FLUID MECHANICS OF ENVIRONMENTAL INTERFACES, SECOND EDITION

Carlo Gualtieri
To all my loved ones

Dragutin T. Mihailović
To Lady N who understood my dreams

Fluid Mechanics of Environmental Interfaces, Second Edition

Editors

Carlo Gualtieri
*Hydraulic, Geotechnical and Environmental Engineering Department,
University of Napoli Federico II, Napoli, Italy*

Dragutin T. Mihailović
Faculty of Agriculture, University of Novi Sad, Novo Sad, Serbia

CRC Press
Taylor & Francis Group
Boca Raton   London   New York   Leiden

CRC Press is an imprint of the
Taylor & Francis Group, an **informa** business

A BALKEMA BOOK

First issued in paperback 2018

CRC Press/Balkema is an imprint of the Taylor & Francis Group, an informa business

© 2013 Taylor & Francis Group, London, UK

Typeset by MPS Limited, Chennai, India

Library of Congress Cataloging-in-Publication Data

Fluid mechanics of environmental interfaces / editors, Carlo Gualtieri, Hydraulic, Geotechnical and Environmental Engineering Department, University of Napoli Federico II, Napoli, Italy, Dragutin T. Mihailović, Faculty of Agriculture, University of Novi Sad, Novo Sad, Serbia. — Second edition.
    pages cm
  Includes bibliographical references and index.
  ISBN 978-0-415-62156-4 (hardback)
 1. Geophysics–Fluid models. 2. Fluid mechanics. 3. Atmospheric turbulance.
4. Hydrology. 5. Ocean-atmosphere interaction. I. Gualtieri, Carlo, editor of compilation.
II. Mihailović, Dragutin T., editor of compilation.
  QC809.F5F587 2012
  551.5'2–dc23

                                                                    2012034826

Published by: CRC Press/Balkema
               P.O. Box 447, 2300 AK Leiden, The Netherlands
               e-mail: Pub.NL@taylorandfrancis.com
               www.crcpress.com – www.taylorandfrancis.com

ISBN 13: 978-1-138-07427-9 (pbk)
ISBN 13: 978-0-415-62156-4 (hbk)

Table of contents

Part four – Processes at interfaces of biotic systems

Preface

Environmental Fluid Mechanics (EFM) studies the motion of air and water at several different scales, the fate and transport of species carried along by these fluids, and the interactions among those flows and geological, biological, and engineered systems. EFM emerged some decades ago as a response to the need of tools to study problems of flow and transport in rivers, estuaries, lakes, groundwater and the atmosphere; it is a topic of increasing concern for decision makers, engineers, and researchers alike. The 1st edition of the book *"Fluid Mechanics of Environmental Interfaces"* published in 2008 was aimed at providing a comprehensive overview of fluid mechanical processes occurring at the different interfaces existing in the realm of EFM, such as the *air-water interface*, the *air-land interface*, the *water-sediment interface*, and the *water-vegetation interface*. Across any of these interface, mass, momentum, and heat are exchanged through different fluid mechanical processes over various spatial and temporal scales.

Following the positive feedback about the 1st edition of the book from the audience, we decided to offer a new edition. Three are the main objectives that we are willing to achieve with the 2nd edition of *"Fluid Mechanics of Environmental Interfaces"*. First, to allow all the contributors to update their chapters considering recent findings in a fast developing research area as the EFM. Second, to extend the coverage of the book to topics that were not considered in the 1st edition, but are indeed of relevance in the EFM field. Third, to add to each chapter an educational part to assist teachers and instructors who will use the book as a textbook or a supplementary readings in their classes.

As for the 1st edition, the book starts with a chapter introducing the concept of EFM and its scope, scales, processes and systems. Then, the book is structured in three parts with fifteen chapters, five more than in the 1st edition. Part one, which is composed of four chapters, covers the processes occurring at the interfaces of the atmosphere with deserts and seas. Part two deals in five chapters with the fluid mechanics at the air-water interface at small scales and sediment-water interface. Finally, part three discusses in six chapters the processes at the interfaces between fluids and biotic systems. Most of the chapters existing in the 1st edition were carefully updated and in some cases also deeply revised and re-organized, such as for chapters 5 and 14.

As already pointed out, five new chapters were added. Chapter 8, by F. Bombardelli and P. Moreno, presents the exchanges at the interface between bed sediments and the overlying waters. These interactions have a tremendous importance for diverse natural and man-made processes such as fining and armouring in rivers, erosion/sedimentation in estuaries, and the cycling of different contaminants in water bodies at large. In the chapter, the characteristics of sediment transport, the concept of incipient motion and the mass balance of solids at the interface are first introduced. Then predictors of diverse variables needed for the mass balance such as bed load flow rates, entrainment functions, and the settling velocity, and the theory of suspended sediment and of bed load are presented. Moreover, the problem of sediment-laden transport of contaminants in water bodies is addressed. Chapter 9 by D. Tonina deals with the hyphoreic exchange. This term means the continuous mixing between surface waters and groundwater due to spatial and temporal variations in channel characteristics. The significance of hyporheic exchange in affecting surface and subsurface water quality and linking fluvial geomorphology, groundwater, and riverine habitat for aquatic and terrestrial organisms has been emerging in recent decades as an important component of conserving, managing, and restoring riverine ecosystems. The chapter presents the concepts, characteristics and environmental effects of hyporheic exchange, and we review the methods for measuring and predicting its characteristics, i.e. hyporheic flux and hyporheic residence time. Chapter 10 by H. Chanson treats EFM aspects of tidal bores. A tidal bore is a hydrodynamic shock propagating upstream as the tidal flow turns to rising. The tidal bore passage is associated with large fluctuations in water depth and instantaneous velocity components and with intense turbulent mixing, and sediment scour and advection in a natural system. Hence the occurrence of a tidal bore is critical to the environmental balance of the estuarine zone in a river and issues such as the sedimentation of the upper estuary, the impact on the reproduction and development of native fish species, and the sustainability of unique eco-systems should be considered. In the chapter both theoretical considerations related to the application of continuity and momentum principles in the analysis of a tidal bore and field observations are presented. The complex interactions between tidal bores and human society are also shortly discussed. Chapter 13 by Y. Tanino describes flow and mass transport under conditions relevant to surface water systems with emergent vegetation. Vegetated surface waters are modelled as homogeneous arrays of discrete, rigid, two-dimensional plant elements. First, typical field conditions are summarized. Then, the standard mathematical formulation for flow through an array of elements is presented and turbulence and mass transport within a homogeneous canopy are described. Finally, the flow at the interface between an emergent canopy and open water is considered. Chapter 16 by D.T. Mihailović and I. Balaz presents maps serving as the combined coupling between interacting environmental interfaces and their behavior in the presence of dynamical noise. Many physical and biological problems, in addition to environmental problems, can be described by the dynamics of driven coupled oscillators. The dynamics of two maps acting as the combined coupling (diffusive and linear) is discussed using methods of nonlinear dynamics, such as bifurcation diagram, Lyapunov exponent, sample and permutation entropy.

As above explained, the third reason for this 2nd edition was the willing of the editors to add at the end of each chapter an educational part. This part is structured in four sections: a synopsis of the chapter, a list of keywords that the reader should have encountered in the chapter, a list of questions and a list of unsolved problems related to the topics covered by the chapter.

Overall, the unique feature of this book to consider all the topics from the point of view of the concept of environmental interface was maintained in this 2nd edition while the coverage of the book was significantly enlarged. As for the 1st edition, the team of the involved contributors is mostly formed by researchers highly experienced in the topics they are covering.

As for the 1st edition, the book is aimed at graduate students, doctoral students as well as researchers in civil and environmental engineering, environmental sciences, atmospheric sciences, meteorology, limnology, oceanography, physics, geophysics and applied mathematics. The book can be adopted as a textbook or supplementary reading for courses at the graduate level in Environmental Fluid Mechanics, environmental hydraulics, hydraulics, open channel flows, physics of the atmosphere, water quality modeling, air quality modeling, atmospheric turbulence and bio-fluid mechanics.

The editors wish to thank all the chapter authors for their continuous and dedicated effort that made possible the realization of this book. The editors also thank the anonymous reviewers of the project for their suggestions and the colleagues, namely F. Bombardelli, A. Bordas, S.T. Rao, and K. Zamani, who presented the 1st edition of the book on international journals such as *Environmental Fluid Mechanics*, *Idojaras* and *Environmental Modelling and Software*, providing thoughtful and detailed remarks that were considered in improving the coverage, the contents and the presentation of this 2nd edition. The editors finally acknowledge with gratitude the assistance of the Editorial Office of CRC Press/Balkema and, especially, of Dr. Janjaap Blom and Ms. José Van der Veer.

October 2012
Carlo Gualtieri
Dragutin T. Mihailović

Preface of the first edition

The field of Environmental Fluid Mechanics (EFM) abounds with various interfaces, and it is an ideal place for the application of new fundamental approaches leading towards a better understanding of interfacial phenomena. In our opinion, the foregoing definition of an environmental interface broadly covers the unavoidable multidisciplinary approach in environmental sciences and engineering also includes the traditional approaches in sciences that are dealing with an environmental space less complex than any one met in reality. An environmental interface can be also considered as a biophysical unit lying between the environment and the organization having the following major functions: (1) to prevent the harmful signals from being injected into the system directly and attacking the valuable structures and channels; (2) to unify the various directions from sub-systems and recursive operations towards the environment; and (3) to fully utilize the internal resources by resolving external variables. The wealth and complexity of processes at this interface determine that the scientists, as it often seems, are more interested in a possibility of non-linear dislocations and surprises in the behavior of the environment than in a smooth extrapolation of current trends and a use of the approaches close to the linear physics. In recent times, researches on fluid mechanics processes at the environmental interfaces have been increasingly undertaken but within different scientific fields and with various applicative objectives.

The aim of the book is to present a comprehensive overview of fluid mechanical processes at the several environmental interfaces. Hence, the matter collected in the book can be considered as a part of the broader context of Environmental Fluid Mechanics in which strong emphasis is placed on the processes involving the exchange of momentum, mass and heat across an environmental interface. The book is aimed at graduate students, doctoral students as well as researchers in civil and environmental engineering, environmental sciences, atmospheric sciences, meteorology, limnology, oceanography, physics, geophysics and applied mathematics. The book can be adopted as a textbook or supplementary reading for courses at the graduate level in Environmental Fluid Mechanics, environmental hydraulics, physics of the atmosphere, water quality modeling, air quality modeling, atmospheric turbulence and bio-fluid mechanics.

Previous books within the EFM field covered only partially the topics presented here. In fact, books on atmosphere dynamics or on air pollution cover only the chapters in the Part 1 of the book. Also, existing books on water quality issues deals only partially with the processes at the environmental interfaces of the hydrosphere. Furthermore, some topics treated in this book, such as momentum and mass-exchange in vegetated open channels, could be found only in papers published on scientific journals. It should be stressed that the book has the unique feature to cover a broad range of scientific knowledge where all the topics are considered from the point of view of the concept of environmental interface. Finally, the team of the involved authors is mostly formed by researchers with many years of experiences in the topics they are covering.

The book is organized in three parts with an introductory chapter by B. Cushman-Roisin, C. Gualtieri and D.T. Mihailović, where scope, scales, processes and systems of EFM are described and discussed together with an overview of EFM processes at environmental interfaces and of challenges to be expected in the next future.

Part one deals with the processes at the atmospheric interfaces. First, the chapter by B. Rajković, I. Arsenić and Z. Grsić covers some theoretical aspects, including molecular and turbulent diffusion, and several areas of modeling of atmospheric dispersion of a passive substance for a point source, such as *Gaussian* and *puff* models. Following this, the chapter by V. Djurdjević and B. Rajković introduces the basic concepts of the air–sea interactions, also discussing the influence of boundary layers on both sides of the air-water interface, and presents the most common approaches to air-sea exchange modeling together with results of sea surface temperature (SST) simulation for the Mediterranean sea obtained by a coupled model with specific modeling of fluxes. The next chapter, by D.T. Mihailović and D. Kapor is devoted to the modeling of flux exchanges between heterogeneous surfaces and the atmosphere. The three approaches commonly applied for calculating the transfer of momentum, heat and moisture from a grid cell comprised of heterogeneous surfaces are discussed. This begs for a combined method and highlights the uncertainties in the parameterization of boundary layer processes when heterogeneities exist over the grid cell. Part one ends with a chapter by G. Kallos that covers the matter related to transport and deposition of dust, the cycle of which is important in the atmosphere and ocean, since dust particles can have considerable impacts on radiation, clouds and precipitation. In this chapter, the state of the art for modeling dust production are reviewed and the impacts on atmospheric and marine processes are discussed.

Part two of the book covers some fluid mechanics processes at the interface between the atmosphere and inland free surface waters. The chapter by C. Gualtieri and G. Pulci Doria deals with gas-transfer at an unsheared free surface, which can have significant impacts on water quality in aquatic systems. First, the effects of the properties of the gas being transferred and of turbulence on gas-transfer rate are discussed. Then, conceptual models are proposed to calculate the gas-transfer rate, including recent developments resulting from both experimental and numerical methods. The next chapter by H. Chanson covers advection-diffusion of air bubbles in turbulent water flows. Herein, air bubble entrainment is defined as the entrainment or entrapment of undissolved air bubbles and air pockets by the flowing waters. After a review of the basic mechanisms of air bubble entrainment in turbulent water flows, it is shown that the void fraction distributions may be represented by analytical solutions of the advection-diffusion equation for air bubbles. Later the microstructure of the air–water flow is discussed, and it is argued that the interactions between entrained air bubbles and turbulence remain a key challenge.

Part three of the book deals with fluid mechanical processes at the interface between water or atmosphere and biotic systems. The chapter by D.T. Mihailović presents transport processes in the system comprised of the soil vegetation and lower atmosphere. The chapter shortly describes the interaction between land surface and atmosphere, such as interaction of vegetation with radiation, evaporation from bare soil, evapotranspiration, conduction of soil water through the vegetation layer, vertical movement in the soil, run-off, heat conduction in the soil, momentum transport, effects of snow presence, and freezing or melting of soil moisture. The chapter also includes a detailed description and explanation of governing equations, the representation of energy fluxes and radiation, the parameterization of aerodynamic characteristics, resistances and model hydrology. The next chapter by B. Lalić and D.T. Mihailović covers turbulence and wind above and within the forest canopy and is focused on forest architecture and on turbulence produced by the friction resulting from air flow encountering the forest canopy. An overview of different approaches oriented towards their parameterization (forest architecture) and modeling (turbulence) is presented. The chapter by P. Gualtieri and G. Pulci Doria deals with vegetated flows in open channels. Particularly, the equilibrium boundary layer developing on a submerged array of rigid sticks and semi-rigid grass on the vegetated bed is characterized based on experimental results carried out by the authors. The last chapter, by G. Nishihara and J. Ackerman discusses the interaction of fluid mechanics with biological and ecological systems. Transport processes in aquatic environments are considered for both pelagic and benthic organisms (those respectively within the water column and at the bottom). The particular issues related to mass transfer to and from benthic plants and animals are considered in detail.

The editors wish to thank all the chapter authors for their continuous and dedicated effort that made possible the realization of this book. The editors also thank the anonymous reviewers of the project for their thoughtful and detailed suggestions that have improved both the contents and presentation of this book. The editors finally acknowledge with gratitude the assistance of the Editorial Office of Taylor & Francis and, especially, of Dr. Janjaap Blom and Richard Gundel.

Carlo Gualtieri
Dragutin T. Mihailović

Biographies of the authors

Josef Ackerman is a Professor in the Department of Integrative Biology at the University of Guelph where he conducts research on the physical ecology of aquatic plants and animals, as well as environmental issues. Most of this research is focused on small-scale fluid dynamic and ecological processes. He holds adjunct faculty positions in the School of Engineering, and was formally Associate Dean of the Faculty of Environmental Sciences, an interdisciplinary faculty that served all academic units at the university. Before coming to Guelph, he was a faculty member at the University of Northern British Columbia, where he played a leading role in founding the university's environmental science and environmental engineering programs and held the Canada Research Chair in Physical Ecology and Aquatic Science. He is currently the Editor in Chief of *Limnology and Oceanography: Fluids and Environments*, an interdisciplinary journal focusing on the interface between fluid dynamics and biological, chemical, and/or geological processes in aquatic systems, and an Associate Editor of *Aquatic Sciences*. Professor Ackerman is the editor of two books and three special issues of journals, former Associate Editor of *Limnology & Oceanography*, and the author of over 60 peer-reviewed publications.

Ilija Arsenić is the Assistant professor of the Meteorology, Physics and Biophysics at the Department for Field and Vegetable Crops, faculty of Agriculture, University of Novi Sad, Serbia. He teaches courses and conducts laboratory classes in Meteorology to the students of various courses and exercises in Biophysics to the students of Veterinary medicine at the Faculty of Agriculture. Additionally, at the Faculty of Sciences he teaches courses in Atmospheric turbulence, Micrometeorology, Air Pollution Modelling, Dynamic Meteorology and Numerical Methods in Weather Forecast. He received a B.S in Physics, M.Sc. in Agrometeorology and defended his Ph.D. thesis in Meteorology and Environmental Modelling at the University of Novi Sad. His main research interest is the numerical modelling of air pollution transport, turbulent processes and processes connected to the dynamic meteorology. Great part of his activities is connected to the constructing and programming HPC cluster computers at the OS level and parallelizing numerical models. Also, he has experience in constructing meteorological measurement sensors and devices and programming appropriate computer programs for them.

Igor Balaž received a B.Sc. in Biology and M.Sc. in Microbiology at the University of Novi Sad, Serbia and defended his Ph.D. thesis in Physics at the Association of Centers for Multidisciplinary and Interdisciplinary Studies (ACIMSI) of the University of Novi Sad. Currently he is a postdoc at the Faculty of Sciences, Department of Physics, University of Novi Sad. His main research interest are artificial life, biosemiotics and general organization of living systems (algebraic and logical properties). He published 6 articles in international peer-reviewed journals; 3 book chapters and 8 articles in international conference proceedings.

Fabián A. Bombardelli holds a degree in Hydraulic Engineering from the National University of La Plata, a Master degree in "Numerical simulation and control" from University of Buenos Aires, Argentina, and a Ph.D. from the University of Illinois, Urbana-Champaign,

USA. Since 2004, he has been Professor (currently Associate with tenure) at the University of California, Davis. His research program focuses on the development of novel theoretical and numerical models and techniques to address multi-phase, environmental problems. His work has been published in peer-reviewed journals such as *Physical Review Letters*, *Physics Fluids*, *International Journal Multiphase Flow*, *Geophysical Research Letters*, *Journal of Geophysical Research*, *Water Resources Research*, etc. He received paper awards from the *American Society of Civil Engineers* (ASCE) and the *American Society of Mechanical Engineers* (ASME) and he received the Best Reviewer Award from *International Association for Hydro-Environment Engineering and Research* (IAHR). He is a member of the Editorial Board of *Environmental Fluid Mechanics*; he serves as member of the Hydraulic Structures Section of IAHR and as Vice-Chair of the Sub-committee for Model Verification and Validation of ASCE.

Hubert Chanson is a 13th Ippen award (IAHR)Professor in Civil Engineering, Hydraulic Engineering and Environmental Fluid Mechanics at the University of Queeensland, Australia. His research interests include design of hydraulic structures, experimental investigations of two-phase flows, applied hydrodynamics, hydraulic engineering, water quality modelling, Environmental Fluid Mechanics, estuarine processes and natural resources. He has been an active consultant for both governmental agencies and private organisations. He authored the textbook *The Hydraulics of Open Channel Flows: An Introduction* (1st edition 1999, 2nd edition 2004) currently used in over 50 universities worldwide and translated into Spanish and Chinese. In 2003, the IAHR presented him with the 13th *Arthur Ippen* Award for outstanding achievements in hydraulic engineering. The American Society of Civil Engineers, Environmental and Water Resources Institute (ASCE-EWRI) presented him with the 2004 award for the Best Practice paper in the *Journal of Irrigation and Drainage Engineering*. Hubert Chanson was invited to deliver plenary keynote lectures in several international conferences and he lectured short courses in Australia, Asia and Europe. He is also member of the *International Association of Hydraulic Engineering and Research* (IAHR) and of the *Institution of Engineers, Australia* (MIEng.Aust.). He chaired the Organisation of the 34th IAHR World Congress held in Brisbane, Australia between 26 June and 1 July 2011. His Internet home page is http://www.uq.edu.au/~e2hchans/.

Benoit Cushman-Roisin is Professor of Engineering Sciences at Dartmouth College, where he teaches a series of courses in Environmental Engineering and Fluid Mechanics in the Thayer School of Engineering. He received his B.Sc. in Engineering Physics at the University of Liège, Belgium, and his doctorate in Geophysical Fluid Dynamics at the Florida State University, where he also taught Physical Oceanography. He later moved to Dartmouth College to teach Fluid Mechanics and Environmental Engineering, the intersection of which is Environmental Fluid Mechanics. He is the author of the first introductory textbook on Geophysical Fluid Dynamics (Prentice Hall, 1994, 2nd Edition for Academic Press, 2011 co-authored with J.M. Beckers) and the lead author of a monograph on the physical oceanography of the Adriatic Sea (Kluwer, 2001). He has authored a number of research articles on various aspects of numerical analysis, physical oceanography and fluid dynamics. He is also the founding and former chief editor of *Environmental Fluid Mechanics*, a peer-reviewed journal published by Springer since 2001. His current research is devoted to the variability of coastal waters (with particular focus on the mesoscale variability in the Adriatic Sea), fluid instabilities, turbulent dispersion, and particle entrainment by jets. Aside from his academic position, Cushman-Roisin also advices various groups and companies on topics related to environmental quality, fluid mechanics and alternative energies.

Vladimir Djurdjević was born in Belgrade on 24th August 1975, Serbia. He graduated from Faculty of Physics, Dept. of Meteorology in 1998 with the average grade of 9.80, the highest average in the history of the department. In 2002 he defended his master thesis *The air-sea interaction in the Mediterranean area*. In 2011 he defended his PhD thesis *Simulations of climate and climate change over South East Europe region using a regional climate model*. His Professional interests are in Numerical modeling of the atmosphere and ocean, and especially regional climate modeling. Since 1999 he is employed at the Institute of Meteorology, Faculty of Physics, Belgrade University, Serbia. As an expert in numerical modeling he participated in several international and national projects. International projects are: *Adriatic integrated coastal areas and rivers basin management system project*, (ADRICOSM-EXT), under IOC-UNESCO, *Simulations of climate change in the Mediterranean Area, (SINTA)*. In summer 2004 he spend summer as an invited expert in numerical modeling of the ocean at the University of Lisbon, Department of Oceanography. For the National meteorological service he had project: *Implementation of coupled Ocean-Atmosphere model* used for medium range integrations in the region. The other project are two domestic projects: *Studying climate change and its influence on environment: impacts, adaptation and mitigation* and *Meteorological extremes and climatic change in Serbia*. Currently he is involved in a European project and Med-Cordex a international indicative. Selected publications include papers on several international conferences and workshops. He has published 5 papers in the international journals from the fields of Micrometeorology and Oceanography. He was a co-author of several chapters in three books.

Zoran Grsić graduated from Faculty of Physics, Dept. of Meteorology at Belgrade University in 1981. In 1992 he defended his master thesis *Critical analysis of Gaussian atmospheric diffusion models*. His professional interests are in Air pollution modelling, Automated systems for real time air pollution distribution assessment and determining zones of influence of chemical and radiation sources, Air quality monitoring networks. Since 1989 he was employed in the Institute of Nuclear Sciences *Vinca* Radiation and Environmental Protection Department in the position of the Research Fellow as the Head of the Group for Meteorology, Radiation and Environmental Protection Department. Currently he is employed as at the National Agency for the radioactive materials. His position is meteorological expert overlooking monitoring and modeling of possible accidents in the nuclear facilities in Serbia. At national level he was in several projects related to the modelling and monitoring of chemical and radioactive pollution. Currently he is the head of service for research of radioactive contamination at the Nuclear facilities of Serbia, the national agency for all facilities using or storing radioactive material. He was leading meteorologist in three national projects. The first one is in the project of the decommissioning of the nuclear reactor in Vinca, near Belgrade, the second one is: *Automated air quality system in industrial zone of Pancevo* and the third one: *Design of continual observing system for the assessment the influence of thermo power plant Nikola Tesla*, which is located in the vicinity of Belgrade. He has four papers in the international journals, 27 papers presented at the various international conferences. He is member of *Air Protection Society of Serbia, Meteorological Society of Serbia* and *Balkan Environmental Association* (B.EN.A).

Carlo Gualtieri is currently Assistant Professor in Environmental Hydraulics at the Hydraulic, Geotechnical and Environmental Engineering Department (DIGA) of the University of Napoli *Federico II*. He received a B.Sc. in Hydraulic Engineering at the University of Napoli *Federico II*, where he received also a M.Sc. in Environmental Engineering and finally a Ph.D. in Environmental Engineering. Prof. Gualtieri has 102 peer-reviewed scientific papers, including 25 publications in scientific journals, 53 publications in conference proceedings, and 24 other refereed publications in subjects related to environmental hydraulics and computational environmental fluid mechanics, with over 80 papers, experimental investigations of two-phase flows, water supply networks management and environmental risk. He is also co-author of 2 textbooks on Hydraulics and author of a textbook on Environmental Hydraulics. He co-edited the books *Fluid Mechanics of Environmental Interfaces (Taylor & Francis*, 2008) and *Advances in Environmental Fluid Mechanics* (*World Scientific*, 2010). He is since 2006 member of the Editorial Board of *Environmental Modelling and Software (Elsevier)* and since 2008 of *Environmental Fluid Mechanics (Springer)*. He contributed as reviewer in several scientific journals (e.g. *Environmental Fluid Mechanics, Environmental Modeling and Software, Journal of Environmental Engineering ASCE, Journal of Hydraulic Engineering ASCE, Journal of Hydraulic Research, Experiments in Fluids, Advances in Water Resources, Water Resources Research,* etc.) and as external examiner for Ph.D. thesis in foreign countries. He co-organized since 2004 the Environmental Fluid Mechanics session at the *International Environmental Modelling & Software Society (iEMSs)* biennal conferences. He is also active as expert reviewer for research funding agencies in several countries. He is member of the *International Association for Hydro-Environment Engineering and Research* (IAHR) and of the iEMSs.

Paola Gualtieri is currently Associate Professor in Hydraulics and Hydraulic Measurements at the Hydraulic, Geothecnical and Environmental Engineering Department of the University of Napoli *Federico II*. She received a B.Sc. in Hydraulic Engineering at the University of Napoli *Federico II*, where she received also a M.Sc. in Environmental Engineering. She finally received a Ph.D. in Hydraulic Engineering at the University of Napoli *Federico II*. She is member of Ph.D. Program Board at the University of Napoli *Federico II*. Prof. Gualtieri produced more than 60 scientific papers, including 2 textbooks, 4 chapters in international scientific books, 8 papers in international journals, 25 papers in international conferences, 14 papers in national conferences in subjects, in the whole related to hydraulic measurements, turbulence in uniform and boundary layer flows, air entrainment at hydraulics structures, and, finally, environmental hydraulics with 28 papers. She contributed as reviewer for international conferences and as expert reviewer for research funding agencies. She is member of the *International Association of Hydraulic Engineering and Research* (IAHR) and of the *American Society of Civil Engineers* (ASCE).

George Kallos is Professor in the Division of Environment and Meteorology, Physics Department, University of Athens and group leader of the Atmospheric Modeling Group in the same Department. He is also a Senior Research Associate at the SUNY/ASRC, Albany, USA, and a Research Professor at the Chapman University, California, USA. Professor Kallos received his Ph.D. in Atmospheric Sciences by the School of Geophysical Sciences, Georgia Institute of Technology, USA, in 1985. The main fields of his scientific and professional activities focus on Atmospheric Dynamics (Regional and Mesoscale Modeling), Air Pollution Modeling (Dispersion, Diffusion, Chemistry), Wave modeling and Marine Meteorology, Regional Climate, Data assimilation, Agricultural Meteorology, Severe Weather Phenomena and Wind Energy Applications. He has published 106 papers in peer reviewed journals, 221 papers in conference proceedings, 9 books and dissertations and more than 160 contributions to other scientific forums. This work has been cited in more than 2000 other research papers. Prof. Kallos is an Associate Editor of the scientific journals *Environmental Fluid Mechanics* and *Journal of Atmosphere-Ocean System*. Prof. Kallos has supervised 10 Ph.D. students and 19 M.Sc theses so far. He has participated in 74 externally funded projects in 60 of which he was the Principal Investigator.

Darko Kapor is Professor of the Theoretical Physics at the Department of Physics, Faculty of Sciences, University of Novi Sad, Serbia. He received a B.S. in Physics at the University of Novi Sad, his M.Sc. in Theoretical Physics at the University of Belgrade, Serbia and defended his Ph. D. Thesis in Theoretical Physics at the University of Novi Sad. He teaches various Theoretical and Mathematical Physics courses and History of Physics to Physics students and a course on Atmospheric Radiation to the students of Physics and Meteorology. Along with his teaching activities in Physics, he also teaches at the multidisciplinary studies of the Center for Meteorology and Environmental Modelling (CMEM) which is the part of the Association of Centers for Multidisciplinary and Interdisciplinary Studies (ACIMSI) of the University of Novi Sad. His main research interest is the Theoretical Condensed Matter Physics, where he is the head of the project financed by the Ministry for Science of the Republic of Serbia. During the last 15 years, he has developed an interest in the problems of theoretical meteorology and worked with the Meteorology group lead by Prof. D.T. Mihailović. He is engaged in transferring the concepts and techniques of the theoretical physics to meteorology, for example to the problem of aggregated albedo for heterogeneous surfaces and more recently the problem of chaos.

Petros Katsafados is an Assistant Professor at the Department of Geography, Harokopio University of Athens, Greece since 2007. In his present affiliation, he is also head of the Atmosphere and Climate Dynamics Group (ACDG) having a Lecturer, a post doctoral researcher and a number of postgraduate students as active members. The entire research and operational activities of the group are available through the website http://meteoclima.gr. He originally studied Mathematics but then he switched scientific field to Atmospheric Physics and Dynamics and he finally completed his PhD thesis from the School of Physics at the National and Kapodistrian University of Athens (NKUA) on 2003. On 2001 he took position as a senior research associate in the School of Physics at NKUA. From that period up today he participated in a number of EU and national funded projects mainly related with air-sea interactions, data assimilation and atmospheric modeling. Since 2002 he has 24 articles in international peer-reviewed journals and more than 60 announcements and publications in conference proceedings in subjects related to atmospheric physics and dynamics.

Branislava Lalić is Associate Professor of the Meteorology, Physics and Biophysics at the Department for Field and Vegetable Crops, Faculty of Agriculture, University of Novi Sad, Serbia. She conducts laboratory classes in Meteorology to the students of various courses and in Biophysics to the students of Veterinary medicine at Faculty of Agriculture. Additionally, at the Faculty of Sciences she teaches courses on General meteorology, Instrumental techniques, Applied meteorology and Agrometeorology to the students of Physics, Meteorology and Environmental Modelling. She received a B.Sc. in Physics at the University of Novi Sad, her M.Sc. in Meteorology at the University of Belgrade, Serbia and defended her Ph.D. Thesis in Meteorology and Environmental Modelling at the University of Novi Sad. Along with her teaching activities at Faculty of Agriculture and Faculty of Sciences she is involved in establishment and organization of the multidisciplinary studies at the Center for Meteorology and Environmental Modelling (CMEM) which is the part of the Association of Centers for Multidisciplinary and Interdisciplinary Studies (ACIMSI) of the University of Novi Sad. Her main research interest is the modeling of physical processes describing biosphere – atmosphere interaction and turbulent transfer between ground and the atmosphere.

Dragutin T. Mihailović is Professor of Meteorology and Biophysics at the Department of Vegetable and Crops, Faculty of Agriculture, University of Novi Sad, Serbia. He is also

Professor of the Modelling Physical Processes at the Department of Physics, Faculty of Sciences at the same university and Visiting Professor at the State University of New York at Albany, NY (USA). He teaches various theoretical and numerical meteorology courses to Physics and Agriculture students. He received B.Sc. in Physics at the University of Belgrade, Serbia, his M.Sc. in Dynamic Meteorology at the University of Belgrade and defended his Ph.D. Thesis in Dynamic Meteorology at the University of Belgrade. He is head of the Center for Meteorology and Environmental Modelling (CMEM) which is the part of the Association of Centers for Multidisciplinary and Interdisciplinary Studies (ACIMSI) of the University of Novi Sad where he has teaching activities. He is also head of the Center for Meteorology and Environmental Predictions, Department of Physics, Faculty of Sciences, University of Novi Sad. His main research interest are the surface processes and boundary layer meteorology with application to air pollution modelling and agriculture. Recently, he has developed an interest for (i) analysis of occurrence of the deterministic chaos at environmental interfaces and (ii) modeling the complex biophysical systems using the category theory and nonlinear dynamics. He has over 60 articles in international peer-reviewed journals; 405 citations in SCI journals; edited 4 books (*Taylor & Francis, World Scientific, Nova Science Publishers*); 15 invited lectures in worldwide institutions and universities and 4 plenary talks.

Patricio Moreno is currently a Ph.D. student at the Department of Civil and Environmental Engineering at the University of California, Davis. He obtained a B.S. in Civil Engineering from Universidad de Santiago, Chile, and a M.Sc. in Environmental Engineering from the University of Tennessee at Knoxville, USA. He also holds a faculty position in the Civil Engineering Department at the Universidad Diego Portales, Chile. His research focuses on the numerical modeling of sediment transport processes and its interaction with turbulent flows.

Gregory Nishihara is an Associate Professor at the Institute for East China Sea Research, Nagasaki University. He received a B.Sc. in Civil Engineering at the University of Hawaii at Manoa and was awarded a Ph.D. in the Science of Marine Resources from the United Graduate School of Agricultural Sciences at Kagoshima University. He received advanced training in the Physical Ecology Laboratory at the University of Guelph, and later at the Laboratory of Marine Botany at Kagoshima University (Japan Society for the Promotion of Science Fellow) where he was a postdoctoral scientist. At Nagasaki, he is working to elucidate the mechanisms involving mass transport and fluid dynamic processes on the primary productivity of seagrass meadows and sargassum forests.

Guelfo Pulci Doria is currently Full Professor in Fluid Mechanics and Hydraulics at the Hydraulic, Geothecnical and Environmental Engineering Department of the University of Napoli *Federico II*. He received a B.Sc. in Electronic Engineering at the University of Napoli *Federico II* in year 1966. From 1993 to 1999 he was head of the Hydraulic and Environmental Engineering Department *G.Ippolito* and in 1994–95 he was president of Napoli Water Supply Authority. He also was 9 times head and 19 times member of the Board of the Ph.D. Program at the University of Napoli *Federico II* and 5 times external reviewer of Ph.D. Program at the University of Roma 3. Now he is also the head of Civil Engineering Doctorate School of University of Napoli *Federico II*. Since 1988 he was national coordinator in 10 National Research Groups funded by the Scientific Research National Department and in 1997 he was co-coordinator of EU funded research program *Waternet*. Prof. Pulci Doria produced about 180 scientific papers, including 4 textbooks, 5 chapters in international scientific books, 7 papers in international journals, 3 invited lectures in conferences, over 40 papers in international conferences and yet over 40 papers in national conferences. He contributed as reviewer for scientific journals and international conferences. His research interests encompass turbulence, hydraulic models and measurements, cavitation, air entrainment at hydraulic structures and, finally, environmental hydraulics with over 30 papers. He is member of the *International Association of Hydraulic Engineering and Research* (IAHR).

Borivoj Rajković is Associate Professor in Institute of Meteorology of Faculty of Physics, University of Belgrade, where he teaches courses in Dynamic Meteorology, Micrometeorology, Numerical and Physical modeling in the Atmosphere. He graduated at Belgrade University, Department for Mathematics, Mechanics and Astrophysics and Ph.D. from Princeton University, program in Geophysical Fluid Dynamics. His professional interests are in Numerical modeling of the atmosphere and ocean, Micrometeorology and Parameterization of physical processes in the Atmosphere. For several years he was joint professor at the University of Novi Sad where he is currently at the Center for Environmental Modeling and Ecological Studies (CIMSI). As an expert in numerical modeling he participated in several international and national projects. International projects are: *Adriatic integrated coastal areas and rivers basin management system project*, (ADRICOSM-EXT), under IOC-UNESCO, *Simulations of climate change in the Mediterranean Area* (SINTA). For the National meteorological service he had project: *Implementation of coupled Ocean-Atmosphere model* used for medium range integrations in the region. The other domestic project is: *Studying climate change and its influence on environment: impacts, adaptation and mitigation*. Currently he is involved in an international initiative Med-Cordex. During 1993–95 he was visiting scientist at the World Laboratory in *LAND-3 project: Protection of the Coastal Marine Environment in the Southern Mediterranean Sea*. Currently he participates in the regional climate project, *Simulation of the Balkan climate in the 21st century* (SINTA) together with scientists from Italy. Selected publications include papers on several international conferences and workshops. He has published 18 papers in the international journals from the fields of Micrometeorology, Oceanography. He is the author of the textbook *Micrometeorology* and co-author of several chapters in three books.

Christos Spyrou is a member of the Atmospheric Modelling and Weather Forecasting Group of the University of Athens since April 2005 as an assistant researcher and PhD candidate. He received his PhD on the Dust Feedback on Radiative Transfer in 2011 and continues to work for the University of Athens. He has more than 6 years of experience in Atmospheric Modelling, Air Pollution Modelling, Weather Forecasting, Energy, Climate and Aerosol Studies. He has worked in the framework of several Projects funded by the EU and the Greek Government as a Senior Researcher (MFSTEP, ESPEN, INSEA, CIRCE, MARINA, POW WOW). For the past 4 years he supervises the operational weather forecasting and data processing of the SKIRON model (limited area and desert dust forecasting) at the University of Athens. Over the past 6 years he has 7 publications in peer reviewed Scientific Journals and 13 papers in Conference Proceedings related to atmospheric physics.

Yukie Tanino is currently a research associate in the Department of Earth Science and Engineering at Imperial College London. Previously, she was a postdoctoral researcher at Laboratoire Fluides, Automatique et Systèmes Thermiques in Orsay, France. She holds a B.S. in Environmental Engineering Science, a M.S. in Civil and Environmental Engineering, and a Ph.D. in Environmental Fluid Mechanics from Massachusetts Institute of Technology. Her main research interests are in the fluid dynamics of, and mass transport in, obstructed flows. Her present research focuses on multiphase flow through porous media, with emphasis on phenomena relevant to geological carbon sequestration and enhanced oil recovery. Key projects are focused on the impact of pore architecture and connectivity on two-phase flow, and capillary trapping under mixed-wet conditions. For her doctoral research, she studied drag, turbulence, and lateral dispersion in random cylinder arrays under conditions relevant to vegetated surface waters.

Daniele Tonina is currently an Assistant Professor with the Center for Ecohydraulics Research at the University of Idaho. He received engineering degrees from the University of Trento (BS, MS, 2000) and the University of Idaho (PhD, 2005). His research interests are in the field of ecohydrology, where he focuses on identifying and modeling linkages between physical processes and biological systems. He is leading a research team evaluating new airborne sensors to acquire both terrestrial and aquatic topographies. He received grants to conduct his research from the US Federal and state agencies, such as the National Science Foundation, the US Bureau of Reclamation and the US Forest Service. He teaches environmental hydrodynamics, aquatic habitat modeling, management of in-channel vegetation and engineering sedimentation. He is a member of the *International Association for Hydro-Environment Engineering and Research* (IAHR) and the *American Geophysical Union* (AGU).

Part one
Preliminaries

CHAPTER ONE

Environmental Fluid Mechanics: Current issues and future outlook

Benoit Cushman-Roisin
*Thayer School of Engineering, Dartmouth College,
Hanover, New Hampshire, USA*

Carlo Gualtieri
*Hydraulic and Environmental Engineering Department,
University of Napoli, Napoli, Italy*

Dragutin T. Mihailović
Faculty of Agriculture, University of Novi Sad, Novi Sad, Serbia

ABSTRACT

All forms of life on earth are immersed in natural fluids, such as the air in the atmosphere and the water in surface and underground systems. The knowledge of natural fluids motions is therefore very important and lead to the implementation of a new discipline, termed Environmental Fluid Mechanics (EFM). EFM is the scientific study of naturally occurring fluid flows of air and water on our planet Earth, especially of those flows that affect the environmental quality of air and water.

In this chapter EFM is introduced. First commonalities and differences between EFM and its cousin disciplines, such as Fluid Mechanics, Hydraulics and Geophysical Fluid Dynamics, are described pointing out their specific purpose and scales. Second, the concepts of stratification and turbulence, which are two essential ingredients of EFM, are introduced. Third, scales, processes and systems within EFM are presented. The concept of environmental interface is defined introducing the EFM processes occurring across the main four environmental interfaces. The chapter ends with a discussion about the challenges facing EFM scientists in the next decades.

1.1 FLUIDS IN THE ENVIRONMENT

All forms of life on earth are immersed in a fluid or another, either the air of the atmosphere or the water of a river, lake or ocean; even, soils are permeated with moisture. So, it is no exaggeration to say that life, including our own, is bathed in fluids. A slightly closer look at the situation further reveals that it is the mobility of fluids that actually makes them so useful to the maintenance of life, both internally and externally to living organisms. For example, it is the flow of air that our lungs that supplies oxygen to our blood stream. The forced air flow created by our respiration, however, is not sufficient; without atmospheric motion around us, we would choke sooner or later in our own exhaust of carbon dioxide.

Likewise, most aquatic forms of life rely on the natural transport of water for their nutrients and oxygen. Our industrial systems, which release pollution on a continuing basis, would not be permissible in the absence of transport and dilution of nearly all emissions by ambient motions of air and water.

In sum, natural fluid motions in the environment are vital, and we have a strong incentive to study the naturally occurring fluid flows, particularly those of air in the atmosphere and of water in all its streams, from underground aquifers to surface flows in rivers, lakes, estuaries and oceans.

The study of these flows has received considerable attention over the years and has spawned several distinct disciplines: meteorology, climatology, hydrology, hydraulics, limnology and oceanography. Whereas the particular objectives of each of these disciplines, such as weather forecasting in meteorology and design of water-resource projects in hydraulics, encourage disciplinary segregation, environmental concerns compel experts in those disciplines to consider problems that are essentially similar: the effect of turbulence on the dispersion of a dilute substance, the transfer of matter or momentum across an interface, flow in complex geometries, the rise of a buoyant plume, and the impact of flow over a biotic system.

The study of environmental flows is also fully integrated in the contemporary emphasis on environmental impacts and sustainable life on planet Earth. According to physicists, the world scientific community will be occupied during the 21st century in large part by problems related to the environment, particularly those stemming from the concern over climate change (Rodhe *et al.*, 2000) as well as many other problems spanning a wide range of spatial and temporal scales. This marks the first time in the history of science that environmental problems lie at the forefront of scientific research.

The following chapters of this book are illustrative of a number of these problems. The common points encourage interdisciplinarity to a degree that is increasing in proportion to the acuity of our environmental problems. This overlap between the various disciplines concerned with the environmental aspects of natural fluid flows has given rise to a body of knowledge that has become known as Environmental Fluid Mechanics. The interdisciplinary aspects become especially manifest in the study of processes at the interfaces between environmental systems.

1.2   SCOPE OF ENVIRONMENTAL FLUID MECHANICS

In the light of the preceding remarks, we can propose a definition: *Environmental Fluid Mechanics* (EFM) is the scientific study of naturally occurring fluid flows of air and water on our planet Earth, especially of those flows that affect the environmental quality of air and water. Scales of relevance range from millimeters to kilometers, and from seconds to years.

According to the preceding definition, EFM does not extend to fluid flows inside organisms, such as air flow in lungs and blood flow in the vascular system, although these can be classified as natural. Rather, these topics more properly belong to specialized biological and medical sciences, which have little in common with studies of outdoor fluid flows.

The preceding definition also distinguishes EFM from classical fluid mechanics, the latter being chiefly concerned with artificial (engineered) fluid motions: flows in pipes and around airfoils, in pumps, turbines, heat exchangers and other machinery that utilizes fluids. In so doing, it treats many different types of fluids and under vastly different pressures and temperatures (Munson *et al.*, 1994). By contrast, EFM is exclusively concerned with only two fluids, air and water, and moreover under a relatively narrow range of ambient temperatures and pressures. Ironically, while classical fluid mechanics tends to view turbulence as a negative element, because it creates unwanted drag and energy loss, EFM accepts turbulence as beneficial, because it favors rapid dispersion and dilution.

The objective of EFM also differs from that of hydraulics, which deals exclusively with free-surface water flow (Chow, 1959; Sturm, 2001). Traditionally, problems in hydraulics have addressed the prediction and control of water levels and flow rates, but the realm of hydraulics has recently been shifting considerably toward environmental concerns (Singh and Hager, 1996; Chanson, 2004). This situation has arisen because it has now become equally important to estimate the effect of turbulent mixing, erosion and sedimentation, and their effects on water quality as it has been to calculate pressures against structures and predict floods. Because of its similarities with other natural fluid flows, the environmental component of hydraulics is incorporated in EFM.

Geophysical fluid dynamics, which studies the physics of atmospheric and oceanic motions on the planetary scale (Cushman-Roisin, 1994), is another branch of fluid mechanics that overlaps with EFM. In geophysical fluid dynamics, however, the strong effect of planetary rotation relegates turbulence to secondary status. Put another way, the two main ingredients of geophysical fluid dynamics are stratification and rotation, whereas those of EFM are stratification and turbulence.

Other cousin disciplines are limnology (study of lakes; ex. Imberger, 1998) and hydrology (study of surface and subsurface water; ex. Brutsaert, 2005). Table 1.1 recapitulates the commonalities and differences between EFM and its cousin disciplines highlighting their purpose, possibility of human control and the role of turbulence within them.

Table 1.1. Topical comparison between Environmental Fluid Mechanics and related disciplines.

	Environmental Fluid Mechanics	Fluid Mechanics	Geophysical Fluid Dynamics	Hydraulics	Hydrology
Air example	Sea breeze	Airfoil	Storm	–	–
Water example	Danube River	Pump	Gulf Stream	Dam	Watershed
Turbulence	Beneficial (Dilution)	Detrimental (Drag)	Secondary importance	Secondary importance	Unimportant
Human control	Limited	Dominant	Nil	Dominant	Limited
Purpose	Prediction & Decision	Design & Operation	Prediction & Warnings	Design & Operation	Prediction & Decision

Finally, it is worth situating the purpose of EFM among that of the other disciplines. Because no one can affect in any direct way the flow of air and water on planetary scales, geophysical fluid dynamics, meteorology and oceanography aim solely at the understanding and prediction of those flows. In contrast, the primary objectives of traditional fluid mechanics and hydraulics are design and operation. Environmental Fluid Mechanics finds its purpose between those extremes; like hydrology and limnology, it is aimed at prediction and decision. Indeed, typical problems in EFM concern the prediction of environmental-quality parameters that depend on natural fluid flows, such as bedload transports and pollution levels. EFM also extends into decision making. Decisions in the realm of EFM, however, do not address how natural fluid flows can be controlled or modified, but rather how inputs from human activities can be managed as to minimize their impact downstream. A typical example is the design of a smokestack (with decisions regarding its location, height, diameter and rate of output) in order to avoid certain levels of ground pollution within a certain radius around its base. Another pertinent example is the management of a lake that is used as a drinking water reservoir but is unfortunately contaminated by methyl tertiary-butyl ether (MTBE). This contaminant. which is an oxygenated compound that has been added to gasoline in the USA, is released in the lake by recreational vehicles. Since gas-transfer, that is volatilization, is believed to be the main removal process of MTBE from the lake, the assessment of MTBE volatilization rate is a critical point for the use of the lake for water supply (Gualtieri, 2006). This example points out another feature of EFM, namely that EFM processes often involve exchange processes between the boundaries of different systems, such as the interface between a water body and the atmosphere or between the atmosphere and the land surface. An overview of these processes will be proposed later in Section 1.5.

EFM thus considers only two fluids, air and water, and each within a relatively narrow range of values, never far from ambient temperatures and pressures, one may then be tempted to ask: Shouldn't such study be relatively straightforward? Why should an entire discipline be devoted to such a narrow object of inquiry? The answers to these questions lie in the several complexities which EFM needs to confront. First, the domain size is typically very large, large enough to enable a number of distinct processes to play simultaneous roles, and it is not uncommon to encounter a hierarchy of processes embedded into one another. For example, sea breeze near the seashore is a larger-scale manifestation of convection and at the same time a smaller-scale component of the local meteorology. Second, the geometry is typically complex, with irregular topography and free surfaces. Third, processes at interfaces, the particular subject of this book, often play a controlling role in the entire system, and details matter for the whole. Fourth, fluid turbulence, although an incompletely known subject of physics, is central to friction, dispersion and dilution in environmental fluids.

1.3   STRATIFICATION AND TURBULENCE

Stratification and turbulence are two essential ingredients of EFM. Stratification occurs when the density of the fluid varies spatially, as in a sea breeze where masses of warm and cold air lie next to each other or in an estuary where fresh river water flows over saline seawater. Such situations with adjacent masses of lighter and denser fluid create buoyancy forces that strongly control the flow by either generating or restricting vertical motion.

1.3.1   Stratification

Stratification is to be distinguished from compressibility. Compressibility, or the variation of density under changing pressure, is responsible for the propagation of sound waves. Intuitively, it is evident that the propagation of sound waves (acoustics) is not relevant to environmental fluid motions. This is because the typical speeds associated with movements of air and water in nature are much less than the sound speed; *i.e.* the Mach number (ratio of fluid velocity to sound speed) is much less than one. In contrast to compressibility, stratification arises because density varies with temperature through what is commonly called thermal expansion: heat dilates the fluid[1], so that warm fluid expands and cold fluid contracts. This effect is often important in natural fluid systems because thermal contrasts across the system create buoyancy forces that may not be negligible, imparting to the fluid a tendency to arrange itself vertically with the denser fluid sinking to the lowest places and the lighter fluid floating on top. Such layering of the fluid according to density, from the heaviest at the bottom to the lightest at the top, is what is properly called stratification. But, the word stratification has been enlarged to encompass any situation in which density differences are important, regardless of whether they occur in the vertical or the horizontal

---

[1]With the exception of fresh water below 4°C.

or both, and whether they are caused by heat or another agent such as salinity (in seawater), moisture (in atmosphere), or suspended matter (in turbid water).

Although a certain degree of stratification is always present in environmental systems, its dynamical effects are not necessarily important in every single instance. There are indeed cases, such as shallow-river flows, where buoyancy forces exert a negligible effect among the other forces at play. To ascertain the importance of density stratification in a particular situation, we can use the following rule. Under the action of gravity, fluid masses of different densities tend to flow so that the heavier ones occupy the lower portion of the domain and the lighter ones the upper portion. In the absence of mixing along the way and of other forces besides gravity, the ultimate result would be a vertical arrangement of horizontal layers with density increasing monotonically downward, which corresponds to a state of least potential energy. The action of other forces, however, create motions that disturb such equilibrium, tending to raise heavier fluid and lower lighter fluid against their respective buoyancy forces. The result is an increase of potential energy at the expense of a portion of the kinetic energy contained in the motion. Therefore, the dynamical importance of stratification can be estimated by comparing the levels of potential and kinetic energies present in the system under consideration.

In most environmental applications, fluid parcels (air or water) undergo only very moderate density variations. For example, a water parcel on the surface of a lake when subjected to solar heating that increases its temperature by 10°C (which almost never occurs) has its density reduced by less than 0.3%! By contrast, we think of the air in the atmosphere as being very compressible, and it is so, but nonetheless the compressibility of air is unimportant in most environmental situations, because air parcels traveling with winds remain within a narrow range of pressures and temperatures and experience density variations that are usually less than 5%. With this in mind, we can write the density ρ of the fluid (mass per volume, in kg/m^3), as the sum of two terms:

$$\rho = \rho_0 + \rho' \tag{1.1}$$

where ρ_0 is a constant and ρ' a variable but small perturbation. For ρ_0, we can adopt the following values:

- for air at standard temperature (15°C) and pressure (101.33 kPa): $\rho_0 = 1.225$ kg/m^3;
- for freshwater at standard temperature (15°C) and atmospheric pressure $\rho_0 = 999$ kg/m^3;
- for seawater at standard temperature (10°C) and salinity (35 ppt) $\rho_0 = 1027$ kg/m^3.

If the density perturbation ρ' changes by a value $\Delta\rho$ over a height H of the fluid (height over which vertical excursions take place), so that a fluid parcel at some level z has a density equal to $\rho_0 + \Delta\rho/2$ and one at level $z + H$ a density equal to $\rho_0 - \Delta\rho/2$ (Figure 1.1), an exchange of volume V between those two parcels causes a rise in potential energy of the heavier one by $mgH = (\rho_0 + \Delta\rho/2)VgH$ and a simultaneous drop in potential energy of the lighter parcel by $(\rho_0 - \Delta\rho/2)VgH$. The net change in potential energy is $\Delta\rho VgH$. On the other hand, the kinetic energy is on the order of $mU^2/2$ per parcel, where U is a measure of the fluid velocity in the system (such as a velocity at some inlet). For the pair of parcels, this adds to $(\rho_0 + \Delta\rho/2)VU^2/2 + (\rho_0 - \Delta\rho/2)VU^2/2 = \rho_0 VU^2$. A comparison of potential energy to kinetic energy leads to forming the ratio:

$$Ri = \frac{gH\Delta\rho}{\rho_0 U^2} \tag{1.2}$$

after division by V. This ratio is called the *Richardson number*.

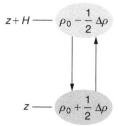

Figure 1.1. Exchange between fluid parcels of different densities and at different heights. Because each displacement is performed either against or with the force of gravity, the exchange causes a modification in potential energy.

The value of the dimensionless ratio Ri permits to determine the importance of stratification in a given system. If Ri is on the order of unity (say $0.1 < Ri < 10$, customarily written as $Ri \sim 1$), a significant perturbation to the stratification can consume a major part of the available kinetic energy, thereby modifying the flow field significantly. Stratification is then important. If Ri is much greater than unity ($Ri >> 1$, or in practice $Ri > 10$), then there is insufficient kinetic energy to perturb the stratification in any significant way, and the latter greatly constrains the flow. But, on the other hand, when Ri is much less than unity ($Ri << 1$, or in practice $Ri < 0.1$), potential-energy variations created by vertical excursions of the fluid against their buoyancy forces cause a negligible drop in kinetic energy, and the stratification is easily erased by vertical mixing. In sum, stratification effects are negligible whenever $Ri << 1$ and important otherwise.

1.3.2 Turbulence

Turbulence is the term used to characterize the complex, seemingly random motions that continually result from instabilities in fluid flows. Turbulence is ubiquitous in natural fluid flows because of the large scales that these flows typically occupy. (The only significant exception is the subsurface flow in porous soils where motion is very slow.) By vigorously stirring the fluid, turbulence is an extremely efficient agent of dilution. This is a major advantage in environmental systems. On the other hand, turbulence comes with a substantial handicap: The complex motions that it generates are beyond any easy description, even by a statistical approach. Some specific types of turbulent flow, such as homogeneous turbulence and shear turbulence, can be described by limited theories and modeled with a good dose of empiricism, but a complete theory of turbulence has not yet been formulated.

The level of turbulence in a fluid system is estimated by comparing the amount of kinetic energy and the work of viscous forces. If ρ_0 is again the average density value in the system, U a typical velocity value, L a characteristic length of the domain (such as its width or height), and μ the viscosity of the fluid, then a measure of the kinetic energy per unit volume is $\rho_0 U^2/2$, while the dissipative work done by viscous forces per unit volume is $\mu U/L$. The ratio of these two quantities is (after removal of the factor 2 which is inconsequential in a definition):

$$Re = \frac{\rho_0 UL}{\mu} \tag{1.3}$$

This is the *Reynolds number*, ubiquitous in fluid mechanics. When Re is large, there is ample kinetic energy and comparatively weak viscous dissipation; the fluid flows relatively freely and is thus apt to exhibit complex spatial patterns and much temporal variability.

This is the case of turbulence. Hence, turbulence occurs whenever the Reynolds number is large. There is rarely a precise value of the Reynolds number below which the flow is simply structured (laminar flow) and above which turbulence occurs, but the transition typically occurs at a Reynolds number of a few thousands. In environmental systems, with large values of L and small values of μ [$\mu = 1.8 \times 10^{-5}$ kg/m·s for air and 1.0×10^{-3} kg/m·s for water], the value of Re almost invariably exceeds 10^6, and the flow is turbulent. The questions that arise are how strong is the turbulence and what is its nature. Environmental fluid turbulence can be broadly divided into two types: *shear turbulence* and *convective turbulence*. Each type is characterized by a turbulent velocity scale, which can then be compared to the mean flow velocity.

In shear turbulence (also called wall turbulence), the turbulent velocity scale is the friction velocity u^*, defined as:

$$u^* = \sqrt{\frac{\tau}{\rho}} \tag{1.4}$$

where ρ is the fluid density and τ is the stress occurring at the boundary (Pope, 2000, page 269). The greater the stress against the boundary, the greater the shear in the mean flow, and the greater its capacity to create turbulent eddies.

In convective turbulence, the turbulent velocity scale, usually denoted w^* because it measures the vertical velocity of rising or sinking thermals, is given by:

$$w^* = (\kappa \alpha g h\, Q)^{1/3} \tag{1.5}$$

where $\kappa = 0.41$ is the Von Kármán constant, α is the thermal expansion coefficient, g the earth's gravitational acceleration, h the height of the system, and Q the *kinematic* heat flux (actual heat flux divided by the fluid's density and heat capacity) (Cushman-Roisin, 1994, page 165). Which among u^* and w^* is largest and how the latter compares to the mean flow directly affect the importance of turbulence in an environmental flow.

The two ingredients of EFM, stratification and turbulence, act generally in competition with each other. Oftentimes, the buoyancy forces of stratification tend to quench turbulence, because vertical movements against buoyancy forces consume kinetic energy to increase potential energy. On the other hand, turbulent motions are capable of mixing the fluid and therefore of reducing the density differences that create stratification. An exception to the rule is convection, which occurs when an unstable, top-heavy stratification releases potential energy that feeds turbulent kinetic energy.

1.4 SCALES, PROCESSES AND SYSTEMS

Environmental problems appear different at different scales, requiring various approaches for their investigation and solution. Likewise, Environmental Fluid Mechanics takes different forms depending on the scale of investigation.

The shortest relevant length scale is that of the smallest turbulent eddy, called the Kolmogorov scale, where viscosity quenches turbulence. It is typically less than a millimeter in environmental fluid flows. Computer models cannot resolve this scale, but it is nonetheless important because it is near this scale that molecular diffusion occurs inside the flow and skin effects take place on the interfaces.

The next scale characterizing EFM motions is usually the local level, where the smaller geometrical dimensions of the system come into play, such as the overall roughness of a

Figure 1.2. A smokestack plume. Note the turbulent billowing inside the plume, which is the cause of its gradual dispersion in the ambient atmosphere. (Photo by the first author).

vegetated surface, the shape of buildings in an airshed, or the structure of a river channel. At this level, the focus is usually on resolvable details of the flow or the concentration field in the vicinity of a single source, such as the jet caused by the discharge of an industrial waste in a body of water or the plume originating from a release of hot gases from a smokestack (Figure 1.2). The understanding of such phenomena proceeds from studies of specific processes. The same process is likely to be present in different environmental systems under almost identical forms. For example, shear-flow instability occurs in the lower atmosphere, in estuaries and also in the near-surface circulation of a lake. Likewise, convective motions driven by top-heavy stratification follow similar dynamics regardless whether they occur in air or water. The same mathematical formulation will therefore be useful in more than one application.

At the next larger level, one considers entire systems, such as a stretch of river, an entire lake, an aquifer, or an urban airshed. In those systems, fluid motions result from several processes acting simultaneously. For example, lake dynamics are characterized by a mix of wind-driven currents, gravity waves, thermal stratification, and winter convection. As one proceeds toward longer scales, one begins to encounter systems of systems, for example, a hydrologic network consisting of multiple river branches and lakes, or the meteorology over a heterogeneous land area.

Table 2 lists the typical length, velocity, and time scales of the most common environmental fluid processes and systems. Not surprisingly, larger systems evolve on longer time scales, with the exception of ocean tides. Depending on the size of the system under consideration, the spatial scale can be regional, continental or even global. As the scale increases, some processes may yield precedence to others. For example, as one approaches continental and global scales, turbulence becomes increasingly less important, and planetary rotation becomes dominant. At the limit of the entire globe, mass budgets (ex. of greenhouse gases) also become important because there is (almost) no escape from the earth.

Table 1.2. Length, velocity and time scales of environmental fluid processes and systems.

	Horizontal Length Scale L	Vertical Length Scale H	Velocity Scale U	Time Scale T
Processes:				
Microturbulence	1–10 cm	1–10 cm	1–10 cm/s	few seconds
Shear turbulence	0.1–10 m	0.1–10 m	0.1–1 m/s	few minutes
Water waves	0.1–10 m	1–100 cm	1–10 m/s	seconds to minutes
Convection	10–1000 m	1–1000 m	0.1–1 m/s	hours, days or seasons
Atmospheric systems:				
Urban airshed	1–10 km	100–1000 m	1–10 m/s	hours
Sea breeze	1–10 km	100–1000 m	1–10 m/s	hours
Thunderstorms	1–10 km	100–5000 m	1–10 m/s	hours
Mountain waves	10–100 km	10–1000 m	1–10 m/s	days
Tornado	10–100 m	100–1000 m	100 m/s	minutes to hours
Hurricane	100–1000 km	10 km	100 m/s	days to weeks
Weather patterns	100–1000 km	10 km	1–10 m/s	days to weeks
Climatic variations	Global	50 km	1–10 m/s	decades and beyond
Water systems:				
Aquifers	1–1000 km	10–1000 m	1–10 m/s	seasons to decades
Wetlands	10–1000 m	1–10 m	1–10 m/s	days to seasons
Small stream	1–10 m	0.1–1 m	1–10 m/s	seconds to minutes
Major river	10–1000 m	1–10 m	1–100 cm/s	minutes to hours
Lakes	1–100 km	10–1000 m	1–10 m/s	days to seasons
Estuaries	1–10 km	1–10 m	0.1–1 m/s	hours to days
Oceanic tides	basin size	basin depth	0.1–10 m/s	hours
Coastal ocean	1–100 km	1–100 m	0.1–1 m/s	few days
Upper ocean	10–1000 km	100–1000 m	1–100 cm/s	weeks to decades
Abyssal ocean	global	basin depth	0.1–1 cm/s	decades and beyond

1.5   EFM PROCESSES AT ENVIROMENTAL INTERFACES

In Section 1.2, EFM was defined as the scientific study of naturally occurring fluid flows of air and water on our planet Earth, especially of those flows that affect the environmental quality of air and water. In fact, these flows carry various substances that can modify environmental quality or be considered as indicators of environmental quality. These substances of concern may be gases, solutes or solids, and they can be naturally present or be produced by human activities. Anthropogenic contaminants can often create severe hazards for both human and environmental health.

There are two primary modes of transport that fall under the scope of EFM:

- *advection*, which is the transport by the flow of the fluid itself;
- *diffusion*, which is the transport associated with random motions within the fluid. These random motions occur at the molecular scale producing *molecular diffusion* or are caused by turbulence, causing *turbulent diffusion*. Molecular diffusion tends to be

important in the close vicinity of interfaces, regulating for example the passage of a soluble gas between air and water, while turbulent diffusion tends to act mostly within the body of the system.

Moreover, a large number of substances of environmental concern are simultaneously subjected to various transformation phenomena:

- *physical transformation*, caused by physical laws, such as radioactive decay;
- *chemical transformation*, produced by chemical reactions, such as hydrolysis and photolysis;
- *biochemical transformation*, due to biological processes, such as the uptake of nutrients by organisms and oxidation of organic matter.

When they reduce the level of contamination or the pollution hazard, transformation phenomena are beneficial to the environment. There are occasions, however, when the transformation creates a new substance that has adverse effects, called a secondary pollutant. A most important example of this is the formation of tropospheric ozone from nitrogen oxides by photochemical reactions.

Both transport and transformation processes investigated by EFM can occur either within one of the environmental fluid systems (atmosphere, hydrosphere) or at the interface with the lithosphere or biosphere. An *environmental interface* can be defined as a surface between two either abiotic or biotic systems that are in relative motion and exchange mass, heat and momentum through biophysical and/or chemical processes. These processes are fluctuating temporally and spatially. The study of interfaces is a crucial prerequisite toward a better understanding of the environment, but it is enormously complex and it is expected to occupy scientists for some significant time in the future (Mihailovic and Balaz, 2007).

In EFM, four main environmental interfaces need be considered, which are: air-water, air-land, water-sediment, and water-vegetation interfaces. They are affected by the following processes:

- The *air-water interface* of streams, rivers, lakes, estuaries, seas and oceans is subjected to momentum, heat and mass transfer. The main actor in *momentum transfer* is the shear stress exerted as the result of a difference between wind speed and direction in the air and the surface velocity in the water. The shear stress generates a wave field, part of which goes to creating surface drift currents. The accompanying *surface heat transfer* represents a relevant source or sink of heat in producing the thermal structure of a water body. Finally, several chemicals are transferred upward to the air or downward to the water depending on the substances involved and departure from equilibrium (Henry's Law). This process is termed *gas-transfer*. Hence, gas transfer of a volatile or semi-volatile chemical is a two-way process involving both dissolution by the water and volatilization into the air across an air-water interface. Furthermore, *air-entrainment* is the entrapment of undissolved air bubbles and air pockets by the flowing water (Chanson, 2004). Finally, the fate of *sea-salt aerosols*, once they are injected into the atmosphere from the ocean source, is governed by a series of physical processes such as transport, coagulation, dry and wet removal and chemical transformation;
- The *air-land interface* is a complex one that connects non-liquid terrestrial surfaces with the atmosphere. Examples are bare soil, desert, rocky land, ice, vegetative cover, buildings, and their non-homogeneous combinations. The physical state of the atmosphere is defined by its *temperature, humidity, wind speed*, and *pressure*. The question is: How does the atmosphere evolve its physical state? To answer this question we must determine the fluxes of *heat, energy* and *momentum* into and out of the air-land interface. A particular type of interface is the *biosphere*, which introduces characteristics of

living organisms. The rates at which trace gasses and energy are transferred through the air-biosphere interface depend upon a *complex* and *non-linear* interplay among physiological, ecological, biochemical, chemical and edaphic (soil) factors as well as meteorological conditions. Note, finally, that the surface fluxes of some gases, such as ammonia, mercury, and certain volatile organic compounds, can be upward into the air as well as downward to the surface and therefore should be studied as *bi-directional fluxes*;

- The *water-sediment interface*, which is very difficult to define precisely, is subjected to several complicated physical and chemical processes responsible for exchange of solids and solutes between the water column and the sediment bed. The physical processes involving the solids are *settling, sedimentation* and *resuspension*. Settling is the downward movement of sediment particles due to their negative buoyancy. Sedimentation occurs once the settled particles reach the bottom and join the bed sediments, while resuspension is the process by which particles of the bed are entrained upward into the water column, usually by shear flow. The processes involving the exchange of fluid and solutes between the water column and the bed sediments are termed *hyporheic flows*. The *hyporheic zone*, where groundwater and stream water mix, has hydrodynamic, physicochemical and biotic characteristics different from those of both the river and the subsurface environments. The effects of the hyporheic exchange processes are twofold. First, the hyporheic zone acts as a storage zone or a dead zone, which temporarily traps stream-transported solutes and subsequently releases them after some time. Second, the metabolic activity of the hyporheic microorganisms significantly alters the in-stream concentration of chemicals, both at the reach scale and at the basin scale. Furthermore, *diffusive exchanges*, either molecular or turbulent and including *adsorption/desorption*, can occur between the water column and the sediment bed. Also, the bed solutes can be subjected to advection and diffusion. *Bioturbation* is the mixing of sediment by small organisms, usually worms, living in the upper layers of the sediment;

- The *water-vegetation interface* is a relatively new subject of study, which considers the interaction between the flowing waters and submerged and/or emerged vegetation. Besides the transfer of substances between vegetation and water, the problem is complicated by the fact that the vegetation can deform under the passage of the water flow. Finally, vegetation also affects the transport of solutes within the flow. Besides classical turbulent diffusion, vegetated flows exhibit *mechanical dispersion* (Nepf et al, 1997). In fact, as in porous media flow, streamlines are circuitous as they bend and branch around the plant stems. Two particles that begin together may travel different winding paths, taking different times to travel the same longitudinal distance and thus they are dispersed. Mechanical and turbulent diffusion are independent and their contribution to the total diffusivity are additive. Another dispersive mechanism may be associated with the backflow region (*dead-zone*) within the plant stem wake, where the contaminants are trapped and separated from the main cloud to be after released enhancing longitudinal dispersion.

The previous overview points to the number and complexity of EFM processes occurring at the interfaces among environmental systems and explain why theoretical, laboratory, field and numerical studies have only begun recently to investigate EFM processes at environmental interfaces and to elucidate their role and effects on environmental quality.

1.6   CHALLENGES OF ENVIRONMENTAL INTERFACES MODELING

As previously outlined, the field of EFM abounds with various interfaces and can serve as an ideal platform for the application of new and fundamental approaches leading towards a better understanding of interfacial phenomena. The preceding definition of an environmental interface broadly covers the requisite multidisciplinary approach so necessary in environmental sciences and yet permits approach by well established scientific methods that have been developed to study the environment within approximations and assumptions designed to alleviate the complexity of the problems. Nonetheless, we anticipate that the next generation or two of EFM scientists will be confronted by the following challenges.

First is the seemingly perpetual problem of fluid turbulence. Without hoping for a miraculous new theory for all forms of fluid turbulence, EFM scientists are asked to continue forging new methods to deal effectively with its effects on environmental processes, particularly shear flow, convection, instabilities, and contaminant dispersion.

On the field side of the discipline, there is a strong need for observational techniques, including new instrumentation, to measure concentrations and fluxes in the very proximity of interfaces. This is particularly challenging not only because interfaces tend to be ill-defined at close range but also because instrumental probes run the risk of interfering with the situation that one is trying to observe in its natural manifestation. In that respect, remote sensing offers a unique advantage.

This leads us to another and relatively profound question: In which circumstances should we view the environmental interface as a fractal surface? And, if such is the case, how can this be accomplished most clearly and effectively in our models?

It goes without saying that computer models are ever more powerful. However, the time when a computer exists that will permit the simulation of an environmental system down to its micro-level (ex. urban-scale airshed model down to the size of an individual sediment particle or river model down to the size of an air bubble) is still in the distant future. Parameterization techniques will continue to be necessary for the undetermined future. Yet, these techniques are not stagnant; they need to evolve as the shortest resolved scale diminishes in the numerical models and as our discoveries and understanding of the factors at play demand the inclusion of evermore more processes in the models.

One particular need for in-depth inquiry, which arises in the context of environmental remediation, is the study of particle-particle interaction inside of a flowing fluid. The current state of the art remains largely empirical, and serious efforts need to be made to move gradually toward a science-based approach to the related processes.

At the opposite end of the spectrum, on the very largest spatial and temporal scales, EFM scientists are called to be ever more conscious of planetary limits and climatic implications.

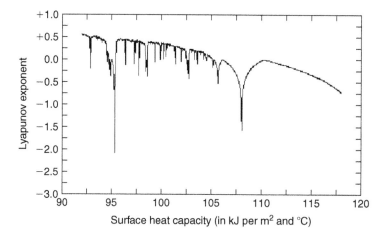

Figure 1.3. Dependence of Lyapunov exponent on soil surface heat capacity. The spectrum is obtained from dimensionless temperature as solution of the energy balance equation for the interface between land and lower atmosphere when the energy is exchanged by all three known mechanisms. Positive values correspond to temporal growth and hence chaotic behaviour.

Acute questions concern the sustainability of water resources and the capacity of environmental systems (atmosphere, hydrosphere, lithosphere and biosphere) to assimilate our waste.

EFM modellers base their calculations on mathematical models for the simulation and prediction of different processes, which are most often non-linear, describing relevant quantities in the field of consideration (Monteith and Unsworth, 1990). Many investigators have proved that complex dynamical evolutions lead to chaotic regime. A small *tuning* of initial conditions may lead the numerical model to instability if the system is a chaotic one. The aforementioned instabilities can be generated in temporal fluctuations on all space-time scales ranging from turbulence to climate. These kinds of uncertainties tend to take place at the interface between two environmental media. The land-air interface of the lower atmosphere and many other environmental interfaces are illustrative examples of the occurrence of irregularities in the temporal variation of some geophysical quantities (Figure 1.3).

APPENDIX A – LIST OF SYMBOLS

	List of Symbols	
Symbol	Definition	Dimensions or Units
C	Chézy coefficient	
H	vertical length scale	[L]
J_b	channel bed slope	
L	characteristic lenght scale	[L]
Q	kinematic heat flux	$[K\,L\,T^{-1}]$
Re	Reynolds number	
Ri	Richardson number	
T	time scale	[T]
U	fluid velocity	$[L{\cdot}T^{-1}]$
V	volume exchanged	$[L^3]$
g	gravitational acceleration constant	$[L\,T^{-2}]$
h	system height	[L]
m	mass	[M]
u^*	shear or friction velocity	$[L\cdot T^{-1}]$
z	vertical coordinate	[L]
$\Delta\rho$	change in density value	$[M\,L^{-3}]$
α	thermal expansion coefficient	$[K^{-1}]$
κ	Von Kármán constant	
μ	fluid dynamic viscosity	$[M\,L^{-1}\,T^{-1}]$
ρ	fluid density	$[M\,L^{-3}]$
τ	shear stress	$[M\,L^{-1}\,T^{-2}]$

APPENDIX B – SYNOPSIS

Environmental Fluid Mechanics (EFM) is the scientific study of naturally occurring fluid flows of air and water on our planet Earth, especially of those flows that affect the environmental quality of air and water. EFM has several points of contacts with more traditional cousin disciplines dealing with fluids, but also its own different and specific features and purposes. Moreover, in EFM applications, the domain size is typically very large, large enough to enable a number of distinct processes to play simultaneous roles, and it is not uncommon to encounter a hierarchy of processes embedded into one another. Stratification and turbulence are two essential ingredients of EFM, which are parameterized through the Richardson number and the Reynolds, respectively. EFM is characterized by several processes occurring at different scales within the natural fluids systems. Among them, the processes across the environmental interfaces are the focus of this book. Four are the main environmental interfaces: air-water interface, air-land interface, water-sediment interface and water-vegetation interface. Finally, it could be stressed that several research challenges, such as the problem of turbulence, the need for more detailed observational techniques and more powerful computer models, are still open for the next generation of EFM scientists.

APPENDIX C – KEYWORDS

By the end of the chapter you should have encountered the following terms. Ensure that you are familiar with them!

Environmental Fluid Mechanics	Shear turbulence	Environmental interface
Stratification	Convective turbulence	Air-water interface
Richardson number	Advection	Air-land interface
Turbulence	Diffusion	Water-sediment interface
Reynolds number	Transformation	Water-vegetation interface

APPENDIX D – QUESTIONS

What is the Environmental Fluid Mechanics?
Which are turbulence and stratification?
What is an environmental interface?
Which are the transport and transformation processes within EFM?
What is the hyporheic zone?

APPENDIX E – PROBLEMS

E1. Define Environmental Fluid Mechanics (EFM) and describe commonalities and differences between EFM and its cousin disciplines, such as Fluid Mechanics, Hydraulics, Geophysical Fluid Dynamics and Hydrology, highlighting their purpose, possibility of human control and the role of turbulence within them. List some examples of application of EFM in the decision-making context.

E2. Define the concepts of stratification and turbulence. Define Richardson number and Reynolds number. Discuss on the role that stratification and turbulence play within EFM. Give an example of a water system in which the presence of stratification significantly reduces the level of turbulence.

E3. A 10 m/s wind blows around a tower that is 15 m wide. The ambient density and viscosity of air are respectively $\rho_0 = 1.20 \, \text{kg/m}^3$ and $\mu = 1.8 \times 10^{-5} \, \text{kg/(m·s)}$. Show that the flow must be turbulent. How weak should the velocity be to make the Reynolds number fall below 1000? Is such value realistic? What can you conclude about the state of the flow in the wake of the tower on any day of the year?

E4. In first approximation, the depth-average velocity in a river is given by $U = C(gHJ_b)^{0.5}$, where C is called the Chézy coefficient, g is the gravitational acceleration, H the local water depth, and J_b the bottom slope of the river. A default value for C is 18. If the critical value of the Reynolds number for the onset of turbulence by shear along a boundary is $Re = 5 \times 10^5$ and if the bottom slope is 1 m per kilometer, what are the minimum water depth and water velocity that will cause the river flow to be turbulent? Are these values realistic? What can you conclude about the level of turbulence if the water depth is 0.7 m?

E5. Describe the four main environmental interfaces and list the processes occurring across these interfaces discussing commonalities and differences among these processes.

REFERENCES

Brutsaert, W., 2005, *Hydrology: An Introduction*, Cambridge University Press, 618 p.

Chanson, H., 2004, *Environmental Hydraulics of Open Channel Flows*, Elsevier Butterworth-Heinemann, 430 p.

Chow, V. T., 1959, *Open-Channel Hydraulics*, McGraw-Hill, 680 p.

Cushman-Roisin, B., 1994, *Introduction to Geophysical Fluid Dynamics*, Prentice Hall, 320 p.

Gualtieri C., 2006, Verification of wind-driven volatilization models. *Environmental Fluid Mechanics*, **6**, pp. 1–24.

Imberger, J., ed., 1998, *Physical Processes in Lakes and Oceans*, American Geophysical Union, 668 p.

Mihailovic, D., and Balaz, I., 2007, An essay about modelling problems of complex systems in Environmental Fluid Mechanics. Idojaras, 111(2–3), 209–220.

Monteith, J. L., and Unsworth, M., 1990, *Principles of Environmental Physics, Second Edition*. Elsevier, 304 p.

Munson, B. R., Young, D. F., and Okiishi, T. H., 1994, *Fundamentals of Fluid Mechanics*, 2nd ed., John Wiley & Sons, 893 p.

Nepf, H.M., Mugnier, C.G., and Zavistoki, R.A., 1997, The effects of vegetation on longitudinal dispersion, *Estuarine, Coastal and Shelf Science*, **44**, 675–684.

Pope, S. B., 2000. *Turbulent Flows*, Cambrige University Press, 771 p.

Rodhe, H., Charlson, R. J., and Anderson, T., 2000, Avoiding circular logic in climate modeling. An editorial essay, *Clim. Change*, **44**, 409–411.

Singh, V. P., and Hager, W. H., eds., 1996, *Environmental Hydraulics*, Kluwer Academic Pub., 415 p.

Sturm, T. W., 2001, *Open Channel Hydraulics*, McGraw-Hill, 493 p.

Part two
Processes at atmospheric interfaces

CHAPTER TWO

Point source atmospheric diffusion

Borivoj Rajković
Faculty of Physics, Institute for Meteorology, University of Belgrade, Belgrade, Serbia

Ilija Arsenić
Faculty of Agriculture, Institute for Vegetables and Crops, University of Novi Sad, Novi Sad, Serbia

Zoran Grsić
Institute of Sciences Vinča, Belgrade, Serbia

ABSTRACT

This chapter covers some theoretical aspects and several areas of modelling of atmospheric dispersion of a passive substance. After the introduction there is a section containing fundamentals about molecular diffusion. It has a derivation of Fick's law including sinks and sources of a passive substance. Some simple cases of sources and sinks are presented and their physical meaning discussed. At the end, we examine the point source substance diffusion in the case of a constant wind.

After the molecular mechanism of diffusion, we look at its generalization, the turbulent diffusion, how it arises and its problems from the modelling point of view. Finally, we present some results such as Taylor's theorem and Richardson's approach.

The second part of the chapter covers the basic models for point source diffusion. The starting point is the Gaussian model. First, we give a derivation of the concept and the several variants that are most common. Next, we discuss some of the limitations that are inherent to this approach, and present an example where one gets quite nice results in spite of all possible criticism of the Gaussian approach. The standard Gaussian model has serious problems in two situations, when the wind is changing either in time or in space, or if the size of the domain is large. In order to address these problems modellers have taken the next step creating the concept of Puff models. Instead of a single puff and its advection downwind, together with the appropriate lateral spreading, now there is a series of such puffs, which are consequently released. Spreading and advection of each puff is done according to its position and the moment of release; thus, such a model is able to take into account possible changes and variations both in time and space. We present the concept and its basic characteristics and then we offer some idea of its potential. Finally, we show several examples where this approach had been used.

Whether we have Gaussian or Puff-type models, in any case we still have to be able to calculate the amount of the deposited substance on the ground at a given location. So, this chapter ends with a subsection about the parameterizations of wet and dry deposition.

2.1 FOREWORD

It is clear that in the era of massive pollution of air, water and land, there is a great need for a reliable method of calculating the spreading of various substances that are constantly injected into the atmosphere. The nature of the flow in the lowest part of the atmosphere makes this task quite a complicated one. So, we set an additional condition that the method of calculation should have some degree of efficiency even if we have to sacrifice some of the features of the problem. Fortunately, a combination of empirical experience and theory that has been advanced in the last 100 years, and the rapid progress in computer power, make it possible to approach the problem and have a decent level of success.

The usual starting point in the problem of diffusion of a passive substance is the so-called "point" source which may be either instantaneous where we have a single "puff" emitted or continuous with a release that lasts for some time. From the methodological point of view, the starting point can be molecular diffusion. After we have introduced basic concepts and given some results we can start with so-called turbulent diffusion. This concept tries to take into account the turbulent nature of the atmospheric flow. That turned out to be, and still is, a very complex problem yet unsolved. Some of the basic parameters, such as the variance of the substance concentration both in the direction of the wind and in the lateral directions, are still not expressed in terms of the velocity fields. This is the well known problem of the "closure" of the equations of motion. There are several approaches in solving this problem but none are a complete solution of the problem. Fortunately, from the large accumulation of measurement data, values of these basic parameters are known with sufficient accuracy. In combination with some theory, they constitute an acceptable tool in solving the problem.

The class of models thus formed are Gaussian and later Puff-type models. They are a combination of empirical experience and a classical Fick's approach to the problem of diffusion. From the pure theoretical point of view we have two important results/concepts in treatment of the turbulent diffusion, Taylor's theorem and Richardson's formulation of the problem. Taylor's theorem explains why turbulent diffusion is a scale-dependent problem and even makes a prediction of the spreading of a "cloud" of a passive substance at the very beginning and at the final stage. The beginning and the final stage are measured relative to the integral time scale. Interestingly, at about the same time, Richardson developed a theory that offered a radical new approach to the solution of the problem. He substitutes a new variable, the so-called *distance–neighbour function* that depends only on the scale of the spreading cloud for density distribution in the x, y and z directions. From the mathematical point of view we are solving a partial differential equation by introducing an integral transformation which leads to a new equation of the same form as Fick's equation but with a *variable* coefficient of "viscosity." He managed to derive the form of the new mixing coefficient using all available empirical data. Unfortunately, the theory does not contain the "inverse" transformation from the distance–neighbour function to the normal distribution of passive substance in 3D (x, y, z) space. There is an alternative, at least in theory. We might seek the solution in the framework of a full three-dimensional prognostic model, very much like the ordinary problem of weather forecasting. The problem is that usually we do not have sufficient knowledge about the starting wind field structure, and even less, about the changes that occur at the boundaries of the domain in which we are trying to make the prediction.

So, for the time being, if efficiency of method is of paramount importance one would still work with a Gaussian-type model with all its enlargements that will account for some of its deficiencies. If computer power is not an issue and the problem's setup allows, we can use the Puff-type model. In the end, we should mention the inverse modelling techniques, such as those based on the Bayesian statistics or Kalman filtering.

2.2 DIFFUSION IN THE ABSENCE OF WIND

Diffusion is a term generally used for *molecular* dispersion of a passive substance consisting of gasses or very small particles. The basic quantity is concentration of the substance, χ, which can be either the number of particles in a unit volume having dimension $[L^{-3}]$, or the amount of mass in one kilogram of air expressed in non-dimensional units, or the volume of gas in a unit volume of air. The assumption that particles are very small allows us to neglect the influence of gravity and effectively treat the substance as a gas.

In relatively calm weather, the diffusion goes down the gradient of its concentration, that is, from the region of higher concentration to the regions of smaller concentration. The relation between flux, which is a mass of substance that is transported through the unit area in one second, and the gradient of concentration, can be expressed by Fick's law of diffusion. The basic assumption is that this transport is *proportional* to the gradient of the concentration. Let the sides of the elementary volume be along the coordinate axes, then the flux through the unit area orthogonal to the x-direction is

$$F(x) = -D\frac{\partial \chi}{\partial x}. \tag{2.1}$$

The constant of proportionality D $[L^2 T^{-1}]$ can be derived from the molecular considerations within the framework of an ideal gas. Its value is about 10^{-7}–10^{-5} m^2/s depending on the kind of gas. For the air we have the number $D_{air} \approx 10^{-5}$ m^2/s. The minus sign in Equation (2.1) denotes that the transport is *down* the gradient of concentration. Convergence of that flux gives the rate of change of χ,

$$\frac{d\chi}{dt} = \frac{dF}{dx} \tag{2.2}$$

which, under Fick's assumption, becomes:

$$\frac{d\chi}{dt} = \frac{d}{dx}\left(D\frac{d\chi}{dx}\right). \tag{2.3}$$

In three dimensions we have

$$\frac{d\chi}{dt} = \nabla(D\nabla\chi). \tag{2.4}$$

The constant D is kept "behind" the differential operator for the more general case of variable D. That is the case in turbulent diffusion when the flow is turbulent. Finally, if we have sources or sinks, with known rates *Src* and *Snk* the diffusion equation becomes:

$$\frac{\partial \chi}{\partial t} = \nabla(D\nabla\chi) + Src + Snk. \tag{2.5}$$

The *Src* measures the amount of gas being formed in a chemical transformation or the amount of pollutant that is emitted from a chimney or some other point or dispersed source, etc. The same goes for the *Snk* term. In order to avoid terminological confusion, we should note that in the equations of motion the whole diffusion term is viewed as the *Snk* term. So, solving Equation (2.5) means calculation of the time evolution of spatial distribution for χ given source(s) and sink(s) with appropriate initial and boundary conditions.

Both source and sink terms may represent quite complicated processes, so we have to make smaller or larger simplifications, which is usually referred to as parameterization. For instance, in the case of a sink term, it is common that the rate of change is *proportional to the amount of the present passive substance*. This is often the case in chemical transformations. Its mathematical form is:

$$Snk = -\sigma\chi, \tag{2.6}$$

where σ is a constant whose meaning will soon be apparent.

2.2.1   Sink term, no diffusion, point source

In order to get a better understanding of the physical meaning of this assumption, we will examine the time evolution of χ in the windless case and no diffusion. In that case, the one-dimensional version of the Equation (2.5) reduces to:

$$\frac{d\chi}{dt} + \sigma\chi = 0 \tag{2.7}$$

which has the solution

$$\chi_h(t) = \chi_0 \exp(-\sigma \cdot t) \tag{2.8}$$

presented in Figure 2.1, upper panel. So, this form of the sink term gives the exponential decay of concentration with e^{-1} folding time of $\tau = 1/\sigma$, i.e. after τ seconds the concentration of substance roughly halves.

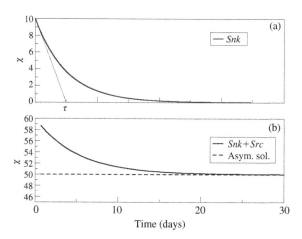

Figure 2.1. The upper panel shows the solution of Equation (2.7) while the lower one of Equation (2.9).

Coming back to Equation (2.5), in its dimensional version for the *Src* term, we start with the simplest case of the constant source whose strength is equal to f, while the *Snk* term is still of the form in Equation (2.6). These assumptions give

$$\frac{d\chi}{dt} + \sigma\chi = f. \tag{2.9}$$

Having in mind the solution from the previous case, we seek the solution in the form:

$$\chi(t) = \chi_h(t)g(t) \tag{2.10}$$

with $\chi_h(t)$ as

$$\chi_h(t) = \chi_0 \exp(-\sigma \cdot t). \tag{2.11}$$

After inserting this in the equation we get:

$$\chi_h(t) = \chi_0 \exp(-\sigma \cdot t) + \frac{f}{\sigma} \tag{2.12}$$

presented in Figure 2.1, lower panel.

The solution has two terms. The first term is the transitional part of the solution and decays with time. For long periods of time or more precisely for time, $t \approx \tau = 1/\sigma$ emerges a balance between source and sink terms:

$$\sigma\chi \approx Src. \tag{2.13}$$

That will always happen no matter how weak or strong is the source since the sink term is parameterized as proportional to the existing amount of passive material, and is always able to "catch up" with the increase of material given by *Src*. But the most problematic aspect of Equation (2.13) is that all "material" released stays very close to the point of release. So, a mechanism that will spread χ is still missing. The spreading is done by the second derivative, the "diffusion" term. To show that, we add the diffusion term while for the source term we choose the point source whose strength is Q. One of the ways to represent point source is through Dirac's delta function. With respect to time we will still restrict ourselves to the steady case, that is,

$$\sigma\chi = v\frac{d^2\chi}{dx^2} + Q\delta(x - x_0). \tag{2.14}$$

2.2.2 Sink term, with diffusion and point source

This is a non-homogeneous equation that can be solved using the Green's function approach. Away from the source we have

$$\sigma\chi_\pm = v\frac{\partial^2\chi_\pm}{\partial x^2}. \tag{2.15}$$

This is a *homogeneous* differential equation with boundary conditions $\chi_\pm \to 0$ for x to $\pm\infty$. Since the coefficients are constant we can immediately write solutions in the form:

$$\chi_\pm = C_\pm \exp\left[\pm\sqrt{\frac{\sigma}{v}}(x - x_0)\right]. \tag{2.16}$$

The non-homogeneous solution is a superposition of the solutions with the continuity condition for $\chi(x)$ at $x = x_0$. The condition for the first derivative at the point $x = x_0$ we can get if we integrate Equation (2.14) around that point,

$$\sigma \int_{x_0+\frac{\varepsilon}{2}}^{x_0+\frac{\varepsilon}{2}} \chi \, dx = v \frac{d\chi}{dx}\bigg|_{x_0+\frac{\varepsilon}{2}} - v \frac{d\chi}{dx}\bigg|_{x_0-\frac{\varepsilon}{2}} + Q. \tag{2.17}$$

If ε is very small there is a balance:

$$v \frac{d\chi}{dx}\bigg|_{x_0+} - v \frac{d\chi}{dx}\bigg|_{x_0-} + Q = 0. \tag{2.18}$$

Together with the continuity of χ we get

$$\chi(x) = \frac{Q}{\sqrt{\sigma \cdot v}} \begin{cases} \exp\left[-\sqrt{\sigma/v}(x - x_0)\right], & x > x_0 \\ \exp\left[-\sqrt{\sigma/v}(x_0 - x)\right], & x < x_0 \end{cases}. \tag{2.19}$$

This solution, shown in Figure 2.2, is symmetric on both sides of x_0 since we have the constant coefficients problem. So, in the case of molecular diffusion and sink term whose "activity" is proportional to the amount of the passive substance we get again exponential decay, but now in space, away from the point source. The width of the distribution is expressed through the ratio of σ/v. As before $\sigma \cdot \chi$ term keeps the passive substance close to the source while the diffusion term spreads it away from the source.

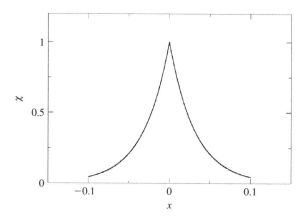

Figure 2.2. Graphical representation of the solution of Equation (2.14). Note the *symmetry* in the x-direction.

The relative strength of those two terms will decide how wide/narrow is the cloud of released material.

2.3   DIFFUSION IN THE PRESENCE OF WIND

Now we introduce motion into the problem, that is, of *advection* of a passive and conservative substance. A passive substance is a substance whose presence does not influence motion but is only carried around by the wind. For instance, water vapour can be viewed as such until condensation occurs. Smoke is another example; in small concentrations it can also be regarded as a passive substance, but in the situation of a large volcanic eruption, it can block the sun and therefore influence not only the winds but in the extreme event even the whole climate. In the conservative case we have

$$\frac{d\chi}{dt} = 0,$$

(2.20)

or explicitly

$$\frac{\partial\chi}{\partial t} + v \cdot \nabla\chi = 0.$$

(2.21)

If the velocities in the problem are much smaller then the speed of sound, we can assume that we have an incompressible fluid for which the continuity equation assumes a quite simple form:

$$\nabla \cdot (v\chi) = 0.$$

(2.22)

This allows us to write the conservation equation in the flux form as:

$$\frac{\partial\chi}{\partial t} + \nabla \cdot (v\chi) = 0.$$

(2.23)

Next we show (prove) that with the appropriate initial and boundary conditions the conservation equation with *Snk* as in Equation (2.6) and general *Src* term:

$$\frac{\partial\chi}{\partial t} + \nabla \cdot (v\chi) + \sigma\chi = f$$

(2.24)

has a unique solution.

Let us consider a cylindrical region G bounded by sides with area S, and at the top and bottom by surfaces St and Sb, respectively. We will denote the initial conditions with χ_0 and the boundary conditions with χ_s, valid at the sides of the cylinder S. For the velocity field we will assume the *no inflow condition*, that is, the normal velocity component is zero at S and that vertical velocity is also zero at the bottom and top of the cylinder

$$u_n = 0 \quad \text{at } S$$
$$w = 0 \quad \text{at } z = 0; z = H.$$

(2.25)

First we multiply Equation (2.24) with χ and get

$$\frac{\partial\chi^2}{\partial t} + \nabla \cdot (v\chi^2) + \sigma\chi^2 = f\chi.$$

(2.26)

If we integrate it over the domain V, over time $0 < t < T$ we get

$$\int_V \frac{\chi^2}{2} dV \bigg|_{t=T} - \int_V \frac{\chi^2}{2} dV \bigg|_{t=0} + \int_{t=0}^T dt \int_V \nabla\left(\frac{v\chi^2}{2}\right) + \sigma \int_{t=0}^T dt \int_V \chi^2 dV$$

$$= \int_{t=0}^T dt \int_V f\chi dV.$$

(2.27)

We then apply Gauss–Ostrogradsky's theorem, the transformation of the volume integral into a surface integral:

$$\int_V \nabla \cdot \left(\frac{v\chi^2}{2} \right) dV = \int_S \frac{u_n \chi^2}{2} ds \tag{2.28}$$

and get

$$\int_V \frac{\chi^2}{2} dV \bigg|_{t=T} - \int_V \frac{\chi^2}{2} dV \bigg|_{t=0} + \int_{t=0}^T dt \int_S \frac{u_n \chi^2}{2} ds + \sigma \int_{t=0}^T dt \int_V \chi^2 dV$$

$$= \int_{t=0}^T dt \int_V f \chi \, dV. \tag{2.29}$$

Now let us introduce new variables u^+ and u^- defined as

$$u^+ = \begin{cases} u_n, & u_n > 0 \\ 0, & u_n < 0 \end{cases} \tag{2.30}$$

and

$$u^- = u_n - u^+. \tag{2.31}$$

With these definitions, Equation (2.27) can be rewritten in the form

$$\int_V \frac{\chi^2}{2} dV \bigg|_{t=T} + \int_{t=0}^T dt \int_S \frac{u_n^- \chi^2}{2} ds + \sigma \int_{t=0}^T dt \int_V \chi^2 dV$$

$$= \int_V \frac{\chi^2}{2} dV \bigg|_{t=0} - \int_{t=0}^T dt \int_S \frac{u_n^- \chi^2}{2} ds + \int_{t=0}^T dt \int_V f \chi dV. \tag{2.32}$$

Now suppose that there are two, *different*, solutions, χ_1, χ_2. In that case, due to the linearity of the governing equation, their difference is also a solution, i.e.

$$\frac{\partial(\chi_2 - \chi_1)}{\partial t} + \nabla v(\chi_2 - \chi_1) + \sigma(\chi_2 - \chi_1) = 0. \tag{2.33}$$

If we introduce a new variable ξ, defined as

$$\chi_2 - \chi_1 = \xi, \tag{2.34}$$

we have

$$\frac{\partial \xi}{\partial t} + \nabla \cdot v\xi + \sigma\xi = 0 \tag{2.35}$$

while the boundary conditions

$$u_n = 0 \quad \text{at } S; \text{ with } u_n < 0 \tag{2.36}$$

now become

$$\xi = 0 \quad \text{at } S; \text{ with } u_n < 0 \tag{2.37}$$

and the integral Equation (2.33) becomes

$$\int_V \frac{\xi^2}{2} dV \Big|_{t=T} + \int_{t=0}^{T} dt \int_S \frac{u_n^+ \xi^2}{2} ds + \sigma \int_{t=0}^{T} dt \int_V \xi^2 dV = 0. \tag{2.38}$$

Since all integrands are positive definite, the above relation is true only if $\xi = 0$, which means that

$$\chi_2 = \chi_1. \tag{2.39}$$

With that we have proved the uniqueness of the solution of the diffusion equation.

Next we analyze the wind case, with the point source, in the same way that we analyzed the windless case. If we denote wind speed with u, the governing equation is

$$u \frac{d\chi}{dx} + \sigma\chi = v \frac{d^2\chi}{dx^2} + Q\delta(x - x_0). \tag{2.40}$$

Away from the source we have the homogeneous equation(s)

$$u \frac{d\chi_\pm}{dx} + \sigma\chi_\pm = v \frac{d^2\chi_\pm}{dx^2} \tag{2.41}$$

with the same boundary conditions as in the windless case $\chi_\pm \to 0$ for $x \to \pm\infty$. We seek particular solutions of Equation (2.41) in the form

$$\chi_\pm(x) = C_\pm \exp[\pm\lambda(x - x_0)] \tag{2.42}$$

which, upon substitution, leads to the quadratic equation for λ

$$\lambda^2 + \frac{u}{v}\lambda - \frac{\sigma}{v} = 0 \tag{2.43}$$

with roots

$$\lambda_\pm = -\frac{u}{2v} \pm \sqrt{\frac{\sigma}{v} + \frac{u^2}{4v^2}}. \tag{2.44}$$

Due to the condition $\chi_+ \to 0$ as $x \to +\infty$ and because

$$\frac{u}{2v} < \sqrt{\frac{\sigma}{v} + \frac{u^2}{4v^2}} \tag{2.45}$$

we discard the λ_- solution.

From the continuity of $\chi(x)$ and its first derivative we finally get

$$\chi(x) = \frac{Q}{\sqrt{\sigma \cdot \nu}} \begin{cases} \exp\left[-\left(\sqrt{\dfrac{\sigma}{\nu} + \dfrac{u^2}{4\nu^2}} - \dfrac{u}{2\nu}\right)(x - x_0)\right], & x > x_0 \\[4mm] \exp\left[-\left(\sqrt{\dfrac{\sigma}{\nu} + \dfrac{u^2}{4\nu^2}} + \dfrac{u}{2\nu}\right)(x_0 - x)\right], & x \le x_0 \end{cases} \tag{2.46}$$

Depending on the sign of u (here we take $u > 0$), typical forms of these solutions are presented in the Figure 2.3.

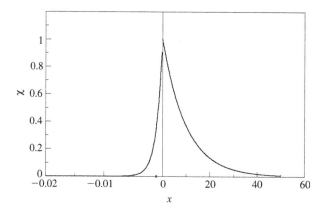

Figure 2.3. Graphical representation of the solution of Equation (2.26). Note the *asymmetry* in the x-direction.

Unlike the case of solution of Equation (2.14) this solution exhibits space *asymmetry* which is a consequence of the presence of wind. Upwind we have "narrowing" of the distribution while downwind "broadening" occurs.

2.4   TURBULENT DIFFUSION

So far we have had diffusion (spreading) of a passive substance by the molecular processes only. Due to the fact that the diffusion coefficient is in that case a constant, mathematical treatment of that problem is relatively easy. But, if a passive substance is released into the atmosphere, most likely close to the ground, measurements show that spreading is much stronger by several orders of magnitude than the calculations for the molecular diffusion suggest. The reason for that is that flow near the ground is always turbulent. The main characteristic of such flows is that they consist of a large number of eddies with very different sizes, which constantly develop and decay. In the case of steady-state turbulence, the distribution of the number of eddies is approximately constant. Its shape depends on several parameters. The basic one is the amount of turbulent kinetic energy (TKE), and the next most important is viscosity. The size of TKE depends on the wind shear and local stability near the ground. The existence of eddies means that instead of molecular movement we have a large number of bigger and smaller vortices that carry around passive substances, that is, we have an extremely complicated pattern of advection field resulting in very efficient diffusion. The biggest eddies are of the order of several hundreds of meters while the smallest ones are small enough so that viscous dissipation is sufficient to transform all TKE into heat. This prevents the formation of even smaller vortices and we are referring to the size of these smallest elements as Kolmogorof's scale. Because of such a large difference from the ordinary diffusion, a new name has been introduced: turbulent diffusion.

Even such a short description of turbulent diffusion is sufficient to indicate that its mathematical treatment must be extremely difficult. The spread of a cloud of a passive substance results from the nonlinear interaction of the turbulent elements of the surrounding air and eddies of a passive substance. The nature of the nonlinear interaction is that it is local. To show that, let us assume that at a particular moment our cloud is very small relative to the turbulent element so it is embedded in it. In that case the cloud will be carried around but without changes in its dimensions. This is depicted in the left sketch in Figure 2.4. Grey is the cloud while in white we have an air eddy. The opposite would be that we have very small eddies of air impinging on a relatively large cloud (the right part of the same figure, where in white are turbulent elements of the air while in grey is the cloud). Air will just mix better the material inside the cloud but again without significant change in the cloud's overall size. But, if we have interaction of the turbulent elements of roughly the same size as the cloud's (the central part of the same figure), then the "left"/"right" edge of the cloud will be extended by the eddy there, thus roughly doubling the size of the cloud.

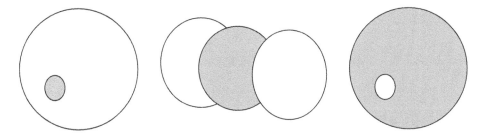

Figure 2.4. Sketch of the three possible situations in relative scales between turbulent elements (white) and cloud (grey).

The same conclusion comes from a simple analysis of the nonlinear $(u\partial_x u)$ term[1] which will give somewhat more precise result as the above intuitive/graphical reasoning. The second fact that we must take into account is that larger elements have larger velocities. The locality of the interaction together with the velocity dependence on the size of the elements then explains the increase in diffusion rate.

2.5 TAYLOR'S THEOREM

In the introduction to the problem of calculating turbulent diffusion we have highlighted the fact that a basic difficulty lies in the fact that the rate of expansion of a cloud of a passive substance depends on the "size" of that cloud at that moment. That fact almost prevents us from the Fickian approach in the diffusion calculations. There is a beautiful explanation/picture of that situation which is encompassed in the so-called Taylor's theorem. Let us first clarify several concepts that have been so far loosely defined or have been surmised intuitively.

---

[1] Let us have two components with respective wave numbers k_1 and k_2. Then $(u\partial_x u)$ will create $\sin k_1 x \cdot \cos k_2 x \cdot \sin(k_1 + k_2)x + \sin(k_1 - k_2)x$ which means that we have two *new* components with wave numbers $(k_1 + k_2)$ and $(k_2 - k_1)$. If $k_1 >> k_2$ then $k_1 + k_2 \sim k_1$ and nothing new happens. The same goes for $k_2 - k_2 \sim k_1$. Only if $k_2 \sim k_1$ then we get $k_1 + k_2 \sim 2k_1$, i.e. creation of the new wave number (smaller eddy).

The first is the "size" of a cloud of a passive substance. Let us, for the sake of clarity, reduce the geometry to one dimension and let at $x = 0$ be a source of a passive substance that continuously releases particles. Further let us assume that their size is very small like smoke, fine dust, pollen etc., so light that we will assume that these particles float in the surrounding air. Now as time passes the released particles will spread away from each other. A possible definition of the cloud size would be the distance from the furthest particle on the left to the furthest particle on the right. But that is not a very clever choice since we know that in every gas such as air we have Maxwell's distribution of velocities and these furthest particles could be very far away but in negligible concentration. The more practical choice is through the following mathematical definition:

$$S = \overline{x_i^2}. \tag{2.47}$$

The advantage of this definition is that it takes into account the concentration as well as the distance of particles in the cloud. The mean wind (in the sense of Reynolds's decomposition, has only large-scale variations larger than the size of the expanding cloud) will not influence the size of the cloud but rather carry it downstream without changes in its geometry. That can be taken into account by introducing the movement of the centre of the cloud as the position of its median, i.e.

$$S = \overline{(x_m - x_i)^2} \tag{2.48}$$

with

$$x_m = \frac{1}{N} \sum_{n=1}^{N} x_i. \tag{2.49}$$

Since this is a trivial extension of the windless case we will return to the zero wind case and analyze the case given by Equation (2.47). The question is how *fast* does a cloud spread. We will define the "speed" of the increase in size as

$$\frac{dS}{dt} = \frac{d}{dt} \overline{x_i^2}. \tag{2.50}$$

Since differentiating and averaging are commutative operations we have

$$\frac{dS}{dt} = \overline{\frac{d}{dt} x_i^2} \tag{2.51}$$

or

$$\frac{dS}{dt} = \overline{\frac{x_i}{2} \frac{dx_i}{dt}} = \frac{1}{2} \overline{x_i v_i}. \tag{2.52}$$

If we express the distance of the i-th particle through the integral of its velocity, from the beginning of the release till time t, we can write

$$\frac{dS}{dt} = \frac{1}{2} \overline{v_i \int_0^t v_i(\tau) d\tau}. \tag{2.53}$$

Since velocity at the time t is independent of the sequence of the integration and again integration and averaging are interchangeable operations we have

$$\frac{dS}{dt} = \frac{1}{2} \int_0^t \overline{v_i(t)v_i(\tau)} d\tau. \tag{2.54}$$

This is the increase of S at the moment t relative to the *beginning* of the release. But, of greater interest is what is happening relative to this moment, that is, we would like to change the frame of reference, from the moment $t = 0$ to the moment t (see Figure 2.5). The time ξ in this new frame is related to τ as

$$t = \tau + \xi. \tag{2.55}$$

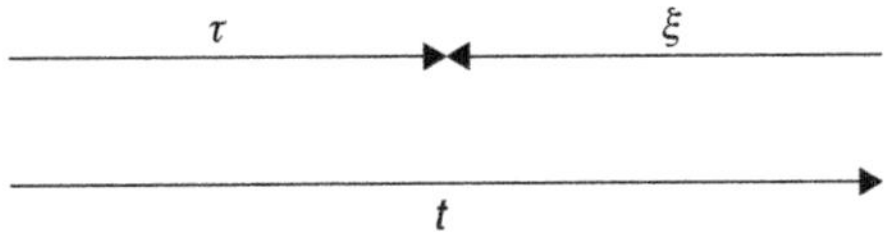

Figure 2.5. Sketch explaining the relation between time relative to the beginning of the release (τ) and time relative to this moment (ξ).

The Equation (2.54) then becomes

$$\frac{dS}{dt} = \frac{1}{2} \int_0^t \overline{v_i(t)v_i(t + \xi)} d\xi. \tag{2.56}$$

Using the definition of the auto-correlation function:

$$R(t, \xi) = \frac{\overline{v_i(t)v_i(t + \xi)}}{\overline{v_i(t)^2}} \tag{2.57}$$

and concentrating on the case of the homogeneous turbulence for which

$$R(t, \xi) = R(\xi), \tag{2.58}$$

we finally get

$$S(t) = \frac{1}{2} \overline{v_i^2} \int_0^t dt' \int_0^{t'} R(\xi) d\xi. \tag{2.59}$$

This relation constitutes Taylor's theorem (Taylor, 1921). Provided that we know the shape of the auto-correlation function, we can calculate the size of the cloud at any moment. Unfortunately, it is even more difficult to get the form of $R(\xi)$, as it is obvious from its definition. So, it seems that we have not gained much. We have expressed the unknown S with another, perhaps even more complicated variable R. Well, if we wanted an operational relation, we didn't get one but there are several very important points that are hidden in this result. Let us first concentrate on the very beginning of the cloud growth. If the time is really short, i.e. ξ is very small, we can assume that $R(\xi \approx 0) \approx 1$ which immediately gives the result:

$$S(t) = \frac{\overline{v_i^2}}{2}t^2 = const \cdot t^2. \tag{2.60}$$

Actually if we want something that has dimensions of length, we should introduce[2]

$$D = \sqrt{x_i^2} = const \cdot t. \tag{2.61}$$

These two relations are an exact derivation and/or confirmation of the experimental fact that the cloud's expansion rate, in the early stages of expansion, increases with time. Now let us look at the other extreme, a very "long" time after the start of the diffusion. What is very long is not yet clear, but it will soon become clear. One of the global parameters that characterizes every auto-correlation function is its integral time scale defined as

$$\int\limits_0^\infty R(\xi)d\xi = T. \tag{2.62}$$

So, if t (or more precisely t') in Equation (2.61) is much larger than T the inner integral's value is close to T. That gives us as the result for D:

$$D = \sqrt{x_i^2} = const' \cdot \sqrt{t} \tag{2.63}$$

Equations (2.61) and (2.63) are telling us that at the beginning of diffusion, the cloud's size grows linearly in time and as the process goes on its growth *slows* down and for the $t >> T$ reduces to the square root of time. The explanation of this result comes from the structure of turbulent flows which is the cause of the spreading. As we have explained earlier, the turbulent character of the flow means that flow consists of many eddies of different sizes. Besides the distribution in size, of even more importance is the distribution in speed. The fact that we must take into account is that *larger* elements have *larger velocities*. The locality of the interaction together with the velocity dependence on the size of the elements explains the *increase* in diffusion rate. At the beginning small elements are responsible for the turbulent diffusion. As the cloud grows larger, larger and faster elements are widening the cloud. This is seen as the increase of the diffusion rate. Once the cloud is comparable and bigger than the size of the elements with the largest kinetic energy, the diffusion rate slows down since there are no more *new* elements faster than the previous one to take over further spreading.

2.6 RICHARDSON'S THEORY

Starting point of Richardson's (Richardson, 1926) theory was also the fact that diffusion depends on the scale of the cloud. Therefore he introduced a new variable, the so-called *distance–neighbour function q(l)*, defined as:

$$q(l) \equiv \frac{1}{N}\int\limits_{-\infty}^{\infty} \chi(x)\chi(x + l)dx, \tag{2.64}$$

---

[2] The variable D can serve as the definition of the cloud size.

with

$$N = \int_{-\infty}^{\infty} \chi(x)dx \tag{2.65}$$

being the number of the particles in the cloud, which we assume is *constant* in time. The name for $q(l)$ becomes its definition:

- $\chi(x)$: is the number of particles on dx
- $(\chi(x)dx)/N$: is that number relative to the total number of particles
- $\chi(x+l)$: is the number of particles, on *unit* length at distance l meters away
- $(\chi(x)dx/N) \cdot \chi(x+l)$: is the *relative* number of *neighbours* of all particles from the section whose length is dx and is l meters away.

When we *add* them all we get a number of neighbours of each particle in a cloud at the relative distance of l meters. For a better understanding let us consider a simple distribution $\chi(x)$ with constant concentration χ_0 over an interval d starting at $x = a$ and 0 elsewhere:

$$\chi(x) = \begin{cases} 0, & x < a \\ \chi_0, & a \le x \le a+d. \\ 0, & x > a+d \end{cases} \tag{2.66}$$

From the definition of $q(l)$ and using the translation $x \to x+l$ we can show that $q(l)$ is an *even* function and therefore it is sufficient to calculate it only for $l > 0$. From Equation (2.66) we get

$$\chi(x)\chi(x+l) = \begin{cases} 0, & x < a \\ \chi_0^2, & a \le x \le a+d-l. \\ 0, & x > a+d-l \end{cases} \tag{2.67}$$

If we insert this into Equation (2.64) we get

$$q(l) = \begin{cases} 0, & -d < l \\ \chi_0(1+l/d), & -d > l > 0 \\ \chi_0(1-l/d), & 0 < l < d \\ 0, & d < l \end{cases}. \tag{2.68}$$

Both, $\chi(x)$ and its $q(l)$, are shown in Figure 2.6.

The main advantage of $q(l,t)$ over $\chi(x,t)$ is that it depends on l and not on x, that is, the scale of the cloud is the only spatial variable in the problem.

If we want to switch to the new framework of $q(l)$ instead of $\chi(x,t)$ two questions arise. The first is can we develop the equation for the time *evolution* of $q(l,t)$? Given that we are successful in that, we face the second problem, can we create the *methodology* with which we can get the *inverse* $\chi(x,t)$ from $q(l,t)$?

To get the prognostic equation for $q(l,t)$ we start with Equation (2.64) by differentiating it

$$\frac{\partial q(l)}{\partial t} = \frac{1}{N} \int_{-\infty}^{\infty} \frac{\partial}{\partial t}(\chi\chi_l)dx = \frac{1}{N} \int_{-\infty}^{\infty} \chi_l \frac{\partial}{\partial t}\chi + \chi \frac{\partial}{\partial t}\chi_l dx. \tag{2.69}$$

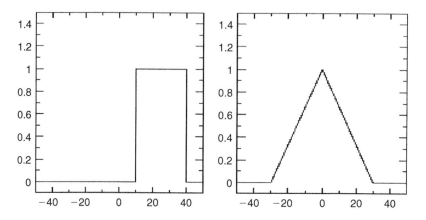

Figure 2.6. On the left, the concentration distribution in arbitrary units. On the right is the distance–neighbor function for that distribution.

Now, if the process is of the Fickian type then

$$\frac{\partial \chi}{\partial t} = K \frac{\partial^2 \chi}{\partial x^2} \tag{2.70}$$

and analogously

$$\frac{\partial \chi_l}{\partial t} = K \frac{\partial^2 \chi_l}{\partial x^2}. \tag{2.71}$$

Noticing that differentiation over x and over l are the same, Equation (2.71) can be rewritten as

$$\frac{\partial \chi_l}{\partial t} = K \frac{\partial^2 \chi_l}{\partial l^2}. \tag{2.72}$$

So, Equation (2.69) becomes:

$$\frac{\partial q(l)}{\partial t} = \frac{1}{N} \int_{-\infty}^{\infty} \left(\chi_l \frac{\partial^2 \chi_l}{\partial x^2} + \chi \frac{\partial^2 \chi_l}{\partial l^2} \right) dx. \tag{2.73}$$

If we transform the integrand using several identities:

$$\chi_l \frac{\partial^2 \chi}{\partial x^2} + \chi \frac{\partial^2 \chi_l}{\partial l^2} = \frac{\partial^2}{\partial x^2}(\chi_l \chi) - 2\frac{\partial \chi}{\partial x}\frac{\partial \chi_l}{\partial l} - 2\chi \frac{\partial^2 \chi_l}{\partial l^2} + 2\chi \frac{\partial^2 \chi_l}{\partial l^2}, \tag{2.74}$$

$$2\frac{\partial \chi}{\partial x}\frac{\partial \chi_l}{\partial l} - 2\chi \frac{\partial^2 \chi_l}{\partial l^2} = \frac{\partial^2}{\partial x \partial l}(\chi \chi_l) \tag{2.75}$$

and

$$2\chi\frac{\partial^2 \chi_l}{\partial l^2} = \frac{\partial^2}{\partial l^2}(\chi\chi_l),$$

(2.76)

we finally get

$$\frac{\partial q(l)}{\partial t} = 2K\frac{\partial^2}{\partial l^2}q(l).$$

(2.77)

The meaning of this is that for the molecular mechanism of diffusion both descriptions, the one using $\chi(x)$ and the other using $q(l, t)$, are equally good. Next we generalize, Equation (2.77) in the form:

$$\frac{\partial q(l)}{\partial t} = \frac{\partial}{\partial l}\left[K(l)\frac{\partial}{\partial l}q(l)\right].$$

(2.78)

Can we find (form) $K(l)$? To do that Richardson analyzed all the data available to him at that time covering a very wide range of scales from the synoptic ones to the smallest, molecular, scales (Figure 2.7). From these data he proposed that $K(l)$ should be

$$K(l) = 0.2l^{3/4}.$$

(2.79)

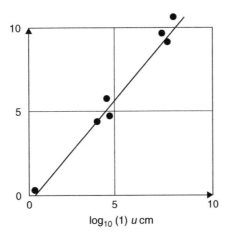

$\log_{10}(1)\ u\ cm$

Figure 2.7. Deduced diffusion coefficient from measurements, observations of the processes from the synoptic scale to molecular one (black dots) and suggested linear interpolation of those data. Linear form for the logarithmic scales indicates power function for $K(l)$.

With this relation, Equation (2.78) is complete and ready to serve as the equation for evolution in space and time for the variable q. The procedure would be as follows, for a given concentration distribution we make an integral transformation, defined in Equation (2.64), to form $q(l, 0)$ and then integrate Equation (2.78) to get $q(l, t)$. What about the second step, the inversion procedure? Unfortunately, he was not as successful in that as he was in the first part of the theory. Maybe that was the reason why he left this problem for over 25 years

(Richardson, 1952), when he showed that for a limited class of concentration distributions he was able to perform the inversion part. Due to the fact that inversion for any $q(l, t)$ has not be found, the general solution of the turbulent diffusion problem using Richardson's approach remains still an open problem.

2.7 THE GAUSSIAN MODEL FOR A POINT SOURCE

From both Taylor's theorem and Richardson's theory, we know that turbulent diffusion of a passive substance has difficulties in dealing with Fick's equation. But none of them offers an operational framework that can give an estimate of, for instance, how big is the concentration of a passive pollutant around, say, a factory chimney or some other quasi-point source. There are two different situations regarding the manner in which the material is released. If we have emissions with relatively short duration we talk about a puff. If, on the other hand, we have a continuous source then we call it a plume.

In our highly industrial era the number of sources is very large and we are forced to come up with some approach that is relatively easy to handle, and yet sufficiently accurate to answer the question of the spatial distribution of concentration from a source that emits a pollutant into windy and unstable/stable atmosphere. The only possible approach is a combination of theory and experiment. The hope is that elements of the dispersion theory can be parameterized using the field measurements and the rest of it supplied from the Fick's equation. To fulfil that, in England in 1925 near the city of Porton a series of field experiments (Pasquill and Smith, 1983) were conducted in which a smoke was released and its concentration was measured. The purpose of the experiment was to find the spatial distribution of the released substance. The atmosphere was close to neutral with the wind of about $7 \, [LT^{-1}]$. Concentration was measured downwind, and in the direction perpendicular to that direction, roughly every 100 meters. From these data an approximate concentration distribution was deduced in the form of the exponential function

$$\chi(x_0, y) = \chi_0 \exp(-ay^r) \tag{2.80}$$

where x_0 is a point in the downwind direction, while y is horizontal distance perpendicular to the x axes. Following these preliminary results from various other experiments, Brahman *et al.* (1952) have analyzed the New Mexico experiments, Crozier and Seely (1955) have analyzed Australian experiments from 1953, Pasquill (1955, 1956) used data from another experiment at Porton, etc. From most of the experimental results general shape of the plume could be expressed as:

$$\chi(x, y, z) = Q \exp[-(by)^r - (cz)^s] \tag{2.81}$$

where x, y, z are distances relative to the source. Parameters b and c depend on the size of the plume in the respective directions. The constant Q is a measure of the rate of emission. If we assume that the wind is constant throughout the considered period, the concentration takes the form of a plume whose main axis is downwind, with lateral spread in both directions. The effect of the wind is that it dilutes the concentration, which means the stronger the wind the smaller the concentration. The amount of the material that is diffused is determined by the strength of the source. Concentration is inversely proportional to the wind's strength,

$$\chi \approx \frac{1}{U}. \tag{2.82}$$

Like in Taylor's theorem for the measure of lateral spread we take:

$$\sigma_i^2 = \frac{\int\limits_0^\infty x_i^2 \chi \, dx_i}{\int\limits_0^\infty \chi \, dx_i}, \quad i = 2, 3.$$ (2.83)

Let us note that from the conservation of mass we have:

$$\iint\limits_{y,z} U\chi \, dy \, dz = Q.$$ (2.84)

Now, the expression using the above relations can be rewritten in the following way

$$\chi(x, y, z) = \frac{Q}{B_1 \sigma_y \sigma_z} \left\{ -\left[\left(\frac{\Gamma(3/r)}{\Gamma(1/r)}\right)^{r/2} \left(\frac{y}{\sigma_y}\right)^r + \left(\frac{\Gamma(3/s)}{\Gamma(1/s)}\right)^{s/2} \left(\frac{y}{\sigma_z}\right)^s \right] \right\},$$ (2.85)

where

$$\frac{1}{B_1} = \frac{rs}{4U} \frac{[\Gamma(3/r)\Gamma(3/s)]^{1/2}}{[\Gamma(1/r)\Gamma(1/s)]^{3/2}}.$$ (2.86)

with $r = s = 2$ and using the relations:

$$\Gamma(n+1) = n\Gamma(n)$$ (2.87)

and

$$\Gamma(1/2) = \sqrt{\pi}$$ (2.88)

we get the so-called standard Gaussian form of the plume

$$\chi(x_0, y, z) = \frac{Q}{\sqrt{2\pi}\sigma_y \sigma_z} \exp\left[-\frac{1}{2}\left(\frac{y^2}{\sigma_y^2} + \frac{z^2}{\sigma_z^2}\right) \right].$$ (2.89)

In the end we want to have a concentration relative to a fixed point, the beginning of the x-axis, the usual position of the source. Then, taking into account the wind we have

$$\chi(x, y, z) = \frac{Q}{\sqrt{2\pi}U\sigma_y \sigma_z} \exp\left[-\frac{1}{2}\left(\frac{y^2}{\sigma_y^2} + \frac{z^2}{\sigma_z^2}\right) \right].$$ (2.90)

All this is valid for ground sources. If the height of the source is at H we have:

$$\chi(x, y, z) = \frac{Q}{\sqrt{2\pi}U\sigma_y \sigma_z} \exp\left[-\frac{1}{2}\left(\frac{y^2}{\sigma_y^2} + \frac{(z-H)^2}{\sigma_z^2} + \frac{(z+H)^2}{\sigma_z^2}\right) \right].$$ (2.91)

The second term in the z-direction comes from the fact that with time the cloud will spread so much that it will reach the ground. In that case, we can imagine a second source that is a mirror image of the original, positioned at $-H$ below the ground so that its contribution to the points above the ground starts exactly at the point where the original cloud touched the ground. In the case of the short release time (puff) we have

$$\chi(x,y,z,t) = \frac{q}{\sqrt{(2\pi)^3}\,\sigma_x\sigma_y\sigma_z}\exp\left\{-\left[\frac{(x-Ut)^2}{2\sigma_x^2} + \frac{y^2}{2\sigma_y^2}\right]\right\}$$

$$\times\left\{\exp\left[-\frac{(z-H)^2}{2\sigma_z^2}\right] + \exp\left[\frac{(z+H)^2}{2\sigma_z^2}\right]\right\}.$$

(2.92)

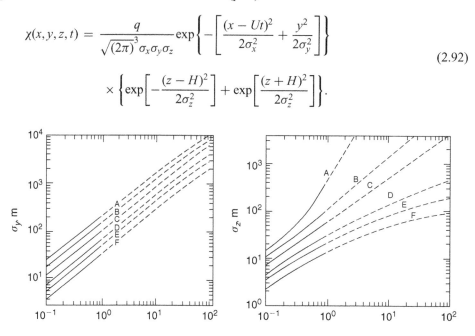

Figure 2.8. On the left panel we have downwind variation of the lateral diffusion coefficient σ_y while on the right we have the same for vertical coefficient σ_z.

Even though we formally differentiate σ_x and σ_y the usual assumption is that there is isotropy in x and y. Obviously, parameters σ_i, $i = 1$, 2, 3 are at the centre of the Gaussian approach and most of the "meteorology" is hidden in them. They should reflect the local stability and the parameters that characterize turbulent flow. To express all that with a single number (two numbers) seems a difficult problem. Here again we insert as much of the empirical experience as we can. Actually our starting point, Equation (2.92), has the fact that flow is turbulent and therefore is characterized by lateral spread of a passive substance (σ_y and σ_z). Taking into account the stability of the atmosphere requires an additional effort. The first attempt was made by Pasquill (1961), later modified by Gifford (1961), and referred to as the Pasquill–Gifford (P–G) stability class. This collective work of several researchers in the interpretation of the available measurements resulted in formation of nomograms (Turner, 1969), shown in Figure 2.8, that have dependence of the σ's in the y and z directions for quite a wide range of distances. The dashed parts of the curves are actually *extrapolations* of the measured data. The whole range of possible stability states, that is, possible values of $\partial\Theta/\partial z$, where $\Theta(z)$ is potential temperature, were divided into seven categories, labelled as A–F. The next step is to determine the category (class) using only the standard meteorological data, 2 meters temperature, 10 meters wind and cloud cover. Pasquill and Smith (1983) devised such a scheme, presented in Table 2.1. The question of stability was covered only with the position of the Sun. The idea is that a high Sun means a warmer part of the day and warmer season in which we should expect an unstable regime within the PBL. In the next table we show how these categories are determined.

Table 2.1. Determination of categories from wind speed, solar radiation and cloud cover data available from routine measurements.

Surface (10 m) wind speed ms^{-1}	Daytime Incoming solar radiation			Night-time Cloudiness	
				>=4/8	<=3/8
<2	A	A–B	B	–	–
2–3	A–B	B	C	E	F
3–5	B	B–C	C	D	E
5–6	C	C–D	D	D	D
>6	C	D	D	D	D

Later, Briggs (1973) turned these graphs into analytical relations thus making them operational for computers. At this moment we must once again state the assumptions and the validity of the results given so far. First from the measurements done over relatively small domains and therefore for short periods, the concentration distribution in the directions normal to the wind direction was approximated by the exponential curves. These measurements and consequent fits have by their nature some spread. So in order to get formal similarity with the Fickian picture we set values of r and s to 2 because in that case we have *Gaussian* distributions. Beside the spatial variation we have also the question of the time averages. The shorter the time average the closer we are to the actual situation. So we have 3 minutes, 10 minutes or hourly σ's. A parameterization has been proposed (Gifford and Hanna, 1973) that takes into account that for the longer times σ should increase:

$$\sigma(t > 10) = \sigma_{10}\left(\frac{t}{t_{10}}\right)^q \tag{2.93}$$

with σ_{10} are denoted values of σ for 10 minutes. Factor q has two values depending on the length of the time interval. Up to an hour $q = 0.2$ while for the longer time, 1 hour $< t < 100$ hours $q = 0.25$.

In spite of the obvious crudeness of the calculation this approach has the advantage of being very straight forward and needs practically one number, wind at the point of release. The Sun's height can be estimated from the astronomy. If there is additional data, in particular temperature gradient near the ground, we can refine the expressions for the two basic parameters σ_y and σ_z. The concept of Pasquil-Gifford-Turner that σ is the only parameter describing the diffusion process was later paralleled by the similarity approach. The group of models based on that concept of similarity has been proposed by several authors: Golder (1972), Horst (1979), Nieuwstad (1980) and Briggs (1982) among others. The starting point of the theory is the well known Monin-Obukhov's theory with its length scale

$$L = \frac{\Theta_0 u_*^3}{\kappa w \theta_0}. \tag{2.94}$$

The next step then is to relate σ_z to various parameters connected to the Monin–Obukhov theory:

$$\sigma_y = \sigma_\theta U F_y\left(u_*, w_*, \frac{z}{L}, z_i\right) \tag{2.95}$$

and

$$\sigma_y = \sigma_\varphi U F_z \left(u_*, w_*, \frac{z}{L}, z_i \right). \tag{2.96}$$

In order to accomplish that, an extensive re-examination of almost all data from the field experiments was done. The basic problem comes from the formulation of Pasquil-Gifford-Turner concept that does not take into account either sensible and latent heat flux nor z_0. Instead they have insolation alone. Golder, in his 1972 paper, was able to produce nomograms, which though made subjectively, relate on the one side the pair z_0, L^{-1} to Pasquill-Gifford-Turner categories (A–F).

As an example of calculations that are based on the Gauss model we present estimates of possible pollution coming from a point source for the period of one year in Figure 2.9. The wind is measured at the height of 40 meters which is close to the height of the chimney, which is the possible source of pollution. The wind was averaged on an hourly basis, which was taken into account when choosing the appropriate σ.

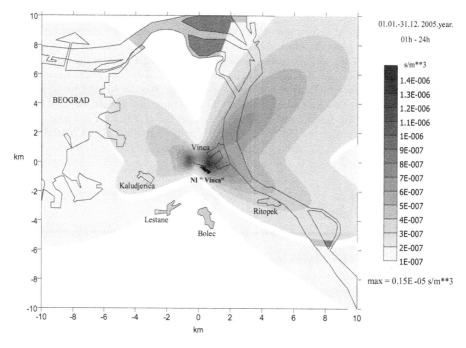

Figure 2.9. The annual concentrations for the 2005. year of a continous point source for the nuclear facility Vinca near Belgrade, Wind data are the standard hourly averaged wind with the direction of the prevailing wind. Wind was measured at the level of 40 meters, approximately the height of the possible source of pollution.

Due to the fast development of the 3-D models capable of calculating turbulent mixing coefficients from the prognostic equation for turbulent kinetic energy, and which are therefore considered as being adequate in calculating (forecasting) change of concentration in time and space, it is possible to test the Gaussian model against their predictions. We made a comparison between concentrations calculated with such a model (actually a 2-D version in the x–z plane) and the Gaussian model whose results we have already shown in the previous example (Grsic, 1991). In a nutshell, he shows that the largest difference between the two models was not greater than 50%, being most of the time between 25% and 35%.

2.8 THE PUFF MODEL FOR A CONTINUOUS POINT SOURCE

As it was pointed out in the introduction, the Puff model is an attempt to generalize the Gaussian concept for non-stationary releases or spatially non-homogeneous wind or for both. The continuous releases are treated as time–series consecutive instantaneous releases, or puffs. The amount of substance q allocated to each puff *is* the release rate Q multiplied by the time interval Δt between two consecutive releases. So, as time passes, the *number* of puffs that have been released is growing. Each puff is carried around by the wind valid for that particular time interval. Beside changes in the position of the centre, the size of each puff also *increases* due to the turbulent diffusion. Figure 2.10 shows a sketch of the actual meandering of the plume (upper panel) and its approximation by the series of puffs consecutively released from the point source located at S (lower panel). Beside puffs, we have a grid of cells spanning the space in which we want to calculate the concentration distribution. These cells are usually of constant volume. Concentration in a cell is the *sum* of the contribution of all puffs released up to that moment. If the index of a receiving cell is denoted by ic and index of puffs as ipf, then the contribution of that puff is

$$\chi_{ic}(x_{ic}, y_{ic}, z_{ic}, n \cdot \Delta t) = \frac{Q \cdot \Delta t}{\sigma_{ipf}} \exp\left\{-\left[\frac{(x_{ic} - x_{ipf})^2}{2\sigma_x^2} + \frac{(y_{ic} - y_{ipf})^2}{2\sigma_y^2}\right]\right\}$$
$$\times \left\{\exp\left[-\frac{(z_{ic} - z_{ipf})^2}{2\sigma_z^2}\right] + \exp\left[\frac{(2z_{inv} - z_{ipf})^2}{2\sigma_z^2}\right]\right\}, \tag{2.97}$$

where σ_{ipf} is defined as:

$$\sigma_{ipf} \equiv (2\pi)^{2/3}\sigma_{x,\,ipf}\sigma_{y,\,ipf}\sigma_{z,\,ipf}. \tag{2.98}$$

It is clear that computational effort in this approach could be several orders of magnitude bigger than in the case of the Gaussian plume approach. Two parameters are involved, telling us how often we release each puff and how high is the spatial resolution of the grid in which

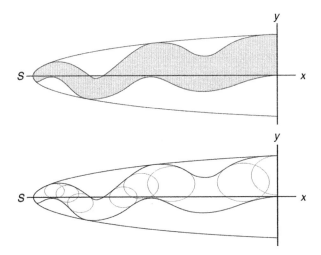

Figure 2.10. Top panel has the accrual shape of the cloud and the lower panel represents it using several puffs. Actual number of puffs is usually larger than on the picture where we have reduced their number for better visibility. Also the edge of the real cloud on the top panel is smoothed near its borders.

we calculate the concentration of a passive substance. Logistically (regarding coding) it is also much more difficult. The model has to keep track of the position of each puff and as time passes these can be large in number. Even when a puff leaves the domain it can come back due to changes of wind direction. On the other hand, the linear nature of the cell concentration calculation makes this problem easy to parallelise and so run on a cluster rather than on a single processor machine.

Next we show an example of the puff approach to the calculation (Grsic and Milutinovic, 2000) of possible contamination by a continuous point source near the city of Novi Sad. Wind data has been reanalyzed from the anemograph tapes and 10-minute averages were made. The stability of the atmosphere was characterized with the temperature gradient between temperature at 5 centimetres and 2 meters. Wind was measured at the standard height of 10 meters. The heights of the possible source, the petrochemical plant chimney, were at a much greater height so we had to perform the vertical extrapolation of the wind data. Following Holstag and Ulden (1983) and Holstag (1984), Beljars (1982), Beljars and Holstag (1991) and using the Monin–Obukhov approach with the necessary modification for the strongly stable situations, we extrapolated winds to 50 meters height.

We have also looked into the differences in the extrapolation results if other methods are used, namely if one has only standard 2 meters temperature. The main goal of the Holstag and Ulden and Holstag papers was exactly that: how, from standard measurements which have only 2 meters temperature, one can *estimate* heat fluxes and therefore use again the Monin–Obukhov approach. Figure 2.11 shows annually averaged diurnal cycle for the measured 10 meters wind, black curve, wind extrapolated at 50 meters using temperature gradient, (grey curve) and wind extrapolated using heat flux estimate, (light grey curve). Based on these winds, we have made an *estimate* of the possible zones of influence. Our runs were 3 hours long and we made calculations twice a day. To estimate the influence of the averaging period for the wind, 10 minutes versus 1 hour, we made a comparison of those two averages. This was repeated for all four seasons, 15th January and 15th March (Figure 2.12), 15th July and 15th September (Figure 2.13). The year was 1998, for which we have the data of both the

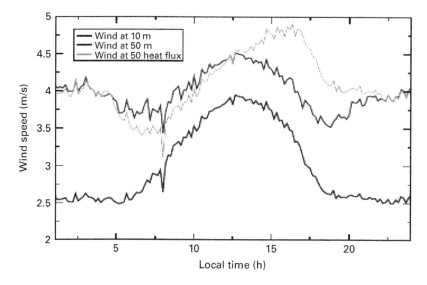

Figure 2.11. The annually averaged diurnal cycle for the measured 10 meters wind, black curve, wind extrapolated at 50 meters using temperature gradient, (grey curve) and wind extrapolated using heat flux estimate, (light grey curve).

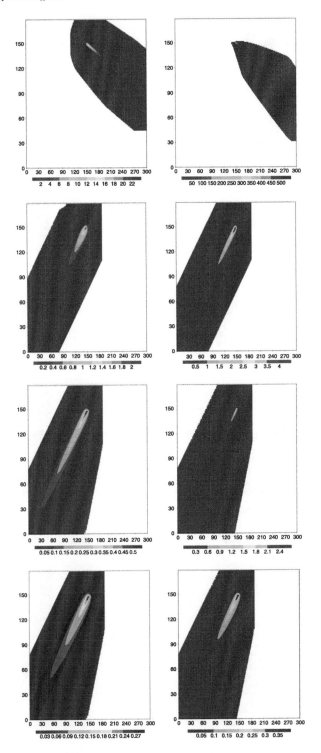

Figure 2.12. On the left, concentration after 3 hours of continuous release. The upper two panels are for the 15th of January. The top panel is for midnight and the one below is for noon of the same day. The lower two panels are for the 15th of March again the upper for midnight and the lower one for noon. The winds are hourly averages. On the right, the same except for the winds which are 10 minutes averages.

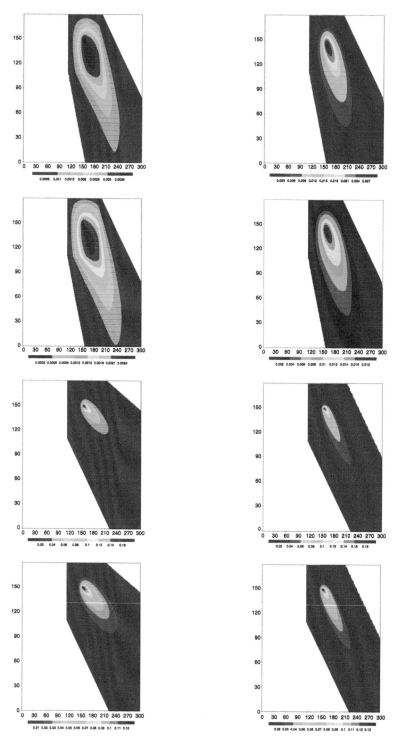

Figure 2.13. On the left, concentration after 3 hours of continuous release. The upper two panels are for the 15th of June. The top panel is for midnight and the one below is for the noon of same day. The lower two panels are for the 15th of September again the upper for midnight and the lower one for noon. The winds are hourly averages. On the right, the same except for the winds which are 10-minutes averages.

wind and the temperature. The source strength was the same for all runs so the differences come from variations in wind and variations in the local stability between day and night and from their seasonal variations. As we would expect, the spatial spread is larger in the case of the 10-minute average. The seasonal variations of concentrations are presumably strongly influenced by local stability rather than by the wind intensity variations.

2.9 DRY AND WET DEPOSITIONS

So far, we have assumed that the amount of passive substance was not changing except for the emissions. However, there are many other mechanisms that might change the quantity of the pollutant. We can have a chemical transformation deposition on the ground by both dry and wet deposition, etc. We will concentrate only on the dry and wet depositions and their parameterizations.

The dry deposition occurs when turbulent eddies hit the ground so that any material they carry sticks to it. The amount of the material that is deposited can be parameterized as

$$\chi_d = V_d \cdot \chi(x, y, z \approx 0) \tag{2.99}$$

where V_d is the so-called deposition velocity, $[L\,T^{-1}]$. Its typical magnitude is about ~ 1 mm/s. In the simplest case, it depends only upon the friction velocity u_* and the mean wind (Thykier–Nielsen and Larsen, 1982)

$$V_d = \frac{u_*}{U}. \tag{2.100}$$

In a more general case, there could be included the so-called aerodynamic resistance (r_a), the resistance representing the viscous sub-layer (r_v) and the resistance representing characteristics of the ground, bulk resistance (r_b). Then Equation (2.100) has three terms:

$$V_d = \frac{1}{r_a + r_v + r_b}. \tag{2.101}$$

There are two possibilities in treating the removed material. The first is the so-called *source–depletion*, where we add all depositions downwind and subtract them from the source. The other, called *surface–depletion*, calculates the flux of material downwind and is represented as a negative source. The second one is more realistic but is computationally more complex. We should also take into account whether we have vapours (gases) or particles. In any case different materials have different deposition rates on different surfaces. We can find deposition parameters by direct measurements at the site and then use those numbers through some interpolation procedure. This is of course the best approach but is expensive in both time and money.

Rain (snow) is a very successful removal mechanism for both gases and particles. A simple parameterization is with the introduction of the washout rate, W_r. It relates the removed concentration of the rain droplets C_0 to the concentration in the rain χ_0, at some reference height

$$W_r = \frac{C_0}{\chi_0} \tag{2.102}$$

(Misra *et al.*, 1985). With the knowledge of W_r and χ_0, the flux of effluent to the surface due to the precipitation is

$$F_{prec} = \chi_0 W_r P, \tag{2.103}$$

where P is the equivalent rainfall in, for instance, mm/hr. From Equation (2.103) we can define, in an analogous way to the dry deposition velocity, the wet deposition velocity as

$$w_r = \frac{F_{prec}}{\chi_0} = W_r P. \tag{2.104}$$

APPENDIX A – LIST OF SYMBOLS

<table>
<tr><td colspan="3" align="center">List of Symbols</td></tr>
<tr><td>Symbol</td><td>Definition</td><td>Dimensions
or Units</td></tr>
<tr><td>B_1</td><td>an arbitrary constant</td><td></td></tr>
<tr><td>C_0</td><td>concentration in rain droplets</td><td>$[\text{m}^{-3}]$</td></tr>
<tr><td>D</td><td>an arbitrary constant or a measure of cloud size</td><td>$[\text{m}]$</td></tr>
<tr><td>D_{air}</td><td>molecular diffusion of air</td><td>$[\text{m}^2\,\text{s}^{-1}]$</td></tr>
<tr><td>F</td><td>flux of substance through unit area orthogonal to x-direction</td><td>$[\text{s}^{-1}]$</td></tr>
<tr><td>F_{prec}</td><td>flux of effluent to surface due to precipitation</td><td>$[(\text{ms})^{-1}]$</td></tr>
<tr><td>G</td><td>cylindrical region</td><td></td></tr>
<tr><td>H</td><td>top of cylinder G, height of source</td><td>$[\text{m}]$</td></tr>
<tr><td>K</td><td>diffusion coefficient</td><td>$[\text{m}^2\,\text{s}^{-1}]$</td></tr>
<tr><td>L</td><td>Monin-Obukhov's length scale</td><td>$[\text{m}]$</td></tr>
<tr><td>N</td><td>number of particles in a cloud</td><td></td></tr>
<tr><td>P</td><td>equivalent rainfall</td><td>$[\text{m}\,\text{s}^{-1}]$</td></tr>
<tr><td>Q</td><td>release rate or strength of point source</td><td>$[\text{kg}\,\text{kg}^{-1}\,\text{s}]$</td></tr>
<tr><td>R</td><td>auto-correlation function</td><td></td></tr>
<tr><td>S</td><td>area that by side bounds cylindrical region G, measure of a cloud size</td><td>$[\text{m}^2]$</td></tr>
<tr><td>Snk</td><td>sink of a substance</td><td>$[\text{kg}\,\text{kg}^{-1}\,\text{s}]$</td></tr>
<tr><td>Src</td><td>source of a substance</td><td>$[\text{kg}\,\text{kg}^{-1}\,\text{s}]$</td></tr>
<tr><td>St, Sb</td><td>top and bottom surfaces that bound cylindrical region G</td><td>$[\text{m}^2]$</td></tr>
<tr><td>T</td><td>integral time scale</td><td>$[\text{s}]$</td></tr>
<tr><td>U</td><td>wind strength, mean wind</td><td>$[\text{m}\,\text{s}^{-1}]$</td></tr>
<tr><td>V</td><td>domain of integration</td><td>$[\text{m}^3]$</td></tr>
<tr><td>V_d</td><td>deposition velocity</td><td>$[\text{m}\,\text{s}^{-1}]$</td></tr>
<tr><td>W_r</td><td>washout rate</td><td></td></tr>
<tr><td>a, b, c</td><td>arbitrary constants</td><td></td></tr>
<tr><td>$const$</td><td>an arbitrary constant</td><td></td></tr>
<tr><td>f</td><td>source strength</td><td>$[\text{kg}\,\text{kg}^{-1}\,\text{s}]$</td></tr>
<tr><td>i</td><td>index of a particle</td><td></td></tr>
<tr><td>ic</td><td>index of receiving cell</td><td></td></tr>
<tr><td>ipf</td><td>index of a puff</td><td></td></tr>
<tr><td>k_1, k_2</td><td>wave numbers</td><td>$[\text{m}^{-1}]$</td></tr>
<tr><td>l</td><td>distance between any two particles in a cloud</td><td>$[\text{m}]$</td></tr>
<tr><td>q</td><td>distance-neighbor function, amount of substance</td><td>$[\text{m}^{-1}]$</td></tr>
<tr><td>r, s</td><td>constants in the distribution function</td><td></td></tr>
<tr><td>r_a</td><td>aerodynamic resistance</td><td>$[\text{s}\,\text{m}^{-1}]$</td></tr>
<tr><td>r_b</td><td>bulk resistance</td><td>$[\text{s}\,\text{m}^{-1}]$</td></tr>
</table>

(Continued)

List of Symbols

Symbol	Definition	Dimensions or Units
r_v	resistance representing viscous sub layer	$[\mathrm{s\,m^{-1}}]$
t	time	$[\mathrm{s}]$
Δt	time interval	$[\mathrm{s}]$
u	x-component of velocity	$[\mathrm{m\,s^{-1}}]$
u_n	component of velocity normal to the surface	$[\mathrm{m\,s^{-1}}]$
u_*	friction velocity	$[\mathrm{m\,s^{-1}}]$
u^+	arbitrary variable	
u^-	arbitrary variable	
v	y-component of velocity	$[\mathrm{m\,s^{-1}}]$
$\boldsymbol{v}$	wind vector	$[\mathrm{m\,s^{-1}}]$
v_i	i-th particle velocity	$[\mathrm{m\,s^{-1}}]$
ν	diffusion coefficient	$[\mathrm{m\,s^{-1}}]$
w	z-component of velocity	$[\mathrm{m\,s^{-1}}]$
w_r	wet deposition velocity	
$w\theta_0$	mean vertical flux of heat	$[\mathrm{m\,K\,s^{-1}}]$
x, y, z	distances, coordinates	$[\mathrm{m}]$
x_0	position of the source of a passive substance	$[\mathrm{m}]$
x_i	position of the i-th particle relative to the source position	$[\mathrm{m}]$
x_{ic}, y_{ic}, z_{ic}	coordinates of the ic-th cell	$[\mathrm{m}]$
x_m	median of a cloud of particles	$[\mathrm{m}]$
z_0	aerodynamic length	$[\mathrm{m}]$
Γ	Gamma function	
Θ	mean potential temperature	$[\mathrm{K}]$
Θ_0	mean potential temperature of the basic state	$[\mathrm{K}]$
δ	Dirac's delta function	
ε	small interval	
κ	Von Kármán constant	
λ	an arbitrary constant	
ξ	difference between χ_2 and χ_1	$[\mathrm{kg\,kg^{-1}}]$
ξ	relative time	$[\mathrm{s}]$
π	number pi	
σ	arbitrary constant	$[1\,\mathrm{m^{-2}}]$
σ_i	measure of lateral spread	$[1\,\mathrm{m^{-2}}]$
$\sigma_x, \sigma_y, \sigma_z$	diffusion coefficients in x, y and z direction	$[1\,\mathrm{m^{-2}}]$
σ_{10}	values of σ, averaged over 10 minutes	$[1\,\mathrm{m^{-2}}]$
σ_{ipf}	measure of the lateral spread of a cloud	$[1\,\mathrm{m^{-2}}]$
τ	time constant	$[\mathrm{s}]$
χ	concentration of a substance	$[\mathrm{kg\,kg^{-1}}]$
χ_s	boundary conditions for χ	$[\mathrm{kg\,kg^{-1}}]$
χ_0	initial conditions	$[\mathrm{kg\,kg^{-1}}]$
χ_1, χ_2	two different solutions for χ	$[\mathrm{kg\,kg^{-1}}]$
χ_{ic}	concentration of a passive substance for the ic-th cell	$[\mathrm{kg\,kg^{-1}}]$
χ_d	deposited material	$[\mathrm{kg\,kg^{-1}}]$
χ_h	homogenous part of solution	$[\mathrm{kg\,kg^{-1}}]$
χ_0	initial concentration of a substance	$[\mathrm{kg\,kg^{-1}}]$

APPENDIX B – SYNOPSIS

Starting with simple conceptual models for calculation of the concentration of a passive substance, more complex models were presented. In the atmosphere, concentration is strongly influenced by the nature of air movement. Since passive substances are usually released near the ground where motion is turbulent, this has to be taken into account when we want to estimate cloud dispersion. The Richardson's theorem does that for short and long time scales compared to the integral time scale. To simulate 3-D fields of concentration we can use a Gaussian model, which is a combination of empirical results and the Fickian diffusion equation. Finally for a time-dependent picture we have to use the Puff extension of time-independent Gaussian models.

APPENDIX C – KEYWORDS

Passive substance concentration
Richardson's theorem
Pasquill-Gifford-Turner
categories

Boundary layer
Molecular diffusion
Puff model

Gaussian model
Turbulent diffusion
Dry and wet deposition

APPENDIX D – QUESTIONS

What is the difference between molecular and turbulent diffusion?
What are the advantages and disadvantages of Gaussian type models?
Is it possible to calculate wind and temperature profile within a mixed layer of the PBL.
Why do we have to do that?

APPENDIX E – PROBLEMS

E1. Find the distance–neighbour function for the distribution of the concentration in the shape of the hat that follows a parabolic distribution. Put the beginning of the coordinate system at the maximum of the concentration.

E2. Derive the power law for the $K(l)$ in Richardson's theory using Buckingham's (PI) theorem.

E3. Derive equation (2.56).

REFERENCES

Beljars, A. C., 1982, The derivation of fluxes from profiles in perturbed areas. *Boundary–Layer Meteorology*, **24**, pp. 35–55.
Beljars, A. C., and Holstag, A. A. M., 1991, Flux parameterization over land surfaces for atmospheric models. *Journal of Applied Meteorology*, **30**, pp. 327–341.
Braham, R. R., Seely, B. K., and Crozier, W. D., 1952, A technique for tagging and tracing air parcels. *Transactions, American Geophysical Union*, **33**, pp. 825–833.
Briggs, G. A., 1982, Similarity forms for ground source surface layer diffusion. *Boundary–Layer Meteorology*, **23**, pp. 489–502.
Briggs, G. A., 1973, *Diffusion Estimates for Small Emissions*, (Washington: U.S. Environmental Protection Agency).

Crozier, W. D. and Seely, B. K., 1955, Concentration distributions in aerosol plumes three to twenty two miles from a point source. *Transactions, American Geophysical Union*, **36**, pp. 42–50.

Gifford, F. A., 1961, Use of routine meteorological observations for estimating atmospheric dispersion. *Nuclear Safety*, **2**, pp. 47–51.

Gifford, F. A. and Hanna, S. R., 1973, Modelling urban air pollution. *Atmospheric Environment*, **7**, pp. 131–136.

Golder, D., 1972, Relations among stability parameters in the surface layer. *Boundary–Layer Meteorology*, **3**, pp. 47–58.

Grsic, Z., 1991, *Critical analysis of Gaussian diffusion models (master thesis, in Serbian)*, (Belgrade: Institute for Meteorology, Belgrade University).

Grsic, Z., and Milutinovic, P., 2000, Automated meteorological station and the appropriate software for air pollution distribution assessment. In *Air Pollution Modelling and Its Application, XIII*, edited by Gryning, S. E. and Batchvarova, E.

Holstag, A. A. M., 1984, Estimates of diabatic wind speed profiles from near-surface weather observations. *Boundary–Layer Meteorology*, **29**, pp. 225–250.

Holstag, A. A. M., and Van Ulden, P., 1983, A simple scheme for daytime estimates of the surface fluxes from routine weather data. *Journal of Applied Meteorology*, **22**, pp. 517–529.

Horst, T., 1979, Lagrangian similarity modeling of vertical diffusion from ground-level sources. *Journal of Applied Meteorology*, **18**, pp. 733–740.

Misra, P. K., Chan, W. H., Chung, D. and Tang, A. J. S., 1985, Scavenging ratios of acidic pollutants and their use in long-range transport models. *Atmospheric Environment*, **19**, pp. 1471–1475.

Nieuwstad, F. T. M., 1980, Application of mixed-layer similarity to the observed dispersion from a ground-level source. *Journal of Applied Meteorology*, **19**, pp. 733–740.

Pasquill, F., 1961,The estimation of the dispersion of windborne material. *Meteorological Magazine*, **90**, pp. 33–49.

Pasquill, F., 1956, Meteorological research at Porton. *Nature*, **177**, pp. 1148–1150.

Pasquill, F., 1955, Preliminary studies of the distribution of particles at medium range from a ground-level point source. *Quarterly Journal of Royal Meteorological Society*, **81**, pp. 636–638.

Pasquill, F. and Smith, F. B., 1983, *Atmospheric Diffusion*, (New York: John Wiley and Sons).

Richardson, R. L., 1952, Transforms for the eddy–diffusion of clusters. *Proceedings of the Royal Society*, **A 214**, pp. 1–20.

Richardson, R. L., 1926, Atmospheric diffusion shown on a distance–neighbor graph. *Proceedings of the Royal Society*, **A 110**, pp. 709–737.

Taylor, G. I., 1921, Diffusion by Continuous Movements, *Proceedings of the London Mathematical Society*, **20**, pp. 196–212.

Thykier–Nielsen, S. and Larsen, S. E., 1982, *The Importance of Deposition for Individual and Collective Doses in Connection with Routine Releases from Nuclear Power Plants*, (Roskilde: RISO National Laboratory).

Turner, D. B., 1969, Workbook of atmospheric dispersion estimates. In *Office of Air Programs Publications*, **AP 26**, (Washington D.C.: Public Health Service Publ.), pp. 26–84.

CHAPTER THREE

Air–sea interaction

Vladimir Djurdjević & Borivoj Rajković
*Institute for Meteorology, Faculty of Physics,
University of Belgrade, Belgrade, Serbia*

ABSTRACT

This chapter will cover the basic concepts of air–sea interaction. It has three sections. After the introduction about the importance of the phenomenon, its two-way nature and the scales (time and space) on which it is important, there is a section on exchange of momentum, energy and mass between the atmosphere and the ocean. Part of this section addresses some of the aspects of modeling approaches used in a variety of problems which are connected to or influenced by air–sea interaction. Air–sea exchanges are strongly influenced by the structures of both media near the atmosphere–ocean interface, notably boundary layers that are present in both media. Therefore, we give a brief discussion of boundary layer structures and we examine in particular the role of the viscous sub-layer in the atmosphere. Then we present the most common approaches to the modeling of these exchanges. We start with some relatively simple concepts such as "Bulk" formulae and then present some more complex approaches. It is difficult to evaluate the quality of a particular model. We usually look into the effects of flux calculation and then, indirectly, we judge about the quality of a particular scheme or approach. Therefore, we present calculations of the sea surface temperature (SST) for the Mediterranean Sea obtained by a coupled model with particular modeling of fluxes. Comparing observed and calculated SST's we offer some ideas about the quality of modeling in that case.

3.1 FOREWORD

The atmosphere and the ocean are interacting mutually over an area that covers five sevenths of the planet's surface. Just from this basic fact alone, we can expect that knowledge of this interaction is important if we wish to understand dynamical characteristics of both entities. This is really the case and the dynamical state of both atmosphere and ocean are in large part determined by the interaction. This interaction works both ways and is determined by their dynamical and physical properties. First we point out the difference in densities of the atmosphere and the ocean. The typical density of the ocean is about 1025 kg/m^3, while the density of the air is roughly 800 times smaller, from 1.2 to 1.3 kg/m^3. A direct consequence of this large difference in densities is that the interaction occurs mostly over the surface where they are in contact. The second physical characteristic that strongly influences the nature of the interaction is the heat capacity. The heat capacity of the ocean is about four times larger than the heat capacity of the air, so the total heat capacity of the unit area column of air through the entire atmosphere is equal to the heat capacity of the unit area layer of the ocean whose depth is only about 2.5 meters. Or putting it differently, the heat that is needed to warm a column of air by one degree can be obtained just by cooling 2.5 meters of water by one degree. Another difference between the atmosphere and the ocean is the absorption of the incoming short-wave radiation, which is the fundamental external source of energy that drives the whole atmosphere–ocean system. The basic difference comes from the fact that the atmosphere is quite a weak absorber in that part of the solar spectrum (about 16%) while the ocean typically absorbs about 80% of the short-wave radiation within its first 10 meters (Jerlov, 1976). That is why the ocean surface appears very dark on satellite pictures. On the other side, the main source of energy for the atmosphere is the long-wave radiation that comes from both the ocean and the earth's surface which radiate as almost black bodies at the respective surface temperatures. These differences in heat and absorption characteristics play a dominant role in the way the ocean influences the state of the atmosphere. Large differences in the heat capacities of the land and the ocean are the main reason why temperature variations over oceans are much smaller compared to those over land. The vast heat capacity of the ocean makes it an efficient store of heat in the summer part of the year. In that part of the year the net energy balance at the ocean surface is such that more energy is gained than lost to the atmosphere. During the rest of the year the accumulated energy is then available for the additional heating of the atmosphere while the ocean cools down, because the energy balance at the surface is negative. The evaporation from the ocean's surface is also an important part of the energy balance of the ocean–atmosphere system. It takes part of the ocean's energy which then becomes available for the atmosphere first through convection and then finally through condensation in the clouds. As a by-product we have moistening of the atmosphere which greatly influences its radiation properties and therefore its temperature. The atmosphere's influence on the ocean works through two mechanisms. The first is the already-mentioned energy exchange involving exchange of radiation and heat through sensible and latent heat fluxes. The second influence is through mechanical forcing due to friction between the surface wind and the ocean.

Thus distribution of the surface winds decisively influences the structure of the ocean–surface circulation. But again, due to the large differences in densities, velocities in the ocean are only about 10% of the velocities in the air, measured at a reference height of 10 meters. Due to such a large density of water, the ocean "carries" more easily the amount of momentum handed over by the atmosphere, and so effectively the atmosphere sees the ocean as a motionless surface. So, close to the ocean surface a large wind shear develops which in turn leads to the fully developed turbulent regime. The ocean surface layer has its source of momentum confined to its very top and that leads to the same consequence, as in the case of the atmosphere, that there is a fully developed turbulent regime. There is yet another way for the atmosphere to influence the ocean, through the precipitation created in the clouds. This influence is twofold, through the local increase of mass thus creating a barotropic component of the pressure gradient force. In the past this was viewed as an important contributor to the ocean currents but now we know that this effect is about 30 times weaker than the effect of the surface winds. The second effect is that of diluting/salting depending on the difference in precipitation–evaporation. If this difference is positive, local salinity decreases thus reducing buoyancy in the top layer of the ocean, otherwise we have increase of buoyancy there. Both of these effects locally influence the pressure field and therefore change the existing pressure gradient force, thus influencing the ocean circulation.

The air–sea interaction, in some degree, influences the whole spectrum of time and space scales in the atmosphere and in the ocean. Generally, the longer the time or space scales, the larger the influence of the interaction. But there are some phenomena, relatively small in size, that owe their existence completely to the air–sea interaction. The first example is the land–sea breeze, a forced circulation due to the temperature contrast across the land–sea interface. Also there is a weak feedback coming from the shape of the wind stress in the vicinity of the land–sea boundary (Mellor, 1986, Rajkovic and Mellor, 1988). The next example is the case of the formation and evolution of tropical cyclones. At the other end of the time and space spectra are the ENSO–El-Nino phenomena, and the seasonal variability of the Somali jet and monsoons. For these time scales and beyond, years or decades, a, merging of the two components (atmosphere and ocean) into one inseparable system is inevitable.

In both media, in the vicinity of the mutual interface, there are well-developed boundary layers, so the interaction must "go through" them and is therefore strongly influenced first by the molecular and then by the turbulent nature of the motion there. The turbulent regime in the atmosphere is formed due to the existence of the strong velocity gradient as mentioned before. Buoyancy flux is the second contributor to the turbulent kinetic energy (TKE). In the case of the ocean, the vertical gradient of the surface currents, caused by the "import" of the momentum flux from the atmosphere, is one of the sources of TKE. Buoyancy flux works in the same way as in the case of the atmosphere except that here precipitation or evaporation can also decrease/increase it. There is another source of buoyancy. That is solar short-wave radiation and its absorption with depth. Due to the seasonal difference in the absorption of solar radiation, there is a seasonal shift in the sign of the buoyancy flux. It is often said that winter for ocean is like summer for the atmosphere. The atmospheric boundary layer (ABL), or as often referred to, the planetary boundary layer (PBL), extends from several hundred meters for high latitude regions in winter to several kilometres in the summer season and the tropics. As we pointed out earlier, the vertical temperature gradient and diurnal amplitude are much smaller over the ocean, so the height of the ocean's BL can be several times smaller than the corresponding ABL over land. Atmospheric boundary layers are well defined in space, having relatively sharp upper boundaries. This is clearly visible from the vertical soundings in the potential temperature but also in other fields (wind and humidity). In all these fields we see large vertical gradients which mark the end of ABL and the beginning of the so-called "free atmosphere." Both atmospheric and oceanic boundary layers have a double structure with the so-called surface layers in the vicinity of the interface and well-mixed layers further away. These surface layers are usually referred to as constant flux layers. Surface layers occupy about 10% of the whole boundary layer and are characterized by large vertical gradients of almost all variables. In the immediate vicinity of the interface, on both sides, there exist viscous sub layers with molecular transports as dominant mechanisms of momentum, heat and even mass (water vapour) transfer. Mixed layers, on the other, hand are characterized by small vertical gradients. Exchange between boundary layers and the rest of the atmosphere/ocean is greatly reduced by the existence of the strong gradients in density at their tops, especially in the case of the oceans with much colder water below the picnocline (region of steepest density gradient region). The ocean counterpart for the ABL is the thermocline layer. There are several ways of defining its depth. The simplest and most often used way is to specify the depth where the temperature gradient exceeds some predefined value. This is usually between 0.5 and 1 degree. Thermocline depth varies from 1000 meters to 50 meters. Note that in oceanographic practice the term "surface layer" can have another meaning as the layer of water that in the past had been influenced by the atmosphere. That layer is usually somewhat deeper than the boundary layer itself.

In modelling air–sea interaction, the processes inside the viscous sub-layer have proved to be an important factor for the evolution of the whole ABL by influencing fluxes near the ocean surface (Janjic, 1994; Liu *et al.,* 1979). Regimes that develop in the viscous sub-layer are determined by a single parameter, the friction velocity. With weak winds and therefore small friction velocity, viscous mechanisms are important and should be taken into account. With the increase in wind and consequently development of waves, the influence of the viscous sub-layer reduces. Other mechanisms developed, such as direct exchange of momentum from the local pressure gradient forces exerted on the waves. In a very strong wind regime we can have direct transfer of water into the atmosphere from the wave spray which leads to the complete collapse of the viscous sub-layer.

3.2   EXCHANGE OF THE MOMENTUM FLUXES

Consideration of the momentum exchange starts with the condition of continuity of fluxes across any surface, including the boundary surface between the atmosphere and the ocean. If we assume that the boundary surface that separates two fluids is smooth and well defined as in Figure 3.1, then in the immediate vicinity of the interface we have

$$M_a = M_o, \tag{3.1}$$

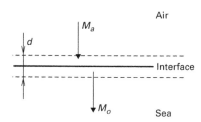

Figure 3.1. Schematic presentation of the air–sea boundary region. M_a is the momentum flux in the atmosphere while M_o is the momentum flux in the in the ocean.

where M_a is the momentum flux in the atmosphere while M_o is the momentum flux in the ocean. As was mentioned several times, because water density is several times larger than air density, the velocity of the sea currents needed to "carry" all the momentum that comes from the atmosphere is much smaller than the velocity of the air that "carries" the same amount of momentum. This fact is well known from the measurements. Typical air velocity is of the order of meters/second while typical surface current velocity is of the order of centimetres/second. So from the atmosphere's perspective, the ocean is a motionless surface very much like the land, while from the ocean's perspective the momentum flux that comes from the atmosphere is a considerable source of kinetic energy and is a key factor in the formation of surface currents. Based on these considerations, for the atmosphere we impose the lower boundary condition of zero velocity while for the ocean the upper boundary condition is in terms of the momentum flux and is set equal to the atmosphere's momentum flux as in Equation (3.1). Based on the measurements in neutral flows, oceanographers developed formulae that relate momentum flux to the wind strength, usually to its value at 10 meters. The coefficient that appears in those formulae is the so-called *drag coefficient* and whole concept is known by that name. Large and Pond (1981) developed a simple algorithm consisting of a bulk formula for calculating the drag coefficient using only the wind velocity:

$$C_D = \begin{cases} 1.2 \cdot 10^{-3}, & 4 \le U \le 11\ [LT^{-1}] \\ (0.49 + 0.065U) \cdot 10^{-3}, & 11 \le U \le 25\ [LT^{-1}]. \end{cases} \tag{3.2}$$

The other well-known formula that takes into account the SST in addition to the wind velocity is the Hellerman and Rosenstein (1983) formula

$$C_D = \alpha_1 + \alpha_2 U + \alpha_3(T_a - T_s) + \alpha_4 U^2 + \alpha_5(T_a - T_s)^2 + \alpha_6 U(T_a - T_s)^2 \tag{3.3}$$

The vertical profile of the wind due to friction must increase upward from the land or sea and is characterized with strong vertical gradients, or shear. They decrease with height and eventually the wind acquires a velocity close to the geostrophic. The existence of the strong shear near the sea/land surface is the main reason why the flow is *turbulent* there. That is the fundamental characteristic of that region which strongly influences all its transports, momentum, energy and any passive substance present there. Further away from this surface region, we still observe a turbulent regime but the generation of turbulence is of a different nature there. The main source of turbulent kinetic energy is the local convective instability or buoyancy production. The whole layer of the atmosphere with turbulent regime is called the Planetary Boundary Layer, PBL in the text below. Its vertical scale is of the order of 1 km with strong diurnal, seasonal and north–south variations. The north–south variability comes from two factors. The first is the position of the Sun resulting in larger surface heating, while the second is the value of the Coriolis parameter. Both of these contribute to the generation of higher PBL at low latitudes. The PBL turbulent fluxes are several orders of magnitude larger than the corresponding molecular fluxes and so there they are the dominant mechanism in transporting the momentum, energy, passive substances, etc.

Due to the difference in densities of water and air, there is no direct mixing between them, i.e. air cannot go "through" water and vice versa. This has a very strong impact on the scale and the mechanism of the exchange, which has to be completely *molecular* in the immediate vicinity of the surface. What are the scales of the relevant variables, length, velocity and momentum? How far up are these viscous fluxes important and where do turbulent fluxes eventually take over and become dominant for the rest of the PBL?

Let us examine the simplest possible case of the flow over a homogeneous flat surface with very large horizontal extension in comparison with the vertical extension. Further we assume constant pressure gradient force (PGF) and no rotation. In that case there exists a steady-state solution in which PGF is balanced by the surface friction. The domain of interest is very close to the ground (that is why we can neglect the influence of the Coriolis term) so we neglect vertical advection by the mean wind. Due to the assumed homogeneity in x and y, horizontal advection and divergence of Reynolds stresses are negligible in comparison with their vertical divergence. With all these assumptions made, the x-component of the equation of motion reduces from

$$\frac{\partial U}{\partial t} + U\frac{\partial U}{\partial x} + V\frac{\partial U}{\partial y} + W\frac{\partial U}{\partial z} - fV$$
$$= PGF + \frac{\partial}{\partial x}\left(v\frac{\partial U}{\partial x} - \overline{uu}\right) + \frac{\partial}{\partial y}\left(v\frac{\partial U}{\partial y} - \overline{uv}\right) + \frac{\partial}{\partial z}\left(v\frac{\partial U}{\partial z} - \overline{uw}\right) \tag{3.4}$$

to

$$\frac{\partial}{\partial z}\left(v\frac{\partial U}{\partial z} - \overline{uw}\right) = -PGF \equiv A. \tag{3.5}$$

As is traditional in the case of turbulent flows, capital letters denote mean values while lower case letters denote deviations from these mean variables. Equation (3.3) expresses the balance between the acceleration due to the PGF and the deceleration due to the convergence of the sum of turbulent x-momentum flux and viscous momentum flux, $v\partial U/\partial z$. If we integrate Equation (3.5) from the surface up to a level z we get

$$-\overline{uw} + v\frac{\partial U}{\partial z} = Az + B, \tag{3.6}$$

where B is the constant of integration. The boundary condition

$$\overline{uw} = 0 \tag{3.7}$$

at $z = 0$ leads to

$$B = v\left[\frac{\partial U}{\partial z}\right]_{z=0}. \tag{3.8}$$

The dimensions of B are $[L^2T^{-2}]$, that of squared velocity, and since it is a consequence of the friction, its square root is called "friction velocity" and is denoted either by u_* or by u_τ. Using this notation and having in mind that we are very close to the surface, balance relation (3.6) reduces to

$$-\overline{uw} + v\frac{\partial U}{\partial z} = u_*^2.$$

(3.9)

Due to the fact that the sum of two momentum fluxes has approximately a constant value, region where this approximation is valid is called the *constant flux layer*. The most important consequence of Equation (3.9) is that we have only two parameters in the problem. One is absolute constant, viscosity v, while the other one is a *dynamical* variable, the friction velocity u_*, velocity scale in the problem. From these two and dimensional arguments we can form the length scale of the problem

$$z_0 = \frac{v}{u_*}.$$

(3.10)

With u_* and z_0 we can rewrite Equation (3.9) in the *non-dimensional* form

$$\frac{-\overline{uw}}{u_*^2} + v\frac{\partial (U/u_*)}{\partial (z/z_0)} = 1.$$

(3.11)

The above relation is illustrated in Figure 3.2 showing relative magnitudes of non-dimensional viscous flux versus non-dimensional turbulent flux in the vicinity of the wall. For $z \leq 0.3z_0$ the viscous mechanism is dominant, while for $z \geq 0.3z_0$ the turbulent mechanism prevails.

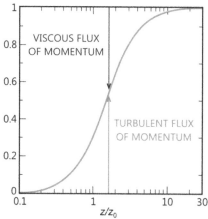

Figure 3.2. The relative magnitudes of non-dimensional viscous flux versus non dimensional turbulent flux, in the vicinity of the wall.

All this is valid for a very smooth and hard surface like a large area covered with ice. For the boundary layers over the water, the most commonly used relation for the value of z_0 is the one proposed by Charnock (1955)

$$z_0 = 0.0156\frac{u_*^2}{g},$$

(3.12)

where g is the gravitational constant.

Calculation of the fluxes above the surface sub-layer has to take into account an additional factor, that of the local stability and, related to it, the existence of heat and humidity fluxes. There are several approaches to this problem but we will concentrate on the so-called Monin–Obukhov theory (M–O in the text below) following the presentation of Janjic (1995). Let us denote the vertical turbulent flux of momentum with M and assume that it depends on the vertical gradient of the mean wind, that is, assume the eddy viscosity concept

$$\overline{uw} = M = -K_M \frac{dU}{dz}, \tag{3.13}$$

where K_M is turbulent diffusivity coefficient for momentum. If we integrate Equation (3.13) from level z_1 to level z_2 with the assumption that M is *constant* within $[z_1, z_2]$ we get

$$U_2 - U_1 = M \int_{z_1}^{z_2} \frac{dz}{K_M} \frac{dU}{dz}. \tag{3.14}$$

Further if we define the *bulk* mixing coefficient of momentum as:

$$\frac{z_2 - z_1}{K_{MB}} \equiv \int_{z_1}^{z_2} \frac{dz}{K_M}, \tag{3.15}$$

Equation (3.14) can be written as

$$M = K_{MB} \frac{U_2 - U_1}{z_2 - z_1}. \tag{3.16}$$

The starting point of the M-O theory is the theory valid for the *neutral* fluid, namely the law of the wall, which says that very close to the wall, in the region where turbulent momentum flux is constant (Equation (3.9)), mean velocity gradient is inversely proportional to the distance from the wall

$$\frac{\partial U}{\partial z} = \frac{u_*}{kz}. \tag{3.17}$$

This is a similarity law that says: if we *scale* velocity properly, then all the possible profiles of the velocity, close enough to the wall, will collapse to a single profile, and the shear of all these profiles is given by the above relation. The quantity u_* ensures that the slope of the first derivative of $U(z)$ profile at $z = 0$ is such that $U(z=0) = 0$. So the complete dynamics of the neutral flow, in the law of the wall region, is determined by that single quantity. For instance, the appropriate length scale $z_0 = u_*/v$ is also derivative of the friction velocity. As we said, both the concept and form of the profiles (gradients) for the neutral flow were the starting point for the M-O theory whose generalization

$$\frac{\partial U}{\partial z} = \frac{u_*}{kz} \varphi_m \left(\frac{z}{L}\right) \equiv \frac{u_*}{kz} \varphi_m (\zeta) \tag{3.18}$$

was proposed by Obukhov (1946) and Monin and Obukhov (1953). The first part comes from the neutral case, while with the introduction of the new, universal function φ_m we take into account all the new moments of stratified flows. There are *two* new moments, heat flux between the surface and the air above it $w\theta_0$, and stability parameter $\beta = 1/\Theta_0$ with $\Theta_0 \approx$ 273°K. From these three u_*, $w\theta_0$ and β the M-O theory suggests the new vertical length scale L

$$L = \frac{u_*^3}{\kappa \beta g w \theta_0},$$ (3.19)

where κ is Von Kármán's constant while g is the gravitational constant. The reason why we have *one* unknown function φ_m of *one* variable $\zeta = z/L$ is that out of four variables, parameters z, u_*, $w\theta_0$ and β, only three are dimensionally independent. In that case, from the Buckingham's Pi theorem, the form of relation must be that of an unknown function, with *one* dimensionless variable. Once we have relation, we must show that it is, indeed, the universal function covering all possible ranges of winds (u_*) and all possible stable and unstable regimes ($w\theta_0$). Following the formulation of the M-O theory, a great deal of effort went into its verification. The best known is the so-called Kansas experiment (Businger *et al.*, 1971).

From the way that M-O theory was initiated, it is clear that we must have, as a boundary condition:

$$\lim_{\zeta \to 0} \varphi_m(\zeta) = 1,$$ (3.20)

From the definition of ζ, we see that such a condition can be achieved either for $z \approx 0$ or for the close to neutral case with $w\theta_0 \approx 0$. With the boundary condition met, we get for the wind profile the logarithmic profile, which is always observed when the atmosphere is close to the neutral one. Now having the value for the gradient of the mean wind, we can rewrite Equation (3.14) as

$$U_2 - U_1 = \frac{u_*}{\kappa} \int_{z_1}^{z_2} \varphi_m(\zeta) \frac{d\zeta}{\zeta}.$$ (3.21)

Integrand of Equation (3.21) is singular for $\zeta = 0$ so we will add "suitable chosen zero" $\varphi_m(0) - \varphi_m(0)$ and group it as:

$$U_2 - U_1 = \frac{u_*}{\kappa} \left(\int_{z_1}^{z_2} (\varphi_m(\zeta) - \varphi_m(0)) \frac{d\zeta}{\zeta} + \varphi_m(0) \int_{z_1}^{z_2} \frac{dz}{z} \right)$$ (3.22)

or with condition (3.20) we have

$$U_2 - U_1 = \frac{u_*}{\kappa} \left(\int_{z_1}^{z_2} (\varphi_m(\zeta) - 1) \frac{d\zeta}{\zeta} + \ln\left(\frac{z_2}{z_1}\right) \right).$$ (3.23)

If we denote

$$\Psi_m(\zeta) \equiv \int_{z_1}^{z_2} (\varphi_m(\zeta) - 1) \frac{dz}{z}$$ (3.24)

then we finally have

$$U_2 - U_1 = -\frac{u_*}{\kappa}\left[\Psi_m(\zeta) + \ln\left(\frac{z_2}{z_1}\right)\right] \tag{3.25}$$

Functions $\varphi_m(\zeta)$ or $\Psi_m(\zeta)$ are the core of the M-O theory. The only way that these functions can be determined is through measurements and then looking for their best fit. Correspondence between measurements and their mathematical expression is not one-to-one, and so there are several formulations for φ_m or Ψ_m. They are divided into two groups. The one is for the unstable stratification and the other one is for the stable stratification. From the already mentioned analysis of the Kansas experiment Businger et al. have proposed one set of Ψ_m. Here we present the form suggested by Mellor (2004)

$$\varphi_m(\zeta) \cong \begin{cases} (1 + a_m\zeta)^{-1/3} & \text{for the unsatble case} \\ 1 + 5\zeta & \text{for the stable case} \end{cases} \tag{3.26}$$

where a_m has a value of 11.5. After definition

$$x \equiv (1 - a_m\zeta)^{1/3} \tag{3.27}$$

Equations (3.26) become

$$\Psi_m(\zeta) \cong \begin{cases} -\frac{3}{2}\ln(x^2 + x + 1) + \sqrt{3}\arctan\frac{2x+1}{\sqrt{3}} + c & \text{for the unstable case} \\ 5\zeta & \text{for the stable case} \end{cases} \tag{3.28}$$

With the explicit form for Ψ, we can calculate K_{MB} and related fluxes provided that we know the values of U_1 and U_2 at levels z_1 and z_2, u_* and L. But L depends, beside u_*, on the surface heat flux $w\theta_0$, which makes the problem both implicit and transcendental. Since we have two unknowns (u_*, and $w\theta_0$) we must create another similar relation, but for the potential temperature Θ. That will be done later in this section.

Apart from the problem of solving for K_{MB} and its counterpart for heat $K_{\Theta B}$, we must analyse the possible positions of levels z_1 and z_2. The upper one must be within the surface layer where the M-O theory applies. This may sometimes be a problem, when that is the height of the lowest level of a numerical model, which can have a relatively low vertical resolution and has in its domain points deep in the North (South), where the whole PBL is much shallower, hence its surface layer. If we are working with the standard measurements that is, two-metre temperatures and ten-metre winds, then we are well within the region of applicability of M-O theory. With the lower level, z_1, the situation is more complicated. From the geometry the lower boundary condition should be at $z = 0$, but due to singularity at $z = 0$, we usually set the lower boundary condition at some height z_0 above the surface. The idea is that below z_0 fluxes will remain constant. We had a similar quantity when we looked at the case of the neutral stratification. Over land z_0 is dominated by the form of the surface, local irregularities that most of the time are much higher than the one that z_0 has for the smooth surface. Beside the mathematical problems with the lower boundary condition we must remember that in the foundations of M-O theory lies the assumption that fluxes are due to completely chaotic turbulent movement. In the case of the boundary layers over water, z_0, is very small which means that turbulent fluxes become comparable to the viscous one. Further, in the case of the weak winds, z_0, is so small that viscous fluxes completely take over. We will come back to this later, when we develop the theory that takes into account the existence of the viscous sub-layer. That this really should be taken into account comes from experience with numerical weather prediction models (Janjic, 1994, 1996; Chan et al., 1996).

We start the viscous sub-layer theory with the Liu et al. 1979 paper. According to this, very close to the surface we have the following relation:

$$U_1 - U_S = D_1 \left[1 - \exp\left(-\frac{z_1 u_*}{D_1 v} \right) \right] \left(\frac{M}{u_*} \right) \qquad (3.29)$$

where the subscript S stands for the surface while index 1 stands for the top of the viscous sub-layer. We will come back to the parameter $D_{1,v}$, the viscosity-related coefficient and M is the momentum flux *above* the viscous sub-layer. In deriving Equation (3.29) Liu et al. have explicitly set the condition of continuity of fluxes across the boundary between the viscous sub-layer and the turbulent layer above. If we introduce the definition

$$\xi = -\frac{z_1 u_*}{D_1 v} \qquad (3.30)$$

and since its value is very small in the sub-layer, we have

$$1 - \exp(-\xi) \approx \xi \qquad (3.31)$$

so the relation (3.29) becomes

$$U_1 - U_S = \frac{z_1 M}{v}, \qquad (3.32)$$

where

$$z_1 = \frac{\xi v D_1}{u_*}. \qquad (3.33)$$

The last relation effectively defines the viscous sub-layer height. From the combination of Equations (3.16) and (3.32) we get

$$v \frac{U_1 - U_S}{z_1} = K_{MB} \frac{U_2 - U_1}{\Delta z} \qquad (3.34)$$

or

$$U_1 = \frac{1}{1 - \dfrac{z_1 K_{MB}}{v \Delta z}} U_S - \frac{\dfrac{z_1 K_{MB}}{v \Delta z}}{1 - \dfrac{z_1 K_{MB}}{v \Delta z}} U_2. \qquad (3.35)$$

This relation states that velocity at the interface can be viewed as a weighted mean of the surface velocity and the velocity at the height z_2. So if we know parameters D_1 and ξ parameterization of the surface layer is complete. In most cases the value of U_1 is set to zero or is negligible relative to the value U_2, but not always. For instance, in the Gulf Stream there are regions with surface currents up to 2 [LT^{-1}], so that U_1 and U_2 are of the same order. According to Janjic (1995) there are three possible regimes (which need to be taken into account while calculating fluxes) regarding the existence of the viscous sub-layer. Furthermore, he proposes that the number which separates these regimes is the Reynolds number for z_0

$$R_e = \frac{z_0 u_*}{\nu} \tag{3.36}$$

with limits

$$z_0 = \max\left(0.018\frac{u_*}{g}, 1.59 \cdot 10^{-15}\right). \tag{3.37}$$

Regarding the momentum flux, if R_e is smaller than R_{e1} corresponding to the value for $u_{*1} = 0.225$ [LT^{-1}], we do include the viscous layer in the calculations. We will call this the *smooth* regime. If the friction velocity is greater than u_{*1} we neglect the influence of the viscous sub layers. That regime is referred to as the *rough* regime. The idea is that having larger u_* the sea surface becomes wavy and there is pressure force upon the surface of the water, enhancing the momentum exchange and thus surpassing the limits that viscosity imposes.

Regarding the value for the constant D_1, Liu suggests parameterization in the form

$$D_1 = GR_e^{1/4}, \tag{3.38}$$

where G is a constant that depends on the flow regime. For the smooth regime Liu gives the value around 30 while for the other two regimes the value of $G \approx 10$ is the best fit to the Mangarella *et al.* (1973) data. This approach has been successfully implemented in NCEP's limited area model and in the version of that model which is fully coupled with the POM (Princeton Ocean Model).

3.3 EXCHANGE OF THE HEAT FLUX

The problem of the heat flux exchange between the ocean and the atmosphere has some similarity with the problem of the momentum flux exchange, but there are also some differences. The geometry of the problem is depicted in Figure 3.3 showing various components of the energy exchange between the ocean and the atmosphere.

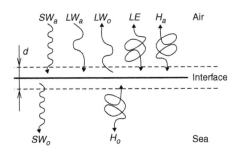

Figure 3.3. Various components of the energy exchange between the ocean and the atmosphere. SW_a is the solar short-wave radiation, LW_a is the atmosphere's long-wave radiation, LW_o is the ocean's long-wave radiation, LE and H_a are latent and sensible heat fluxes, SW_o is the solar short-wave radiation that is absorbed by the ocean and H_o is the ocean's sensible heat flux.

Assuming that we have a balance of all energy components in the layer, whose depth is d, we can write the heat balance as:

$$H_o = (LW_a - LW_o) + H_a + LE + (SW_a - SW_o). \tag{3.39}$$

The terms in the brackets are the net long-wave and short-wave components of the radiation fluxes respectively. Analogous to the case of the momentum flux, we assume that sensible and latent heat fluxes can be expressed as:

$$H = \overline{w\theta} = -K_H \frac{d\Theta}{dz} \tag{3.40}$$

and

$$LE = \overline{wq} = -K_H \frac{dq}{dz}. \tag{3.41}$$

We also assume that the mixing coefficient is the same in both fluxes. Variable q is the specific humidity. Again as in the case of momentum, integrating we get

$$H = -K_{HB} \frac{\Theta_2 - \Theta_1}{z_2 - z_1} \tag{3.42}$$

and

$$LE = -K_{HB} \frac{q_2 - q_1}{z_2 - z_1}. \tag{3.43}$$

The M-O theory for the heat fluxes has

$$\frac{\partial \Theta}{\partial z} = -\frac{\Theta_*}{\kappa z} \varphi_h(\zeta) \tag{3.44}$$

and

$$\frac{\partial q}{\partial z} = -\frac{q_*}{\kappa z} \varphi_h(\zeta) \tag{3.45}$$

with the scales for heat and humidity defined as $\Theta_* \equiv H/u_*$ and $q_* \equiv LE/u_*$. Again by integrating Equation (3.45) from z_1 to z_2 we obtain for Θ and q

$$\Theta_2 - \Theta_1 = -\frac{\Theta_*}{\kappa} \left[\Psi_h(\zeta) + \ln\left(\frac{z_2}{z_1}\right) \right] \tag{3.46}$$

and

$$q_2 - q_1 = -\frac{q_*}{\kappa} \left[\Psi_h(\zeta) + \ln\left(\frac{z_2}{z_1}\right) \right]. \tag{3.47}$$

As in the case of momentum, viscous sub-layer fluxes are introduced, for heat as

$$\Theta_1 - \Theta_S = D_2 \left[1 - \exp\left(-\frac{z_{1T} u_*}{D_2 \chi}\right) \right] \left(\frac{H}{u_*}\right) \tag{3.48}$$

and for the specific humidity as

$$q_1 - q_S = D_3 \left[1 - \exp\left(-\frac{z_{1T} u_*}{D_3 \lambda} \right) \right] \left(\frac{LE}{u_*} \right), \tag{3.49}$$

where χ and λ are molecular, heat and humidity viscous coefficients, Θ_S is sea surface temperature (henceforth SST) and q_S is specific humidity just above the water surface, which is assumed to have its saturation value. Using the assumption $z_{1q} u_* / D_1 \lambda \approx 1$, Equations (3.46) and (3.48), and definitions of bulk coefficients we evaluate Θ_1 and q_1 as

$$\Theta_1 = \Theta_S \frac{1}{1 - \dfrac{z_{1T} K_{HB}}{\chi \Delta z}} - \Theta_2 \frac{\dfrac{z_{1T} K_{HB}}{\chi \Delta z}}{1 - \dfrac{z_{1T} K_{HB}}{\chi \Delta z}} \tag{3.50}$$

and

$$q_1 = q_S \frac{1}{1 - \dfrac{z_{1q} K_{HB}}{\lambda \Delta z}} - q_2 \frac{\dfrac{z_{1q} K_{HB}}{\lambda \Delta z}}{1 - \dfrac{z_{1q} K_{HB}}{\lambda \Delta z}}, \tag{3.51}$$

with definitions

$$z_{1T} = \frac{\xi \chi D_2}{u_*} \tag{3.52}$$

and

$$z_{1q} = \frac{\xi \lambda D_3}{u_*}. \tag{3.53}$$

The saturation value can be calculated either from the Clausius–Clapeyron relation or from some of the empirical relations that will cover wider range of validity in terms of temperature, like Teten's formula

$$e_{sat}(T) = 0.618 \exp\left(\frac{17.2T}{T + 237.3} \right), \tag{3.54}$$

where T is temperature of the air (water) in deg C for the standard pressure of 1000 mb. D_2 and D_3 can be expressed, like in the momentum case, as

$$D_2 = GR_e^{1/4} P_r^{1/2} \tag{3.55}$$

and

$$D_3 = GR_e^{1/4} S_c^{1/2}, \tag{3.56}$$

where G is constant whose value depends on the regime. Again we have three regimes with limiting R_e or corresponding u_* values. The first regime is the same one that we had for momentum, i.e. $u_* < u_*$ with $u_{*1} = 0.225$ [LT^{-1}]. The second regime is for $u_{*1} < u_* < u_{*2}$ where $u_{*2} = 0.7$ [LT^{-1}] while the third regime has $u_* > u_{*2}$. In the first two regimes we have viscous sub layer with temperature and humidity at the top of it as in Equations (3.5) and (3.51). In the third regime, *rough with spray* we neglect the viscous calculations with the idea that with such strong winds waves have spray causing direct injection of water into the air.

Mellor (2004) has slightly different formulae for differences in potential temperature and humidity, analogous to

$$\Theta_2 - \Theta_1 = -\frac{\Theta_*}{\kappa}\left[\Psi_h(\zeta) + \ln\left(\frac{z_2}{z_1}\right)\right] + F_{YK}\left(\frac{z_0 u_*}{v}, P_r\right)$$

$$q_2 - q_1 = -\frac{q_*}{\kappa}\left[\Psi_h(\zeta) + \ln\left(\frac{z_2}{z_1}\right)\right] + F_{YK}\left(\frac{z_0 u_*}{v}, S_c\right). \tag{3.57}$$

The two extra terms are corrections for the viscosity whose parameterization, according to the laboratory results (Yaglom and Kader, 1974), is

$$F_{YK} = 3.14\left(\frac{u_* z_0}{v}\right)^{1/2}\left(P_r^{2/3} - 0.2\right) + 2.11. \tag{3.58}$$

Numbers P_r and S_c are Prandtl's turbulent number and Schmidt's number respectively.

The energy flux exchanges between the atmosphere and the ocean illustrate nicely the two-way (or circular) nature of the energy flux as depicted in the sketch given in Figure 3.4. Part of the energy coming from the atmosphere represents the forcing factor for the ocean while on the other side the ocean is also one of the sources of energy for the atmosphere.

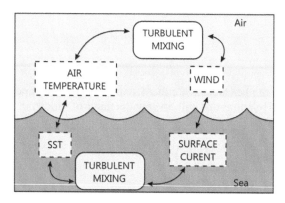

Figure 3.4. Sketch depicting the two-way nature of the atmosphere and the ocean interaction.

Beside these energy exchanges, momentum input from the atmosphere is a very important contributor to the formation of the surface currents while the energy fluxes are dominant contributors to the SST. Apart from the energy fluxes which operate *in situ* the other mechanism that influences the value of SST is advection by the ocean currents through the advection process. For models of the atmosphere, the SST is either the lower boundary condition itself or determines (together with the air temperature) surface heat fluxes which again are the lower boundary conditions for other models. So for the SST forecast we need atmospheric fluxes while these atmospheric fluxes are in turn dependent on the SST. A very similar situation is with the mass where evaporation and precipitation, and increase/decrease in the salinity of the sea, are two connected processes. We will come back to this in the next section.

We turn now to the radiation fluxes. In situations without knowledge of the surface long-wave radiation fluxes, oceanographers use empirical formulae with the net radiation $LW_a - LW_o$ being the most frequently calculated quantity. That can be done as suggested by May (1986)

$$LW = [\sigma \cdot T_a^4(0.4 - 0.05e_a^{1/2}) + 4\sigma \cdot T_a^3(T_S - T_a)](1 - 0.75C^{3.4}), \tag{3.59}$$

where $\sigma = 5.6 \cdot 10^{-8}$ is the Boltzmann's constant, T_a is the two-metre temperature [deg C], e_a is water vapour's partial pressure [mb's], T_S is water temperature [deg C] and C *is* cloud cover [%]. For the flux of the short-wave radiation we can use (Reed, 1977)

$$SW_a = Q_{TOT}(1 - 0.62C + 0.0019\beta)(1 - \alpha), \tag{3.60}$$

where C is again cloud cover, β is solar noon altitude in degrees and α is the albedo of the ocean. Q_{TOT} is defined as the sum of solar direct Q_{DIR} and diffuse Q_{DIFF} radiation, i.e.

$$Q_{TOT} = Q_{DIR} + Q_{DIFF}, \tag{3.61}$$

where

$$Q_{DIR} = Q_0 \tau \exp[-\sec(z)] \tag{3.62}$$

and

$$Q_{DIFF} = \frac{(1 - A_a)Q_0 - Q_{DIR}}{2}. \tag{3.63}$$

with $Q_0 = 1370$ [ML^{2}T^{-3}] being short-wave flux at the top of the atmosphere, with τ as transmission coefficient of the atmosphere with the value of 0.7 while $A_a = 0.09$ is the absorption coefficient of the combined effect of the water vapour and the ozone (Rosati and Miyakoda, 1988 and Castellari *et al.*, 1997). Part of the incoming short-wave radiation will partly penetrate the water and will be absorbed there. According to Paulson and Simson (1977), the depth variation of that flux, due to the attenuation, can be calculated as

$$SW_o(z) = SW_a(re^{-z/a_1} + (1 - r)e^{-z/a_2}), \tag{3.64}$$

with SW_a as short-wave flux at the ocean's surface while r, a_1 and a_2 are constants related to the optical properties of the water that, according to Jerlov (1976), can be classified into five groups. Values of these parameters depending on the group are given in Table 3.1.

Table 3.1. Values for the coefficients r, a_1 and a_2 for different types of sea water.

Jerlov type	I	Ia	Ib	II	III
r	0.58	0.62	0.67	0.70	0.78
a_1	0.35	0.60	1.0	1.5	1.4
a_2	23.0	20.0	17.0	14.0	7.9

The typical depth that sunlight can penetrate varies from 25 to 50 meters.

If we have values for SW_a and LW_a from measurements or from an atmospheric model as in the case of a numerical weather prediction model, LW_a can be treated as an independent term rather than part of the net radiation term. The ocean is also a source of long wave radiation that can be calculated according to the Stephan–Boltzmann's law for a grey body

$$LW_o = \varepsilon\sigma \cdot T_S^4, \tag{3.65}$$

with constant ε close to 1, $\varepsilon = 0.985$ (Gill, 1982). Temperature T_s is the SST in [deg K].

3.4   THE MASS AND SALINITY FLUXES

The question of mass and salinity fluxes, as part of the air–sea interaction, can be regarded as a single question because changes of the salinity of the sea can be viewed as the flux of the fresh water to/from the ocean. The diagram of that is shown in Figure 3.5.

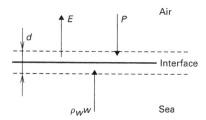

Figure 3.5. Diagram of mass flux. Flux of mass into the ocean, represented by the precipitation rate P, flux of mass out of the ocean, represented by the evaporation rate E, and flux of mass into the ocean $\rho_w w$ that balances the difference of the first two.

The balance of these fluxes means

$$\rho_w w = E - P \tag{3.66}$$

where ρ_w is the density of the water, w is the vertical velocity in the ocean while E and P are fluxes of water vapour and liquid water (in precipitation) from the atmosphere. On the other hand, the salt balance can be depicted as in Figure 3.6.

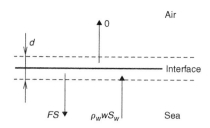

Figure 3.6. Fluxes contributing to the salt balance. Letter 0 denotes that atmospheric salt flux is zero. In the ocean we have diffusion of salt, FS and vertical flux of salt, $\rho_w w S_w$ due to vertical advection.

leading to

$$FS + \rho_w w S_w = 0, \tag{3.67}$$

where FS is diffusion of salt, $\rho_w w S_w$ is the vertical flux of salt due to vertical advection, and S_w is salinity at the sea surface. Combining equations for the fluxes of mass and salinity we get

$$FS = -(E - P)S_w. \tag{3.68}$$

3.5 SIMULATIONS OF AIR-SEA INTERACTION OVER THE MEDITERRANEAN AREA

Starting with climate modeling, the air–sea interaction was introduced as the basic factor in the large-scale and longer-term simulations. With the introduction of regional climate modeling, the spatial scale reduces but the time scale remains the same. That means that there is still a need for the air–sea interaction. Finally, with the extension of weather forecast periods beyond 5–7 days, the air–sea interaction found its place in the models for weather prediction (Miyakoda, 2002).

To approach such a problem, we have created a coupled air–sea interaction model for a limited area (Djurdjevic and Rajkovic, 2002) by coupling NCEP's Eta meso-scale atmospheric model (Janjic, 1984, 1994; Mesinger *et al.*, 1988), as the atmospheric component, with POM (Princeton Ocean Model) (Mellor and Yamada, 1982; Mellor and Blumberg, 1985), as the ocean component. Exchanges of fluxes and SST were done interactively, after every physical time-step in the atmospheric model (~360s). For this exchange we made a special coupler. Surface fluxes from Eta E-grid were interpolated on the POM C-grid using bilinear interpolation. The SST from the C-grid was set on the E-grid using simple averaging of all points that are inside the area of corresponding E-grid point (the resolution of the ocean model was about two times larger than the atmospheric model resolution). How good is such a model depends on the success of the coupling, which means how good are the fluxes of energy and momentum that are exchanged between the two components of the model. That is not so easy to verify against direct observations, so one can look at the SST as a variable most directly dependent on these exchanges.

Air–sea interaction in the Mediterranean area was analysed. The length of simulation was one year (2002). It is important to emphasize that the run was uninterrupted for the whole year, which means: start with a single initial field for both the atmosphere and the ocean, and then only updating at the boundaries. The ocean part was initialized from the MODB data set, which is the monthly climatology of the Mediterranean Sea. For the atmosphere part the German meteorological service, (Deutshen Wetterdienst or DWD) data were used both for the initial and for the boundary conditions. The atmospheric boundary conditions were updated every six hours. The boundaries for the ocean were kept constant, i.e. no exchange through the boundaries.

The main topic is to verify the quality of computed fluxes. This will be done indirectly through verification of the SST. In Figure 3.7 we show time evolution of the mean SST for the whole Mediterranean Sea. We can see that the annual variation was reproduced with remarkable accuracy. Even on shorter time scales, model was able to follow short scale variations of SST.

To infer the influence of coupling on various results we have compared coupled and uncoupled runs (Figure 3.8). For the uncoupled run we had specified the climatological SST (Reynolds climatology).

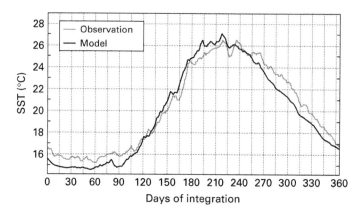

Figure 3.7. Mean SST for the Mediterranean Sea.

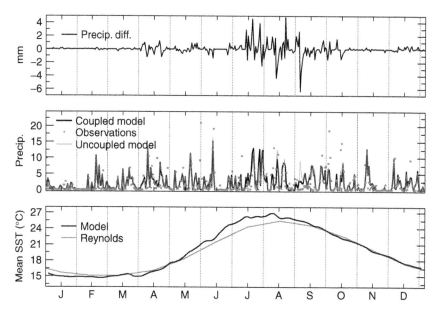

Figure 3.8. The bottom panel shows mean SST from the coupled run (black line) and prescribed climatological SST (gray line). The middle panel shows precipitation (cumulative diurnal) from the coupled run, black line, the same for the run with the climatological SST, gray line, and dots show observations of cumulative precipitation. The top panel shows differences in cumulative precipitation between coupled and uncoupled runs.

First we look into the coupled versus the uncoupled SST, which is presented in the bottom panel of the figure. It is clear that the differences are season-dependent. They are stronger during the summer season. Since the atmosphere gets part of its moisture from the sea we looked into the area-averaged diurnal accumulations of precipitation as well, (middle panel in the same figure). The precipitation data covers almost the whole of Serbia (the south-east of the Balkan peninsula), the area for which we had the data for that particular year. In general, both runs had surprisingly good precipitation forecasts. The annual accumulation for the observations was 721 mm, for the coupled model it was 750 mm and for the uncoupled it was 746 mm.

Differences, in the top panel, were concentrated over the June, July and August period, which was also true for the diurnal averages. In comparison with the observations there is some scatter but, overall, the coupled model does slightly better than the run with the climatological SST. This, of course, depends crucially on how far or on how close is the actual SST to the climatological one. Presumably, the fact that differences in the SST calculations lead to differences in the precipitation forecasts indicates that these differences come from the differences in the latent heat fluxes, so we have another indication of quality of flux calculations.

Using the coupled air–sea model, the annual variation in average SST for the whole Mediterranean Sea was reproduced with remarkable accuracy. That skill was maintained even on shorter time scales. Runs with prescribed climatological SST had also surprisingly good precipitation forecasts. Errors in the annual accumulation were less than 25 mm and 20 mm for coupled and uncoupled models respectively. Differences were concentrated over the June, July and August period. The same was valid in the case of diurnal accumulations.

3.6 AIR–SEA INTERACTION IN LAND-SEA BREEZE

Due to the very different heat capacities of water bodies (ocean, sea, lake or even large rivers) and adjacent land, heating by the Sun forms two different air masses near the surface. The one over land is warmer, the so-called "thermal low". The air mass over water is cooler and shallower. The border between the two air masses is called the sea breeze front, and due to the increasing pressure gradient the system moves inland. Colder air causes rising of the warm air over land. Aloft the circulation is opposite. As compensation, the air above land moves over the water and eventually sinks, thus forming a local closed circulation. During the night the opposite circulation is formed. This phenomenon, although very local, is very important for the pollution problems that large cities on the coast have. The best-known examples are Los Angeles in the USA and Athens in Greece. The land-sea breeze is very important factor in the determination of the state of pollution and is a major source of its relief. Aside from its role in the reduction of the pollution level, the land-sea breeze influences the climatology of the coastal regions. In the absence of it the sea's influence would be reduced to the part of the coast in the vicinity of the coastline. Of course these local circulations are superimposed on the larger scale circulations present in the area. The Pacific coast of the USA is one such example. Also one should take into account that the land-sea surface thermal gradient is solely created by local heating by the Sun. The sate of the ocean can be, and often is, influenced by the local and sometimes distant currents. The sate of possible up-welling or down-welling is a strong factor in determining the sea surface temperature. We will analyse this in more detail later on.

Numerical modelling of the land-sea breeze has been going on since meso-scale models were introduced as a par excellence example of a very local circulation system. Even though the land-sea breeze is a non-hydrostatic phenomenon it has been modelled quite success-fully in all its features quite successfully, its initiation, duration, the height up to which it propagates and its extent over water and land. Here we will concentrate on the possibility of sir-sea interaction within it. The material that will be present here is largely from (Mellor, 1986; Rajković, 1986; Rajković and Mellor, 1988).

We will start with ocean part of the problem. In Figure 3.9 we have a schematic presentation of the coastal waters, and it the movements due to the wind stress that the the atmosphere exerts on the ocean. The top of the ocean is stably stratified due to the sun's heating and would remain still in the absence of the coast. From the equations of motion applied to the water's surface we can see that offshore-onshore variation of the wind stress is the relevant factor. Right at the coast we always have its abrupt reduction to zero but it can change (reduce) in a wider region away from the coastline. In the case as presented in Figure 3.9 we

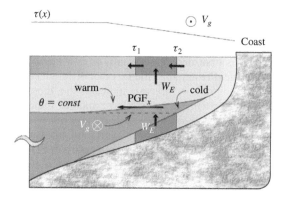

Figure 3.9. Schematic presentation of processes in the coastal is of the ocean when offshore-onshore wind stress, $\tau(x)$, is variable. There are two zones with Ekman pumping. One is near the surface, but there is another near the bottom. This secondary upwelling is responsible for the rising of the isopycnal surface leading to formation of the geostrophic current parallel to the coast.

can have divergence of the surface water and local upwelling. If we have the alongshore wind component, it will induce a water moment away from the coast. This will be compensated by the upwelling, which will bring colder waters from below (this is depicted as the green area next to the coast). If further wind stress increases away from the coast this will create divergence of the surface water and cause upwelling there as well. This upweling will raise isopycnal surfaces (Θ = constant) and create a local pressure gradient which will then create an along-shore geostrophic current. A more detailed analysis by Mellor (1986) has shown that depending on the onshore-offshore *shape* of the wind stress we can have either *strong* upwelling and *poleward* coastal flow or quite the opposite situation with *equatorward* flow and quite weak upwelling. In his analysis Mellor starts from the equations relevant for the long-term averages (neglecting the nonlinear terms).

$$-\int_{-H}^{0} v\,dz = -\frac{gH}{f}\partial_x \eta - \frac{1}{f}\int_{-H}^{0}\int_{z}^{0}\partial_x b\,dz'\,dz + 0 - \frac{\tau_x^b}{f} \tag{3.69}$$

$$\int_{-H}^{0} u\,dz = 0 = -\frac{gH}{f}\partial_y \eta - \frac{1}{f}\int_{-H}^{0}\int_{z}^{0}\partial_y b\,dz'\,dz + \frac{\tau_y^0}{f} - \frac{\tau_y^b}{f}. \tag{3.70}$$
$$\quad\;\;\text{(A)}\qquad\qquad\qquad\text{(B)}\qquad\qquad\text{(C)}\;\;\text{(D)}$$

From these it follows that, in the steady state, a balance must exist between the offshore-onshore pressure gradient force (sum of A and B), the Ekman transport (term C) and the bottom Ekman transport (term D). The sign of the A + B + C determines the coastal flow direction.

In addition to the influence of the large-scale wind, a land-sea breeze may also influence the shape of the wind stress in the vicinity of the coast. The question is how much. We will be looking at long-term averages, that is, the quasi-steady state of both atmosphere and ocean. To simulate that we have done quasi-coupling between atmosphere and ocean by consecutive runs of atmospheric and oceanic models for relatively log period, long enough so that time changes become very small. It took about 40 days for that. The models were 2D ones with POM as the ocean model, but it turned out that by a slight change in definition of buoyancy

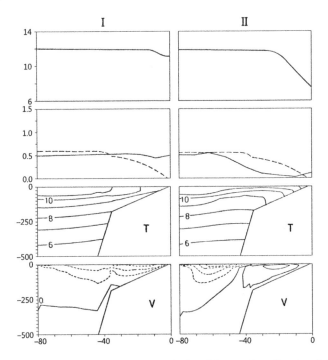

Figure 3.10. Two, very different, starting profiles of SST in the offshore-onshore direction. Column I has a weak SST variation, which gave the wind stress profile in the panel below. That wind stress will form temperature filed in the ocean given in the next panel below and alongshore current, the lowest panel. Column II is the same starting point with a strong SST variation. Note that current directions are opposite for the two SST distributions.

POM can be transferred into an atmospheric model as well. Two, very different, scenarios were designed regarding the possible SST; its cross-shore direction variations. In column I of Figure 3.10 we have the case with weak SST gradient, presumably resulting from weak upwelling. The other case, column II would be a result of much stronger upwelling. These two SST's were then set as lower boundary conditions for the temperature field, and were imposed on the atmospheric model. The large-scale pressure gradient force was inferred from the surrounding large-scale wind field, in the form of the steady geostrophic wind and was the same for both runs. The temperature field was also induced from the local climatology with slightly lower PBL top over the sea. The land temperature was variable in time with sine shape and with the observed amplitude. In the absence of the radiation module temperature was returned to the morning level using the nudging technique.

In the second row, from top down we show the time average of the produced wind stresses in the cross-shore direction (full line). The time average was over the previous 7 days. We can see that the two profiles are very different and according to Mellor's analysis should produce very different responses in the near-coast circulation. Indeed when these two wind forcing were applied to the ocean model they resulted in the cross-section profiles depicted in the two lowest panels, temperature and long-shore current respectively Both ocean runs started from temperature fields that had only vertical variation, the ones inferred from the local climatological temperature fields. These pictures show that the two resulting SST variations, in the cross-shore direction, are in very good agreement with the initial ones. All this is a confirmation that, at least local, structures of the atmospheric wind field and ocean temperature and long-shore currents are mutually dependant.

APPENDIX A – LIST OF SYMBOLS

<table>
<tr><td colspan="3" align="center">List of Symbols</td></tr>
<tr><td>Symbol</td><td>Definition</td><td>Dimensions
or Units</td></tr>
<tr><td>A</td><td>an arbitrary constant</td><td>$[\mathrm{m\,s^{-2}}]$</td></tr>
<tr><td>A_a</td><td>absorption combined coefficient of water
vapour and ozone</td><td></td></tr>
<tr><td>B</td><td>constant of integration</td><td>$[\mathrm{m^2\,s^{-2}}]$</td></tr>
<tr><td>C</td><td>cloud cover</td><td>%</td></tr>
<tr><td>D_1</td><td>a constant</td><td></td></tr>
<tr><td>D_2</td><td>a constant</td><td></td></tr>
<tr><td>D_3</td><td>a constant</td><td></td></tr>
<tr><td>E</td><td>flux of water vapour</td><td>$[\mathrm{kg\,m^{-2}s^{-1}}]$</td></tr>
<tr><td>F_{YK}</td><td>surface layer correction term for viscosity part of fluxes</td><td></td></tr>
<tr><td>FS</td><td>salt flux at sea surface</td><td>$[\mathrm{kg\,m^{-2}s^{-1}}]$</td></tr>
<tr><td>G</td><td>a constant</td><td></td></tr>
<tr><td>H</td><td>heat flux in the vertical</td><td>$[\mathrm{K\,m\,s^{-1}}]$</td></tr>
<tr><td>H_a</td><td>heat flux in the vertical in atmosphere</td><td>$[\mathrm{K\,m\,s^{-1}, W\,m^{-2}}]$</td></tr>
<tr><td>H_o</td><td>heat flux in the vertical in ocean</td><td>$[\mathrm{K\,m\,s^{-1}, W\,m^{-2}}]$</td></tr>
<tr><td>K_M</td><td>turbulent diffusivity coefficient for momentum</td><td>$[\mathrm{m^2\,s^{-1}}]$</td></tr>
<tr><td>K_{MB}</td><td>bulk turbulent diffusivity coefficient for momentum</td><td>$[\mathrm{m^2\,s^{-1}}]$</td></tr>
<tr><td>K_H</td><td>turbulent diffusivity coefficient for heat</td><td>$[\mathrm{m^2\,s^{-1}}]$</td></tr>
<tr><td>K_{HB}</td><td>bulk turbulent diffusivity coefficient for heat</td><td>$[\mathrm{m^2\,s^{-1}}]$</td></tr>
<tr><td>[L]</td><td>dimension of length</td><td></td></tr>
<tr><td>L</td><td>Monin–Obukhov length</td><td>$[\mathrm{m}]$</td></tr>
<tr><td>LE</td><td>latent flux in the vertical</td><td>$[\mathrm{kg\,kg^{-1}ms^{-1}}]$</td></tr>
<tr><td>LW</td><td>net long-wave radiation flux</td><td>$[\mathrm{K\,m\,s^{-1}, W\,m^{-2}}]$</td></tr>
<tr><td>LW_a</td><td>atmospheric long-wave radiation flux</td><td>$[\mathrm{K\,m\,s^{-1}, W\,m^{-2}}]$</td></tr>
<tr><td>LW_o</td><td>oceanic long-wave radiation flux</td><td>$[\mathrm{K\,m\,s^{-1}, W\,m^{-2}}]$</td></tr>
<tr><td>M</td><td>momentum flux in the vertical</td><td>$[\mathrm{m^2\,s^{-2}}]$</td></tr>
<tr><td>M_a</td><td>momentum flux in the atmosphere</td><td>$[\mathrm{m^2\,s^{-2}}]$</td></tr>
<tr><td>M_o</td><td>momentum flux in the ocean</td><td>$[\mathrm{m^2\,s^{-2}}]$</td></tr>
<tr><td>P</td><td>flux of liquid water (precipitation)</td><td>$[\mathrm{kg\,m^{-2}s^{-1}}]$</td></tr>
<tr><td>PGF</td><td>pressure gradient force</td><td>$[\mathrm{m\,s^{-2}}]$</td></tr>
<tr><td>P_r</td><td>Prandtl's turbulent number</td><td></td></tr>
<tr><td>Q_0</td><td>solar short-wave flux at the top of the atmosphere</td><td>$[\mathrm{W\,m^{-2}}]$</td></tr>
<tr><td>Q_{TOT}</td><td>total solar short-wave radiation at surface</td><td>$[\mathrm{W\,m^{-2}}]$</td></tr>
<tr><td>Q_{DIFF}</td><td>diffuse part of solar short-wave radiation</td><td>$[\mathrm{W\,m^{-2}}]$</td></tr>
<tr><td>Q_{DIR}</td><td>direct part of solar short-wave radiation</td><td>$[\mathrm{W\,m^{-2}}]$</td></tr>
<tr><td>R_e</td><td>Reynolds number</td><td></td></tr>
<tr><td>S_c</td><td>Schmidt number</td><td></td></tr>
<tr><td>S_W</td><td>salinity at the sea surface</td><td>$[\mathrm{psu}]$</td></tr>
<tr><td>SW_a</td><td>short-wave radiation flux at surface</td><td>$[\mathrm{K\,m\,s^{-1}, W\,m^{-2}}]$</td></tr>
<tr><td>SW_o</td><td>part of short-wave radiation flux in ocean</td><td>$[\mathrm{K\,m\,s^{-1}, W\,m^{-2}}]$</td></tr>
<tr><td>[T]</td><td>dimension of time</td><td>$[T]$</td></tr>
<tr><td>T</td><td>temperature</td><td>$[\mathrm{K}]$</td></tr>
</table>

(Continued)

List of Symbols

Symbol	Definition	Dimensions or Units
T_a	air 2 m temperature	[°C]
T_s	sea surface temperature	[°C]
U	mean wind velocity in x direction	[m s^{-1}]
V	mean wind velocity in y direction	[m s^{-1}]
W	mean wind velocity in z direction	[m s^{-1}]
a_1	constant related to the optical properties of water	[m]
a_2	constant related to the optical properties of water	[m]
e_a	saturation water vapour partial pressure	[h Pa]
e_{sat}	saturation water vapour pressure	[h Pa]
f	Coriolis parameter	[s^{-1}]
g	gravitational acceleration constant	[m s^{-2}]
q	specific humidity (of water vapour)	[kg kg^{-1}]
q_s	surface specific humidity (of water vapour)	[kg kg^{-1}]
r	constant related to the optical properties of water	
t	time	[s]
u_*	friction velocity	[ms^{-1}]
uu	kinematic flux of U-momentum in x direction	[m^2 s^{-2}]
uv	kinematic flux of U-momentum in y direction	[m^2 s^{-2}]
uw	kinematic flux of U-momentum in z direction	[m^2 s^{-2}]
w	vertical velocity	[m s^{-1}]
$w\theta$	kinematic flux of heat in the vertical	[K m s^{-1}]
wq	kinematic flux of latent heat in the vertical	[kg kg^{-1} m s^{-1}]
z_o	aerodynamic length based on friction velocity	[m]
z_{1T}	viscous sub layer height for temperature	[m]
z_{1q}	viscous sub layer height for humidity	[m]
Θ	potential temperature	[K]
Θ_0	constant, characteristic potential temperature in the surface layer	[K]
Θ_S	surface potential temperature	[K]
Ψ	integral of surface layer stability correction terms	
α	albedo of the ocean surface	
β	stability parameter, solar noon altitude	[k^{-2}], [rad]
ε	emissivity of sea surface	
ζ	dimensionless height in the surface layer	
η	sea elevation	
κ	Von Kármán constant	
λ	heat molecular viscosity coefficient	[m^2 s^{-1}]
υ	kinematic molecular viscosity	[m^2 s^{-1}]
ξ	non dimensional height	
ρ_w	density of water	[kg m^{-3}]
σ	Boltzmann's constant	[W m^{-2}K^{-4}]
τ	atmospheric transmission coefficient	
ϕ_h	surface layer stability correction term for heat	
ϕ_m	surface layer stability correction term for momentum	
χ	humidity molecular viscosity coefficient	[m^2 s^{-1}]

APPENDIX B – SYNOPSIS

The mechanism of air–sea interaction was explained giving the physical mechanisms involved in it. Exchange of momentum, heat and salinity/water fluxes were analysed in detail. The importance of the viscous sub-layer was emphasized and a model of it presented. After that, the results of a coupled regional climate model were presented for the Mediterranean Sea. The results show a very high level of correspondence between simulations and observed SST. Note that the coupling strategy was to exchange fluxes in *every* time step of the atmospheric model. This led to a very accurate diurnal signal of SST. It has been recently argued that it may be important even at much longer time scales. Finally the land-sea breeze has been given as another example of air-sea interaction. This was confirmed by running a semi-coupled 2D atmosphere-ocean system and examining its steady state solution, which confirmed the notion that atmospheric wind field, ocean temperature field and long-shore currents are mutually dependant.

APPENDIX C – KEYWORDS

Energy exchange	Water (mass) flux	Land-sea breeze
Momentum flux	Sea surface temperature	Upwelling
Heat flux	Coupled models	

APPENDIX D – QUESTIONS

Why do we have to assume the continuity of momentum over an interface?
What is bulk formula and in which context it is used?
What is the reason for introducing the Monin-Obukhov theory?
The Monin-Obukhov theory is expressed as a system of two coupled transcendental implicit
    equations. How can we solve it?

APPENDIX E – PROBLEMS

E1. Write a computer program (using mat-lab or Fortran language) that calculates momentum and heat fluxes for a given temperature gradient between surface and some height (usually 2 meters) and wind intensity (usually at 10 meters) using the Monin-Obukhov theory. Take into account that different similarity functions should be applied for unstable and stable stratification.

REFERENCES

Businger, J. A., Wyngaard, J. C., Izumi, Y. and Bradley, E. F.,1971, Flux-profile relationship in the atmospheric boundary layer. *Journal of Atmospheric Science*, **28**, pp. 181–189.
Castellari, S., Pinardi, N. and Leaman, K., 1998, A model study of air–sea interactions in the Mediterranean Sea. *Journal of Marine Systems*, **18**, pp. 89–114.
Charnock, H., 1955, Wind stress on a water surface. *Quarterly Journal Royal Meteorological Society*, **81**, pp. 639–640.
Chen, F., Mitchell, K., Janjic, Z. I. and Baldwin, M., 1996, Land–surface parameterization in the NCEP Mesoscale Eta Model. In *Activities in Atmospheric and Oceanic Modelling*, No. 23, 4.4, (Geneva: CAS/JSC WGNE).

Djurdjevic, V. and Rajkovic, B., 2002, Air–sea interaction in Mediterranean area. In *Spring Colloquium on the Physics of Weather and Climate "Regional weather prediction modelling and predictability"*, (Trieste: ICTP).

Gill, A. E., 1982, *Atmosphere–Ocean Dynamics*. (New York: Academic Press).

Janjic, Z. I., 1996, The surface layer parameterization in the NCEP Eta Model. In *Research Activities in Atmospheric and Oceanic Modeling*, 4.16–4.17, (Geneva: CAS/JSC WGNE).

Janjic, Z. I., 1995, The surface layer parameterization in Eta model. In Summer *School in Meteorology: Hydrological Cycle in Atmospheric Models*, Doc. 8, (Belgrade: Federal Hydro–meteorological Institute).

Janjic, Z. I., 1994, The step-mountain Eta coordinate model: Further developments of the convection, viscous sub-layer and turbulence closure schemes. *Monthly Weather Review*, **122**, pp. 927–945.

Janjic, Z. I., 1984, Non-linear advection schemes and energy cascade on semi-staggered grids. *Monthly Weather Review*, **112**, pp. 1234–1245.

Jerlov, N. G., 1976, *Marine Optics*, (Amsterdam: Elsvier Scientific Publishing Company).

Liu, W. T., Katsaros, K. B. and Businger, J. A., 1979, Bulk parameterization of air–sea exchanges of heat and water vapour including the molecular constraints at the interface. *Journal of Atmospheric Science*, **36**, pp. 1722–1735.

Mangarella, P. A., Chambers, A. J., Street, R. L., and Hsu, E. Y., 1973, Laboratory studies of evaporation and energy transfer through a wavy air–water interface. *Journal of Physical Oceanography*, 3, pp. 93–101.

May, P.W., 1986, A brief explanation of Mediterranean heat and momentum flux conditions. NORDA Code 322, NSTL, MS 39529.

Mellor, G. L., 2004, *Users guide for a three-dimensional, primitive equation, numerical ocean model*, (Princeton: Princeton University).

Mellor, G. L., 1986, Numerical simulation and analysis of the mean coastal circulation of California. *Continental Shelf Research*, **6**, pp. 689–713.

Mesinger, F., Janjic, Z. I., Nickovic, S., Gavrilov, D., and Deaven, D. G., 1988, The step-mountain coordinate: model description and performance of alpine lee cyclogenesis and for a case of an alpine redevelopment. *Monthly Weather Review*, **116**, pp. 1493–1518.

Miyakoda, K., 2002, Strategy for regional Seasonal Forecast. In *Ocean Forecasting: Conceptual Basis and Applications*, edited by Pinardi, N. and Woods, J. D., (Berlin: Springer), pp. 179–199.

Monin, A. S. and Obukhov, A. M., 1954, Basic laws of turbulent mixing in the atmosphere near the ground. *Trudy Geophizicheskogo Instituta*, **24**, pp. 1963–1987.

Obukhov, A.M., 1946, Turbulence in thermally inhomogeneous atmosphere. *Trudy Geophizicheskogo Instituta*, **1**, pp. 95–115.

Paulson, C. A. and Simpson, J., 1977, Irradiance measurements in the upper ocean. *Journal of Physical Oceanography*, **7**, pp. 952–956.

Rajković, B. and Mellor, G. L., 1988, Coastal ocean response to atmospheric forcing. In *International Colloquium on Ocean Hydrodynamics, 19th–Small-Scale Turbulence and Mixing on the Ocean*, (Amsterdam: Elsevier Science Publishers), pp. 141–149.

Reed, R. K., 1977, On estimating insolation over the ocean. *Journal of Physical Oceanography*, **17**, pp. 854–871.

Rosati, A. and Miyakoda, K., 1988, A general circulation model for upper ocean simulation. *Journal of Physical Oceanography*, **18**, pp. 1601–1626.

Yaglom, A. M. and Kader, B. A., 1974, Heat and Mass Transfer between a rough wall and turbulent fluid flow at high Reynolds and Peclet numbers. *Journal of Fluid Mechanics*, **62**, pp. 601–623.

CHAPTER FOUR

Modelling of flux exchanges between heterogeneous surface and atmosphere

Dragutin T. Mihailović
Department for Field and Vegetable Crops, Faculty of Agriculture, University of Novi Sad, Novi Sad, Serbia

Darko Kapor
Department of Physics, Faculty of Sciences, University of Novi Sad, Novi Sad, Serbia

ABSTRACT

In numerical models of atmospheric flow it is necessary to consider the properties of boundary-layer flow as averaged over the grid cell of the model. "Flux aggregation" is the process by which an effective horizontal average or aggregate of turbulent fluxes is formed over heterogeneous surfaces. The aggregated flux differs from spatial average of equilibrium fluxes in an area, due to nonlinear advective enhancement associated with local advection across surface transitions. Aggregated fluxes can be related to vertical profiles only above the blending height. The concept of so-called blending height has become frequently used approach to the parameterization of areally averaged fluxes over heterogeneous surfaces. There are three approaches commonly taken for calculating the transfer of momentum, heat and moisture from a grid cell comprised of heterogeneous surfaces. They are: (a) "parameter aggregation", where grid cell mean parameters such as roughness length, albedo, leaf area index, stomatal resistance, soil conductivity, etc., are derived in a manner which attempts to incorporate in the best way the combined non-linear effects of each of different relatively homogeneous subregions ("tiles") over the grid cell; (b) "flux aggregation", where the fluxes are averaged over the grid cell, using a weighted average with the weights determined by the area covered by each tile; and (c) a combination of the "flux aggregation" and "parameter aggregation" methods. However, if large differences exist in the heterogeneity of the surfaces over the grid cell, then a combined method has to be applied. In "parameter aggregation" and "flux aggregation", numerical modellers usually either use the dominant type for the grid cell or make a simple linear average to determine grid cell averages of certain parameters. Both these methods lead to uncertainties in the parameterization of boundary layer processes when heterogeneities exist over the grid cell. In this chapter we describe: (1) the concept of the blending height, (2) an approach for aggregation of aerodynamic surface parameters, (3) an approach for aggregation of albedo and (4) a combined method for calculating the surface temperature and water vapour pressure over heterogeneous surface.

4.1 FOREWORD

The effect of land surface heterogeneity on the atmosphere and on the surface energy balance has attracted widespread interest because understanding of this effect is fundamental to a comprehensive knowledge of regional and global hydrometeorological processes. Moreover, many investigators are concerned that inadequate treatment of heterogeneity may weaken confidence in large-scale models, which do not resolve heterogeneity at scales smaller than the model grid. Several technical advances have spurred interest in heterogeneity further; not the least of which is the availability of satellite data. Remote sensing technology offers high-resolution data to quantify regional and global heterogeneity and make areal-average measurements representing the effective areal-average value of surface parameters. Computational advances and increased interest in climate, and therefore in the modelling of land-atmosphere interactions, have also promoted interest in surface heterogeneity (Michaud and Shuttleworrth, 1997).

"Aggregation" generally refers to spatial averaging of some heterogeneous surface variable such as albedo, soil hydraulic properties, soil moisture, fraction of vegetation cover, surface temperature, surface reflectance, sensible heat flux, latent heat flux, surface resistance, aerodynamic resistance, or aspects of topography; or it refers to spatial averaging of some near-surface meteorological field such as temperature, humidity or precipitation. There is the question of how to "average" (arithmetically or logarithmically being two of the ways), and how to determine the size of the region over which averaging should be performed. This size depends on the degree of heterogeneity and whether there is a causal relationship between the variable being averaged and the quantity to be calculated in the model. There is a possibility that aggregation will fail when a heterogeneous variable has a nonlinear relationship with some other variable of interest. Moreover, aggregation strategies may be dependent on model formulation. Aggregation is a more limited enterprise than "scaling", because scaling seeks to find a basis for relating a phenomenon at one scale to an analogous phenomenon at other scales. Michaud and Shuttleworrth (1997) emphasized that interest in the "aggregation problem" is motivated largely either by the desire to make efficient use of highly resolved spatial data, or by the desire to proceed confidently without utilisation of detailed data. In other words, it seeks to address the question, "How can we model variable processes spatially using a grid cell which is coarse enough to be economical, yet fine enough that results are not affected by sub-grid-scale variability?" However, the topic of aggregation is equally pertinent to the question of adequate spatial resolution of measurements; hence there is a need to investigate the effect of spatial resolution on the accuracy of remotely sensed measurements.

In numerical models of atmospheric flow it is necessary to consider the properties of boundary-layer flow as averaged over the grid cell of the model. "Flux aggregation" is process by which an effective horizontal average or aggregate of turbulent fluxes is formed over heterogeneous surfaces. The aggregated flux differs from spatial average of equilibrium fluxes in an area, due to nonlinear advective enhancement associated with local advection across surface transitions. Aggregated fluxes can be related to vertical profiles only above the blending height. The concept of so-called blending height has become frequently used approach to the parameterization of areally averaged fluxes over heterogeneous surfaces (e.g. Wieringa, 1986; Mason, 1988; Claussen, 1990, 1991, 1995). For above the blending height modifications of air flow owing to changes in surface conditions will not be recognisable individually, but an overall stress or heat flux profile will exist, representing the surface conditions of a large area. This concept should be applicable to variation in surface conditions at scales considerably smaller than 10 km, i.e., for so-called disorganised or Type A landscapes (Shuttleworrth, 1988). At these scales. the concept of blending height has been tested by microscale models (e.g. Mason, 1988; Claussen, 1991). In flow over terrain inhomogeneities at scale larger than 10 km, i.e. over so-called organised or Type B landscapes, blending takes place essentially above the surface layer where Coriolis effect must not be ignored. Moreover, in Type B landscapes, secondary circulations may develop which mix momentum and energy throughout the planetary boundary layer efficiently and presumably affect surface fluxes (Claussen, 1995).

Until the middle of the last decade, the hydrologists and meteorologists have invested a large effort in making the theoretical and modelling background related to the aggregation of fluxes and parameters. Those efforts and results reached are comprehensively elaborated by Michaud and Shuttleworrth (1997) through the Tucson Aggregation Workshop summary findings that will be given in this chapter in exactly the same form as it was done in the paper by aforementioned authors. They can be summarised as follows:

- Aggregation of land surface properties appears to be successful to within an accuracy of about 10% in many, but not all, circumstances. Stated more precisely, effective parameter values representing the areal averages of land surface properties in models of surface-atmosphere interactions have been calculated successfully from simple aver-aging rules, with the form of the latter being related to the nature of the variable being averaged (e.g. Shuttleworrth, 1991). Patch-scale and meso-scale simulations show that energy fluxes calculated from these effective (aggregated) parameters can be within 10% of energy fluxes obtained from higher-resolution simulations (Dolman and Blyth, 1997; Noilhan *et al.*, 1997; Sellers *et al.*, 1997).

  Using a combination of wind tunnel experiments, theoretical analysis, and simula-tion, Raupach and Finnigan (1997) showed that the regional energy balance is insensitive to the presence of hills of moderate size, providing that the nature of the vegetation and soil at the surface and the soil water available to the vegetation are uniform. Aggregation of near-surface meteorology considered in isolation is likely to be successful for slopes up to 20%.

  The above successes are encouraging, but additional work is needed in (1) the aggre-gation of soil hydraulic properties, (2) lateral near-surface water and groundwater flow, and (3) examination of the effect of distinct lateral changes in vegetation height. In addition, some, but not all, researchers point to the need for additional work in the aggregation of soil moisture. Although there has been substantial progress in under-standing scaling of ecohydrologically relevant soil parameters at plot and field scales $(1-10\,000\,m^2)$ (Kabat *et al.*, 1997), this progress has been little recognised by the large-scale meteorological modelling community; the applicability of these scaling procedures at large scales remains under-explored. In terms of soil moisture, several researchers (Wood, 1997; Sellers *et al.*, 1997) have shown that neglecting small-scale moisture variability may compromise coarse-grid simulations of areal-average evaporation, though Sellers *et al.* (1997) view this as of secondary significance.
- Meso-scale heterogeneity in land surface properties is now known to be capable of gen-erating meso-scale circulations, which can have a significant effect on vertical energy transfers within the atmosphere. Parameterization of this phenomenon, which would allow general circulation models to accommodate these additional sub-grid-scale atmo-spheric transport processes, is a topic of active research (Pielke *et al.*, 1997). However, some researchers (Noilhan *et al.*, 1997) view the need to provide such parameterization with less urgency, drawing attention to the moderating effect of winds.
- The purpose of many aggregation studies is to provide information to refine or stim-ulate regional and global models of the interactions between soil, vegetation, energy, and water. The basic tools for regional ecohydrological modelling have already been developed and applied in mountainous terrain (Thornton *et al.*, 1997). Adequate speci-fication of finely resolved near-surface meteorology, particularly precipitation, is one of the difficulties that needs to be addressed, but there is currently no universally accepted procedure for doing this.
- Remotely sensed vegetation indices contain useful information on the bulk stomatal resistance and photosynthetic uptake of vegetation (Sellers *et al.*, 1992), but the roles of vegetation type and nutrition on the interpretation of these indices require further investigation.

● Aggregation of remotely sensed measurements in sparse canopies can be accomplished with little error in some circumstances (such as aggregation of surface temperature from $1 \, m^2$ to $1 \, km^2$) but not others (such as aggregation of sensible heat to $1 \, km^2$ (Moran et al., 1997).

4.2   THE CONCEPT OF BLENDING HEIGHT

In the studies of the heterogeneous terrain, Wieringa (1986) suggested averaging momentum fluxes at a blending height. He interpreted the blending height as a height above which modifications of air flow owing to changes in surface conditions will not be recognised individually, but overall stress or heat flux exist, representing the surface conditions of a large area. Mason (1988) more explicitly defined the blending height l_b [L] as a scale height at which the flow is approximately in equilibrium with local surface and also independent of horizontal position. Using the latter definition, the momentum flux $-\overline{(uw')}$[L^2T^{-2}] on average over a heterogeneous surface is

$$\left[-\overline{(uw')} = \sum_i \left[-\overline{(uw')}_i \right] \right] = \kappa^2 U^2(l_b) \sum_i \frac{\sigma_c^i}{\ln\left(\frac{l_b}{z_0^i} \right)^2} \tag{4.1}$$

where square brackets denote a horizontal average, σ_c^i is fractional area covered by a patch i with the roughness length z_0^i [L]. U [LT^{-1}] is mean wind speed and κ the Von Kármán constant (here $\kappa = 0.4$). An aggregated roughness length z_{0a} [L] can also be defined from Eq. (4.1), as

$$\frac{1}{\ln\left(\frac{l_b}{z_{0a}} \right)^2} = \sum_i \frac{\sigma_c^i}{\ln\left(\frac{l_b}{z_0^i} \right)^2}. \tag{4.2}$$

Mason (1988) provided a heuristic model, which indicates that

$$\frac{l_b}{L_c} \left(\ln \frac{l_d}{z_0} \right)^2 \approx 2k^2 \tag{4.3}$$

where L_c [L] is horizontal scale of roughness variations, and from Eq. (4.3) one can conclude that $l_b/L_c \approx O(10^{-2})$. Claussen (1991) deduced the blending height from numerical simulations of air flow over a surface with randomly varying roughness. He found that the sum of errors owing to the assumptions of horizontal homogeneity and equilibrium with the local surface attains a minimum at a height, which is roughly as large as the diffusion height scale l_d

$$\frac{l_b}{L_c} \left(\ln \frac{l_d}{z_0} \right) \approx c_i k \tag{4.4}$$

where the constant c_i should be $O(1)$. Claussen (1990) found $c_i = 1.75$. Using either estimate of blending height, Eq. (4.3) or (4.4), one obtains reasonably accurate estimates of an aggregated roughness length. Differences between estimates are small particularly when considering the inaccuracy in determining L_c. From simulations of air flow over randomly varying surface conditions, Claussen (1991) inferred that L_c is the length scale at which on average the surface conditions change over a larger fetch.

4.2.1 Parameter aggregation

Provided L_c and σ_c^i are known, then the blending height and the aggregated roughness length can be obtained from Eqs. (4.2) and (4.3) or (4.4). The average momentum flux is finally computed from the aggregated roughness length. The computation of areally averaged fluxes from aggregated parameters will be called "parameter aggregation" in the following. Formally, an areally averaged flux $\langle \Phi \rangle$ is

$$\langle \Phi \rangle = f(\psi_a, ...) \tag{4.5a}$$

where the vector of aggregated surface parameters is a function of surface parameters of each land type,

$$\psi_a = f(\psi_i). \tag{4.5b}$$

For example, z_{0a} is given by Eq. (4.2), (4.3) or (4.4), but for an aggregated albedo α_a, $\alpha_a = \sum_i \sigma_c^i \alpha_i$.

4.2.2 Flux aggregation

In stratified flow, it has been proposed (e.g. Wood and Mason, 1991; Noilhan and Lacarrère, 1992) to apply the method of "parameter aggregation" also to estimation of areally averaged heat fluxes, i.e. by defining proper values of aggregated albedo, aggregated leaf area index, or aggregated stomatal resistances. However, "parameter aggregation" will fail if surface conditions vary strongly. For example, definition of an aggregated soil heat conductivity is cumbersome in the presence of water and soil. The heat flux into soil is predominantly conductive, whereas water advection or thermoclinic circulation could influence the heat flux into water. Likewise, it has been shown (e.g. Claussen, 1990; Blyth *et al.*, 1993) that an aggregated stomatal resistance is impossible to find if the local resistances vary strongly.

A second complication arises as a result of the nonlinear relationship between turbulent fluxes and vertical mean profiles. For example, the vertical gradient of potential temperature can be positive on average over larger area, whereas the averaged heat flux is upward, because strong turbulence in small regions of unstable stratification can dominate the averaged heat flux, resulting in an averaged flux opposite to the averaged vertical gradient of potential temperature. This process is important in the winter polar zones (e.g. Stössel and Claussen, 1993; Claussen, 1995). To circumvent these problems, Claussen (1991) suggested computing momentum and heat fluxes at the blending height for each land-use type, which can be identified in the area under consideration. Consequently, the averaged surface fluxes are obtained by the average of surface fluxes on each land-use surface weighted by its fractional cover σ_c^i. This method is called "flux aggregation" in the following. Formally,

$$\langle \Phi \rangle = \sum_i \sigma_c^i \Phi_i \tag{4.6a}$$

where

$$\Phi_i = f(\psi_i, ...). \tag{4.6b}$$

Fluxes Φ_i also depend on turbulent transfer coefficients, which in turn are functions some of the components ψ_i. The requirement of computing the surface fluxes for each land type at the blending height leads to a revised formulation of turbulent transfer coefficients which differs from the conventional formulation (Claussen, 1991).

4.3 AN APPROACH FOR AGGREGATION OF AERODYNAMIC SURFACE PARAMETERS OVER HETEROGENEOUS SURFACE

Numerical modellers usually either use the dominant surface type over the grid cell or a simple linear average to determine grid cell averages of surface parameters. Both methods have problems in parameterising the surface layer processes when large heterogeneities exist over the grid cell (Mason, 1988; Claussen, 1995; Hess and McAvaney, 1998). However, it is possible to make aggregation of some surface parameters over the grid cell in a more physical way as it is done by Mihailovic et al. (2002). They suggested approaches for: (1) calculating the exchange of momentum between the atmosphere and heterogeneous surface, (2) deriving the equation for the wind speed profile in a roughness sublayer under neutral conditions, and (3) derivation of the aggregated roughness length and displacement height over the grid cell.

4.3.1 Mixing length and momentum transfer coefficient

We derive first an expression for the momentum transfer coefficient K_m [L^{2}T^{-2}] and the wind profile, under neutral conditions above a heterogeneous grid cell consisting of patches of vegetation, solid part (e.g. bare soil, rock, urban tile), and water. The non-uniformity of the vegetative part is expressed by the surface vegetation fractional cover σ_i representing the i type of vegetation cover filling the grid cell. Their sum takes values from 0 (when only solid surface or water are present) to 1 (when the ground surface is totally covered by plants). The non-uniformity of solid (solid parts of urban area, rock solid and bare soil) and liquid portions (sea, river, lake, water catchments) of the grid cell will be denoted by symbols δ_i and v_i, representing the surface solid and water fractional cover respectively. The total sum of all these fractional covers must be equal to unity. A realistic surface of a grid cell is rather porous with patches of solid material, vegetative portions and free air spaces inside and around it, which can produce quite different modes of turbulence in comparison with a uniform underlying surface. Also, the designed underlying surface in the grid cell is a mosaic of patches of various sizes and different aerodynamic characteristics. Presumably, this mosaic will produce microcirculations with possible flow separations at leading and trailing edges, setting up a highly complex dynamic flow. In this section, we will not address the consequences of such non-uniformity of the vegetation part of the underlying surface. Instead, following calculations are based on the assumption that the underlying surface is a combination of the only three portions consisting of a vegetative portion, characterised with total fractional cover σ, a solid portion, characterised with total δ, and a liquid portion having total fractional cover $v = 1 - \sigma - \delta$.

As suggested by Mihailovic et al. (1999), who introduced an expression for the mixing length over a grid cell consisting of vegetated and non-vegetated surface, the aggregated mixing length l_m^a [L] at level z [L] above a grid cell consisting of a heterogeneous surface as defined above, might be represented by some combination of their single mixing lengths. If, as a working hypothesis, we assume a linear combination weighted by fractional cover, according to mixing length theory we can define l_m^a as

$$l_m^a = \kappa \left[\sum_{i=1}^{K} \sigma_i \varsigma_i (z - d_i) + \sum_{i=1}^{L} \delta_i z + \sum_{i=1}^{M} v_i z \right], \tag{4.7}$$

where σ_i, δ_i and ν_i are partial fractional covers for vegetation, solid part, and water surface, with K, L, and M as the maximum number of patches in the grid cell respectively, while d_i is zero displacement height for the ith vegetative part in the grid cell. Parameter ς_i is the dimensionless constant introduced by Mihailovic *et al.* (1999) that depends on morphological and aerodynamic characteristics of the vegetative cover whose values vary according to the type of vegetative cover. The functional form of the parameter ς, considered as a function of leaf drag coefficient C_d and leaf area index LAI, was derived empirically by Lalic (1997) and Lalic and Mihailovic (1998). They analysed the wind profiles in the sub-layer above a broad range of vegetation [i.e. short grass (Morgan *et al.*, 1971), tall grass (Jacobs and van Boxel, 1988) and forest (De Bruin and Moore, 1985)], using the maximum and minimum values of LAI for 20 types of vegetation listed in Delage and Verseghy (1995). Comparison of model simulations with observations showed a good agreement with the expression $\varsigma^2 = \sqrt{2}\,(C_d LAI)^{1/10}$ for short grass, $\varsigma^2 = 2\,(C_d LAI)^{1/5}$ for tall grass, and $\varsigma^2 = 4\,(C_d LAI)^{1/2}$ for forest.

The momentum transfer coefficient K_m for the non-homogeneous vegetative cover is

$$K_m = l_m^a\, u_m^a \tag{4.8}$$

here u_*^a [LT^{-1}] is a friction velocity above non-homogeneously covered grid cell. Replacing l_m^a, in Eq. (4.8), by the expression (4.7), we get

$$K_m = \kappa \left[\sum_{i=1}^{K} \sigma_i \varsigma_i (z - d_i) + \sum_{i=1}^{L} \delta_i z + \sum_{i=1}^{M} \nu_i z \right] u_*^a. \tag{4.9}$$

4.3.2 Wind profile

Using the assumption that the friction velocity u_*^a is equal to $l_m^a\, du/dz$ yields

$$u_*^a = \kappa \left[\left(\sum_{i=1}^{K} \sigma_i \varsigma_i + \sum_{i=1}^{L} \delta_i + \sum_{i=1}^{M} \nu_i \right) z - \sum_{i=1}^{K} \sigma_i \alpha_i d_i \right] \frac{du}{dz}. \tag{4.10}$$

This equation can be integrated to

$$u(z) = \frac{u_*^a}{\kappa} \frac{1}{\displaystyle\sum_{i=1}^{K} \sigma_i \varsigma_i + \sum_{i=1}^{L} \delta_i + \sum_{i=1}^{M} \nu_i} \ln \left[\left(\sum_{i=1}^{K} \sigma_i \xi_i + \sum_{i=1}^{L} \delta_i + \sum_{i=1}^{M} \nu_i \right) z - \sum_{i=1}^{K} \sigma_i \varsigma_i d_i \right] + C_i \tag{4.11}$$

where C_i is an integration constant. If we introduce the following notations

$$\Lambda = \sum_{i=1}^{K} \sigma_i \varsigma_i + \sum_{i=1}^{L} \delta_i + \sum_{i=1}^{M} \nu_i \tag{4.12}$$

and

$$\Gamma = \sum_{i=1}^{K} \sigma_i \varsigma_i d_i,$$ (4.13)

then Eq. (4.11) can be written in a concise form

$$u(z) = \frac{u_*^a}{k} \frac{1}{\Lambda} \ln(\Lambda z - \Gamma) + C_i.$$ (4.14)

The constant C_i can be found if we introduce the assumption that the extrapolation of the wind profile given by Eq. (4.14) produces zero wind velocity at some height z_k [L] defined as

$$z_k = Z_0 + D$$ (4.15)

where

$$Z_0 = \frac{z_0}{\Lambda}$$ (4.16)

and

$$D = \frac{\Gamma}{\Lambda}$$ (4.17)

The last two expressions can be considered as aggregated roughness length and displacement height over a non-homogeneous surface in the grid cell as in Mihailovic *et al.* (1999) for the case of a surface consisting only of bare soil and vegetation patches.

The above condition can then be written as

$$0 = \frac{u_*^a}{\kappa} \frac{1}{\Lambda} \ln(\Lambda z_k - \Gamma) + C_i$$ (4.18)

After substituting the expressions (4.15), (4.16) and (4.17) into Eq. (4.18), we find that the constant C_i is given by

$$C_i = \frac{u_*^a}{\kappa} \frac{1}{\Lambda} \ln z_0$$ (4.19)

Finally, combining the expressions (4.14) and (4.18), we derive a wind profile in the roughness sublayer above the non-uniform surface in the grid cell under neutral conditions, which can be written in the form

$$u(z) = \frac{u_*^a}{\kappa \Lambda} \ln \frac{z - D}{Z_0}$$ (4.20)

In this wind profile, Z_0 and D, defined by Eqs. (4.16) and (4.17), represent the aggregated roughness length and displacement height above the grid cell, respectively. Note that the aerodynamic properties of different types of vegetation, expressed through the vegetation-type dependent parameter ς, are incorporated into the expressions for Λ and Γ and, thus, Z_0, D and $u(z)$.

4.3.3   Parameterization of roughness length and displacement height

Eq. (4.14) can be used in numerical modelling of atmospheric processes above built-in urban areas and forest canopies since their dynamics exhibits many similarities as well as dissimilarities (Fernando *et al.*, 2001). This wind profile can be also successfully applied to modelling processes above an urban grid cell (Mihailovic *et al.*, 2005; Lazic *et al.*, 2002). In the parameterization of the aggregated roughness length given by Eq. (4.16), it seems that a suitable choice would be to separate the vegetative, z_{0v} [L], and non-vegetative, z_{0n} [L], parts of the grid cell. Bearing in mind that the non-vegetative part includes solid and liquid fraction with roughness lengths z_{0s} [L] and z_{0l} [L] respectively, the aggregated roughness length may be written in the form

$$Z_0 = \frac{1}{\Lambda} \frac{\sigma z_{0v} + \delta z_{0s} + \nu z_{0l}}{\sigma + \delta + \nu} \qquad (4.21)$$

Since the sum of total fractional covers is equal to 1, the last expression can be simplified

$$Z_0 = \frac{\sigma z_{0v} + \delta z_{0s} + \nu z_{0l}}{\Lambda} \qquad (4.22)$$

For roughness length of solid and water fraction, we use a simple average having the form

$$z_{0s} = \frac{\sum\limits_{i=1}^{L} \delta_i z_{0s}^i}{\sum\limits_{i=1}^{L} \delta_i} \qquad (4.23)$$

and

$$z_{0l} = \frac{\sum\limits_{i=1}^{M} \nu_i z_{0l}^i}{\sum\limits_{i=1}^{M} \nu_i} \qquad (4.24)$$

However, for the roughness length of the vegetative part, we will use a simple average in combination with the expression for the generalised roughness length (Mihailovic *et al.*, 1999). In that case, we obtain

$$z_{0v} = \frac{\sum\limits_{i=1}^{K} \frac{\sigma_i \varsigma_i^m}{\sigma_i(\varsigma_i-1)+1} z_{0v}^i}{\sum\limits_{i=1}^{K} \sigma_i} \qquad (4.25)$$

where ς_i^m is a parameter for ith part of a vegetative cover in the grid cell, while m is a parameter that has a value of 2 according to Mihailovic *et al.* (1999). The use of this parameter in the expression for the wind profile in the roughness layer gives systematically better results above the broad range of plant communities than the classical logarithmic wind profile (Mihailovic *et al.*, 1999).

Substituting (4.23), (4.24) and (4.25) into Eq. (4.16), we obtain the expression for the roughness length Z_0 as

$$Z_0 = \frac{1}{\Lambda}\left[\sum_{i=1}^{K} \frac{\sigma_i \varsigma_i^m}{\sigma_i(\varsigma_i - 1) + 1} z_{ov,i} + \frac{\sum_{i=1}^{L} \delta_i z_{0s}^i}{\sum_{i=1}^{L} \delta_i +} + \frac{\sum_{i=1}^{M} v_i z_{ol,i}}{\sum_{i=1}^{M} v_i} \right] \qquad (4.26)$$

According to Eqs. (4.12), (4.13) and (4.17), the aggregated displacement height D has the form

$$D = \frac{\sum_{i=1}^{L} \varsigma_i \alpha_i d_i}{\sum_{i=1}^{K} \sigma_i \varsigma_i + \sum_{i=1}^{L} \delta_i + \sum_{i=1}^{M} v_i} \qquad (4.27)$$

Mihailovic *et al.* (2002) have performed numerical tests comparing the aforementioned expressions for aggregated aerodynamic characteristics with some earlier approaches (Kondo and Yamazawa, 1986; Claussen, 1995). It was done by comparison of the wind profiles using the observations obtained in an urban area. They found that (1) there exists a better physical justification of the derivation of aggregate aerodynamic characteristics than in the case when aggregation is made by a simple averaging method, (2) in numerical experiments with different fractions of grid cell components the aggregated aerodynamic parameters show more realistic reproduction of the behaviour of observed features, and (3) the wind profile above the urban area obtained by Eq. (4.20) simulates more correctly the wind speed than the two other methods.

4.4 AN APPROACH FOR AGGREGATION OF ALBEDO OVER HETEROGENEOUS SURFACE

In the grid-based environmental models, numerical modellers usually make a simple averaging to determine the albedo as the grid cell-average albedo, a key variable in the parameterization of the land surface radiation and energy budgets (Wetzel and Boone, 1995; Jacobson, 1999, Hu *et al.*, 1999). Recently, attempts go towards the calculation of the net shortwave radiation by combining the net albedo from different patches (Walko *et al.*, 2000). However, a physics-based analysis indicates that there is a significant deviation of the albedo above such a heterogeneous surface from that calculated by simple averaging, seriously affecting the calculated values of quantities describing surface biophysical processes like land surface energy budgets, canopy photosynthesis and transpiration, urban area physics and snow melt, among others (Mihailovic and Kallos, 1997; Delage *et al.*, 1999). It is, therefore, important to understand the general behaviour and limitations of the approaches used for aggregating the albedo over a heterogeneous grid cell in current land surface models. With these issues in mind, this section considers a new approach for aggregating the albedo over a very heterogeneous surface in land surface schemes for use in grid-based environmental models following Kapor *et al.* (2002) and Mihailovic *et al.* (2003). More precisely, they introduced a method for accounting for the effect of different height levels and nature of the surfaces present in a given grid cell.

4.4.1 Definition of the loss coefficient

This procedure, although transparent, is rather cumbersome, so that we shall demonstrate it using a situation with rather simple geometry, i.e. a two-patch grid-cell with a simple geometrical distribution and different heights of its components. We start with a discussion of the basic assumptions of the approach, and then derive a general expression for the aggregated albedo. The derived expression for the albedo of this particular grid-cell is compared with the conventional approach, using a common parameterization of albedo over the same grid cell (Oke, 1987).

First of all let us state the basic assumptions. We suppose that the basic constituent of the albedo, coming from the grid-cell, describes the diffuse, homogeneous, isotropic single scattering of incoming radiation from a given surface. This simplifying assumption neglects the multiple scattering effect and the dependence of the albedo on the zenith angle of the incident radiation. Apparently, within this approach, the geometry plays an essential role. In our approach, a part of the radiation reflected from the lower surface is completely absorbed by the lateral sides of the surface lying on a higher level. Consequently, the idea is to calculate the ratio of the reflected energy lost in this manner by calculating the solid angle within which these lateral sides are seen from each point of the lower surface. It is important to stress here another assumption that differs this work from one of Schwerdtfeger (2002). We assume here that the observer (measuring instrument) is sufficiently high so that the whole grid-cell is seen under a small angle and the influence of height could be neglected.

To calculate the radiant energy flux dE/dt, we introduce the total intensity of radiation I obtained from the monochromatic intensity by integrating it over the entire range of the spectrum. Taking into account that within our approach I is a constant, we can write down our basic expression following Liou (2002)

$$\left(\frac{dE}{dt}\right) = I \, dS \cos\theta \, d\Omega, \tag{4.28}$$

where dS is the infinitesimal element of surface on which radiation comes or reflects from, $\cos\theta$ describes the direction of the radiation stream, while $d\Omega = \sin\theta \, d\theta \, d\varphi$ is the element of solid angle within which our differential amount of energy is confined to.

After stating our basic assumptions, we shall explain our analytic treatment for the most general case. Let us concentrate on the average albedo of the properly chosen grid-cell of the area S as presented in Fig. 4.1.

For simplicity we assume that this region consists of two surface types, with different albedos and heights. Accordingly, we assume that this grid-cell is divided into two subregions having the areas S_1 and S_2 with corresponding albedos α_1 and α_2 respectively, while the relative height of the higher surface is h. In order to define the position of a particular point we have to use the global (x,y,z) reference frame as well as the local reference frame (x',y',z') assigned to each point, which is used for the calculation of the solid angle under which the vertical boundary between two surfaces is seen from the given point (Fig. 4.2). Let us note that the local axes are parallel to the corresponding global ones. According to the conventional approach, the average albedo $\bar{\alpha}_c$ over the grid-cell of an arbitrary geometry is given as

$$\bar{\alpha}_c = \alpha_1 \sigma_1 + \alpha_2 \sigma_2 \tag{4.29}$$

in terms of the fractional covers $\sigma_i = S_i/S$ $(i=1,2)$, where $S = S_1 + S_2$ is the total grid cell area.

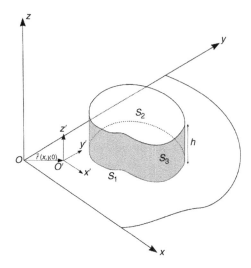

Figure 4.1. Schematic representation of the grid-cell of an arbitrary geometry consisting of two surfaces of the relative height h. Notation follows the text.

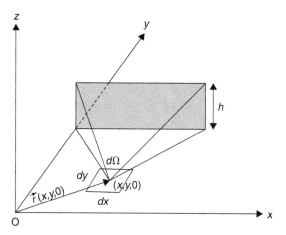

Figure 4.2. Schematic representation of the differential solid angle used in definition of $(dE/dt)_1$.

Our idea is to introduce the "loss coefficient" k_l ($0 < k_l \leq 1$), which measures the relative radiant flux lost from the reflected beam from the lower surface due to their non-zero relative height. We must emphasise that our basic assumption is that the flux of radiation that reaches the vertical boundary surface of the area S_3 (which lies in the plane orthogonal to the surfaces S_1 and S_2) is completely lost. This means that we are not taking into account the contribution of the radiation reflected from the surface S_3 to the total reflected flux of radiation. In that way we calculate the average albedo $\bar{\alpha}_n$ of this grid-cell as

$$\bar{\alpha}_n = (1 - k_l)\,\alpha_1\sigma_1 + \alpha_2\sigma_2. \tag{4.30}$$

One way of accounting the possible reflection from the vertical boundary would be to add the term including the albedo of the vertical boundary which, however, need not be equal to

α_2 at all, which poses an additional problem. Finally, our definition of the loss coefficient brings us to the following relation

$$k_l = \frac{\left(\dfrac{dE}{dt}\right)_l}{\left(\dfrac{dE}{dt}\right)_h} \tag{4.31}$$

where $(dE/dt)_h = IS_1\pi$ is the amount of flux which the land surface of area S_1 emits into the upper half-space (Liou, 2002), while $(dE/dt)_l$ is the part of the total energy coming from surface S_1 towards surface S_3. Our definition of the loss coefficient is conceptually analogous to the idea of the sky-view factor introduced by Oke (1987). More precisely, his sky-view factor would be represented as $1 - k_l$ for the infinite obstacle case. This concept is currently used in some urban models for estimation of the trapping of solar radiation and outgoing longwave radiation flux by the urban street canyon system (Masson, 2000). Let us note that in our approach we are interested in aggregating the albedo so we do not consider the particular fluxes that are in the focus of the foregoing studies.

4.4.2 Calculation of the loss coefficient

The amount of emitted flux reaching the vertical boundary is calculated as the sum of all infinitesimal amounts of radiant flux emitted from the infinitesimal surface element $dxdy$ (centred around the point with position vector $\vec{r}$), confined in the solid angle $d\Omega$ under which the element $dxdy$ "sees" the surface S_3 (Fig. 4.2). Let us note that we have chosen the lower surface to have $z = 0$ so it is omitted in the calculations.

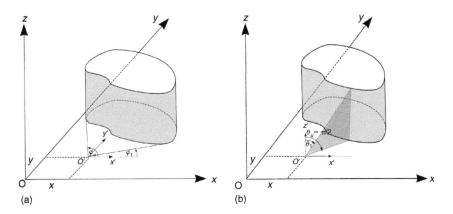

Figure 4.3. Definition of the boundaries for the integration over the local (a) azimuthal and (b) zenithal angle for the grid-cell of an arbitrary geometry.

Figures 4.3a and 4.3b show boundaries for the integration over the azimuth (φ_1, φ_u) and zenith (θ_1, θ_u) angles in terms of the global coordinates (x, y) of the given point, where the subscripts l and u denote the lower and upper boundary respectively. Accordingly, we can write down the following relation

$$\left(\frac{dE}{dt}\right)_l = I \iint_S dx\,dy \int_{\varphi_l(\vec{r})}^{\varphi_u(\vec{r})} d\varphi \int_{\theta_l(\vec{r},\varphi)}^{\theta_u(\vec{r},\varphi)} \cos\theta \sin\theta\,d\theta. \tag{4.32}$$

Combining Eqs. (4.31) and (4.32), we can evaluate the loss coefficient k needed for calculating the average albedo given by Eq. (4.30).

In order to demonstrate this procedure, we shall apply this analytic treatment to a particular situation consisting of the square grid-cell with the edge size L presented in Fig. 4.4. For the sake of simplicity, we assume that this grid-cell is divided into two subregions having rectangular form. These two subregions have areas $S_1 = L \times l$ and $S_2 = L \times (L - l)$, with corresponding albedos α_1 and α_2 respectively, while the relative height of the higher surface is h. Now $(dE/dt)_h = ILl\pi$, (while $(dE/dt)_l$ given by Eq. (4.32) becomes

$$\left(\frac{dE}{dt}\right)_l = I \int_0^l dy \int_0^L dx \int_{\varphi_l}^{\varphi_u} \int_{\theta_l}^{\theta_u} \cos\theta \sin\theta\,d\theta\,d\varphi \tag{4.33}$$

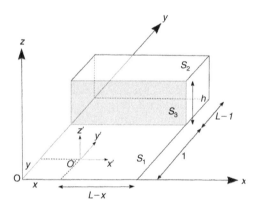

Figure 4.4. Schematic representation of the square grid-cell consisting of two surfaces of the relative height h. Notation follows the text.

with

$$\varphi_l = arctg\frac{l-y}{L-x} \qquad \varphi_u = \frac{\pi}{2} + arctg\frac{x}{l-y} \tag{4.34}$$

and

$$\theta_l = arctg\frac{l-y}{h\sin\varphi} \qquad \theta_u = \frac{\pi}{2} \tag{4.35}$$

as defined in Kapor et al. (2002). Let us note that the expression for θ_l is valid for any $0 \leq \varphi \leq \pi$. Introducing the reduced dimensionless quantities

$$\hat{x} = \frac{x}{L}, \quad \hat{y} = \frac{y}{L}, \quad \hat{l} = \frac{l}{L}, \quad \hat{h} = \frac{h}{L},$$

our final result for the loss coefficient (4.31), as the function of the reduced relative height $\widehat{h}$ and reduced length $\widehat{l}$, after some substitutions can be presented as

$$
\begin{aligned}
k_l(\widehat{l},\widehat{h}) = \frac{1}{\widehat{l}\pi} &\left\{ \widehat{l}\, arctg\frac{1}{\widehat{l}} - \sqrt{\widehat{h}^2 + \widehat{l}^2}\, arctg\frac{1}{\sqrt{\widehat{h}^2 + \widehat{l}^2}} + \widehat{h}\, arctg\frac{1}{\widehat{h}} \right. \\
&+ \frac{1}{4}(1 - \widehat{l}^2)\left[\ln(1 + \widehat{l}^2) - \ln(1 + \widehat{h}^2 + \widehat{l}^2) \right] + \frac{1}{4}\widehat{h}^2 \ln(1 + \widehat{h}^2 + \widehat{l}^2) \\
&+ \frac{1}{4}(1 - \widehat{h}^2)\ln(1 + \widehat{h}^2) + \frac{1}{4}\widehat{l}^2\left[\ln\widehat{l}^2 - \ln(\widehat{h}^2 + \widehat{l}^2) \right] \\
&\left. + \frac{1}{4}\widehat{h}^2\left[\ln\widehat{h}^2 - \ln(\widehat{h}^2 + \widehat{l}^2) \right] \right\}
\end{aligned}
\tag{4.36}
$$

Let us now compare the effect of the two approaches given by expressions (4.29) and (4.30) (i.e. the conventional and proposed approaches) by analyzing some limiting cases. Expression (4.36) behaves as $h/(2l)$ for small h/l so it vanishes identically for $h = 0$. For $l \rightarrow 0$, and consequently $\widehat{l} \rightarrow 0$, it has a finite value equal to 1/2, and since σ_1 vanishes, average albedo tends to α_2, as it should. For further analysis, we have calculated the ratio of the average albedos obtained by the proposed and conventional approaches as

$$
\Gamma = \frac{\overline{\alpha}_n}{\overline{\alpha}_c}.
\tag{4.37}
$$

In the particular case $\alpha_1 = \alpha_2/2$ this ratio becomes

$$
\Gamma = \frac{1 - \left[1 + k_l\left(\widehat{l},\widehat{h}\right) \right]\frac{\widehat{l}}{2}}{1 - \frac{\widehat{l}}{2}}
\tag{4.38}
$$

where $k_l(\widehat{l},\widehat{h})$ is given by expression (4.36). Fig. 4.5 depicts Γ as a function of the reduced length $\widehat{l} = l/L$ and considered the reduced relative height $\widehat{h} = h/L$ as the parameter. The inspection of this plot indicates that the albedo calculated by the proposed approach is always lower than the conventional one, decreasing non-linearly when $\widehat{l}$ increases. So the decrease in albedo is up to 20 percent for $\widehat{l} = 1$ and a reduced relative height of 1. These differences in albedo may have a significant impact on the calculation of the energy budget over the grid-cell. This study depicts another important property: for $l = L$, the loss coefficient does not vanish, but in fact remains finite with a value coming from

$$
\begin{aligned}
\lim_{\widehat{l} \to 1} k_l(\widehat{l},\widehat{h}) = &\frac{1}{4} + \frac{1}{\pi}\left(\sqrt{1 + \widehat{h}^2}\, arctg\sqrt{1 + \widehat{h}^2} - \widehat{h}\, arctg\,\widehat{h} \right) \\
&- \frac{1}{2}\left(\sqrt{1 + \widehat{h}^2} - \widehat{h} \right) - \frac{1}{4\pi}\widehat{h}^2 \ln\frac{(1 + \widehat{h}^2)^2}{(2 + \widehat{h}^2)\widehat{h}^2}.
\end{aligned}
\tag{4.39}
$$

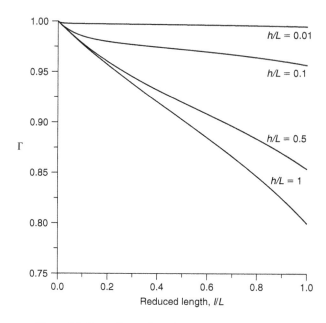

Figure 4.5. Dependence of Γ ratio on the reduced length l/L.

This is the consequence of the fact that a vertical surface at the edge of the grid-cell must have an impact to its albedo. This can be explained in a more extreme case by considering the albedo of a square grid-cell surrounded by vertical boundaries of height h. If we neglect the boundaries, its albedo would be equal to some value α. However, it can be seen very easily that due to the additivity of solid angles, the effective albedo is equal to α multiplied by the factor

$$\Lambda(\widehat{h}) = 1 - 4\kappa(1, \widehat{h}) \tag{4.40}$$

i.e. a factor whose magnitude is between 0 and 1. In fact, for $h = 0$, it is equal to 1, while for a large $\widehat{h}$ it vanishes. One should notice the importance of this effect for the calculations particularly in urban areas, where the height might be close or even larger than the cell size, so $\widehat{h}$ need not be small at all. For the small values of the reduced relative height $\widehat{h}$, the loss coefficient is proportional to $\widehat{h}$ what allows us in practical calculations in environmental modelling to use a rather simplified form of loss coefficient instead of its complete form given by the expression (4.36). In fact dimensional considerations indicate that this must be true in the most general case. More precisely, if there are more than two patches, the loss coefficient for any surface due to the presence of another (higher) surface should be always proportional to its relative height, so in the future work we shall use this approximation to study some practical situations.

To calculate the albedo of urban grid cells using the proposed method by aggregating their albedos over several patches included in the grid cell, we have to suppose that all patches have a rectangular form located to each other. This case can be treated analytically in principle, yet the expression is much more complex. So we decided to treat it by an empirical approach based on the knowledge of the behaviour studied above. Let us study any two patches having contact at some line (denoted by i and j; i,j enumerating patches).

Expanding the expression (4.29) we obtained effective albedo of the lower surface (let us say i) as

$$\bar{\alpha}_n = \bar{\alpha}_{c(i,j)} - k_{l(i,j)}\alpha_i\sigma_i. \tag{4.41}$$

We know that the limiting expression for only two patches and small relative height is $k_{l(i,j)} = (h_j - h_i)/(2L)$. However, in practice the coefficient $k_{i,j}$ between two adjacent surfaces can be estimated by some empirical expressions based on dimensional consideration.

4.4.3 Application of the Monte Carlo ray tracing approach

Our previous studies indicated that in the case of very complicated grid cell geometr preventing the analytical solution for the loss coefficient, one must use some numerical approach. For the evaluation of the integral (4.32). It turned out that most efficient and highly reliable method is a particular form of Monte Carlo calculations, so called Monte Carlo ray tracing (MCRT) method. It was shown that it reproduces the analytical results up to a high precision. Let us first explain the general idea of the calculation procedure.

The main idea of the MCRT method is to follow the path of appropriately chosen ray of light, after it had undergone diffuse, isotropic single scattering from the lower surface S_1 of the grid cell. Our observation of ray's destiny is in the sense of a possibility that ray may be absorbed by the vertical boundary, that is the lateral side of surface lying on a higher level, if it reaches it in accordance with our single scattering assumption.

Averaging in this way the observed behavior over a large number ($N = 10^6$) of the followed light paths, we can conclude about the value of loss coefficient k_l as the origin of radiative flux loss, within a given grid-cell geometry. The details of a particular Monte Carlo procedure strongly depend on the geometry of the grid-cell, so we will illustrate the particular procedure related to the simplest grid-cell geometry shown in Fig. 4.6.

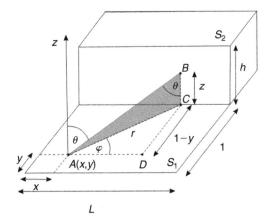

Figure 4.6. Schematic diagram of the procedure for application of the MCRT method for calculating the aggregated albedo.

The point $A(x,y)$, which belongs to the lower surface and represents a point of the intercept of this surface and the incoming beam, is randomly sampled by generating two random numbers r_1 and r_2 uniformly distributed in the interval $(0, 1)$. The area of the lower surface is $S_1 = L \times l$. So we write that $x = r_1 \times L$ and $y = r_2 \times l$. In agreement with our basic assumption – diffusive and single ray scattering, we choose a random direction in the upper half-space (θ, φ) [with $\theta \in (0, \pi/2)$; $\varphi \in (0, 2\pi)$] to simulate the trace of scattered beam. Using the uniform random numbers r_3 and r_4 from the range $(0, 1)$ gives us the way to choose this random direction as $\varphi = r_3 \times 2\pi$ and $\theta = r_4 \times \pi/2$.

Further approach was based on the idea of line-plane intersection, where the reflected beam was treated as a straight line while the vertical area had the role of a plane. The intersection of the line and the plane can be derived using general expression of the analytical geometry (McCrea, 1960), but the following is based on computer oriented exposition by Bourke (1990). The intersection of the line and the plane occurs when

$$N(PP + t(PP1 - PP)) - (N \cdot P0) = 0 \tag{4.42}$$

where points PP and $PP1$ are two known points on the line, $P0$ is the familiar point on the plane, N is the plane normal and t is the line coefficient. The final expression of line coefficient, represented in x, y, z coordinates is

$$t = \frac{Nx(P0x - PPx) + Ny(P0y - PPy) + Nz(P0z - PPz)}{Nx \cos\varphi \sin\theta + Ny \sin\varphi \sin\theta + Nz \cos\theta} \tag{4.43}$$

where subtraction of PP and $PP1$ is performed in terms of azimuthal and zenithal angles. The coordinates of intersection point are found as

$$x = PPx + t \cos\varphi \sin\theta$$
$$y = PPy + t \sin\varphi \sin\theta$$
$$z = PPz + t \cos\theta \tag{4.44}$$

Now, if the point $B(x, y, z)$ lies within the borders of vertical area, then diffusively scattered beam will be absorbed (in the single scattering approximation) and this case is positive for absorption. This procedure was repeated $N = 10^6$ times and the *loss coefficient* was estimated as

$$k_l = N_a \ (number \ of \ cases \ which \ were \ positive \ for \ absorption) /$$
$$N \ (number \ of \ conducted \ numerical \ experiments). \tag{4.45}$$

An example of the previously exposed ideas can be found in (Kapor *et al.*, 2010) where it has been demonstrated how the results obtained can be further incorporated into different complex schemes to evaluate various important parameters, air temperature for example.

4.5 A COMBINED METHOD FOR CALCULATING THE SURFACE TEMPERATURE AND WATER VAPOUR PRESSURE OVER HETEROGENEOUS GRID CELL

In numerical modelling of surface layer processes, as mentioned above, two approaches are commonly taken for calculating the transfer of momentum, heat and moisture from a grid cell comprised of heterogeneous surfaces. They are: (1) "parameter aggregation", where grid cell mean parameters are derived in a manner which attempts to best incorporate the combined non-linear effects of each of different relatively homogeneous subregions ("tiles") over the grid cell and (2) "flux aggregation", where the fluxes are averaged over the grid cell, using a weighted average with the weights determined by the area covered by each tile; according to Hess and McAvaney (1997) and Hess and McAvaney (1998); there is also the third approach as a combination of the "flux aggregation" and "parameter aggregation" methods, the so-called combined method. When the underlying surface over the grid cell is homogeneous, the turbulent transfer physics can be treated as (1) or (2) (Claussen, 1991; Claussen, 1995). If large differences exist in the heterogeneity of the surfaces over the grid cell then a combined method has to be applied. The application of the aggregation method requires a control regarding its sensitivity to chaotic time fluctuations, realisability and proper aggregation of biophysical parameters relevant for calculating turbulent fluxes over the grid cell (Mihailović, 2002). However, when either the "flux aggregation" method or its combination with the "parameter aggregation" is used, then certain anomalies can arise through the "Schmidt paradox", leading to a situation of the occurrence of the counter-gradient transport between the surface and the lowest model level. In this section we will suggest a method that combines the "parameter aggregation" and the "flux aggregation" approaches in calculating the surface temperature of the heterogeneous grid cell.

In the following text we use angular brackets to indicate an average of certain physical quantity A over the grid cell, i.e.

$$\langle A \rangle = \sum_{i=1}^{NP} \xi_i A_i \tag{4.46}$$

where NP is the number of patches within a grid cell and ξ_i is fractional cover for the ith surface type. In "parameter aggregation" approach, the mean sensible heat flux $\langle H_0 \rangle$ and latent heat flux $\lambda \langle E_0 \rangle$, calculated over the grid cell, where λ is latent heat of vaporisation, are found by assuming, for example, the aerodynamic resistance representation, i.e.

$$\langle H_0 \rangle = \rho_p c_p \frac{\langle T_0 \rangle - T_a}{\langle r_a \rangle} \tag{4.47}$$

and

$$\lambda \langle E_0 \rangle = \frac{\rho_p c_p}{\gamma} \frac{\langle e_0 \rangle - e_a}{\langle r_a \rangle}, \tag{4.48}$$

where ρ_p is the air density, c_p is specific heat of air at constant pressure, γ is psychrometric constant, r_a is resistance between canopy air or ground surface and the atmospheric lowest model level, T is temperature and e is water vapour pressure. The subscript a indicates the atmospheric lowest model level and the subscript 0 indicates the surface or environment within the canopy. The $\langle r_a \rangle$ is defined as

$$\langle r_a \rangle = \langle r_s \rangle \delta \mu + \frac{1}{\kappa \langle u_* \rangle} \ln \frac{\langle z_a \rangle - \langle d \rangle (1 - \delta)}{\langle z_b \rangle - \langle d \rangle (1 - \delta)}, \tag{4.49}$$

where r_s is the bare soil surface resistance, δ ($\delta = 1$ for the bare soil, water and solid fraction; $\delta = 0$ for vegetative surface) and μ ($\mu = 1$ for the bare soil fraction; $\mu = 0$ for vegetative surface, water and solid fraction) are parameters, u_* is friction velocity, z_a is height of the lowest atmospheric model level, z_b is a height taking values z_0 and h_c (canopy height) for the barren/solid/water and vegetative part respectively. For r_s is used the empirical expression given by Shu Fen Sun (1982), i.e.

$$\langle r_s \rangle = d_1 + d_2 \langle w_1 \rangle^{-d_3} \tag{4.50}$$

where d_1, d_2 and d_3 are empirical constants (Mihailovic and Kallos, 1997) , while w_1 is the top soil layer volumetric soil moisture content. If the surface "flux aggregation" approach is applied then the mean surface fluxes are given by

$$\langle H_0 \rangle = \rho_p c_p \sum_{i=1}^{NP} \xi_i \frac{T_{m,i} - T_a}{r_{a,i}} \tag{4.51}$$

$$\lambda \langle E_0 \rangle = \frac{\rho_p c_p}{\gamma} \sum_{i=1}^{NP} \xi_i \frac{e_{m,i} - e_a}{r_{a,i}}. \tag{4.52}$$

where the subscript m refers to the single patch in the grid cell (vegetation, bare soil, water urbanised area) whose temperature is calculated under the land surface scheme. However, according to Hess and McAvaney (1998), it seems that averaging temperatures over different patches in the grid cell, rather than the sensible heat flux, can be the source of problems. We will suggest an alternative method for their calculation diagnostically from Eqs. (4.47) and (4.48), when the grid-averaged fluxes are known from Eqs. (4.51) and (4.52). It is done by following the works of Hess and McAvaney (1998) and Mihailovic et al. (2002). Since we have three unknowns, it is necessary to introduce the associated "parameter" and "flux aggregation" equations for momentum

$$\langle u_*^2 \rangle = \left[\frac{\kappa \Lambda}{\ln \dfrac{z_a - \langle D \rangle}{\langle Z_0 \rangle}} \right]^2 \langle F\left(\langle Ri_b \rangle, u_a, \langle T_0 \rangle, T_a\right)\rangle u_a^2, \tag{4.53}$$

and

$$\langle u_*^2 \rangle = \sum_{i=1}^{NP} \xi_i \left[\frac{\kappa \Lambda_i}{\ln \dfrac{z_a - D_i}{Z_{0,i}}} \right]^2 F_i\left[Ri_{b,i}, u_a, T_{m,i}, T_a\right] u_a^2 \tag{4.54}$$

where $\langle Z_0 \rangle$, $\langle D \rangle$ and Λ are given by Eqs. (26), (27) and (12), F represents the nonneutral modification, Ri_b is bulk Richardson number and u_a is wind speed at the lowest model level. Now, the mean averaged momentum flux is calculated from Eq. (4.54). If this value is substituted into Eq. (4.53) the resulting equation can be solved for $\langle F \rangle$. The "parameter aggregation" version of the aerodynamic resistance $\langle r_a \rangle$ can be now determined (since $\langle F \rangle, \langle Z_0 \rangle$, $\langle D \rangle$ and $\langle h_c \rangle$ are all known). Thus,

$$\langle r_a \rangle = \langle r_s \rangle \delta \mu + \frac{\left[\dfrac{\kappa \Lambda}{\ln \dfrac{z_a - \langle D \rangle}{\langle Z_0 \rangle}} \right]}{\kappa \left\{ \sum_{i=1}^{NP} \xi_i \left[\dfrac{\kappa \Lambda_i}{\ln \dfrac{z_a - D_i}{Z_{0,i}}} \right]^2 F_i\left[Ri_{b,i}, u_a, T_{g,i}, T_a\right] \right\}^{1/2}} \ln \frac{z_a - \langle D \rangle(1 - \delta)}{\langle z_b \rangle - \langle D \rangle(1 - \delta)}. \tag{4.55}$$

Hence, the grid-averaged surface values of temperature and water vapour pressure can be found from Eqs. (4.47) and (4.48), i.e.

$$\langle T_0 \rangle = \frac{\langle r_a \rangle \langle H_0 \rangle}{\rho_p c_p} + T_a \qquad \langle e_0 \rangle = \frac{\langle r_a \rangle \gamma \lambda \langle E_0 \rangle}{\rho_p c_p} + e_a \qquad\qquad (4.56)$$

4.6 CONCLUSION

The aim of this chapter was to review various procedures for treating heterogenous grid cells, characterstic for realistic situations. It is shown that this variety of options demands that the choice of the approach should be made by the modeller, depending on the particular situation.

APPENDIX A – LIST OF SYMBOLS

List of Symbols		
Symbol	Definition	Dimensions or Units
C_d	a leaf drag coefficient	
C_i	an integration constant	
D	displacement height above the grid cell	
F	the nonneutral modification	
$\langle H_0 \rangle$	the mean sensible heat flux calculated over the grid cell	
I	the total intensity of radiation	
K, L, M	the maximum number of patches in the grid cell, respectively	
K_m	the momentum transfer coefficient	$[m^2\ s^{-2}]$
L	the edge size of the square grid-cell	
$[L]$	the dimension of length	
LAI	a leaf area index	
L_c	the horizontal scale of roughness variations	$[m]$
NP	the number of patches within a grid cell	
Ri_b	the bulk Richardson number	
S	the total grid cell area	
dS	the infinitesimal element of surface on which radiation comes or reflects from	
S_1, S_2	the areas of the subregions of the grid-cell with corresponding albedos α_1 and α_2 respectively	
S_3	the area which lies in the plane normal to the surfaces S_1 and S_2	
T	the temperature	
$[T]$	the dimension of time	
U	the mean wind speed	$[m\ s^{-1}]$
Z_0	the aggregated roughness length above the grid cell	
a	the subscript which indicates the atmospheric lowest model level	

(Continued)

List of Symbols

Symbol	Definition	Dimensions or Units
c_i	the constant	
c_p	specific heat of air at constant pressure	
d_1, d_2	empirical constants	
d_3	empirical constant	
d_i	the zero displacement height for the ith vegetative part in the grid cell	
dE/dt	the radiant energy flux	
e	the water vapour pressure	
h	the relative height of the higher surface	
h_c	canopy height	
$\widehat{h}$	the reduced relative height	
k_l	the "loss coefficient", which measures the relative radiant flux lost from the reflected beam from the lower surface due to their nonzero relative height	
$\widehat{l}$	the reduced length	
l_b	the blending height as a scale height at which the flow is approximately in equilibrium with local surface and also independent of horizontal position	[m]
l_d	the diffusion height scale	
l_m^a	the aggregated mixing length at level z	[m]
m	a parameter; the subscript refers to the single patch in the grid cell (vegetation, bare soil, water urbanised area)	
r_a	the resistance between canopy air or ground surface and the atmospheric lowest model level	
r_s	the bare soil surface resistance	
u_a	wind speed at the lowest model level	
u_*	friction velocity	
u_*^a	a friction velocity above non-homogeneously covered grid cell	[m s^{-1}]
$-\overline{(uw')}$	the momentum flux on average over a heterogeneous surface	[m^2 s^{-2}]
w_1	the top soil layer volumetric soil moisture content	
z	level above a grid cell	[m]
z_{0a}	an aggregated roughness length	[m]
z_{0l}	an aggregated roughness length over the liquid fraction of the grid cell	[m]
z_{0n}	an aggregated roughness length over the non-vegetative part of the grid cell	[m]
z_{0s}	an aggregated roughness length over the solid fraction of the grid cell	[m]
z_{0v}	an aggregated roughness length over the vegetative part of the grid cell	[m]

(Continued)

List of Symbols

Symbol	Definition	Dimensions or Units
z_a	the height of the lowest atmospheric model level	
z_b	a height taking values z_0 and H (canopy height) for the barren/solid/water and vegetative part, respectively	
$\langle \Phi \rangle$	an areally averaged flux	
Φ_i	the fluxes which in turn are functions some of the components ψ_i	
α_a	an aggregated albedo	
$\bar{\alpha}_c$	the average albedo over the grid-cell of an arbitrary geometry	
$\bar{\alpha}_n$	the average albedo of the grid-cell	
γ	psychrometric constant	
δ	the total fractional cover for a solid portion; parameter	
δ_i	a partial fractional cover for a solid part	
ς	the vegetation-type dependent parameter	
ς_i	the dimensionless constant, parameter	
ς_i^m	a parameter for ith part of a vegetative cover in the grid cell	
ξ_i	the fractional cover for the ith surface type	
θ_1, θ_u	the zenithal angles where the subscripts l and u denote the lower and upper boundary, respectively	
κ	Von Kármán constant	
λ	latent heat of vaporisation	
$\lambda \langle E_0 \rangle$	latent heat flux calculated over the grid cell	
μ	parameter	
ν	the total fractional cover for a liquid portion	
ν_i	a partial fractional cover for a water surface	
ρ_p	the air density	
φ_1, φ_u	the azimuthal angles, where the subscripts l and u denote the lower and upper boundary, respectively	
σ	the total fractional cover for a vegetative portion	
σ_c^i	a fractional area covered by a patch i with the roughness length z_0^i	
σ_i	a partial fractional cover for vegetation	
ψ_a	the vector of aggregated surface parameters	
ψ_i	the surface parameters of each land type	
$d\Omega$	the element of solid angle within which our differential amount of energy is confined to	
0	the subscript which indicates the surface or environment within the canopy	

APPENDIX B – SYNOPSIS

One of the problems in modeling processes at the environmental interfaces is the fact that, in most cases, the grid cell is not homogeneous, i.e., we contend with the rather heterogeneous surface above which the majority of the processes occur. This scenario affects very strongly all of the calculations and demands a specific approach to atmospheric flows in such situations. We address the so-called "flux aggregation", a process by which an effective horizontal average or aggregate of turbulent fluxes is formed over heterogeneous surfaces. One must understand that because of a nonlinear advective enhancement associated with local advection across surface transitions, it differs from the spatial average of equilibrium fluxes in an area. There are three approaches used in practice: "flux aggregation", "parameter aggregation" and the combination of the two. The concept of blending height is introduced so that aggregated fluxes can be related to vertical profiles only above this height. After introducing this concept, we present in this chapter an approach for the aggregation of aerodynamic surface parameters, and then, the aggregation of albedo in case of a surface consisting of parts with differing heights and a combined method for calculating the surface temperature and water vapor pressure over a heterogeneous surface.

APPENDIX C – KEYWORDS

By the end of the chapter you should have encountered the following terms. ensure that you are familiar with them!

Heterogeneous grid-cell	Schmidt paradox	Monte carlo method
Parameter aggregation	Mixing length	Environmental models
Aggregation of fluxes	Turbulent transport	Albedo over urban area
Blending height	Aggregated albedo	Albedo over rock area
Combined method of aggregation	Loss coefficient	Albedo over land

APPENDIX D – QUESTIONS

What are the basic difficulties for calculations within a heterogeneous grid cell?
Describe the concept of aggregation.
What is the difference between aggregation and the "common" averaging of fluxes?
Define the blending height and explain the advantages of this concept.
Why must a mountainous rocky ground with the same type of rock be considered heterogeneous from the perspective of the albedo calculation?

APPENDIX E – PROBLEMS

E1. Explain, in your own words, the principles of operation of Monte Carlo calculations.

E2. Analyze the relation between the blending height and the aggregation roughness length.

E3. Derive the expression (4.36) starting from (4.31), following the lines described in the text.

E4. In the numerical modeling of surface processes, there exist two approaches that are commonly used for calculating the transfer of momentum, heat and moisture from a grid cell consisting of heterogeneous surfaces.
    Count them and describe in more detail.

E5. Explain under what circumstances the "Schmidt paradox" can arise? Describe these circumstances, and then, discuss the occurrence of the counter-gradient transport between the surface and the lowest model level.

REFERENCES

Bourke, P., 1991, Intersection of a plane and a line. http://paulbourke.net/geometry/planeline/

Claussen, M., 1990, Area-averaging of surface fluxes in a neutrally stratified, horizontally inhomogeneous atmospheric boundary layer. *Atmospheric Environment*, **24a**, pp. 1349–1360.

Claussen, M. 1991, Estimation of areally-averaged surface fluxes. *Boundary-Layer Meteorology*, **30**, pp. 327–341.

Claussen, M., 1995, Flux aggregation at large scales: on the limits of validity of blending height. *Journal of Hydrology*, **166**, pp. 371–382.

Delage, Y. and Verseghy, D., 1995, Testing the effects of a new land surface scheme and of initial soil moisture conditions in the Canadian global forecast model. *Monthly Weather Review*, **123**, pp. 3305–3317.

Delage, Y., Wen, L. and Belanger, J.M., 1999, Aggregation of parameters of the land surface model CLASS. *Atmosphere-Ocean*, **37**, pp. 157–178.

De Bruin, H.A.R. and Moore, C.J., 1985, Zero-plane displacement and roughness length for tall vegetation, derived from a simple mass conservation hypothesis. *Boundary-Layer Meteorology* **42**, pp. 53–62.

Dolman, A.J. and Blyth, E.M., 1997, Patch scale aggregation of heterogeneous land surface cover for mesoscale meteorological models. *Journal of Hydrology*, **190**, pp. 252–268.

Fernando, H.J.S., Lee, S.M., Anderson, J., Princevac, M., Paradyjak, E. and Grossman-Clarke, S., 2001, Urban fluid mechanic: Air circulation and contaminant dispersion in cities. *Environmental Fluid Mechanics*, **1**, pp. 107–164.

Hess, G.D. and McAvaney, B.J., 1997, Note on computing screen temperatures humilities and anemometer-height winds in large-scale models. *Australian Meteroogical Magazine*, **46**, pp. 109–115.

Hess, G.D. and McAvaney, B.J., 1998, Realisability constraints for land-surface schemes. *Global and Planetary Change*, **19**, pp. 241–245.

Hu, Z., Islam, S. and Jiang, L., 1999, Approaches for aggregating heterogeneous surface parameters and fluxes for mesoscale and climate models. *Boundary-Layer Meteorology*, **93**, pp. 313–336.

Jacobs, A.F.G. and van Boxel, J.H., 1988, Changes of the displacement height and roughness length of maize during a growing season. *Agricultural and Forest Meteorology*, **42**, pp. 53–62.

Jacobson, M.Z., 1999, *Fundamentals of atmospheric modeling*, Cambridge University Press, The Edinburgh Building, Cambridge.

Kabat, P., Hutjes, R.W.A. and Feddes, R.A., 1997, The scaling characteristics of soil pammews: From plot scale heterogeneity to subgrid parametexization. *Journal of Hydrology*, **190**, pp. 364–397.

Kapor, D., Mihailovic, D.T., Tosic, T., Rao, S.T. and Hogrefe, C., 2002, An approach for the aggregation of albedo in calculating the radiative fluxes over heterogeneous surfaces in atmospheric models. In *Integrated Assessment and Decision Support Proceedings of the First Biennial Meeting of the International Environmental Modelling and Software Society, 24-27 June, Lugano, Switzerland,* Vol. 2, edited by Rizzoli, A.E. and Jakeman, A.J., pp. 448–453.

Kapor, D.V., Cirisan, A.M., and Mihailovic, D.T., 2010, Calculation of Aggregated Albedo in Rectangular Solid Geometry on Environmental Interfaces. In: *Advances in Environmental Fluid Mechanics;* D.T. Mihailovic, C. Gualtieri, Eds.; World Scientific Publishing Co Pte Ltd, New Jersey, USA, 2010, pp. 145–165.

Kondo, J. and Yamazawa, H., 1986, Aerodynamic roughness over an inhomogeneous ground surface. *Boundary-Layer Meteorology*, **35**, pp. 331–348.

Lalic, B., 1997, *Profile of wind speed in transition layer above the vegetation*, University of Belgrade, Masters Thesis, (in Serbian).

Lalic, B. and Mihailovic, D.T., 1998, Derivation of aerodynamic characteristics using a new wind profile in the transition layer above the vegetation. *Research Activities in Atmospheric and Oceanic Modelling, 27*, pp. 4.25–4.26.

Lazic, J., Mihailovic, D.T., Lalic, B., Arsenic, I. and Hogrefe, C., 2002, Land air surface scheme (LAPS) for use in urban modelling. In *Integrated Assessment and Decision Support Proceedings of the First Biennial Meeting of the International Environmental Modelling and Software Society, 24-27 June, Lugano, Switzerland,* Vol. 2, edited by Rizzoli, A.E. and Jakeman, A.J., pp. 448–453.

Liou, K.-N., 2002, *An introduction to atmospheric radiation, 2nd Edition,* Academic Press, Inc., NY.

Mason, P.J., 1988, The formation of areally averaged roughness lengths. *Quarterely Journal of Royal Meteorological Society*, **114**, pp. 399–420.

Masson, V., 2000, A physically-based scheme for the urban energy budget in atmospheric models. *Boundary -Layer Meteorology*, **94**, pp. 357–397.

McCrea, W.H., 1960, *Analytical Geometry of Three Dimensions, 2nd Edition*, Oliver and Boyd Ltd, Edinburgh and London, p. 160

Michaud, J.D. and Shuttleworrth, W.J., 1997, Executive summary of the Tucson Aggregation Workshop. *Journal of Hydrology*, **190**, pp. 176–181.

Mihailović, D.T., 2002, Environmental Modeling of physical, biophysical and chemical processes in the atmosphere plant-soil-interaction: How Nonlynearity Affects the Solutions?, In *Integrated Assessment and Decision Support Proceedings of the First Biennial Meeting of the International Environmental Modelling and Software Society, 24-27 June, Lugano, Switzerland,* Vol. 2, edited by Rizzoli, A.E. and Jakeman, A.J., pp. 383–388.

Mihailović, D.T., Lalic, B., Rajkovic, B. and Arsenic, I., 1999, A roughness sublayer wind profile above non-uniform surface. *Boundary-Layer Meteorology*, **93**, pp. 425–451.

Mihailović, D.T. and Kallos, G., 1997, A sensitivity study of a coupled-vegetation boundary-layer scheme for use in atmospheric modelling.*Boundary-Layer Meteorology*, **82**, pp. 283–315.

Mihailović, D.T., Kapor, D., Hogefre, C., Lazic, J. and Tosic, T., 2003, Parameterization of albedo over heterogeneous surfaces in coupled land- atmosphere schemes for environmental modelling, Part I: Theoretical background. *Environmental Fluid Mechanics*, **4**, pp. 57–77.

Mihailović, D.T., Rao, S.T., Hogefre, C. and Clark, R., 2002, An approach for the aggregation of aerodynamicmic parameters in calculating the turbulent fluxes over heterogeneous surfaces in atmospheric models. *Environmental Fluid Mechanics*,**2**, pp. 315–337.

Mihailović, D.T., Rao, S.T., Alapaty, K., Ku, J.Y., Arsenic, I. and Lalic, B., 2005, A study on the effects of subgrid-scale representation of land use on the boundary layer evolution using a 1-D model. *Environmental Modelling and Software*, **20**, pp. 705–714.

Moran, M.S., Humes, K.S. and Pinter, Jr. P.J., 1997, The scaling characteristics of remotely sensed variables for sparsely-vegetated heterogeneous landscapes. *Journal of Hydrology*,**190**, pp. 338–363.

Morgan, D.L., Pruitt, W.O. and Lourence, F.J., 1971, Analysis of energy, momentum, and mass transfers above vegetative surfaces. *Research and Development Technical Report ECOM 68-G10-F*, Department of Water Science and Engineering, University of California, Davis, U.S.A.

Noilhan, J. and Lacarrère, P., 1992, GCM gridscale evaporation from mesoscale modelling, In: *Proceedings of a workshop held at ECMWF on fine-scale modelling and the development of parameterization schemes, 16–18 September 1991, European Centre for Medium-Range Weather Forecasts*, Reading, UK, pp. 245–274.

Noilhan, J., Lacarrère, P., Dohuan, A.J. and Blyth, E., 1997, Defining area-average parameters in meteorological models for land surfaces with mesoscale heterogeneity. *Journal of Hydrology,* **190**, pp. 302–316.

Oke, T.R., 1987, *Boundary layer climates, 2nd Edition*, Methuen, London, New York.

Pielke, R.A., Zeng, X., Lee, T.J. and Dahr, GA., 1997, Mesoscale fluxes over flat landscapes for use in larger scale models. *Journal of Hydrology*, **190**, pp. 317–337.

Raupach, M.R. and Finnigan, J.J., 1997, The influence of topography on meteorological variables and surface- atmosphere interactions. *Journal of Hydrology*, **190**, pp. 182–213.

Schwerdtfeger, P., 2002, Interpretation of airborn observation of the albedo. *Environmental Modelling and Software*, **17**, pp. 51–60.

Sellers, P.J., Heiser, M.D. and Hall, F.G., 1992, Relations between surface conductance and spectral vegetation indices at intermediate (100 m^2 to 15 km^2) length scales. *Journal of Geophysical Research*, **97(D17)**, pp. 19033–19059.

Sellers, P.J., Heiser, M.D., Hall, F.G., Verma, S.B., Desjardins., R.L., Schuepp, P.M. and MacPherson, J.I., 1997, The impact of using area-averaged land surface properties-topography, vegetation condition, soil wetness in calculations of intermediate scale (approximately 10 km^2) surface-atmosphere heat and moisture fluxes. *Journal of Hydrology*, **190**, pp. 269–301.

Shuttleworrth, J.W., 1988, Macrohydrology-the new challenge for process hydrology. *Journal of Hydrology*, **100**, pp. 31–56.

Shuttleworrth, J.W., 1991, The Modellion Concept. *Reviews of Geophysics*, **29**, pp. 585–606.

Stössel, A. and Claussen, M. 1993, A new atmospheric surface-layer scheme for a large-scale sea-ice model. *Climate Dynamics*, **9**, pp. 71–80.

Sun, S.F., 1982, *Moisture and heat transport in a soil layer forced by atmospheric conditions*. M.S. Thesis, Department of Civil Engineering, University of Connecticut.

Thornton, P.E., Running, S.W. and White, M.A., 1997, Generating surfaces of daily meteorological variables over large regions of complex terrain. *Journal of Hydrology*, **190**, pp. 214–251.

Wetzel, P.J. and Boone, A. 1995, A Parameterization for Atmosphere-Cloud- Exchange (PLACE): Documentation and testing of a detailed process model of the partly cloudy boundary layer over heterogeneous land. *Journal of Climate,* **8**, pp. 1810–1837.

Wieringa, J., 1986, Roughness-dependent geographical interpolation of surface wind speed averages. *Quarterely Journal of Royal Meteorological Society*, **112**, pp. 867– 889.

Walko, R.L., Band, L.E., Baron, J., Kittel, T.G.F., Lammers, R., Lee, T.J., Ojima, D., Pielke, R.A., Taylor, C., Tague, C., Tremback, C. J. and Vidale, P.L., 2000, Coupled atmosphere-biophysics-hydrology models for environmental modeling. *Journal of Applied Meteorology*, **39**, pp. 931–944.

Wood, E.F., 1997, Effects of soil moisture aggregation on surface evaporative fluxes *Journal of Hydrology*, **190**, pp. 398–413.

Wood, N. and Mason, P.J., 1991, The influence of stability on effective roughness lengths for momentum and heat transfer. *Quarterly Journal of Royal Meteorological Society*, **117**, pp. 1025–1056.

CHAPTER FIVE

Desert dust uptake-transport and deposition mechanisms – impacts of dust on radiation, clouds and precipitation

George Kallos

School of Physics, University of Athens, Athens, Greece

Petros Katsafados

Department of Geography, Harokopio University, Athens, Greece

Christos Spyrou

School of Physics, University of Athens, Athens, Greece

ABSTRACT

Desert dust cycle is considered an important factor in the atmosphere and the ocean. Dust particles exert a significant number of impacts on radiative transfer, cloud formation and precipitation. Desert dust can reduce the incoming solar radiation on the surface, while at the same time warm middle tropospheric layers and affect stability and precipitation. Moreover it can assist in the formation of small water droplets and suppress precipitation or, in combination with sea salt and anthropogenic pollutants, form gigantic CCNs that behave like Ice Nucleus (IN) and enhance precipitation. Dust deposition can affect significantly the marine biological processes, by providing nutrients in the sea surface.

The mechanisms for dust production are very complicated and depend on several parameters like friction velocity, soil composition and granulation, soil moisture, vegetation etc. The transport and deposition processes depend mainly on particle size and geometry. Small dust particles can be transported in long distances depending on atmospheric conditions. For example Saharan dust can cross Mediterranean in less than a day while the cross-Atlantic path can last one or two weeks. Almost 10^8 tons of Saharan dust is deposited over the Mediterranean waters and Europe every year.

In this chapter the dust production mechanisms from desert sources will be reviewed and the impacts on the atmospheric and marine processes will be discussed.

5.1 INTRODUCTION

Soil dust produced by aeolian activity is considered a major source of Particulate Matter (PM) in the atmosphere. Dust is extracted from desert, arid and semi-arid regions of the planet under favourable weather conditions and is transported to short and long distances (from a few centimetres to thousands of kilometres). Agricultural and other human activities are also considerable sources of mineral dust, but at a smaller scale than naturally produced dust aerosols. Once airborne dust becomes an important climate modifier as: (1) it affects the backscattering and absorption of solar and terrestrial radiation (Miller and Tegen, 1998; Intergovernmental Panel on Climate Change (IPCC), 2007), (2) it reduces the incoming solar radiation at the earth's surface by a considerable amount (up to 10% under extreme events) and therefore produces a cooling that masks the global warming (Ramanathan *et al.*, 2001, Alpert *et al.*, 1998), (3) it is causing mid tropospheric warming by absorbing of radiation and on that way it stabilizes the lower troposphere and affects the water budget (Levin *et al.*, 2005). Because the dust source areas are near regions with fragile water budget, perturbations in production can affect precipitation and water budget. Since dust production is affected by soil moisture, perturbations in the water cycle in arid and semi-arid regions can affect the dust cycle. The feedback between dust, cloud formation and precipitation is not straightforward but is very complicated (Levin *et al.*, 1996; Solomos *et al.*, 2010). The entire system becomes more complicated when other factors like sea-salt spraying and/or anthropogenic pollutants (aerosols) coexist (Levin *et al.*, 2005). In the past decades, several studies have also indicated a clear connection between suspended particulate matter and health effects (Mitsakou *et al.*, 2008). Microscopic dust is so small that it can sidestep the lungs natural defences (Mitsakou *et al.*, 2005).

Almost one third of the earth's land surface is desert, arid land with sparse vegetation and very small amounts of rainfall. The deserts may be areas covered by sand, rocks gravels and rarely some plants can be found. Mineral deposits like salt can be found in the surface as a result of transport and erosion. Erosion is mainly caused due to strong winds over source areas, but can be initiated due to temperature differentiation, friction between various sizes of stones and water.

Soil dust consists of particle with diameters ranging from submicron levels to tens of microns. The particle size is a function of various parameters related to the way they are created as well as the composition and characteristics of source areas (Tegen and Fung, 1994). The transport and deposition is subject of the particle size and composition as well as turbulence and wind strength.

The estimated global dust-emission rates falls within a range from 1000 to 3000 Tg yr^{-1}, and about 80% of the dust is from the Northern Hemisphere (Goudie and Middleton 2006; Tanaka *et al.*, 2007). The world's largest source of dust is the Sahara Desert with an estimated range of dust emission from 160 to 760 Tg yr^{-1}, ranging from one-third to over half of the total global dust emission. Comparing this amount with the annual production of sulphate aerosols at global scale that is at the range of several hundreds of Mtyr^{-1} someone can see immediately a difference of at least one to two orders of magnitude.

Dust mobilized from the Sahara region can be transported hundreds and thousands of kilometres away towards Mediterranean and Europe as well as towards the Atlantic and Indian Oceans. As Guerzoni and Chester (1996), Kallos *et al.* (2005), and Kallos *et al.* (2007), found it the amount of Saharan dust deposited over the Mediterranean waters is of the order to 10^8 Mtyr^{-1}. Similar amount is crossing the Mediterranean and transported towards Europe. The amounts deposited over the Atlantic Ocean are even higher (Kallos *et al.*, 2006, 2007) especially during the summer seasons. The other deserts known for their high productivity of dust particles are the Gobi desert, the desert of Namibia, Australia, Peru, SW USA and other smaller. Smaller amounts of desert dust are produced from the areas around lakes with specific characteristics, mainly with high amounts of salt and other minerals (e.g. Salt Lake in Utah, southern Aral, the area around the Caspian Sea, Deal Sea).

The impacts of dust in the atmosphere and climate have been briefly mentioned previously (Miller and Tegen, 1998; Andreae, 1996). The impacts of the deposited desert dust on the ocean surface and therefore the marine environments are also considerable (Martin and Fitzwater, 1988; Goudie and Middleton 2001). Desert dust can cause radiative and heat perturbations at the ocean top, it can affect phytoplankton and other kind of marine productivity and of course, it can affect fluxes of important chemical species in the atmosphere like di-methyl-sulfate (DMS). Desert dust in the ocean can trigger various biochemical reactions between dust ingredients and the marine environment. Key elements like iron, phosphorus and other micronutrients (Guerzoni *et al.*, 1999).

Desert dust can affect also fauna and flora. Deposition of dust over plants can affect photosynthesis, evapotranspiration and heat exchange. It can act also as fertilizer.

The urban air quality in many regions around the world is affected from desert dust transport on many ways as described in Rodriguez *et al.* (2001), Papadopoulos *et al.* (2003) and Spyrou *et al.* (2010) among others (e.g. by increasing the PM concentration at levels above the imposed regulations, by reducing visibility, by reducing the incoming solar radiation, by deposited over surfaces in buildings etc and then by resuspension can be in the atmosphere). For example, most of the South European cities cannot meet the imposed European Union air quality standards on PM concentrations (Mitsakou *et al.*, 2008). Health effects are also associated with desert dust outbreaks (Rodriguez *et al.*, 2001; Kallos *et al.*, 2007).

Because of the importance of the dust cycle in the atmosphere, biosphere and hydrosphere, the dust cycle in the atmosphere (production, transport deposition) and its main properties will be further analyzed in the next sections.

5.2 PHYSICAL PROCESSES

The desert dust cycle is considered as a complex geophysical process. It involves soil erosion and atmospheric processes. The impacts of the desert dust on environment and climate are several. They are ranging from modifications on radiative transfer mechanisms (short and long wave), air quality degradation in urban environments, modification of water budget especially in arid and semi- arid regions and is associated with desertification and aridity. The transported and deposited dust material significantly affects the marine environment because it may significantly modify the marine biochemistry after deposition to ocean waters (Martin and Fitzwater, 1988; Prospero *et al.*, 1996). Aerosols interact strongly with solar and terrestrial radiation in several ways (Yoshioka *et al.*, 2007; IPCC, 2007; Kallos *et al.*, 2009a), also known as "Direct Aerosol Effect – DRE" (IPCC, 2007). By absorbing and scattering the solar radiation aerosols reduce the amount of energy reaching the surface (Kaufman *et al.*, 2002; Tegen 2003). Moreover, aerosols enhance the greenhouse effect by absorbing and emitting outgoing longwave radiation (Dufrense, 2001; Tegen, 2003). Effects on construction materials, rain acidification, and visibility degradation have been also been reported. They also pointed out that dust aerosols are an important source of inaccuracies in numerical weather prediction and especially in General Circulation Models (GCMs) used for climate research. It is also worth mentioning that some intense dust storms catastrophically affect the regions in the neighbourhood of dust sources, causing loss of human life and economic damage. Dust plumes can affect remote locations significantly; because they increase the PM concentrations and especially the fine, ones (PM2.5) and therefore they can have significant health effects (Mitsakou *et al.*, 2008). According to Barkan *et al.* (2004) the highest aerosol index values in North Africa and Arabian Peninsula were estimated during June and July while the area around Lake Chad, has demonstrated local maximum values and, contrary to the other sources, is active throughout the year. These estimations were made by analyzing the TOMS instrument data for a period of fourteen years (1979–1992).

Dust mobilization exhibits high seasonal variability of the dust mobilization that depends on the source characteristics as well as the global atmospheric circulation (Ozsoy *et al.*, 2001). The dust production in the highest productive area of North Africa and Arabian Peninsula is subject of seasonal variability and the characteristics of general circulation on the planetary scale. During winter and spring, the Mediterranean region is affected by two upper air jet streams: the polar front jet stream, normally located over Europe, and the subtropical jet stream, which is typically located over northern Africa. The combined effects of these westerly jets in late winter and spring support the propagation of extra tropical cyclones towards the East and Southeast, resulting in dust plume intrusion in the Mediterranean (Figure 5.1).

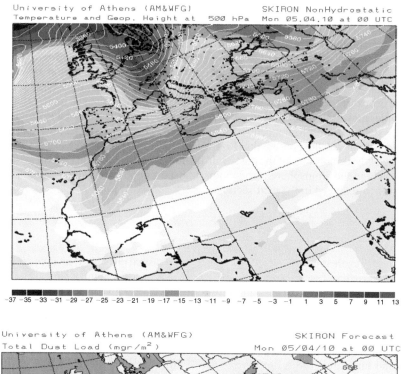

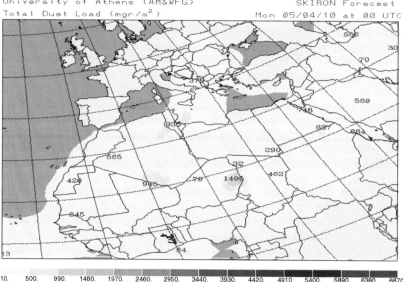

Figure 5.1. Synoptic conditions favour the dust transfer from the North Africa towards Eastern Mediterranean. The plots depict the geopotential height (gpm) with the temperature (C) at 500 hPa and the dust load (in gr/m²). The figure obtained from the Skiron operational cycle and it is valid for 5th April 2010 at 00UTC.

During the transient and cold seasons, most of the dust events that transport significant amounts of dust from Saharan towards the Mediterranean Sea and Europe occur. These seasons are characterized by the low index circulation of the year as described by Papadopoulos *et al.* (2003) and Rodriguez*et al.* (2001). During summer, the amount of produced and transported dust is almost twice as large as in winter (Husar *et al.*, 1997). The highest amounts of mobilized dust in Sahara are transported towards the tropical Atlantic, Caribbean Sea and even North America with the aid of the easterlies (Perry *et al.*, 1997; Kallos *et al.*, 2005).

The dust storm is created by the injection of dust particles in the atmosphere. This injection is function of various parameters like wind shear, the size and the composition of the particles, and the soil moisture. Soil particles can move in three different ways namely creeping, saltation, and suspension:

- Creeping is the rolling and/or sliding of particles along the ground. Creeping is supported by light winds and low particle granulation.
- Saltation is the kind of soil particle movement through a series of jumps or skips. When the particles are lifted into the atmosphere, they start drifting for approximately farther downwind before they fall down again. The horizontal drifting is for approximately four times the vertical lifting. When the particles return to ground, they hit other particles or the ground and then they jump up again and progress forward. Smaller particles can be produced during the impact (Figure 5.2).
- Suspension is the process that occurs when soil particles (usually sediment materials) lifted into the air and remain aloft by winds. If the particles are sufficiently small and the upward air motion is able to support the weight of the individual grains, they will hold aloft. The larger particles settle due to gravitational forcing while the smaller ones remain suspended and transported by turbulence. The amount of the suspended particles is function of wind speed: strong winds can assist in suspension of larger particles. The suspended particles are moving initially by turbulence and later by the organized flow patterns.

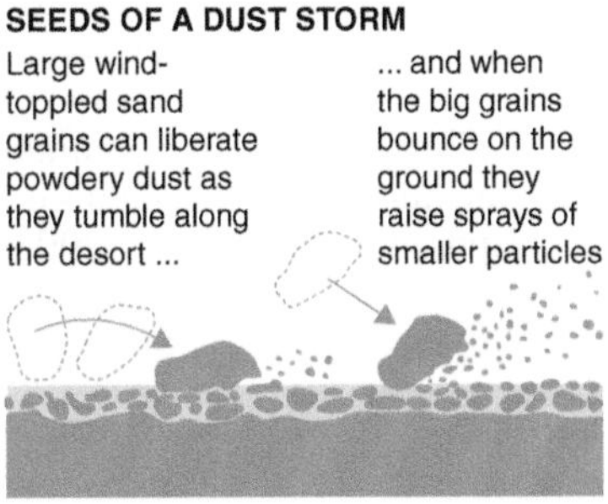

SEEDS OF A DUST STORM

Large wind-toppled sand grains can liberate powdery dust as they tumble along the desort ...

... and when the big grains bounce on the ground they raise sprays of smaller particles

Figure 5.2. Production of small desert particles through saltation. Source: *Geological Society of America.*

The organized strong flow pattern (mesoscale and/or synoptic scale) can lift up the dust particles by thousands of meters and transported horizontally downwind hundreds or thousand of kilometres. Large-scale turbulence or updrafts assist in suspending the soil particles until they settle down by gravitational forcing and/or wet scavenging and deposition processes. Regularly, the smaller particles (usually of the size of PM2.5 and less) are transported the larger distances while particles of size higher than PM10 are deposited faster over smaller distances (a few kilometres to a few hundreds of kilometres).

5.3   PARTICLE SIZE AND SETTLING VELOCITY

Soil dust particles belong to one of the three major types of aerosols namely (a) continental or desert aerosols, (b) industrial aerosols and (c) volcanic aerosols. Soil dust particles (called also continental aerosols) are of a wide range in diameter. Usually they are of diameter of submicron to a few tens of μm. The particulate portion of an aerosol is referred to as Particulate Matter (PM). PM is a collective term used for very small solid and/or liquid particles found in the atmosphere. The geometry, size, composition and in general, physical and chemical properties is varying significantly. Particle size can range from 0.001 to 500 μm (Seinfeld and Pandis, 1998). Particles in the range of 2.5–10.0 μm are called as "coarse" particles; while the others with diameter from 0.1 to 2.5 μm are called as "fine". The smaller particles (less than 0.1 μm) are called as "ultra fine". There are two categories that well known the so-called PM2.5 and PM10 and define particles with diameters less than 2.5 μm and 10 μm respectively. The range of horizontal transport of the particles is function of the size and composition. In general, particles of the category PM2.5 behave as perfect gases because the gravitational settling is negligible. Particles of size PM10 are heavier and therefore the gravitational settling is larger and deposit in relatively small to moderate distances. The particles that are larger than PM10 deposit quickly near the sources.

Since the gravitational settling for soil particles of the category PM2.5 is very small, they are subject for long-range transport. Transport scales of 1000 km are characteristic in such cases. A common phenomenon associated with such kind of transport of dust particles is the "red snow" or "mad rain" encountered in Northern Europe, Asia or even North America. Soil particles larger than PM10 usually are transported in distances ranging from a few meters to a few kilometres. While the transport of such particles is not a subject of long range transport, their effects are significant near the sources and for the production of smaller particles as they collide with others while falling down (saltation). According to Alfaro *et al.* (1997), the size distributions of the aerosols released by silt and clay soil textures have medium respective diameters of 1.6, 6.7 and 14.2 μm. The total mass of released dust depends on particle size distribution. Over the source areas, the mass distribution can be described by the three modal lognormal function of D'Almeida (1987). Although dust production is initiated by the entrainment of sand sized particles ($\sim$60 m in diameter), only smaller particles with radius $r \leq 10$ m reside in the atmosphere long enough to be transported over large distances (Zender *et al.*, 2003). For the long range traveling particles, the minimum (r_{min}), maximum (r_{max}) and effective (r_{eff}) radius of each size bin, number median radius of the distribution (r_n) and geometric standard deviation (σ_g), proposed by Perez *et al.* (2006) is summarized in Table 5.1.

Table 5.1. Main characteristics of typical dust particles (Source: Perez *et al.*, 2006).

B_{in}	r_{min}	r_{max}	r_{eff}	r_n	σ_g
1	0.1	0.18	0.15	0.2986	2
2	0.18	0.3	0.25	0.2986	2
3	0.3	0.6	0.45	0.2986	2
4	0.6	1	0.78	0.2986	2
5	1	1.8	1.3	0.2986	2
6	1.8	3	2.2	0.2986	2
7	3	6	3.8	0.2986	2
8	6	10	7.1	0.2986	2

The wet and turbulent dry deposition processes are the main mechanisms for removal of particles less than 10 μm. Particles larger than 10 μm are basically removed by gravitational settling. The sand particles are large and cannot participate in the longer-term atmospheric transport. Although, their role in dust storms is considerable near the source areas since high amounts of sand mass are lifted and drifted with the turbulence eddies, especially in the area of density current. Such phenomena are responsible for mobilization of large amounts of sand towards areas adjacent to dust sources and hence the expansion of desertification.

The dust particles that are moving within the atmosphere will continue doing it as long as the upward motion is greater than the speed at which the particles fall through air. The relationship between the falling speed (or settling velocity) and the particle size is shown in Figure 5.3. As we can see, particles capable of travelling great distances are these with diameters less than 20 μm since the falling speed is about 0.1 m/s. Particles larger than 20 micrometers in diameter fall disproportionately faster. The PM10 particles fall at about 0.03 m/s. Fine particles fall with very low speeds (~0.001 m/s). Finest clay particles settle very slowly and therefore can be transported very large distances under favourable synoptic weather conditions. This is especially true over oceans under anticyclonic conditions where wet removal processes do not exist.

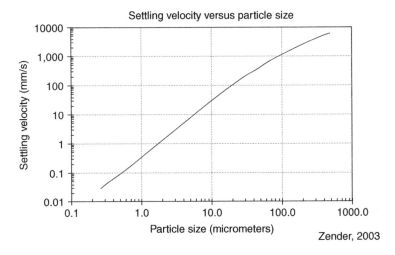

Figure 5.3. The settling velocity, as a function of particle size (source:http://www.meted.ucar.edu).

5.4 SOURCE AREAS

The dust particles encountered in most of the deserts are clay particles with diameters less than 2 μm, silt particles with size ranging from 2 to 50 μm, and sand-size particles that are greater than 75 μm. Therefore, areas that contain soil particles with such characteristics can act as dust sources under the appropriate weather conditions. The source areas favouring the production of fine particles appropriate for long-range transport are these with fine-grained soils, rich in clay and silt. Areas with large soil particles (sand) can act as sources for dust storms of local scale.

A considerable amount of soil dust is taken up by the wind from arid or semi-arid areas around the globe and then transported to smaller or larger distances. Smaller amounts can be produced from other areas and human activities, mainly agricultural areas and/or roads, under certain circumstances. From all dust sources, the Saharan desert is the major production area. The estimates of soil dust emissions exhibit significant variations. The large variations can be attributed to the frequent spatial inhomogeneities in soil properties and the incapability of the existing methodologies to cover such issues on an accurate way. Therefore, the figures provided for the dust production on annual base are subject to assumptions made in the methodology of the calculations such as the surface properties, particle granulation, soil moisture characteristics and f course rate of scavenging. Older estimated emissions are of the range of 500 to 5000 million tones per year. Recent ones suggest the range of 1000 to 3000 million tones per year as more realistic. The dust emissions from Saharan desert are of the range of 130 and 760 million tones per year. The range of emitted dust between 260 and 710 million tonnes per year has been also provided in the literature (Callot *et al.*, 2000, Prospero, 1996, Swap *et al.*, 1992). The dust emission by itself is not an accurate estimate of the phenomenon because someone has to take in the account suspension time scales and range of transport. Most of the emitted dust settles down quickly producing usually producing smaller particles that are emitted later and transported over longer distances.

There is strong relationship between dust production areas and aridity or with low annual rainfall amounts (usually with rainfall < 200–250 mm/year). The so called "dust belt" extends from Western Africa to Middle East, the Arabian Peninsula and East almost up to Himalaya. This is the most "productive" area for dust. The main reason is the small amount of rain, the composition of soil, the daily temperature range and in general the geomorphological characteristics of the area (e.g. ephemeral playa-lakes, rivers, lakes and steams, and in general drainage basins in the proximity of mountains without vegetation). Usually, these ephemeral formations during the wet season collect eroded soils that are exposed to resuspension processes during the dry season (Querol *et al.*, 2002).

Mapping the dust production areas and characterization of their productivity is an important issue due to various implications of dust in the environment, water management and climate. A major effort devoted towards this direction by Prospero *et al.* (2002). In this work, they used satellite data (Total Ozone Mapping Spectrometer TOMS data) to identify the dust regions and their characteristics on global scale. According to Prospero *et al.* (2002), the largest and most persistent sources are located at the latitudes of the subtropical high of the Northern Hemisphere, mainly in a broad "dust belt" that extends from the west coast of North Africa, over the Middle East, Central and South Asia, to China. There are some mountainous regions (e.g., Afghanistan, Iran, Pakistan, and China) that are significant dust sources, especially the valleys between mountain peaks. Considerable amounts of produced dust are also outside of this belt. In particular, there are areas in the Southern Hemisphere with remarkable dust activity as in Namibia, Australia, Peru etc. Other dust production areas associated with human impacts are well documented, e.g., the Caspian and Aral Seas, Tigris-Euphrates River Basin, SW North America, and the loess lands in China. Of course, the largest and most active sources are located in areas where there is little or no human presence.

The most active dust sources are associated with topographic lows or they are in areas with frequent exchange between mountains and valleys of highlands as shown in Figure 5.4. In this figure, a typical desert area in SW Algeria is shown where hills and valleys are in a stripe formation (NASA photo).

The Mediterranean Region is affected by dust storms very often. Every day, there is a region of the Mediterranean Sea where North African dust is deposited. In addition, Europe and especially Southern Europe, receives similar amounts of dust as the Mediterranean Sea. This is especially true during late spring and summer (Guerzoni and Chester, 1996; Prospero, 1996; Moulin *et al.*, 1998). The most important sources of the dust are eastern Algeria, Tunisia, Libya, and Egypt.

The most important dusts sources of the planet have been identified, described, and grouped by Prospero *et al.* (2002). Following the work of Prospero *et al.* (2002), the most important dust sources with their major characteristics are briefly described below:

Mauritania and Western Sahara:

This is an area with important sources that contribute to the production of dust plumes directed towards the Atlantic Ocean. They become active early in the year and remain

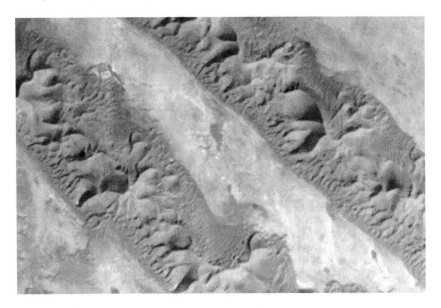

Figure 5.4. Dust uptake areas in NE Algeria. Stripe formation of the uptake areas with sand dunes and dust-salt mixture. Source NASA. Photo taken from the International Space Program, Photo ISS013-E-75141, 2 September 2006. Available from http://earthobservatory.nasa.gov/Newsroom/NewImages

productive until late Fall with peak production during summer months. High productivity is partially due to drainage activities during winter and partially due to trade wind systems.

Mali, Mauritania, Niger and the Ahaggar Mountains:

This area contains some of the most productive sources all over the world. This is due to existence of several sand dunes evident in many locations, the composition of the soil (high granulation), and the absence of precipitation (very seldom) and the presence of trade wind systems in the area, mainly the easterlies. The area is habituated by a very small amount of people with negligible agricultural activities. The most productive period is late spring to late Fall and the suspended dust is primarily directed towards the Atlantic Ocean and secondarily to other directions according to the prevailing weather systems. A certain amount of dust, especially the large particles contribute to the expansion of desertification in the surrounding areas.

Lake Chad Basin and the Bodele Depression:

This is the most productive dust area of the world. It contributes to the dust plumes directed towards West, East and North. The dust areas remain productive during all seasons with minima during late autumn. There is always dust in the air for most of the regions of this large area. The soil consists of sediments that are rich in clay amounts and therefore dust clouds can form easily even with light winds due to high granulation. Production is enhanced also from the activity of drainage formations in many places. Sand dunes that are continuously productive all over the year cover large areas (hundreds of kilometres towards each direction). Most of the sub-regions and especially the Bodele are the main contributors of dust plumes directed towards Atlantic, Gulf of Guinea and also towards Mediterranean.

Tunisia and Northeast Algeria:

This area has some very productive areas at various elevations. The most productive dust areas are in locations where temporal salt lakes are temporarily formed. The dust particles suspended from this area have different hygroscopicity and therefore they affect the cloud formation and precipitation. In addition to high hygroscopicity the dust particles from these areas are of mixed alluvial, silt and clay type. The drainage activity and the formation of the seasonal lakes enhance the dust productivity during the dry period of the year. The most productive period of the year is spring and autumn. The salty-water lakes (called chotts) and the associated dust source regions lie in the lee of the Atlas mountains and therefore receive small amounts of precipitation (approximately 100 mm on annual base) not adequate to keep water during the dry season.

Libyan and Egyptian Desert:

A large area that extends from Eastern Libya to Egypt is dust productive during most of the year, with the most intensive period during spring and autumn. The northern part of this area is a low-lying region where water is drained from the surrounding areas forming the "wadis". These areas are highly productive after the rainy season or temporal rains. The dust productive areas are often broken by the oases. These dust sources are of alluvial type and contribute significantly in the dust storm formation towards the Mediterranean Sea and Europe.

Sudan, Ethiopian highlands and Horn of Africa:

This is a large area with large variability in dust sources. The maximum productivity is from May to July. The productivity is moving towards North at the beginning of summer and then again southward during fall. There are areas with sand dunes while other productive areas at the Sudan highlands with rich in clay soils. Wadi-type formations can be encountered too. Runoff formations in the Ethiopian, Somali and Eritrean high lands turn in high productive regions after drying out. The dust production from these regions is transported towards the Red Sea, the Gulf of Aden and the Indian Ocean. Very often, the dust plumes are mixed with biomass burning that is a seasonal procedure in agricultural or semi-arid areas.

Middle East Deserts:

Middle East is a region where dust sources are too many but with different characteristics. Almost the entire Arabian Peninsula is considered as a dust source. Most of the sources are in low elevation and extend up to the coastal areas. Even the mountainous regions are considered as secondary dust sources mainly due to heavy deposition from major dust storms in the nearby locations or even from East Africa. The Midle East regions alond the Mediterranean coast are not considered as dust sources. Although, this belt is relatively narrow (a few hundreds of kilometres). Dust sources exist in central Turkey but they exhibit a seasonal production cycle with its maximum during summer. The dust source areas of the Northern part of the Middle East region (Mesopotamia region) are associated partly to the agricultural practice along the centuries. Dust storms in the Middle East region are associated with typical weather patterns prevailing in the area mainly during the transition seasons of Spring and Autumn.

West and South West Asia deserts:

The soutwestern Asia dust sources are mainly located in the Iran-Afganistan and Pakistan belt. Several regions of Northwest India are dust mainly during the Spring Season (mainly the Rajasthan area). They extend up to the Caspian Sea region. Most of the active sources are located in valleys between the major mountainous ridges of the area. Dust production exhibits high seasonality in the area with maximum production after the rainy period when the land starts drying and the poor vegetation dries out.

All the area around the Caspian Sea is rich in dust production areas. The dust sources extend until the Aral sea where during the last half century the overuse of water extended the sources to the southern part of the lake. The southern Aral dust-source region is rich in minerals like sea salt.

Dust source regions extend all over the area of Kazakhstan, Azerbaijan, Kyrgyzstan, and Turkmenistan and in general in the area North of Himalaya. Several dust sources in this extended region exhibit high seasonality because during the cold period of the year they cease down due to the fact that the soil is relatively wet and vegetation covers the ground, at least partially. During summer the productivity is relatively high under certain weather types.

Central Asia Deserts:

In central Asia there are very important dust sources like Gobi, Tarim and Takla Makan desrts. Most of these dust sources are active during spring. The geological formations of these dust sources lead to the formation of heavy dust mobilization episodes that have severe consequences in the adjacent urban locations as well as in the agriculture. The consequences extend to big urban conglomerates of China, Korea and Japan. Several dust episodes have been recorded where dust plumes cross the Pacific Ocean and transport dust towards North America.

In the Central Asia there are small dust source regions where salt and other minerals coexist with soil. Such areas are dry lakes or adjacent to lakes where human intervention led to desertification.

The Mongolian steppes are considerable dust sources during the warm period of the year and especially during the transition from green to dry (dry out of shrub and grass).

Australia:

The Central Australian Continent contains several dust sources with persistent activity. The most active season starts during September-October (Australian spring) and peaks in December-February Australian summer).

Southern Africa:

The most known South African dust sources are located in the Namimbia, Botswana and of course the Okawango Delta. Kalahari is a well-known desert of this area. A seasonal variability in dust production is evident but most of the dust sources remain actiuve during all the year. The Namibia desert area extend up to the Southeast Atlantic coastal area. The South African dust sources is a major producer of dust transported towards South Atlantic.

Western Unided States and Mexico:

The largest dust sources in United States are located in the area between the Sierra Nevada and Cascades to the West and Rocky Mountains to the East. Well-known desrts are in Nevada, East California Utah (Salt Lake Region and South), New Mexico and Arizona. Secondary sourcesare located in Western Texas. The dust source regions of Southwestern United States extend to Mexico to the South up to Central America. In general, most of these sources are of low to moderate productivity that is restricted mainly within the boundary layer. Dust productivity in areas of dry lakes is considerable because of the coexistence with minerals and salt. In general, dust production has been increased during the last two centuries due to the human intervention in these regions. The dust productivity in this source regions exhibit high seasonal variability where it's maximum occurs during spring.

Latin America dust sources:

In South America, there a few dust sources. These sources are mainly located in the Altiplano region of Bolivia. They extend in Southern Peru, Argentina and Chile. Most of these dust sources are in highlands where typical elevation is higher than 2000 m above sea level and up to 4000 m. Some of the dust sources are located in dry lake regions and therefore soil coexists with salt.

Minor dust sources exist also in Western Argentina and the Patagonia Region.

5.5 WIND AND TURBULENCE

Having defined the dust sources, one must turn to the characteristics of the wind field, which play a key role in moving and lofting the dust particles. The initial dust and sand particles that will move (at wind speeds of 5–13 m/s) are those whose diameter ranges from 0.08 to 1 mm (80–1000 micrometers). For both larger and smaller particles to move, stronger winds are required. Apparently, the impact created by saltation of the initial particles when lifted can cause the smaller particles to be hurled aloft.

Generally speaking, in order to mobilize dust, winds at the surface need to be 15 knots or greater. The Table 5.2 shows an overview of wind speeds required to lift particles in different source environments.

Table 5.2. Threshold dust-lofting wind speed for different desert environments (source: http://www.meted.ucar.edu).

Environment	Threshold Wind Speed (m/s)
Fine to medium sand in dune-covered areas	4.50–6.70
Sandy areas with poorly developed desert pavement	8.95
Fine material, desert flats	8.95–11.16
Alluvial fans and crusted salt flats (dry lake beds)	13.40–15.60
Well-developed desert pavement	18.90

Once a dust storm starts, even when wind speeds slow to below initiation levels, it can maintain the same intensity. The reason lays in the fact that the bond between the dust particles and the surface is broken and saltation mechanism allows dust to lift. For a perfectly laminar flow, the mobilized particles would move in a thin layer across the desert floor. In order that a dust storm be created, it is necessary to get that dust up in the air. Substantial turbulence in the atmospheric boundary layer is typically required for the lofting of dust.

Typically, the turbulence and horizontal roll vortices that loft the dust up and away from the surface are created by the wind shear. It stands to reason that dust storms will be favoured by an unstable boundary layer, since vertical motions are required to loft the particles. So, a stable boundary layer suppresses updrafts and inhibits dust raise. In similar way, the vertical extent of dust lofting is limited by a low-level inversion. Due to the lack of vegetation, dust-prone regions can experience extreme daytime heating of the ground causing the establishment of an unstable boundary layer, which deepens as the amount of heating increases. Thus, it is the mid-latitude deserts, with their extreme daytime temperatures, which are particularly prone to an unstable boundary layer. On the other hand, dry desert air leads to a wide diurnal temperature cycle. A strong radiative cooling lead to rapid heat loss after sunset, the lowest atmosphere is cooled, resulting in a surface-based inversion with potentially strong effects on blowing dust.

Such inversion suppresses vertical motions in the boundary layer so it becomes hard to lift dust. A 10-knot wind may raise dust during the day, but at night, it may not. However, formation of a surface-based inversion will have little effect to the dust already in suspension higher in the atmosphere. Furthermore, sufficiently strong winds will inhibit formation of an inversion or even remove one that has already formed.

5.6 FRICTION VELOCITY

As it has been previously discussed, the wind strength is not sufficient to lift up dust. The wind field must be sufficiently turbulent to loft dust and in general to have unstable conditions. Friction velocity is the parameter that expresses such conditions on the best and simple way. In general, dust mobilization is proportional to the flux of momentum, or stress, into the ground. A friction velocity of 0.6 m/s is typically required to raise dust. Friction velocity u^* (cm/s) is defined as:

$$u^* = \frac{V_s \cdot \kappa}{\ln(z_s/z_0) - \psi_m \cdot (z_s/L)} \tag{5.1}$$

where:
V_s is the wind speed at the midpoint z_s of the surface layer,
κ is the Von Kármán constant,
z_0 is the surface roughness ($z_0 = 0.01$ for the desert),
ψ_m is the stability parameter for momentum, and
L is the Monin-Obukhov length. For neutral conditions, $z_s/L = 0$ and $\psi_m = 0$.

In daytime, the atmosphere over the desert is usually unstable so that $z_s/L < 0$ and $\psi_m > 0$, and more momentum is transferred to the ground. Table 5.3 presents some typical values of u* for different values of threshold wind velocity (**Vt**) under neutral and unstable conditions (Westphal *et al.*, 1988).

5.7 DIFFUSION EQUATION

The dust cycle in the atmospheric environment is in general, described by a set of K independent Euler-type prognostic continuity equations for dust concentration of the form:

$$\frac{\partial C_k}{\partial t} = -u\frac{\partial C_k}{\partial x} - v\frac{\partial C_k}{\partial y} - (w - v_{gk})\frac{\partial C_k}{\partial z} - \nabla(K_H \nabla C_k) - \frac{\partial}{\partial z}\left(K_Z \frac{\partial C_k}{\partial z}\right)$$

$$+ \left(\frac{\partial C_k}{\partial t}\right)_{SOURCE} - \left(\frac{\partial C_k}{\partial t}\right)_{SINK} \tag{5.2}$$

Table 5.3. Typical values of friction velocity (u^*) for different values of threshold wind velocity (V_t) under neutral and unstable conditions (source: http://www.meted.ucar.edu).

V_t (m/s)	u^* Neutral ($z/L = 0$)	Unstable ($z/L = -2$)
5	29	35
8	46	55
11	64	77

where:
K indicates the number of the particle size bins ($k = 1, \ldots, K$),
C_k is the dust concentration of a k-th particle size bin,
u and v are the horizontal velocity components,
w is the vertical velocity,
v_{gk} is the gravitational settling velocity,
∇ is the horizontal nabla operator,
K_H is the lateral diffusion coefficient,
K_Z is the turbulence exchange coefficient,
$(\partial C_k / \partial t)_{SOURCE}$ is the dust production rate normally over the dust source areas, and
$(\partial C_k / \partial t)_{SINK}$ is the sink term, which includes both wet and dry deposition fractions.
    The total concentration C is a weighted sum of concentrations of K particle size classes used. Usually they are used 4–12 size bins of dust particles.

$$C = \sum_{k=1}^{K} \delta_k C_k; \quad \sum_{k=1}^{K} \delta_k = 1 \tag{5.3}$$

δ_k denotes a mass fraction of the k-th particle category.
    From the time, the dust particles are on the air they are transported higher into the boundary layer but they can be transferred to the ground. The return to the ground after travelling smaller or larger distances (from a few meters to thousands of kilometres) The mobilization of dust in the atmosphere can occur through the:

- dispersion mechanisms,
- gravitational settling of dust particles and
- entrainment of dust in convective activities and deposition with precipitation.

5.8   DISPERSION OF DUST

In general, dispersion is the ensemble of the mechanisms that transfer a dust plume downstream from its source region. It includes a kind of dilution process and the more air is mixed with a plume, the more dilution there will be and the more the plume spreads out and disperses. Dispersion is one way mechanism that does not allow reconstruction of plumes. Figure 5.5 is a schematic drawing of this process. As it is shown, dispersion processes from a point source the concentration has not a uniform pattern throughout the plume. It remains highest in the centre of the plume while it reduces away from the centre. The dispersion pattern of a dust plume is similar to one described in air quality studies.

Dust plume geometry

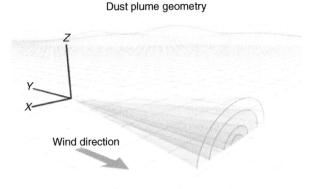

Figure 5.5. Schematic representation of the dust dispersion (source:http://www.meted.ucar.edu).

Turbulence is the primary mechanism that controls dispersion since it mixes ambient air with the plume (more turbulence leads to better dispersion). All three types of turbulence participate in the dispersion procedure: the mechanical turbulence, the turbulence caused by shear, and the turbulence caused by buoyancy.

- Mechanical turbulence is usually produced when the air flows over and/or around obstacles.
- Turbulence from shear can result from the vertical variation of wind speed and/or direction.
- Buoyancy turbulence can be caused by air bubbles raised due to the heating of the lower atmospheric layers (e.g. heating of the ground during day-hours and the formation of unstable conditions Figure 5.6).

Figure 5.6. The buoyancy turbulence (source:http://www.meted.ucar.edu).

It is worth mentioning that dust plumes are not only dispersed by turbulence but also are retained in the air because of it. If turbulence is not present, dust particles are generally settled at a rate of 300 metres per hour. However, this is strongly dependent on synoptic and mesoscale conditions so the establishment of an unstable atmospheric environment will slow down the rate at which the dust settles.

As it was previously mentioned, unstable conditions favour the lofting of dust and the formation of dust storms. Atmospheric stability also has a strong influence on how dust disperses. Figure 5.7 depicts the difference of the dispersion of dust plumes generated under stable and unstable conditions. The plume dispersion is intensified in both horizontal and vertical directions in case of an unstable environment. This effect is significantly more pronounced for vertical dispersion. With stable atmosphere, the dust remains relatively concentrated vertically, compared to dispersion under unstable conditions, while under neutral conditions, the plume will spread roughly equally in horizontal and vertical directions.

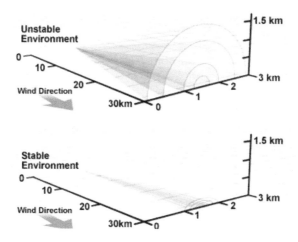

Figure 5.7. Dispersion and atmospheric stability (source: http://www.meted.ucar.edu).

5.9   SETTLING OF DUST

Dry and wet deposition consist the main mechanisms of PM removal from the atmospheric environment. The term "wet deposition" refers to the process where the aerosols are scavenged by precipitation. If they are removed by gravitational settling, it is referred as dry deposition. Different particle sizes are removed by different mechanisms.

When the dust particles are very small (ultra-fine with diameter less than 0.1 μm), are mainly removed by coagulation. The coagulation rate is determined by the mobility of ultra-fine particles and by the mass concentration of the entire aerosol population. Wet deposition is the main removal process for aerosols in the 0.1–10 μm size range. Particles in this size range are most efficient in acting as Cloud Condensation Nuclei (CCN). This range covers also coarser particles. For particles coarser than 10 μm, dry deposition or sedimentation becomes significant. Gravitational forces are important for dust particles of this size. Settling is actually grouping by particle size, with the largest falling out first and the smallest falling out last. Consequently, near the source area there will settle larger and heavier particles, with the smaller ones settling farther away. In general, the sedimentation velocity (v_s) can be obtained by equating the drag force and the weight of particles:

$$v_s = \frac{d_p^2 \rho_p g}{18\eta} \tag{5.4}$$

where η is the dynamic viscosity, d_p is the diameter of the particles, g is the gravitational acceleration ($9.80\,\mathrm{ms}^{-2}$) and ρ_p is the particle density. The sedimentation velocity becomes significant for particles coarser than $10\,\mu\mathrm{m}$. The dry deposition velocity (v_d) is obtained if the sedimentation velocity is divided by the concentration near the surface C_p. The dry deposition rate D_d is then defined as the mass of the PM deposited per surface area unit during the time unit:

$$D_d = v_d M \qquad (5.5)$$

where M corresponds to the mass concentration immediately adjacent to the surface (Meszaros, 1999).

The dust particles are in general hygroscopic. This is the reason why dust particles are in general good Cloud Condensation Nucleus (CCN) for cloud droplet formation. Because of this property, precipitation processes are quite effective for dust removal from the atmosphere. The removal mechanism of particles known as in-cloud scavenging occurs when aerosol particles are removed from the atmosphere by condensation. In addition to these processes, additional particles present in the atmosphere are washed out by precipitation. This process is called below-cloud scavenging (or washout). In the below-cloud mechanism, depending on the particle size, there occur two processes: Fine and ultrafine particles with diameters below 0.5–1 μm are removed by diffusion due to their Brownian motion. Coarser particles are removed by their inertial deposition onto cloud droplets or ice crystals.

Dust in aqueous environment (e.g. clouds and precipitation) can modify significantly precipitation acidity. Precipitation is considered acidic when pH is lower than 5.6 (Granat, 1972). This acid rain damages vegetation, building materials and affects the biogeochemical functioning of ecosystems. Figure 5.8 represents the stages of the wet deposition. More detailed formulation on dry and wet deposition processes will be provided below in the next paragraphs.

Figure 5.8. The wet deposition stages (source: http://www.meted.ucar.edu).

5.10 DESERT DUST FEEDBACK ON RADIATIVE TRANSFER

The presence of aerosols into the atmosphere has a profound effect on the radiative transfer and energy balance of the troposphere. In contrast to greenhouse gases (CO_2, CH_4, etc) which affect only the infrared radiation from the earth's surface, dust particles interact with both "sides" of the energy spectrum. By scattering and absorbing solar radiation they reduce the amount of energy reaching the ground (IPCC 2007, Tegen 2003; Spyrou *et al.*, 2010).

Aerosol radiative effects in the longwave spectrum are usually smaller than in the short-wave. On the infrared dust particles absorb and reemit radiation, thus acting as a greenhouse gas (Tegen 2003; Helmert 2007). While the scattering of solar radiation (direct aerosol effect) tends to cool the atmosphere, the absorption of radiation by aerosols leads to a warming of the atmosphere and to a suppression of cloud formation (semi-direct effect; Stanelle *et al.*, 2010).

The magnitude of the feedback depends strongly on the optical properties of particles (single scattering albedo, asymmetry parameter, extinction efficiency), which in turn depend on the size, shape and refractive indexes of dust particles (Tegen, 2003; Helmert *et al.*, 2007). The mineral composition of the dust source areas (Tegen, 2003), as well as the chemical composition and transformation of aerosols during their transportation (Wang *et al.*, 2005) are all factors on the optical intensity of dust. Due to the complexity of these processes and the rapid changes in the tempo-spatial variability of dust it is very difficult to accurately describe the radiative effects.

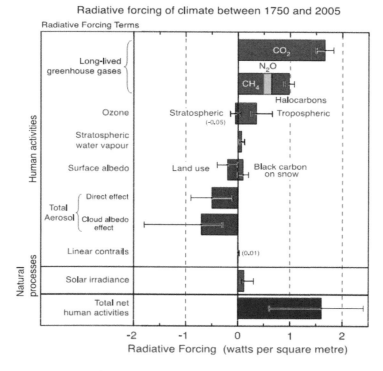

Figure 5.9. Summary of the principal components of the radiative forcing of climate change (source: IPCC 2007).

In Figure 5.9 a summary of the principal components of the radiative forcing of climate change are presented (IPCC 2007). The values represent forcings in 2005 relative to the start of the industrial era (1750). The black lines attached to each bar represents the range of uncertainty for each value. In the case of aerosols the uncertainty is of the same order of magnitude with the actual value, thus proving the difficulty in estimating the dust radiative effect. The large uncertainty is attributed to the global mineralogical variability of dust source areas (Claquin *et al.,* 1998).

Measurements of the direct radiative effect (DRE) in various locations have shown that the local DRE can be very significant. At the framework of the SHADE (Saharan Dust Experiment) project Haywood *et al.* (2003) have measured a reduction in the incoming solar radiation up to $-130\,W/m^2$ at the West African Coast. During ERBE (Earth Radiation Budget Experiment) Hsu *et al.* (2001) have calculated a reduction of $-45\,W/m^2$ at North Africa while at the same time an increase of $+25\,W/m^2$ on the infrared has been observed. Similar results for North Africa were presented by Haywood *et al.* (2005). During the SAMUM campaign (SAharan Mineral DUst ExperiMent) Bierwirth *et al.* (2008) have studied an intense desert dust episode on the 19th of May 2006 at Morocco. For the duration of the event the net surface radiation changed from -19 to $+24\,W/m^2$ in various locations. The interesting part is the increase of the surface temperature observed at Northeastern Morocco, due to aerosol forcing.

Recent studies using General Circulation Models (GCM), observations and data from the AeroCom (Aerosol Comparisons between Observations and Models – Huneeus *et al.,* 2011) project have estimated the global DRE from natural and anthropogenic aerosols (Table 5.4). As far anthropogenic aerosols are concerned the direct effect on radiative transfer has an order of $0.1\,W/m^2$ and smaller (IPCC 2007). Due to the complexity and small contribution of the anthropogenic part it is often excluded from simulations and modeling studies.

Table 5.4. Global radiative effect of aerosols based on model simulations (yellow), observations (green) and AEROCOM data (red). (Sources: IPCC 2007; Bierwirth *et al.,* 2008).

Reference	Shortwave Forcing (W/m^2)	Longwave Forcing (W/m^2)	Net Top Of The Atmosphere
Liao *et al.,* 2004	-0.21	$+0.31$	$+0.1$
Reddy *et al.,* 2005	-0.28	$+0.14$	-0.14
Jacobson 2001	-0.20	$+0.07$	-0.13
Miller *et al.,* 2004	-0.33	$+0.15$	-0.18
Yoshioka *et al.,* 2007	-0.92	$+0.31$	-0.61
Shell & Somerville 2007	-0.73	$+0.23$	-0.5
Myhre and Stordal 2001	-0.53	$+0.13$	-0.4
GISS	-0.75	$+0.19$	-0.56
UIO-CTM	-0.56	$+0.19$	-0.37
LSCE	-0.6	$+0.3$	-0.3
UMI	-0.54	$+0.19$	-0.35

5.11 THE FORMULATION OF THE DUST CYCLE

This paragraph introduces a formulation for the description of the dust cycle in the atmosphere. It consists of a sophisticated scheme for the dust production and concentration which are estimated by a set of K independent Euler-type equations. Equation (5.5) offers the general form of these equations. An advanced parameterization scheme for the dry and wet deposition is also included. The methodology and the formulation described in this chapter can be found in mode details in Spyrou *et al.* (2010). It is an ensemble of state of the art formulation for various processes. As it was mentioned before, the dust cycle in the atmosphere highly depends on meteorological conditions especially at the surface. The mechanism described is embedded in an atmospheric model the SKIRON (Kallos *et al.*, 1997). All the formulation associated with the atmospheric processes is according to one in this atmospheric model. The SKIRON model has been developed at the University of Athens and is based on the well known atmospheric model ETA/NCEP. The dynamics of the model is based on: large-scale numerical solutions controlled by conservation of integral proper-ties (Arakawa, 1966; Janjic, 1977; Janjic, 1984), energetically consistent time-difference splitting (Janjic, 1979; Janjic, 1997), and the step-like mountain representation (Mesinger, 1984; Mesinger *et al.*, 1988). A conservative positive definite scheme (Janjic, 1997) has been applied for horizontal advection of passive substances (including dust concentration). The physics incorporated consists of: the viscous sublayer models over water (Janjic, 1994) and over land (Zilitinkevitch, 1995), the surface layer scheme based on the similarity the-ory (Janjic, 1996b), a turbulence closure scheme based on Kolmogorov-Heisenberg theory (Janjic, 1996a), the Betts-Miller-Janjic deep and shallow moist convection scheme (Betts, 1986; Janjic, 1994), the land surface scheme (Chen *et al.*, 1996), the grid-scale precipitation scheme (Zhao and Carr, 1997), and the radiation scheme (Lacis and Hansen, 1974; Fels and Schwartzkopf, 1975). Recently the RRTMG radiation scheme (Mlawer *et al.*, 1997; Iacono *et al.*, 2000; Iacono *et al.*, 2008) has been implemented with the purpose of studying the radiative feedback of aerosols.

5.11.1 Dust production

The dust particles start moving mainly when the larger particles (with diameters greater then 10 μm) break soil cohesion forces and release zfiner particles into the atmosphere. This is the saltation (bombardment) process (Zender *et al.*, 2006). The near-ground atmospheric conditions, the soil properties and its conditions (e.g. soil moisture, heating conditions, and vegetation cover) regulate the amount of the released dust. The momentum flux from the atmosphere determines the quantity of mobilized dust. The soil properties define the potential and the quantity of released dust.

Usually, the lower boundary condition in dust modeling is either surface fluxes or surface concentration. The "flux" approach is followed in several dust models (Westphal *et al.*, 1987; Marticorena and Bergametti, 1995; Tegen and Fung, 1994). In the SKIRON the second method is followed in order to have consistency with the atmospheric processes as they are modelled.

Two groups of parameters govern the released surface concentration of mobilised particles and the corresponding surface vertical flux. The first group relates to the structure and state of soil, while the second one describes the turbulent state of the surface atmosphere. The flux dependence on friction velocity is a subject where there is no full agreement among different authors. Zender *et al.* (2003) proposed that the vertical dust flux F_S to be represented by:

$$F_{s,j} = T \cdot A_m \cdot S \cdot a \cdot Q_s \cdot \sum_{i=1}^{I} M_{i,j} \tag{5.6}$$

where T a global tuning factor, A_m the fraction of arid or semi-arid soil in each grid cell, S the efficiency with which soil produces dust for a specific meteorological forcing, a the sandblast mass efficiency, Q_s the horizontal flux of large particles that initiate saltation and $M_{i,j}$ the mass overlap between source and transport modes (Zender *et al.*, 2003). The subscript j denotes the particle size categories as described in Table 5.1.

Grid areas which act as desert dust sources in the model are specified using arid and semi-arid categories of the global vegetation data set. This can be done by mapping global vegetation data into the horizontal model grid and then counting numbers of desert points falling into SKIRON model grid boxes. Parameter α which is the fraction of a grid point area covered by desert surface is calculated by:

$$A_m = \frac{number\ of\ dust\ \text{points in model grid box}}{\text{total number of vegetation points in model grid box}} \tag{5.10}$$

The sandblast mass efficiency is defined in the model through the clay content of the soil using the parameterization of Marticorena and Bergametti (1995) and the corresponding texture classes.

$$a = 100\exp\left[(13.4M_{clay} - 6)\ln 10\right] \tag{5.11}$$

5.11.2 Threshold friction velocity

The soil wetness and particle size strongly determine the threshold friction velocity at which the soil erosion starts. Soil water which water resists in the soil due to capillary forces on surfaces of the soil grains, and due to molecular adsorption, increases the threshold friction velocity, therefore reducing the amount of dust injected into the atmosphere.

The soil moisture effects are included in the formulation of u_{*t} following the method of Fecan *et al.* (1999). The maximal amount of the adsorbed water w' is an increasing function of the clay fraction in the soil. Based on empirical data, Fecan *et al.* (1999) estimate w' to be a second order polynomial function of clay fraction in soil:

$$w' = 0.0014\,(\%clay)^2 + 0.17\,(\%clay) \tag{5.14}$$

A combination of this expression with the parameter β_k from Table 5.5 is established between w' and the seven considered texture classes, as given in Table 5.6.

Table 5.5. Correspondence between texture classes and soil types, and relative contributions of clay/sand/silt. (source: Nickovic et al., 2001).

l	*ZOBLER* Texture Classes	*Cosby Soil Types*	*M*			
			Clay	Small Silt	Large Silt	Sand
1	coarse	loamy sand	0.12	0.08	0.08	0.80
2	medium	silty clay loam	0.34	0.56	0.56	0.10
3	fine	Clay	0.45	0.30	0.30	0.25
4	coarse-medium	sandy loam	0.12	0.18	0.18	0.70
5	coarse-fine	sandy clay	0.40	0.10	0.10	0.50
6	medium-fine	clay loam	0.34	0.36	0.36	0.30
7	coarse-medium-fine	sandy clay loam	0.22	0.18	0.18	0.60

Fecan *et al.* (1999) defined the threshold velocity as:

$$
\begin{aligned}
u_{*tk} &= U_{*tk} && \text{for } w \leq w'\ (dry\ soil) \\
u_{*tk} &= U_{*tk}\sqrt{1 + 1.21\,(w - w')^{0.68}} && \text{for } w > w'\ (wet\ soil)
\end{aligned}
\tag{5.15}
$$

Table 5.6. Correspondence between soil texture classes and $w'T'$. (source: Nickovic et al., 2001).

l	*Cosby Soil Types*	$w'T'$ (%)
1	loamy sand	2.5
2	silty clay loam	6.8
3	Clay	11.5
4	sandy loam	2.5
5	sandy clay	10.0
6	clay loam	6.8
7	sandy clay loam	3.5

w corresponds to the ground wetness. According to Bagnold (1941), the threshold friction velocity for dry soil as:

$$U_{*tk} = \begin{cases} \dfrac{0.129 \cdot \left(\dfrac{\rho_p g D_{opt}}{\rho_\alpha}\right)}{(1.928 \ Re \ F^{0.092} - 1)^{0.5}} & 0.03 < Re \ F < 10 \\[4mm] 0.12 \cdot \left(\dfrac{\rho_p g D_{opt}}{\rho_\alpha}\right)\left(1 - 0.0858 \ e^{-0.0617(Re \ F - 10)}\right) & Re \ F > 10 \end{cases} \quad (5.16)$$

where g is gravity, and ρ_p and ρ_α are particle and air densities, respectively. The parameter $D_{opt} = 60 \ \mu m$ the particle diameter for which the threshold friction velocity has its minimum and ReF the Reynolds number.

According to Jackson, (1996) there is still production of dust even below the threshold friction velocity when u_* decreases. In order to avoid underestimation of dust production by parameterizing the inertial effect, the cases shown schematically in Figure 5.10 are treated in the following way:

a) Fluxes start to operate when $u_* < u_{*ts} = 0.9 \times U_{*tk}$ (at time step t_1) increases to the value $u_* > U_{*tk}$ (at time step t_2).

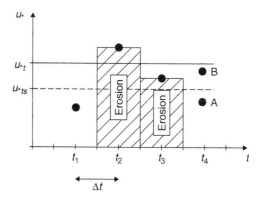

Figure 5.10. Conceptual model describing dust production under sub-threshold friction velocity conditions. At model time step t_1, u_* is bellow u_{*t} and there is no dust production. At t_2, u_* exceeds u_{*t} and dust production starts. At t_3, u_* is bellow the threshold (u_{*t}) but above the sub-threshold value $(u_{*ts} = 0.9 \ u_{*t})$ and there is still dust production driven by inertial forces. At t_4, dust production is ceased if either u_* is bellow u_{*ts} (point A) or even it increases but stays bellow the threshold (point B) (Source: Nickovic *et al.*, 2001).

b)   Fluxes are still operating when $u_* > U_{*tk}$ (at time step t_2) falls to the value $u_{*ts} < u_* < U_{*tk}$ (at time step t_3).

c)   Fluxes are ceased when $u_{*ts} < u_* < U_{*tk}$ (at time step t_3) stays in the interval $[u_{*ts}, U_{*tk}]$ or decreases to $u_* < u_{*ts}$ (at time step t_4).

5.11.3 Viscous sublayer effects

As Janjic (1994) showed, a viscous sub-layer is necessary to be included in the formulation of boundary layer over water surfaces. In the thin viscous sublayer, vertical transport is realised through molecular diffusion. In the boundary layer above the viscous sub-layer, the fluxes are defined by turbulent mixing. The characteristics of the viscous sub-layer are different for various surface turbulent conditions.

The viscous sub-layer formulation is applied to the dust concentration and the dust source term in the concentration equation (5.2), following Janjic (1994) and be in consistence with Chamberlain (1983) and Segal (1990):

$$\left(\frac{\partial C_k}{\partial t}\right)_{SOURCE} = -\frac{F_{Sk}}{\Delta z} \tag{5.17}$$

k is the particle size bin, Δz is the depth of the lowest atmospheric layer and F_{Sk} is the turbulent flux of dust concentration above the viscous sub layer. F_{Sk} is expressed as:

$$F_{Sk} = \nu \frac{C_{0k} - C_{Sk}}{z_C} \tag{5.18}$$

The subscript 0 is used to indicate the values at the interface of the viscous and turbulent layers while S denotes the surface values.

The depth of the viscous sub layer is estimated by:

$$z_C = \frac{0.35 M R r^{0.25} S c^{0.5} \nu}{u_*} \tag{5.19}$$

M is a parameter that is used to define different turbulent regimes.

Two other useful parameters are the roughness Reynolds number Rr and the Schmidt number Sc:

$$Rr = \frac{z_0 u_*}{\nu}, \qquad Sc = \frac{\nu}{\lambda} \tag{5.20}$$

λ is the particle diffusivity, and $z_0 = \max(0.018(u_*^2/g), 1.59 \times 10^{-5})$ according to Zoumakis and Kelessis (1991).

For the dust transport, the three regimes defined in Janjic (1994) are assumed: smooth and transitional, rough, and very rough, depending on Rr (or, equivalently, on u_*). According to his definition, transitions between regimes occur at $u_{*r} = 0.225$ m s^{-1} and $u_{*s} = 0.7$ m s^{-1}. M takes the value of 30 for the smooth regime and $M = 10$ for the others.

According to Businger (1986), the particle diffusivity λ is much smaller than ν. For λ the experimental work of Chamberlain et al. (1984) provides the necessary values for each regime and particle diameter.

Based on the above formulation, expression (18) can be expressed in a simplified form to define turbulence above the viscous sub-layer:

$$F_{Sk} = K_S \frac{C_{LMk} - C_{0k}}{\Delta z} \tag{5.21}$$

K_S is the surface-mixing coefficient for concentration, and LM is used to define the first level of the free atmosphere above the viscous syblayer. K_S is the mixing coefficient for heat and moisture. The Monin-Obukhov method is used to calculate the surface-mixing coefficient as suggested byJanjic (1996b).

The lower boundary condition for concentration is obtained from the requirement that the fluxes (7) and (5.21) are matched at the interface of the viscous and turbulent layers:

$$C_{0k} = \frac{C_{Sk} + \omega C_{LMk}}{1 + \omega} \qquad (5.22)$$

$$\omega = \frac{\left(\frac{K_S}{\Delta z}\right)}{\left(\frac{\lambda}{z_C}\right)} \qquad (5.23)$$

Based on the above, the surface fluxes are calculated as:

$$F_{Sk} = K_S^* \frac{C_{LMk} - C_{Sk}}{\Delta z} \qquad (5.24)$$

Where C_{Sk} is defined by (13), and

$$K_S^* = \frac{1}{1 + \omega} K_S \qquad (5.25)$$

as a corrected similarity-theory mixing coefficient.

5.11.4   Shear-Free Convection Effects

As it was stated by Zilitinkevich *et al.* (1998), overheated soil surfaces may generate strong vertical transport and dust production even without the presence of wind shear. In such cases production follows a different parameterization based on Beljaars (1994) and Janjic (1996b). Beljaars' correction converts the surface buoyancy flux $(\overline{w'T'})_S$ into the turbulent kinetic energy of the near-surface wind by the following fraction:

$$U_B^2 = \left(1.2 \times W^*\right)^2 \qquad (5.26)$$

with

$$W^* = \left[\frac{1}{273} \times gh \left(\overline{w'T'}\right)\right]^{1/3} \qquad (5.27)$$

h is the depth of the convective boundary layer.

In Nickovic *et al.* (2001) the shear-free flow conditions and the appropriate formulation is summarized. According to their formulation, the shear-free convective dust flux is estimated using the 'aerodynamic' mixing coefficient

$$K_{AC} \equiv \frac{F_S}{U_* \Delta C_A} \qquad (5.28)$$

instead of using the bulk coefficient (5.25). This expression is considered as appropriate for conditions where $10^{-10} \le z_{0u}/h \le 10^{-5}$ (z_{0u} is the roughness length for momentum).

Again: according to Nickovic *et al.* (2001), the concentration "aerodynamic increment" is formulated by

$$\Delta C_A \equiv C_{AS} - C_{LM} = \left(C_S - \frac{F_C}{\kappa U_*} \ln \frac{z_{0u}}{z_{0C}} \right) - C_{LM} \tag{5.29}$$

with $\kappa = 0.4$ to be the Von Kármán constant, C_{LM} the lowest atmospheric level concentration, C_S the surface concentration defined by (13), and C_{AS} the concentration extrapolated logarithmically downwards to the level $z = z_{0u}$. By using the formulas (5.28) and (5.29) together with the following Zilitinkevich *et al.* (1998) formulations:

$$\frac{U_*}{W_*} = 0.36 \left(\frac{z_{0h}}{h} \right)^{0.1} \tag{30_1}$$

and

$$K_{AC} = \frac{1}{4.4 \left(\frac{h}{z_{0u}} \right)^{0.1} - 1.5} \tag{30_2}$$

the surface flux at shear-free atmospheric layers should be expressed as:

$$F_S = \frac{0.36 \left(\frac{z_{0u}}{h} \right) W_* \left(C_S - C_{LM} \right)}{\frac{1}{\kappa} \ln \frac{z_{0u}}{z_{0C}} + 4.4 \left(\frac{z_{0u}}{h} \right)^{-1} - 1.7} \tag{31}$$

5.12 DUST SINKS

5.12.1 Dry deposition

There are various dry deposition schemes proposed that are used in various atmospheric composition models. Here the one resulting from the resistance approach and it is similar to that implemented in UAM-AERO (Kumar *et al.*, 1996) is being used. Thus, particle deposition velocity is calculated using the following resistance equation:

$$V_d = V_{sed} + \frac{1}{r_a + r_b + r_a r_b V_{sed}} \tag{32}$$

where $r_a = 1/ku_* [\ln(1/z_0) - \varphi_h]$ and $r_b = 1/u_* (S_c^{-2/3} + 10^{-3/S_t})$ are the aerodynamic and boundary resistances, respectively, k the Von Kármán's constant, z_0 the surface roughness length, φ_h a stability correction term, S_c the Schmidt number, S_t the Stokes number that characterizes the atmospheric air flow and V_{sed} the gravitational terminal settling velocity, as defined by the Stokes' law. Consequently, the scheme is applied separately for each particle size bin and model grid point.

5.12.2 Wet deposition

The wet removal of dust from the atmosphere can be expressed in two stages: one inside the cloud and the other below the cloud bottom. Inside the cloud, the scavenging processes are rather complicated because they depend on various processes taking place that are controlled by the type of hydrometeors, the soil hygroscopicity, the co-existence with other species (e.g. sea salt, sulphates or nitrates, organic particulate matter). The description of such processes is beyond of the scope of this chapter.

However a simplified formulation can be used to express the rate of dust scavenged by precipitation Seinfeld and Pandis (1998):

$$\frac{\partial C}{\partial t} = -\Lambda C \tag{46}$$

where Λ is a scavenging coefficient. For aerosols inside clouds:

$$\Lambda_c = 4.2 \times 10^{-7} \frac{EP}{d_d} \tag{47}$$

where E is the collection efficiency, P is the precipitation rate and d_d is the cloud drop diameter, as calculated by the atmospheric model, for the estimation of the raindrop fall speed at the bottom of the grid boxes. For the wet scavenging below the precipitating clouds, the scavenging coefficient derived by Seinfeld and Pandis (1998) for the collection of cloud droplets is used:

$$E(d_p) = \frac{4}{Re\,Sc} \left(1 + 0.4\,Re^{1/2}Sc^{1/3} + 0.16\,Re^{1/2}Sc^{1/2}\right)$$

$$+ 4\phi \left[\frac{\mu}{\mu_w} + \phi(1 + Re^{1/2})\right] + \left(\frac{St - S^*}{St - S^* + 2/3}\right)^{3/2} \tag{48}$$

where μ and μ_w are the kinematic viscosity of air $(1.8 \times 10^{-5}\,\text{kg m}^{-1}\,\text{s}^{-1})$ and water $(10^{-3}\,\text{kg m}^{-1}\,\text{s}^{-1})$ respectively, $\phi = d_p/d_d$ is the ratio of particle to droplet diameter, Re is the Reynolds number for the droplet, S_c is the Schmidt number for the collected particle and S_t is the Stokes number of the collected particle. Finally, the parameter S^* is given by the following equation:

$$S^* = \frac{1.2 + \ln(1 + Re)/12}{1 + \ln(1 + Re)} \tag{49}$$

APPENDIX A – LIST OF SYMBOLS

List of Symbols		
Symbol	Definition	Dimensions or Units
A_k	function of the particle Reynolds number	
A_m	ratio between areas of small collectors and area of the roughness elements, which depends upon different vegetation types m	
B_{St}	particle surface Stanton number	
C	total concentration	$[\text{kg m}^{-3}]$
C_{AS}	concentration extrapolated logarithmically downwards to the level $z = z_{0u}$	$[\text{kg m}^{-3}]$
C_k	dust concentration of a k-th particle size bin	$[\text{kg m}^{-3}]$

(Continued)

List of Symbols

Symbol	Definition	Dimensions or Units
C_{LM}	lowest atmospheric level concentration	$[\text{kg m}^{-3}]$
C_p	concentration near the surface	$[\text{kg kg}^{-1}]$
C_{Sk}	surface concentration	$[\text{kg m}^{-3}]$
D_d	dry deposition rate	$[\text{kg m}^{-2}\text{ s}^{-1}]$
D_{opt}	the particle diameter for which the threshold friction velocity has its minimum	$[\mu\text{m}]$
F_S	vertical dust flux	$[\mu\text{gr m}^{-2}\text{ s}^{-1}]$
F_{Sk}^{EFF}	effective surface vertical flux	$[\mu\text{gr m}^{-2}\text{ s}^{-1}]$
F_{Sk}	turbulent flux of dust concentration above the viscous sublayer	$[\text{kg m}^{-2}\text{ s}^{-1}]$
G	function which reflects the properties of particles and depositing surfaces	
K	number of the particle size bins $(k = 1, \ldots, K)$,	
K_{AC}	aerodynamic mixing coefficient	
K_H	lateral diffusion coefficient	$[\text{m}^2\text{s}^{-1}]$
K_S	surface-mixing coefficient for concentration	$[\text{m}^2\text{s}^{-1}]$
K_Z	turbulence exchange coefficient	$[\text{m}^2\text{s}^{-1}]$
L	Monin-Obukhov length	$[\text{m}]$
L_*	minimum Monin-Obukhov length	$[\text{m}]$
LM	lowest atmospheric layer	
M	mass concentration immediately adjacent to the surface	$[\text{kg m}^{-3}]$
M_t	parameter varying for different turbulent regimes	
$M_{i,j}$	the mass overlap between source and transport modes	
P	precipitation rate	$[\text{mm h}^{-1}]$
Q_s	the horizontal flux of large particles that initiate saltation	$[\text{kg m}^{-2}\text{ s}^{-1}]$
Re	Reynolds number	
R_k	radius of a k-th particle size class	$[\text{m}]$
Rr	roughness Reynolds number	
S	The efficiency with which soil produces dust for a specific meteorological forcing	
Sc	Schmidt number	
St	Stokes number	
St_a	Stanton number over vegetation surfaces	
T	global tuning factor for dust production	
U_{*tk}	threshold friction velocity for dry soil	$[\text{m s}^{-1}]$
U_B^2	fraction of the surface buoyancy flux converted into the turbulent kinetic energy	$[\text{m}^2\text{ s}^{-2}]$
V_s	wind speed at the midpoint z_s of the surface layer	$[\text{m s}^{-1}]$
V_d	particle deposition velocity	$[\text{m s}^{-1}]$
V_{sed}	the gravitational terminal settling velocity	$[\text{m s}^{-1}]$
W^*	convective scale velocity	$[\text{m s}^{-1}]$
Λ	scavenging coefficient	
Λ_c	scavenging coefficient inside clouds	
a	sandblast mass efficiency	

(Continued)

List of Symbols

Symbol	Definition	Dimensions or Units
c_{dm}	local drag coefficient for vegetation	
c_{vm}	local viscous drag coefficient	
d_p	diameter of the particles	[μm]
d_d	cloud drop diameter	[μm]
f_{B0}	empirical constant, which takes into account effect of the blow-off over the vegetation surfaces	
g	gravitational acceleration constant	[m s^{-2}]
h	depth of the convective boundary layer	[m]
k	size category	
m	number of different vegetation types	
r_a	aerodynamic resistance	
r_b	boundary resistance	
u, v	horizontal velocity components	[m s^{-1}]
u_*	friction velocity	[m s^{-1}]
u_{*t}	threshold value of the friction velocity bellow which dust production ceases	[m s^{-1}]
u_{*ts}	sub-threshold friction velocity	[m s^{-1}]
v_s	sedimentation velocity	[m s^{-1}]
w	vertical velocity	[m s^{-1}]
w_g	ground wetness	[cm^3 cm^{-3}]
w'	volumetric soil moisture	[cm^3 cm^{-3}]
$\overline{(w'T')}_S$	surface buoyancy flux	[m s]
z_C	depth of the viscous sublayer	[m]
z_S	midpoint of the surface layer	[m]
z_{0u}	roughness length for momentum	[m]
z_0	surface roughness	[m]
Δz	depth of the lowest atmospheric layer	[m]
α	fraction of a grid point area covered by desert surface	
β	influence of soil textures	[kg kg^{-1}]
γ	ratio between the mass available for uplift and the total mass of a specific particle size category	[kg kg^{-1}]
γ_k	ratio between the mass available for uplift and the total mass	[kg kg^{-1}]
δ_k	mass fraction of the $k-$th particle category	[kg kg^{-1}]
η	dynamic viscosity	[Pa s]
η_{BD}	collection efficiency for Brownian diffusion	
η_e	efficiency of vegetation to collect the aerosol	
η_{imp}	collection efficiency for impaction	
η_{int}	collection efficiency for interception	
η_s	collection efficiency for interception by smaller vegetation elements	
κ	Von Kármán constant	
λ	particle diffusivity	[m^2 s^{-1}]
ν	molecular diffusivity for momentum	[m^2 s^{-1}]

(*Continued*)

List of Symbols

Symbol	Definition	Dimensions or Units
v_{dep}	dry deposition velocity	$[\text{m s}^{-1}]$
v_{gk}	gravitational settling velocity	$[\text{m s}^{-1}]$
$(v_{gk})^{LM}$	gravitational settling velocity at the lowest model level	$[\text{m s}^{-1}]$
v_{IL}	turbulent deposition velocity at the top of the viscous sublayer z_S	$[\text{m s}^{-1}]$
v_{SL}	turbulent deposition velocity in the layer between z_S and $10\,\text{m}$	$[\text{m s}^{-1}]$
ρ_a	air density	$[\text{kg m}^{-3}]$
ρ	particle density	$[\text{g cm}^{-3}]$
ρ_{pk}	density of a k-th particle size class	$[\text{g cm}^{-3}]$
$(\partial C_k/\partial t)_{SOURCE}$	dust production rate normally over the dust source areas	$[\text{kg m}^{-3}{\cdot}\text{s}^{-1}]$
$(\partial C_k/\partial t)_{SINK}$	sink term which includes both wet and dry deposition fractions	$[\text{kg m}^{-3}{\cdot}\text{s}^{-1}]$
$\partial P/\partial t$	precipitation rate	$[\text{m s}^{-1}]$
ϕ	washout parameter	
ψ_m	stability parameter for momentum	
∇	horizontal nabla operator	

APPENDIX B – SYNOPSIS

Soil dust produced from desert, arid and semi-ardi areas of the planet is considered a major source of Particulate Matter in the atmosphere. Under favourable weather conditions dust mobilized from the Sahara desert area can be transported and deposited hundreds and thousands of kilometres away towards Mediterranean and Europe as well as towards the Atlantic and Indian Oceans. The production of dust is governed by saltation, a process triggered when larger wind-blown particles bounce on the desert soil's surface, releasing smaller particles. The effects of dust particles are numerous, ranging from air quality degradation to radiative feedback and cloud formation. Depending on their size, distribution and optical properties, dust particles reflect and absorb incoming solar and outgoing infrared radiation, thus changing the energy budget of the atmosphere. These phenomena are highly complicated and various parameterisations have been developed for their representation. An ensemble of state of the art formulations is embedded in the atmospheric model SKIRON, based on the well known ETA/NCEP model. The system includes various parameterizations for the description of the desert dust cycle and the dynamics and feedbacks involved with naturally produced aerosols.

APPENDIX C – KEYWORDS

Desert Dust	Source Areas	Sahara Desert
Aerosols	Saltation	Turbulence
Friction Velocity	Dry Deposition	Wet Deposition
Radiative Transfer	Dust Feedback	SKIRON Model

APPENDIX D – QUESTIONS

C1. What is the main mechanism for the production of desert dust particles?
C2. What are the parameters that govern the magnitude of the dust production?
C3. How the friction velocity affects dust production?
C4. What are the major source areas of the Saharan desert?
C5. What are the main dust removal mechanisms?
C6. How do dust particles interact with radiation?

REFERENCES

Alfaro, S.C., Gaudichet, A., Gomes, L. and Maille, M., 1997, Modeling the size distribution of a soil aerosol produced by sandblasting. *Journal of Geophysical Research,* **102**, pp. 11239–11249.

Alpert, P., Kaufman, Y.J., Shay-El, Y., Tanre, D., Da Silva, A., Schubert, S. and Joseph, J.H., 1998, Quantification of dust-forced heating of the lower troposphere, *Nature,* **395**, pp. 367–370.

Andreae, M. O., 1996. Raising dust in the greenhouse, *Nature* **380**: pp. 389–340.

Arakawa, A., 1966, Computational design for long-term numerical integration of the equations of fluid motion: Two dimensional incompressible flow. Part I, *J. Comp. Phys.***1**, pp. 119–143.

Bagnold, R. A., 1941, *The Physics of Blown Sand and Desert Dunes,* pp. 265, Morrow, New York.

Barkan, J., Kutiel, H. and Alpert, P., 2004, Climatology of Dust Sources in North Africa and the Arabian Peninsula, Based on TOMS Data, *Indoor Built Environ 2004*, pp. 13: 407–419.

Beljaars, A.C.M., 1994, The parameterization of surface fluxes in large-scale models under free convection, *Quarterly Journal of the Royal Meteorological Society,* **121**, pp. 255–270.

Betts, A., 1986, A new convective adjustment scheme. Part I: Observational and theoretical basis, *Quarterly Journal of the Royal Meteorological Society,* **112**, pp. 677–693.

Bierwirth Eike, Manfred Wendisch, Andr Eehrlich, Birgit Heese, Matthias Tesche, Dietrich Althausen,Alexander Schladitz, Detlef Muller, Sebastian Otto, Thomas Trautmann, Tilman Dinter, Wolfgang Von Hoyningen-Huene and Ralph Kahn 2008: Spectral surface albedo over Morocco and its impact on radiative forcing of Saharan dust. *Tellus* (2009), **61B**, 252–269. DOI: 10.1111/j.1600-0889.2008.00395.x.

Businger, J.A., 1986, Evaluation of the accuracy with which dry deposition can be measured with current micrometeorological techniques, *Journal of Climatology and Applied Meteorology,* **25**, pp. 1100–1124.

Callot, Y., Marticorena, B. and Bergametti, G., 2000, Geomorphologic approach for modelling the surface features of arid environments in a model of dust emissions: Application to the Sahara desert, *Geodin. Acta,* **13(5)**, pp. 245–270.

Chamberlain, A.C., 1983, Roughness length of sea, sand and snow, *Boundary-Layer Meteorology,* **25**, pp. 405–409.

Chamberlain, A.C., Garland, J.A. and Wells, A.C., 1984, Transport of gasses and particles to surfaces with widely spaced roughness elements, *Boundary-Layer Meteorology,* **24**, pp. 343–360.

Chen F., Mitchell, K., Janjic, Z. and Baldwin, M., 1996, Land-surface parameterization in the NCEP Mesoscale Eta Model, *Research Activities in Atmospheric and Oceanic Modelling, WMO, Geneva,* CAS/JSC WGNE, **23**, 4.4.

Claquin T., M. Schulz, Y. Balkanski and O. Boucher, 1998, Uncertainties in assessing radiative forcing by mineral dust, Tellus, **50B**, 491–505, 1998.

D'Almeida, G.A 1987: Desert aerosol characteristics and effects on climate. In: Leinen, M., Sarnthein, M. (Eds.), Palaeoclimatology and Palaeometeorology: Modern and Past Patterns of Global atmospheric transport. NATO ASI series, C, **282**, pp. 311–338.

Dufrense, J.L., Gautier, C. and Ricchiazzi P. 2001: Longwave Scattering of Mineral Aerosols. *Journal of Atmospheric Sciences*, **59**, pp.1959–1966.

Fecan, F., Marticorena, B. and Bergametti, G., 1999, Parameterization of the increase of the Aeolian erosion threshold wind friction velocity due to soil moisture for arid and semi-arid areas. *Annales Geophysicae*, **17**, pp. 194–157.

Fels, S.B. and Schwartzkopf, M.D., 1975, The simplified exchange approximation: A new method for radiative transfer calculations, *Journal of Atmospheric Science*, **32**, pp. 1475–1488.

Georgi, F., 1986, A particle dry-deposition parameterization scheme for use in tracer transport models, *Journal of Geophysical Research*, **91**, pp. 9794–9806.

Gillette, D.A. and Passi, R., 1988, Modeling dust emission caused by wind erosion, *Journal of Geophysical Research*, **93**, pp. 14233–14242.

Goudie, A.S. and Middleton, N.J. 2006, Desert Dust in the Global System. *Springer*.

Goudie, A.S. and Middleton, N.J., 2001, Saharan dust: Sources and trajectorie, *Trans. Inst. Br. Geogr.* NS, 26, pp. 165–181.

Granat, G., 1972, On the relation between pH and the chemical composition in atmospheric precipitation, *Tellus*, **24**, pp. 550–560.

Guerzoni, S. and Chester, R. (Eds), 1996, *The Impact of Desert Dust Across the Mediterranean*, Kluwer Academic., Norwell, Mass.

Guerzoni, S., Chester, R., Dulac, F., Herut, B., Loye-Pilot, M.-D., Measures, C., Migon, C., Molinaroli, E., Moulin, C., Rossini, P., Saydam, C., Soudine, A. and Ziveri, P. 1999, The role of atmosphere deposition in the biogeochemistry of the Mediterranean Sea, *Prog. Oceanogr*, **44**, pp. 147–190.

Haywood, J.M, *et al.*, 2003, Radiative properties and direct Radiative effect of Saharan dust measured by the C-130 aircraft during SHADE:1. Solar spectrum. *J. Geophys. Res.*, **108**(D18), 8577, doi:10.1029/2002JD002687.

Haywood, J.M., *et al.*, 2005, Can desert dust explain the outgoing longwave radiation anomaly over the Sahara during July 2003. *J. Geophys. Res.*, **110**, D05105, doi:10.1029/2004JD005232.

Helmert J., B. Heinold, I. Tegen, O. Hellmuth and M. Wendisch, 2007, On the direct and semidirect effects of Saharan dust over Europe: A modelling study, *J. Geophys. Res.*, **112**, doi:10.1029/2006JD007444.

Hsu, N.C., J.R. Herman, and C. Weaver, 2000, Determination of Radiative forcing of Saharan dust using combined TOMS and ERBE data. *J. Geophys. Res.*, **105**(D16), 20649–20661.

Huneeus N., M. Schulz, Y. Balkanski, J. Griesfeller, J. Prospero, S. Kinne, S. Bauer, O. Boucher, M. Chin, F. Dentener, T. Diehl, R. Easter, D. Fillmore, S. Ghan, P. Ginoux, A. Grini, L. Horowitz, D. Koch, M. C. Krol, W. Landing, X. Liu, N. Mahowald, R. Miller, J.-J. Morcrette, G. Myhre, J. Penner, J. Perlwitz, P. Stier, T. Takemura, and C. S. Zender 2011: Global dust model intercomparison in AeroCom phase I, *Atmos. Chem. Phys.*, **11**, 7781–7816, 2011, doi: 10.5194/acp-11-7781-2011.

Husar, R.B., Prospero, L.M. and Stowe, L.L., 1997, Characterization of tropospheric aerosols over the oceans with the NOAA advanced very high resolution radiometer optical thickness operational product, *Journal of Geophysical Research*, **102**, pp. 16889–16909.

Iacono, M.J., E.J. Mlawer, S.A. Clough and J.-J. Morcrette, 2000: Impact of an improved longwave radiation model, RRTM. on the energy budget and thermodynamic properties of the NCAR community climate mode, CCM3. *J. Geophys. Res.*, **105**, 14873–14890.

Iacono, M.J., J.S. Delamere, E.J. Mlawer, M.W. Shephard, S.A. Clough, and W.D. Collins, 2008, Radiative forcing by long-lived greenhouse gases: Calculations with the AER Radiative transfer models, *J. Geophys. Res.*, **113**, D13103, doi:10.1029/2008JD009944.

Intergovernmental Panel on Climate Change (IPCC), *Climate Change 2007: The Physical Science Basis*, Cambridge University Press, UK, 2007.

Jackson, D.W.T., 1996, Potential inertial effects in Aeolian sand transport: preliminary results, *Sedimentary Geology*, **106**, pp. 193–201.

Jacobson, M.Z., 2001, Global direct radiative forcing due to multicomponent anthropogenic and natural aerosols. *J. Geophys. Res.*, **106**(D2), 1551–1568.

Janjic Z.I., 1996a, The Mellor-Yamada Level 2.5 turbulence closure scheme in the NCEP Eta Model, in: *Research Activities in Atmospheric and Oceanic Modeling*, edited by Ritchie, H., (WMO, Geneva, CAS/WGNE), **4**, pp. 4.15.

Janjic Z.I., 1996b, The Surface Layer Parameterization in the NCEP Eta Model, in: *Research Activities in Atmospheric and Oceanic Modeling*, edited by Ritchie, H., (WMO, Geneva, CAS/WGNE), **4**, pp. 4.16–4.17.

Janjic Z.I., 1997, Advection scheme for passive substance in the NCEP Eta Model, in: *Research Activities in Atmospheric and Oceanic Modeling*, edited by Ritchie, H., (WMO, Geneva, CAS/WGNE).

Janjic, Z.I., 1977, Pressure gradient force and advection scheme used for forecasting with steep and small scale topography, *Contrib. Atm. Phys.*, **50**, pp. 186–199.

Janjic, Z.I., 1979, Forward-backward scheme modified to prevent two-grid-interval noise and its application in sigma coordinate models, *Contrib. Atm. Phys.*, **52**, pp. 69–84.

Janjic, Z.I., 1984, Non-linear advection schemes and energy cascade on semi-staggered grids, *Monthly Weather Review*, **112**, pp. 1234–1245.

Janjic, Z.I., 1994, The Step-mountain Eta Coordinate Model: Further Developments of the Convection, Viscous Sublayer and Turbulence Closure Schemes, *Monthly Weather Review*, **122**, pp. 927–945.

Kallos G., C. Spyrou, C. Mitsakou, 2009: Short and Long Wave Radiative Forcing from Desert Dust and Impacts on Weather and Climate, *European Geosciences Union General Assembly 2009* , **EGU2009-8867**, Vienna, Austria, 19–24 April 2009.

Kallos, G. and other members of the participating groups at the SKIRON project, 1997: The SKIRON forecasting system: VOL. I–VI: ISBN 960-8468-14-0. Available from the University of Athens and the author.

Kallos, G., Katsafados, P., Spyrou, C. and Papadopoulos, A., 2005, Desert dust deposition over the Mediterranean Sea estimated with the SKIRON/Eta, *4th EuroGOOS Conference*, 6–9 June 2005, Brest, France.

Kallos, G., Papadopoulos, A., Katsafados, P. and Nickovic, S., 2006: Trans-Atlantic Saharan dust transport: Model simulation and results. *Journal of Geophysical Research-Atmosphere*, **111**, doi: 10.1029/2005JD006207.

Kallos, G., Astitha, M., Katsafados, P. and Spyrou, C., 2007, Long-Range Transport of Anthropogenically and Naturally Produced PM in the Mediterranean and North Atlantic:Present Status of Knowledge, *Journal of Applied Meteorology and Climatology*, (in press).

Kaufman, Y.J., D. Tanre, B.N. Holben, S. Mattoo, L.A. Remer, T.F. Eck, J. Vaughan, and B. Chatenet, 2002: Aerosol radiative impact on spectral solar flux at the surface, derived from principal-plane sky measurements, *J. Atmos. Sci.*, 59: 635–646.

Kumar, N., F.W. Lurmann, A.S. Wexler, S. Pandis, and J.H. Seinfeld. 1996. Development and Application of a Three Dimensional Aerosol Model. Presented at the A&WMA Specialty Conference on Computing in Environmental Resource Management, Research Triangle Park, NC, December 2–4, 1996.

Lacis, A.A. and Hansen, J.E., 1974, A parameterization of the absorption of air-solar radiation in the earth's atmosphere, *Journal of Atmospheric Sciences*, **31**, pp. 118–133.

Levin Z., Ganor, E. and Gladstein V., 1996, The Effects of Desert Particles Coated with Sulfate on Rain Formation in the Eastern Mediterranean, *Journal of Applied Meteorology*, 35, pp. 1511–1523.

Levin Z., Teller, A., Ganor, E. and Yin, Y., 2005, On the interactions of mineral dust, sea-salt particles and clouds: A measurement and modeling study from the Mediterranean Israeli Dust Experiment campaign, *Journal of Geophysical Research*, 110, D20202, doi:10.1029/2005JD005810.

Li, X., Maring, H., Savoie, D., Voss, K. and Prospero, M., 1996, Dominance of mineral dust in aerosol light-scattering in the North Atlantic trade winds, *Nature*, **380**, pp. 416–419.

Liao, H., J.H. Seinfeld, P.J. Adams, and L.J. Mickley, 2004: Global radiative forcing of coupled tropospheric ozone and aerosols in a unifi edgeneral circulation model. *J. Geophys. Res.*, **109**, D16207, doi:10.1029/2003JD004456.

Marticorena, B., and Bergametti, G., 1995, Modeling the atmospheric dust cycle: 1. Design of a soil-derived dust emission scheme, *Journal of Geophysical Research*, **100**, pp. 16415–16430.

Martin, J.M. and Fitzwate,r S.E., 1988, Iron deficiency limits phytoplancton growth in the north-east Pacific subarctic, *Nature*, **331**, pp. 341–343.

Mesinger, F., 1984, A blocking technique for representation of mountains in atmospheric models, *Rivista di Meteorologia Aeronautica*, **44**, pp. 195–202.

Mesinger, F., Janjic, Z.I., Nickovic, S., Gavrilov, D. and Deaven, D.G., 1988, The step-mountain coordinate: model description and performance for cases of Alpine lee cyclogenesis and for a case of an Appalachian redevelopment, *Monthly Weather Review*, **116**, pp. 1493–1518.

Meszaros, E., 1999, Fundamentals of atmospheric aerosol chemistry, Akdemiai Kiado.

Miller, R. L., Tegen, I. and Perlwitz, J. 2004. Surface radiative forcing by soil dust aerosols and the hydrological cycle. *J. Geophys. Res.* **109**, doi: 10.1029/2003JD00408.

Miller, R.L. and Tegen, I., 1998, Climate Response to soil dust aerosols, *Journal of Climate*, **11**, pp. 3247–3267.

Mlawer, E.J., S.J. Taubman, P.D. Brown, M.J. Iacono and S.A. Clough, 1997: RRTM, a validated correlated-k model for the longwave. *J. Geophys. Res.*, **102**, 16,663–16,682.

Moulin, C., Lambert, C.E., Dayan, U., Masson, V., Ramonet, M., Bousquet, P., Legrand, M., Bałkanski, Y.J., Guelle, W., Marticorena, B., Bergametti, G. and Dulac, F., 1998, Satellite climatology of African dust transport in the Mediterranean atmosphere, *Journal of Geophysical Research*, **103**, pp. 13137–13144, 10.1029/98JD00171.

Myhre, G., and F. Stordal, 2001: Global sensitivity experiments of theradiative forcing due to mineral aerosols. *J. Geophys. Res.*, **106**, 18193–18204.

Nickling, W.G. and Gillies, J.A., 1989, Emission of fine-grained particulates from desert soils, *Paleoclimatology and Paleometeorology: Modern and Past Patterns of Global Atmospheric Transport,* edited by Leinen, M. and Sarnthein, M., (Kluwer Academic Publishers), pp. 133–165.

Nickovic, S., Kallos, G., Papadopoulos, A. and Kakaliagou, O., 2001, A model for prediction of desert dust cycle in the atmosphere, *Journal of Geophysical Research*, **106**, pp. 18113–18129.

Ozsoy, E., Kubilay, N., Nickovic, S. and Moulin, C., 2001, A hemisphere dust storm affecting the Atlantic and Mediterranean in April 1994: Analyses, modeling, ground-based measurements and satellite observations, *Journal of Geophysical Research*, **106**, pp. 18439–18460.

Papadopoulos A., Katsafados, P., Kallos, G., Nickovic, S., Rodriguez, S. and Querol, X., 2003, Contribution of Desert Dust Transport to Air Quality Degradation of Urban Environments, Recent Model Developments. *26th NATO/CCMS ITM on Air Pollution Modeling and its Application*, ISBN 0-306-48464-1, pp. 279–286.

Perez C., S. Nickovic, J. M. Baldasano, M. Sicard, F. Rocadenbosch, and V. E. Cachorro 2006: A long Saharan dust event over the western Mediterranean: Lidar, Sun photometer observations, and regional dust modelling, *J. Geophys. Res.*, **111**, D15214, doi:10.1029/2005JD006579, 2006.

Perry, K.D., Cahill, T.A., Eldred, R.A. and Dutcher, D.D., 1997, Long-range transport of North African dust to the eastern Union States, *Journal of Geophysical Research*, **102**, pp. 11225–11238.

Prospero, J. M., 1996: The atmospheric transport of particles to the ocean, in *Particle Flux in the Ocean*, edited by V. Ittekkott, S. Honjo, and P. J. Depetris, *SCOPE Rep. 57*, pp. 19–52, John Wiley, New York.

Prospero, J.M., 1996, Saharan dust transport over the North Atlantic Ocean and Mediterranean: An overview, in *The Impact of Desert Dust Across the Mediterranean*, edited by Guerzoni, S. and Chester, R., (Kluwer Acad., Norwell Mass.), pp. 133–151.

Prospero, J.M., Ginoux, P., Torres, O., Nicholson, S.E. and Gill, T.E., 2002, Environmental characterization of global sources of atmospheric soil dust identified with the nimbus 7 total ozone mapping spectrometer (TOMS) absorbing aerosol product, *Rev. of Geophys.*, **40**, doi:10.1029/2000RG000095.

Querol, X., Alastuey, A., Rodriguez, S., Viana, M.M., Artinano, B., Salvador, P., Mantilla, E., Santos, S.G.D., Patier, R.F., Rosa, J.D.L., Campa, A.S.D.L. and Menedez M., 2002, Interpretation de series temporales (1996–2000) de niveles de particulas en suspension en Espana, Ministerio de Medio Ambiente, Madrid.

Ramanathan V., Crutzen, P.J., Lelieveld, J., Mitra, A.P., Althausen, D., Anderson, J., Andreae, M.O., Cantrell, W., Cass, G.R., Chung, C.E., Clarke, A.D., Coakley, J.A., Collins, W.D., Conant, W.C., Dulac, F., Heintzenberg, J., Heymsfield, A.J., Holben, B., Howell, S., Hudson, J., Jayaraman, A., Kiehl, J.T., Krishnamurti, T.N., Lubin, D., McFarquhar, G., Novakov, T., Ogren, J.A., Podgorny, I.A., Prather, K., Priestley, K., Prospero, J.M., Quinn, P.K., Rajeev, K., Rasch, P., Rupert, S., Sadourny, R., Satheesh, S.K., Shaw, G.E., Sheridan, P. and Valero, F.P.J., 2001, Indian Ocean Experiment: An integrated analysis of the climate forcing and effects of the great Indo-Asian haze. *Journal of Geophysical Research*, **106**, pp. 28371–28398.

Reddy, M.S., O. Boucher, Y. Balanski, and M. Schulz, 2005: Aerosol optical depths and direct radiative perturbations by species and sourcetype. *Geophys. Res. Lett.*, **32**, L12803, doi:10.1029/2004GL021743.

Rodriguez, S., Querol, X., Alastuey, A., Kallos, G. and Kakaliagou, O., 2001. Saharan dust inputs to suspended particles time series (PM10 and TSP) in Southern and Eastern Spain, *Atmospheric Environment*, **35/14**, pp. 2433–2447.

Segal, M., 1990, On the impact of thermal stability on some rough flow effects over mobile surfaces, *Boundary-Layer Meteorology*, **52**, pp. 193–198.

Seinfeld, J.H. and Pandis, S.N., 1998, *Atmospheric Chemistry and Physics: From Air Pollution to Climate Change*, (New York: Wiley-Interscience), ISBN 0471178160.

Shao, Y., Raupach, M.R. and Findlater, P.A., 1993, Effect of saltation bombardment on the entrainment of dust by wind, *Journal of Geophysical Research*, **98**, pp. 12719–12726.

Shell, K. and Somerville, R. C. J. 2007: Direct radiative effect of mineral dust and volcanic aerosols in a simple aerosol climate model. *J. Geophys. Res.* **112**, doi:10.1029/2006JD007197.

Slinn, W.G.N., 1982, Prediction for particle deposition to vegetative canopies, *Atmospheric Environment*, 16, pp. 1785–1794.

Spyrou C., C. Mitsakou, G. Kallos, P. Louka, and G. Vlastou, 2010: An improved limited area model for describing the dust cycle in the atmosphere. *Journal of Geophysical Research*, 115, D17211, doi:10.1029/2009JD013682, 2010.

Stanelle T., B. Vogel, H. Vogel, D. Baumer, and C. Kottmeier, 2010: Feedback between dust particles and atmospheric processes over West Africa during dust episodes in March 2006 and June 2007, *Atmos. Chem. Phys.*, **10**, 10771–10788, 2010, doi: 10.5194/acp-10-10771-2010.

Swap, R., Garstang, M., Greco, S., Talbot, R. and Kallberg, P., 1992, Sahara dust in the Amazon basin, *Tellus*, **44**, pp. 133–149.

Tanaka Y. Taichu, Howard Hanson, 2007: "Global dust budget". In: *Encyclopedia of Earth. Eds.* Cutler J. Cleveland (Washington, D.C.: Environmental Information Coalition, National Council for Science and the Environment).

Tegen I., 2003: Modeling the mineral dust aerosol cycle in the climate system. *Quaternary Science Reviews*, **22**, pp. 1821–1834.

Tegen, I. and Fung, I., 1994, Modeling of mineral dust in the atmosphere: Sources, transport and optical thickness, *Journal of Geophysical Research*, **99**, pp. 22897–22914.

Wang Y., Guoshun Zhuanga , Yele Sund, Zhisheng Anb, 2005: Water-soluble part of the aerosol in the dust storm season-evidence of the mixing between mineral and pollution aerosols, *Atmospheric Environment*, **39**, pp. 7020–7029.

Westphal, D.L., Toon, O.B. and Carlson, T.N., 1987, A two-dimensional numerical investigation of the dynamics and microphysics of Saharan dust storms, *Journal of Geophysical Research*, **92**, pp. 3027–3049.

Westphal, D.L., Toon, O.B. and Carlson, T.N., 1988, A case study of Mobilization and transport of Saharan dust, *Journal of Atmospheric Sciences*, **45**, pp. 2145–2175.

White, B.R., 1979, Soil transport by winds in Mars, *Journal of Geophysical Research*, **84**, pp. 4643–4651.

Yoshioka M., Mahowald N.M., A.J. Conley, W.D. Collins, D.W. Fillmore, C.S. Zender, D.B Coleman, 2007: Impact of Desert Dust Radiative Forcing on Sahel Precipitation: Relative Importance of Dust Compared to Sea Surface Temperature Variations, Vegetation Changes, and Greenhouse Gas Warming, *American Meteorological Society*, **20**, pp. 1445–1467, doi: 10.1175/JCLI4056.1.

Zender, C.S., Bian, H. and D. Newman, 2003. Mineral Dust Entrainment and Deposition (DEAD) model: Description and 1990s dust climatology, *J. Geophys. Res.-Atmos.*, **108**, D14, doi:10.1029/2002JD002775.

Zhao, Q. and Carr, F.H., 1997, A Prognostic Cloud Scheme for Operational NWP Models, *Monthly Weather Review*, **125**, pp. 1931–1953.

Zilitinkevich, S.S, Grachev, A.A. and Hunt, J.C.R., 1998, Surface frictional processes and non-local heat/mass transfer in the shear-free convective boundary layer, in *Buoyant Convection in Geophysical Flows*, edited by Plate, E.J. *et al.*, (Kluwer Acad Norwell, Mass), pp. 83–113.

Zilitinkevich, S.S., 1995, Non-local turbulent transport: pollution dispersion aspects of coherent structure of convective flows, in *Air Pollution III – Volume I. Air Pollution Theory and Simulation,* edited by. Power, H., Moussiopoulos, N. and Brebbia, C.A., (Computational Mechanics Publications, Southampton Boston), pp. 53–60.

Zoumakis, N.M., and A. G. Kelessis, 1991, The dependence of the bulk Richardson number on stability in the surface layer, *Boundary Layer Meteorology*, **57**, 407–414.

Part three
Processes at water interfaces

CHAPTER SIX

Gas-transfer at unsheared free-surfaces

Carlo Gualtieri & Guelfo Pulci Doria

*Hydraulic, Geotechnical and Environmental Engineering Dept. (DIGA),
University of Napoli, Napoli, Italy*

ABSTRACT

Transport processes through the gas-liquid interfaces are of paramount importance in Environmental Fluid Mechanics. In environmental systems gaseous substances may be directly exchanged between air and water in either direction across the air-water interface. Gas fluxes being transferred can be upward to the air or downward to the water depending on the substances involved. Nowadays, despite the significant theoretical, laboratory, field and numerical studies, research efforts have not yet achieved a complete understanding of gas-transfer process.

In this chapter gas-transfer is introduced. First it is explained how the physicochemical characteristics of the substance being transferred affect gas-transfer process and they can control which phase governs the process. Second the interaction between turbulence in the bulk liquid and the air-water interface a for a substance being controlled by the liquid phase. Third, gas-transfer in open channel flows is presented in detail discussing the literature conceptual models for the prediction of gas-transfer rate. Then, recent efforts about the numerical simulation of gas-transfer process at an unsheared interface are described. The chapter ends pointing out the challenges to be faced and the needs for future research about gas-transfer.

6.1 INTRODUCTION

Transport processes through the gas-liquid interfaces are of paramount importance in a number of areas of industrial engineering, such as chemical and mechanical engineering, and for geophysical and environmental systems. In such systems, gaseous substances may be directly exchanged between air and water in either direction across the air-water interface. Gas fluxes being transferred can be upward to the air or downward to the water depending on the substances involved. Thus, *gas transfer* is a two-way process involving both gas absorption, i.e. air to water, and volatilization, i.e. water to air, across an air-water interface, for a volatile or semi-volatile chemical. In the Environmental Fluid Mechanics field, for processes at the free surfaces of terrestrial water bodies, early interest related the absorption of atmospheric oxygen in natural waters. This process is also termed as atmospheric reaeration. Since dissolved oxygen (DO) is commonly considered as the main indicator of aquatic ecosystem health, reaeration is one of the most relevant source of DO in the water bodies, whose DO level are depleted by natural causes or the discharge of organic matter (USEPA, 1985; Chapra, 1997). The volatilization of many chemicals, such as mercury, PCBs, PAHs and pesticides, has been widely recognized as an important process determining the transport, fate, and chemical loadings of these contaminants in the atmosphere and in large water bodies, such as lakes, estuaries and oceans (USEPA, 1997). Also, the assessment of volatilization rate of environmentally important compounds of low molecular weight such as benzene, chloroform, methylene chloride, and toluene from rivers and streams contaminated by spills or industrial discharges has been subject of continuing interest. Therefore the estimation of both reaeration and volatilization rate is a key issue in the application of a modeling framework of dissolved oxygen balance or of contaminant transport and fate (Chapra, 1997).

More recently, the exchange of moisture, carbon dioxide (CO_2) and other greenhouse gases between the atmosphere and the oceans or the lakes have become important because of their impact on global warming. It is estimated that approximately 30–40 per cent of man-made CO_2 is taken up by the oceans, but these estimates are significantly affected by the uncertainties in the prediction of gas-transfer rate at the air-water interface (Banerjee and MacIntyre, 2004).

Nowadays, despite the significant theoretical, laboratory, field and numerical studies, research efforts have not yet achieved a complete understanding of gas-transfer process. Also, predictive models currently available are not yet able to predict its rate in all the environmental and hydrodynamic conditions. In the hydraulic and environmental engineering field several empirical equations have been long proposed to estimate both reaeration and volatilization rates, but recent studies have demonstrated that these equations cannot have a general application (Melching and Flores, 1999; Gualtieri *et al.*, 2002; Gualtieri, 2006). Therefore, intensive researches are currently carried on to gain insight into the complex mechanisms of gas-transfer and to develop a physically sound and reliable predictive equation of gas-transfer rate.

First of all, we can define *gas-transfer* as an *interphase mass-transfer process* that occurs at the air-water interface if a non-equilibrium condition between the air phase and the water phase exists for a chemical. The equilibrium or non-equilibrium condition generally depends on chemical potential of the considered species within the phase involved, which is related to concentration, which is simpler to be measured. Thus, the transport of material between phases is controlled by the gradient in concentration across the interface, which represents the *driving force* of the gas-transport process. As a result of this gradient, a flux of the chemical moves through the air-water interface. Also, this flux should be related to the characteristics of transport processes near the air-water interface. These processes can occur at the molecular scale and are also affected by turbulence because the flow in the atmosphere and in the water body is turbulent. Thus, a first qualitative assessment of gas-transfer process would lead to state that a gas-transfer flux J_{g-t} driven by concentration gradient could be generally expressed using Fick's law (Thibodeaux, 1997) as:

$$J_{g-t} = -(D_m + D_t) \cdot \frac{dC}{dz} \tag{6.1}$$

where D_m and D_t are, respectively, the molecular and *turbulent* or *eddy* diffusion coefficient and dC/dz is the concentration gradient of the species being transferred, where z is the vertical coordinate. Notably, the gas being transferred is assumed to be distributed uniformly in the bulk fluid. Also, the magnitude of the eddy diffusion coefficient D_t in the natural environment is usually many times larger than molecular diffusivity D_m.

Equation (6.1) points out that gas-transfer process depends on the physicochemical characteristics of the substance being transferred and on the interaction between turbulence in the atmosphere and/or in the water body, on one hand, and the air-water interface, on the other. The latter feature introduces a second critical point that is related to the relative importance of the gas-phase, i.e. the atmosphere, and of the water-phase, i.e. the water body, on gas-transfer process. It is likely that sometimes one phase can prevail and transport processes occurring within this phase should be better investigated to gain insight into gas-transfer process. Third, another critical point is expected to be related to where the turbulence is produced, i.e. whether close to the air-water interface or far from it, since the interplay between turbulent motions and the interface should be different.

The previous short discussion suggests to divide the subject and to organize the chapter as follows. Section 2 explains how the physicochemical characteristics of the substance being transferred affect gas-transfer process and they can control which phase governs the process. Section 3 provides a discussion on how turbulence generally interacts with the air-water interface a for a substance being controlled by the water phase. This discussion highlights that a more detailed approach requires to consider separately conditions where turbulence is produced far from the air-water interface, that is an unsheared interface, and where turbulence is produced close to the interface, that is a sheared interface. Thus, Section 4 deals with the gas-transfer at an unsheared air-water interface. First of all, dimensional analysis of gas-transfer process is presented to achieve a robust theoretical framework where suitable modelling efforts can be developed. After then, classical and more recent modeling approaches starting from Lewis-Whitman two films theory are discussed. Both approaches based on *global* and *local* properties of turbulence are presented. Moreover, results from both laboratory and field studies together with those coming from numerical simulations are also considered to elucidate physical features of the gas-transfer process and to assess models performances. Finally, conclusive remarks are drawn also highlighting the areas where future research would be useful.

6.2   GAS-TRANSFER – INFLUENCE OF GAS CHARACTERISTICS

The previous short discussion pointed out that gas-transfer process is governed by the interplay of turbulent and molecular transport processes. Hence Equation (6.1) includes molecular diffusivity, which depends on the characteristics of both the gas being transferred and the fluid, air or water, where the transfer occurs. However, there is another important characteristics of the gas involved in the transfer that should be considered. In fact, it is well known that if a vessel of gas-free distilled water is exposed to the atmosphere, gaseous compounds, such as oxygen or carbon dioxide, cross the air-water interface and enter into solution. The process will continue until a fixed level of the gas for a given temperature will be reached. In other words, an equilibrium is established between the partial pressure of the gas in the atmosphere and the concentration in the water phase. This equilibrium can be expressed by Henry's law as:

$$p = H_e C_{sat} \qquad\qquad (6.2)$$

where p is the partial pressure, H_e is Henry's constant and C_{sat} is the saturation concentration of the gas into the water. From equation (6.2) Henry's constant is the ratio of the partial pressure of the gaseous phase to the solubility of the gas in the water phase. Equation (6.2) points out that at a fixed partial pressure of the gas, saturation concentration of the gas and hence its solubility decreases with the increasing value of H_e.

Equation (6.2) could be also presented in dimensionless form using the ideal gas law:

$$p Vol = n_m R T_a \qquad\qquad (6.3)$$

where *Vol* is the volume of the gas, n_m is the number of moles, and T_a is absolute temperature in K. Finally, R is the universal gas constant, which is equal to 8.314. From Equation (6.3), the molar concentration of the gas could be expressed in terms of its partial pressure as:

$$C = \frac{n}{Vol} = \frac{p}{R T_a} \qquad\qquad (6.4)$$

which can be introduced into Equation (6.2) to yield:

$$H = \frac{H_e}{RT_a} = \frac{C}{C_{sat}} \tag{6.5}$$

where H is the dimensionless Henry's constant.

Table 6.1 lists the values of Henry's constants He and H at 25°C for some substances in the field of Environmental Fluid Mechanics.

Table 6.1. Values of Henry's constants H and H_e at 25°C.

Chemical	Source	$M - g$/mole	$H_e - Pa \times m^3 \times mole^{-1}$	H
Aroclor 1016	Chapra, 1997	257.9	3.35E+01	1.35E−02
Aroclor 1242	Chapra, 1997	266.5	3.85E+02	1.55E−01
Aroclor 1248	Chapra, 1997	299.5	3.59E+02	1.45E−01
Aroclor 1254	Chapra, 1997	328.4	1.46E+02	5.91E−02
Aroclor 1260	Chapra, 1997	375.7	7.17E+02	2.89E−01
Mean PCBs	Chapra, 1997	305.6	2.18E+02	8.78E−02
Al drin	Various	364.91	1.67E+00	6.72E−04
Dieldrin	Various	380.91	1.09E+00	4.38E−04
Lindane	Various	290.83	3.33E−01	1.34E−04
Toxaphene	Chapra, 1997	430	5.72E+03	2.31E+00
Benzene	Rathbun, 1998	78.11	5.57E+02	2.24E−01
Naphthalene	Rathbun, 1998	128.2	5.60E+01	2.26E−02
Methylbenzene	Rathbun, 1998	92.14	6.38E+02	2.57E−01
Ethylbenzene	Rathbun, 1998	106.17	7.56E+02	3.05E−01
Chlorobenzene	Rathbun, 1998	112.6	3.58E+02	1.44E−01
Trichloromethane	Rathbun, 1998	257.9	3.91E+02	1.58E−01
Trichloroethylene	NIST, 2000	266.5	1.07E+03	4.34E−01
1,2-Dichloroethane	Rathbun, 1998	299.5	1.14E+02	4.60E−02
MTBE	Various	328.4	6.43E+01	2.59E−02
Mercury	Various	375.7	1.25E+03	5.03E−01

The influence of temperature on He or H and, hence, on C_{sat} was already introduced but the saturation concentration of a gas is affected also by two other parameters, water salinity and partial pressure variations due to elevation. Some empirical equations were developed to predict how these factors influence saturation of dissolved oxygen (Chapra, 1997). These equations point out that saturation concentration of dissolved oxygen decreases as temperature and water salinity increase. On the other hand, saturation concentration increases with the increasing pressure.

The influence of Henry's constant on gas-transfer process can be pointed out considering a volume of fluid across the air-water interface. A qualitative approach shows that the interface due to the surface tension of the fluid could be considered as a semi-solid *wall*. Thus, approaching to the interface, turbulent motions become increasingly damped and molecular transport takes control over turbulent transport. Considering for now only mass transport, it can be expected that a *diffusive* or *concentration* boundary sublayer (CBL) develops on both sides of the interface, while outside these sublayers turbulence governs transport processes (Fig. 6.1). At this point, we limit the discussion to this but important details on the interplay of turbulence and these sublayers will be further provided (Section 1.3).

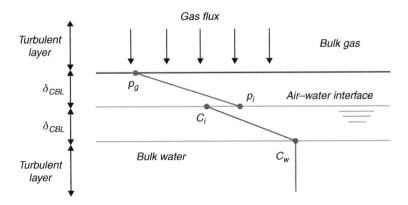

Figure 6.1. Sketch of gas-transfer across the air-water interface.

Fig. 6.1 relates to a flux from the atmosphere to a waterbody, such as in the reaeration process. To enter the bulk water, the gas must cross both the CBLs. Recall that the gas-transfer process is related to a non-equilibrium condition holding between the air phase and the water phase. Thus, we can assume according to Equation (6.1) that the gas flux is proportional to the concentration gradient existing between the interface and the bulk fluid through a coefficient. First, the gas must move through the CBL on the air-side and the gas flux $J_{g-t-gas}$ is:

$$J_{g-t-gas} = k_g(C_g - C_i) \tag{6.6a}$$

where C_g & C_i are gas concentration in the bulk gas and at the air-water interface, respectively, and k_g is the gas-transfer velocity in the CBL on the air-side. Concentrations are related to pressures by Equation (6.4), so Equation (6.6a) yields:

$$J_{g-t-gas} = \frac{k_g}{RT_a}(p_g - p_i) \tag{6.6b}$$

where p_g & p_i are gas pressure in the bulk gas and at the interface, respectively.

Similarly, the gas must cross the CBL on the water-side and the gas flux moving across this CBL is:

$$J_{g-t-water} = k_w(C_i - C_w) \tag{6.7a}$$

where C_w is gas concentration in the bulk water and k_w is the gas-transfer velocity in the CBL on the water-side. Since at the interface equilibrium holds, Equation (6.2) allows to express the concentration at the interface C_i in Equation (6.7a) as a function of the pressure at the interface p_i to yield:

$$p_i = H_e \left(\frac{J_{g-t-water}}{k_w} + C_w \right) \tag{6.7b}$$

while Equation (6.6b) yields:

$$p_i = p_g - \frac{RT_a J_{g-t-gas}}{k_g} \tag{6.6c}$$

Equations (6.6c) and (6.7b) can be equated and solved for the gas flux J_{g-t} as:

$$J_{g-t} = \frac{1}{\frac{1}{k_w} + \frac{RT_a}{H_e k_g}} \left(\frac{p_g}{H_e} - C_w \right) \tag{6.8a}$$

which points out as the gas-transfer process depends on the equivalent concentration gradient existing between the gas phase and the water phase. Equation (6.8a) can be also expressed as:

$$J_{g-t} = K_L \left(\frac{p_g}{H_e} - C_w \right) \tag{6.8b}$$

where K_L, which is equal to:

$$K_L = \frac{1}{\frac{1}{k_w} + \frac{RT_a}{H_e k_g}} = k_w \frac{H_e}{H_e + RT_a(k_w/k_g)} \tag{6.9}$$

is called *gas-transfer coefficient*. Equation (6.9) confirms that the gas-transfer process depends also on gas characteristics, that is the value of Henry's constant H_e. Inspection of (9) highlights that chemicals with high H_e are rapidly purged from the water, whereas chemicals with low H_e tend to stay in solution. Also, we can note that Equation (6.9) shows that the process encounters a resistance moving across the CBLs which is analogous to that of two resistors in series in an electrical circuit. In other words, the total resistance to gas transfer R_{tot} depends on each resistance in the water and gaseous CBL as:

$$R_{tot} = R_g + R_w \tag{6.10}$$

where R_g and R_w are the resistance in the CBLs on the air-side and water-side, respectively:

$$R_g = \frac{RT_a}{H_e k_g} \qquad R_w = \frac{1}{k_w} \tag{6.11}$$

Therefore, depending on the relative magnitudes of H_e, k_g, and k_w, the process may be controlled by the water, the gas, or both CBLs. Particularly, the influence of the water CBL can be quantified as:

$$\frac{R_w}{R_{tot}} = \frac{R_w}{R_g + R_w} = \frac{\frac{1}{k_w}}{\frac{RT_a}{H_e k_g} + \frac{1}{k_w}} = \frac{H_e}{H_e + RT_a(k_w/k_g)} \tag{6.12}$$

Few data are available for gas transfer coefficients k_w and k_g. In the open ocean a value of 8.3×10^{-3} m/s is commonly used for k_g (Rathbun and Tai, 1982). Field and laboratory data show that the gas-film coefficient k_g is typically in the range from 3.00×10^{-3} to 3.00×10^{-2} m/s, whereas k_w lies between is 5.00×10^{-6} to 5.00×10^{-5} m/s (Schwarzenbach et al., 1993). In lakes, k_w varies from 1.16×10^{-6} to 1.16×10^{-4} m/s and k_g from 1.39×10^{-3} to 1.39×10^{-1} m/s (Chapra, 1997). These values correspond to a range from 0.1 to 10 m/day for k_w and from 120 to 12000 m/day for k_g. Thus, the ratio k_w/k_g generally is in the range from 0.001 to 0.01, with the higher values in small lakes due to lower k_g because of wind sheltering (Chapra, 1997).

Table 6.1 showed that H_e can significantly change among different substances and Equation (6.12) demonstrates that the ratio R_w/R_{tot} increases with the increasing value of the Henry's constant. Thus, the higher the Henry's constant, the more the control of gas-transfer process shifts to the CBL on the water-side.

Recently, the values of ratio R_w/R_{tot} for 20 environmental contaminants was evaluated (Gualtieri, 2006). The considered contaminants were 6 different *PCBs*; 4 *pesticides*, aldrin, dieldrin, lindane and toxaphene; 2 *aromatic hydrocarbons*, benzene and naphtalene; 2 *alkyl-benzenes*, methylbenzene and ethylbenzene; 2 *halogenated alkanes*, chloroform and 1,2-dichloroethane; and, finally, chlorobenzene, trichloroethylene (TCE), methyl tertiary-butyl ether (MTBE), and mercury. Mean values of H_e for a temperature of 25°C were applied. Also, three values were considered for the ratio k_w/k_g here. They were obtained coupling maximum, minimum, and mean value for k_w with the minimum, maximum, and mean value for k_g and they are listed in Table 6.2. The percentage resistance to the mass-transfer in the liquid CBL finally was estimated. Results for the *mean* conditions are shown in Fig. 6.2, where the data for some environmentally important gases, such as ammonia, sulfur dioxide, carbon dioxide, nitrogen and oxygen, are also presented. Results showed that lindane, dieldrin, and aldrin are controlled by the gaseous CBL, whereas the remaining chemicals are controlled by the CBL on the water-side. This is the case of sparingly soluble gases such as O_2 and CO_2.

Table 6.2. Values of gas-transfer coefficients k_w, k_g and their ratio k_w/k_g.

	High	Mean	Low
k_w – m/s	1.157E−04	1.157E−05	1.157E−06
k_g – m/s	1.389E−01	1.389E−02	1.389E−03
Ratio k_w/k_g	8.33E−02	8.33E−04	8.33E−06

Noticeably, if a lower value of the ratio k_w/k_g is applied, the control shifts to the liquid CBL. Hence, results in Fig. 6.2, where the ratio k_w/k_g is equal to $k_w/k_g = 8.33 \times 10^{-4}$, are representative of *mean* conditions. If the ratio R_w/R_{tot} is nearly equal to unity, then Equation (6.9) yields that gas-transfer velocity in the CBL on the water-side is equal to gas-transfer velocity, that is:

$$K_L \approx k_w \qquad (6.13)$$

which means that the gas-transfer process is affected only by fluid mechanics processes in the water body.

The forthcoming discussion will be addressed to gas-transfer process for a substance being controlled by the CBL on the water-side, which is a very common condition for the gas-transfer in the Environmental Fluid Mechanics.

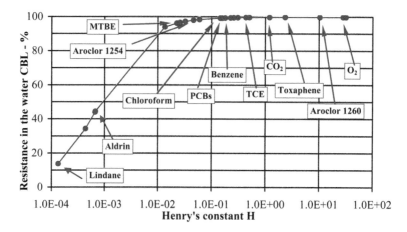

Figure 6.2. Resistance to gas-transfer in the water CBL.

6.3 GAS-TRANSFER – INFLUENCE OF TURBULNCE

Characterizing turbulence influence on gas-transfer across air-water interface has been proved to be difficult since this influence depends on relative phase velocities, roughness of surfaces at the interface, frictional and adhesive forces, surface tensions and several other parameters (Weber and DiGiano, 1996), and complex, anisotropic effects of the free surface on turbulence further complicate the modeling effort as well (Moog and Jirka, 1999). However, to introduce how turbulence generally interacts with the air-water interface for a substance being controlled by the water phase, we can start to consider a stagnant water body, where hydrodynamics processes have negligible effects on gas-transfer, as illustrated in Fig. 6.3 (Socolofsky and Jirka, 2002).

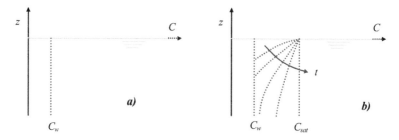

Figure 6.3. Gas-transfer in a stagnant water body.

If the water body has along its depth a uniform initial concentration C_w, which is lower than saturation concentration C_{sat} (Fig. 6.3a), we can define the following initial condition:

$$C(z,0) = C_w \tag{6.14}$$

The air-water interface is then instantaneously exposed to an infinite source of the gas. Since $C_w < C_{sat}$, the gas tends to cross the interface and to dissolve into the water. The process will continue until the water body will reach over all the depth saturation concentration. Dissolution reaction is a fast reaction but the movement of the gas inside the water is controlled by diffusion (Fig. 6.3b). One-dimensional advection-diffusion equation in the vertical direction could be applied to study this case, neglecting advection term since the fluid is stagnant:

$$\frac{\partial C}{\partial t} = D_m \frac{\partial^2 C}{\partial z^2} \qquad (6.15)$$

where Equation (6.14) defines initial condition and the boundary conditions are:

$$
\begin{aligned}
C(-\infty, t) &= C_w \\
C(0, t) &= C_{sat} \\
C(0, 0) &= C_{sat} \\
C(z, 0) &= C_w
\end{aligned}
\qquad (6.16)
$$

Note that the presented case corresponds to that of diffusion in a semi-infinite medium from a constant concentration source. Thus, the solution is:

$$\frac{C(z, t) - C_w}{C_{sat} - C_w} = 1 - erf\left(\frac{-z}{\sqrt{4D_m t}}\right) \qquad (6.17)$$

where the minus sign inside the error function is needed since z is negative downward. Equation (6.17) can be used to derive the flux across the air-water interface. According to Fick's law, the one-dimensional diffusive flux is:

$$J_{g-t-z} = -D_m \left.\frac{\partial C}{\partial z}\right|_{z=0} \qquad (6.18a)$$

Substituting the solution above, the flux becomes:

$$J_{g-t-z}(t) = -(C_{sat} - C_w)\sqrt{\frac{D_m}{\pi t}} \qquad (6.18b)$$

which demonstrates that the flux increases with the molecular diffusivity of the gas into the water and with the gradient existing to saturation. The gas-transfer flux can be also expressed as:

$$J_{g-t-z}(t) = -K_L(C_{sat} - C_w) \qquad (6.19)$$

where the gas-transfer coefficient is given by:

$$K_L = \sqrt{\frac{D_m}{\pi t}} \qquad (6.20)$$

The thickness of the CBL on the water-side can be evaluated, after some algebra, as:

$$\delta_{CBL} = \sqrt{2D_m t} \qquad (6.21)$$

which shows that the CBL in a stagnant water body grows deeper indefinitely in time and with the molecular diffusivity of the exchanged gas. This result holds when turbulence is absent and it should seen as an idealized, very unlikely case. However, it could be considered as a starting point of a discussion about the effects of turbulence on gas-transfer process.

As previously outlined, the interface due to the surface tension of the fluid could be considered as a semi-solid *wall*. Therefore, momentum and mass transport processes are expected to be governed by the interplay between turbulent and molecular transport within an hierarchal structure of layers: the *turbulent layer (TL)*, the *velocity boundary sublayer (VBL)* and the aforementioned *diffusive* or *concentration boundary sublayer (CBL)*.

Far from the interface, in the turbulent layer, both momentum and mass transport is dominated by turbulent motions, that provide full vertical mixing. Thus, the main body of gaseous and liquid phases are assumed to be well-mixed with the gas profile practically uniform at the bulk concentration. In the turbulent layer, momentum and mass transport processes can be related to the turbulent eddy viscosity v_t and to the turbulent eddy diffusivity D_t, respectively. Reynolds analogy allows to consider these parameters having the same order of magnitude, that is:

$$v_t \approx D_t \tag{6.22}$$

They are related by the turbulent Schmidt number Sc_t:

$$Sc_t = \frac{v_t}{D_t} \tag{6.23}$$

which is approximately equal to the unity, that is turbulent momentum and mass transport have the same strength, which is higher than that of transport processes occurring at the molecular scale. In other words, in the natural environment, within the turbulent layer, $v_t >> v$ and $D_t >> D_m$, where v is the kinematic viscosity of the fluid and D_m is the molecular diffusivity of the gas into the fluid. Strictly reasoning, the vertical mass-transport is a combination of molecular and turbulent diffusion and the vertical diffusivity K_v is the sum of molecular D_m and turbulent eddy diffusivity D_t, but we can assume that turbulent diffusion is predominant. Turbulent eddy diffusivity D_t can be also related to the dissipation rate of turbulent kinetic energy ε. In fact, in steady turbulence, the rate of energy transfer from one scale to the next is the same for all scales and it is per unit mass of fluid equal to ε. On the other hand, under certain conditions, assuming a balance between total kinetic energy related to Reynolds stresses and the viscous dissipation, a logarithmic profile structure holds and the dissipation ε could be expressed as (Wüest and Lorke, 2003):

$$\varepsilon = \frac{u^{*3}}{\kappa z} \tag{6.24}$$

where u^* is the friction velocity and κ is Von Kármán constant $\kappa = 0.41$. Typical turbulent layers heights range from several meters to several tens or hundreds of meters in lakes and oceans, respectively, and several hundreds of meters to kilometres in the atmosphere (Lorke and Peeters, 2006).

Approaching to the air-water interface, at scales where viscous forces play a relevant role, turbulent eddies are increasingly damped as they approach closer than their length scale. Thus, turbulent momentum and mass transport mechanisms become weaker and someway increasingly comparable with those occurring at the molecular scale. Both v_t and D_t decrease steeply assuming values which may be comparable with those of v and D_m respectively. We could expect that approaching to the interface, molecular transport takes control over turbulent transport and momentum and mass boundary sublayers develop

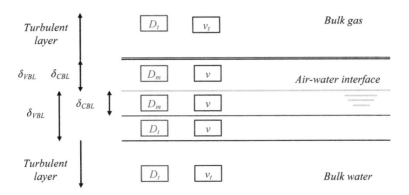

Figure 6.4. Hierarchal structure of layers at the air-water interface.

on both sides of the air-water interface (Fig. 6.4). The first sublayer is termed *velocity boundary sublayer (VBL)* and the second is called *diffusive* or *concentration boundary sublayer (CBL)*. Their thicknesses are δ_{VBL} and δ_{CBL}, respectively. Note that sometimes a difference is underlined between diffusive and concentration sublayers. The latter is related to a concentration gradient only and it is also called *outer concentration sublayer*, the former is the region where that gradient is linear and it is also termed *inner concentration sublayer* (Magnaudet and Calmet, 2006). Hence, the diffusive sublayer would be a component of the CBL. However, in the following discussion we will not distinguish the diffusive sublayer from the CBL. Inside the VBL momentum transport is governed by fluid viscosity. Inside the CBL mass transport is controlled by the molecular diffusivity of the gas in the fluid. The relative thickness of these sublayers is related to the importance of diffusion of momentum and diffusion of mass by molecular transport. This can be expressed through the ratio of the kinematic viscosity to the molecular diffusivity of the gas into the fluid:

$$Sc = \frac{v}{D_m} \tag{6.25}$$

which is the Schmidt number. In other words, the Schmidt number describes the relative intensity of momentum and mass transport processes occurring at the molecular scale. In the air, Sc is close to 1 and the sublayers have about the same thickness. For example, Sc is for CO_2, NH_3 and H_2O equal to 0.83, 0.53 and 0.56, respectively (Jähne and Haußecker, 1998).

The situation is completely different in the liquid phase since within the range of temperature typical of environmental processes, water kinematic viscosity is in the order of 1×10^{-6} m^2/s, whereas molecular diffusivity of a gas into water is in the order of 1×10^{-9} m^2/s resulting in a Sc in the order of 10^3. Therefore in the water phase the diffusion of mass is much more slower than the diffusion of momentum and the mass boundary sublayer is significantly thinner than the viscous boundary sublayer. Also, in contrast to the air phase, Sc depends significantly on temperature, in turn making the gas transport inside the water phase temperature-dependent. Notably, v decreases with temperature, while D_m increases with temperature. Hence Sc decreases with the increasing temperature. For example, the Schmidt number for dissolved oxygen and cyclohexane is in the range from 950 to 440 and from 2223 to 985, respectively if temperature is ranging from 10 to 25°C (Gualtieri, 2005b).

Since we are dealing with the gas-transfer process for a substance being controlled by the CBL on the water-side, further details must be provided about the structure of velocity and concentration boundary sublayers in the water phase. As previously outlined, approaching to the air-water interface from the water side, the velocity boundary layer is first encountered. Inside the VBL the velocity gradient is constant and its thickness δ_{VBL} could be defined as the distance below the interface where D_t equates water kinematic viscosity ν. The height δ_{VBL} could be scaled with the friction velocity u^* in the turbulent layer below as (Lorke and Peeters, 2006):

$$\delta_{VBL} = \frac{11\nu}{u^*} \tag{6.26}$$

and δ_{VBL} is typically $\delta_{VBL} \approx 10^{-3}-10^{-4}$ m.

Inside the VBL, although turbulent diffusion is damped, the rate of strain of scalar tracer concentration fields creates enhanced concentration gradients, which increase transport due solely to molecular diffusion (Lorke and Peeters, 2006). Thus, mixing rates of tracers in the velocity boundary layer are still higher than those occurring at the molecular scale and measured concentration profiles are usually well mixed up to a certain distance from the interface. Approaching further to the air-water interface, turbulent eddy diffusivity D_t decreases down to the molecular diffusivity D_m. This defines the thickness δ_{CBL} of the CBL, where the transport due to the eddies becomes negligible compared to molecular diffusion and a linear concentration gradient holds up to the interface since viscous straining is no more capable to increase mixing above that occurring at the molecular scale. Therefore, it should be expected that the thickness of the concentration boundary sublayer would be related to the level of turbulence in the TL and to the *strength* of both momentum and mass transport mechanisms occurring at the molecular scale. The former using $u*$ was related to δ_{VBL}. The latter is represented from Equation (6.22) by the Schmidt number Sc. Also, D_t was assumed to be dependent from the vertical distance from the interface z. Therefore, δ_{CBL} could be expressed as:

$$\delta_{CBL} = \frac{\delta_{VBL}}{Sc^\alpha} \tag{6.27}$$

where α is a coefficient which is usually assumed to be between $1/3$ and $1/4$ (Wüest and Lorke, 2003).

Equation (6.27) demonstrates that δ_{CBL} is solute-specific and is slightly temperature-dependent, as Sc changes with temperature. If $\alpha = 1/3$, Eq. (6.27) shows that δ_c is for the substances of environmental concern range from $1/13$ to $1/6$ the thickness of the velocity boundary layer δ_{VBL} (Gualtieri, 2005a). Sometimes, since $Sc \approx 10^3$ and $Sc^{1/3} \approx 10$, δ_{CBL} is approximated as $\delta_{CBL} = 0.1\delta_{VBL}$ and it is typically $\delta_{CBL} \approx 10^{-4}-10^{-5}$ m, that is tens or hundreds of microns. Interestingly, the previous discussion holds also at the sediment-water interface, where the same hierarchal structure of turbulent and viscous layers exists and the same key parameters control momentum and mass transport processes (Lorke and Peeters, 2006).

At this point a fundamental question arises: how turbulence interacts with the outlined structure of layers? It may be expected that turbulent eddies moving randomly over the water depth delivering periodically water parcels from the bulk liquid close to the air-water interface. After their arrival at the interface, the effect of the eddies is twofold:

- first, they erode the boundary sublayers structure, thereby limiting the growth of the concentration boundary sublayer thickness, δ_{CBL}. Also, since the concentrations in bulk fluid and at the interface are independent of δ_{CBL}, this effect increases the concentration gradient; hence, according to equation (1), the gas-transfer flux is larger than in the stagnant case;

- second, turbulent eddies moved up to the air-water interface cause motion within the concentration boundary sublayer, thereby increasing the effective diffusivity. thus, the gas-transfer flux is again larger than in the stagnant case.

Furthermore, if, for example, reaeration is the considered gas-transfer process, water parcels carried by the turbulent eddies from the bulk liquid to the air-water interface are characterized by low concentration of dissolved oxygen. Upon their arrival to the interface, they are exposed to dissolved oxygen source and enriched by molecular diffusion until turbulent eddies bring again them down in the bulk water increasing dissolved oxygen levels there.

This brief discussion points out that the general effect of turbulence is to increase gas-transfer flux but also that molecular diffusion is still expected to be a rate-limiting process. However, this general outcome should be considered only as a starting point for a more detailed analysis which requires to consider where turbulence is produced. Three cases can occur:

- *unsheared interface*, also sometimes termed as *bottom-shear generated turbulence*. In fact, if the winds are lights, fluid motions and turbulence that can be observed near the air-water interface are generated elsewhere. This is typical of open channel flows, such as streams and rivers, where turbulence is generated at the bottom wall and is then self-transported towards the free surface. Another case is if turbulence is produced in the shear layer between subsurface currents flowing at different velocities. In both cases, these turbulence structures then impinge at the free surface producing effects as boils that can be easily seen at the surface of rivers. Turbulence can finally produced by heat losses that give rise to natural convective motions on the liquid side (Banerjee and MacIntyre, 2004). Furthermore, the turbulence structure near the free surface can have a close relationship with surface-wave fluctuations and the Froude number of the flow;
- *sheared interface*, also termed as *wind-shear generated turbulence*, which refers to lakes or the sea when the wind blows above an almost still air-water interface. When a significant winds blowing over the free surface, drift currents and wind-waves due to the wind shear across the air-water interface are produced. In this case the turbulence generation occurs at the interface itself giving rise to phenomena that are qualitatively different with regard to the gas-transfer. Also, at moderate wind speed, microbreaking starts changing the structure of turbulence at the air-water interface and affecting gas-transfer rate. Moreover, in lakes, oceans and wetlands, when cooling occurs, turbulence at the air-water interface is induced by heat loss but usually largest heat loss are related to evaporation due to high winds. So at a wind sheared interface, turbulence is also due to convection motions in the water volume and it is termed *buoyant-convective-induced turbulence* (Banerjee and MacIntyre, 2004; Jirka et al., 2010). Finally, at high winds, wave breaking with air entrainment significantly affects gas-transfer;
- *combined wind-stream turbulence* when both air flow and water flow together exist and bed shear and interfacial shear are simultaneously present in the water layer. This conditions typically hold in large rivers and estuaries.

Section 1.4 will discuss in detail *bottom-shear generated turbulence*, proposing a dimensional analysis of the gas-transfer process and presenting a review of experimental results, conceptual models and numerical simulations available in the literature to gain insight into this process and to estimate its rate.

6.4   GAS-TRANSFER AT AN UNSHEARED INTERFACE

In open channel flows, such as streams and rivers, the surface turbulence is mainly generated at the bottom boundary of the streams or in the shear layer between subsurface currents flowing at different velocities. In both cases turbulence structures could then impinge on the air-water interface producing effects as boils that can be easily seen at the surface of rivers and that can affect the gas-transfer process. Thus, since surface turbulence is generated elsewhere, this case can be termed as *bottom-shear generated turbulence* (Nakayama, 2000) and the air-water interface is accordingly called *unsheared interface* or *shear-free interface* (Banerjee and MacIntyre, 2004; Magnaudet and Calmet, 2006).

The structure of fluid motions in the bottom-shear generated turbulence and their effect on the region near the air-water interface have been experimentally investigated in a number of studies, which used different techniques such as laser-Doppler velocimetry and optical probe, digital particle image velocimetry (DPIV) and video cameras, laser-induced fluorescence (LIF) (Rashidi and Banerjee, 1988; Komori *et al.*, 1989; Kumar *et al.*, 1998; Herlina and Jirka, 2004; Jirka *et al.*, 2010). First, Rashidi and Banerjee (1988) using high speed videos observed periodic ejection of intensely turbulent fluid with low streamwise momentum from the wall into the relatively quiescent bulk fluid. Between the bursts and the interface, a high speed region with steep velocity gradient developed. Hence, the motion of bursts toward the interface was forced to slow down and then to turn back to the wall, giving rise to characteristics rolling structures, which rotate clockwise if the flow was viewed as going from left to right (Rashidi and Banerjee, 1988). Komori *et al.*, (1989) observed that large-scale turbulent eddies ejected by bursting from the buffer region of the bottom moved upward to the interfacial region and arrived at the free-surface. Also, they successfully related these bursting motions to the gas-transfer process. In a more detailed study, three types of persistent coherent structures were observed near the air-water interface (Fig. 6.5), that is upwellings, downwellings or downdrafts, and spiral eddies (Kumar *et al.*, 1998). Upwellings, also called splats, were produced by large active structures (bursts) originated in the sheared region at the channel bottom and impinged on the free surface. They moved with it for some time, but the vanishing of the vertical velocity at the surface then forced upwellings to stretch in the horizontal and roll up resulting in the creation of downwelling structures, also termed antisplats, when two neighbouring upwellings collided, which moved back to the flow. At the edges of the upwellings were seen to be generated spiral eddies, typically attached to the free surface. These eddies often merged if rotating in the same direction, and form pairs if rotating in the opposite directions. Spiral eddies persisted for long period and they were finally destructed by merging, by new upwellings impinging on them upward, and by viscous dissipation (Kumar *et al.*, 1998). It should be noted that in

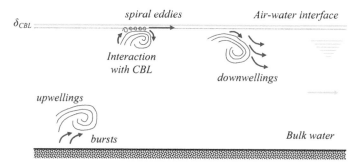

Figure 6.5. Coherent structures involved in the gas-transfer process.

buoyant-convective-induced turbulence sinking of cold plumes and the upwards movement of the warmer water is the dominant mechanism transport (Jirka *et al.*, 2010).

The first step of the proposed study of gas-transfer process at an unsheared interface is to develop a proper dimensional analysis which would consider all the parameters that are likely to be involved in that process. These parameters should reflect both the fluid and gas properties, and the hydrodynamics of the flow. As presented in Subsection 1.4.1, dimensional analysis can provide a general relation for the gas-transfer rate, where this rate is related with parameters describing the hydrodynamics and environmental conditions of the mean flow.

However, Section 1.3 already pointed out that turbulence close to the air-interface where the concentration boundary layer is embedded is a key factor affecting the transport rate across the interface. Also, the aforementioned experimental studies revealed almost at a basic level of understanding the mechanism of interaction between turbulent coherent structures and air-water interface. Hence, several conceptual models were proposed in the literature to relate gas-transfer rate with hydrodynamics parameters representing the turbulence conditions at the interface. These parameters can represent both *global* and *local* properties of turbulence. These models are reviewed and discussed in Subsection 1.4.2 starting from the classical two-film model to the latest models proposed in the literature to account near-surface turbulence characteristics. More recently, numerical methods have been applied to investigate gas-transfer process to overcome difficulties still existing in the experimental techniques. Particularly, both Direct Numerical Simulation (DNS) and Large Eddy Simulation (LES) were applied allowing detailed determination of velocity and concentration fields very near the air-water interface, as described and discussed in Subsection 1.4.3. Finally, Subsection 1.4.4 compares results from both conceptual models and numerical simulations with available experimental data.

6.4.1 Dimensional analysis

Dimensional analysis usually starts with the selection of the parameters affecting the process being modelled. Considering now a channel with wide rectangular section so that the hydraulic radius $R_h \approx 4h$ and the shape factor $\psi = h/W$ is always very low. Gas-transfer process should be affected by the following parameters (Gualtieri *et al.*, 2002; Gualtieri *et al.*, 2006):

- natural constants and fluid properties, such as the gravitational acceleration constant g, the water density ρ, the water dynamic viscosity μ, and the water surface tension T_s;
- gas exchanged properties, such as the molecular diffusion coefficient D_m;
- flow properties, such as the mean depth h, the mean streamflow velocity u, the energy line slope J_e, the channel bed slope J_b and the roughness coefficient of Colebrook-White equation ε_{cw}.

Note that some literature empirical equation contain other parameters, such as Froude number Fr, friction velocity u^*, water discharge Q and kinetic turbulent energy dissipation rate per unit mass ε, which can be all expressed through the listed parameters. Water density ρ and the water dynamic viscosity μ can be combined to form water kinematic viscosity v, that is $v = \mu/\rho$. Also, the water surface tension, T_s, was transformed into a kinematic parameter as $\tau_s = T_s/\rho$. Hence, the process can be considered as kinematic. Thus, it holds:

$$K_L = f_1[D_m, \tau_s, v, g, h, u, J_e, J_b, \varepsilon_{cw}] \tag{6.28}$$

Assuming as fundamental quantities the water mean depth h and the molecular diffusivity D_m, a proper dimensional analysis leads to:

$$Sh = f_2[Sc, We, Re, Fr, J_e, J_b, S] \qquad (6.29a)$$

where Sh, Sc, We, Re, and Fr are the classical Sherwood number, Schmidt number, Weber number, Reynolds number, and Froude number, respectively. Sh, Re and We are defined as:

$$Sh = \frac{K_L \cdot h}{D_m} \qquad (6.30)$$

$$Re = \frac{u \cdot 4h}{\nu} \qquad (6.31)$$

$$We = \frac{u^2 \cdot h}{\tau_s} \qquad (6.32)$$

Finally, the relative roughness S is:

$$S = \frac{\varepsilon_{cw}}{4h} \qquad (6.33)$$

Equation (6.29a) provides the dimensionless gas-transfer rate in an open channel for liquid-controlled chemicals. This equation holds whatever is the gas involved in the gas-transfer. Also, in Equation (6.29a) the temperature influence is directly accounted for through the temperature dependent parameters, such as Sc, We and Re. This represents an advantage respect to the common application of a temperature corrective coefficient, such as the classical θ of Van't Hoff-Arrhenius equation (Chapra, 1997).

Equation (6.29a) can be modified. First of all, assuming that uniform flow conditions hold, the energy line slope J_e and the channel bed slope J_b are equal, that is $J_e = J_b$. Second, the Froude number in Equation (6.29a) could be discarded since it can be expressed using S, Re and J_e (Gualtieri *et al.*, 2002). In fact, classical Darcy-Weisbach equation states:

$$J_e = \frac{f}{4h}\frac{u^2}{2g} = \frac{f}{8}Fr^2 \qquad (6.34)$$

where f is the friction factor, that in a turbulent flow is $f = f(Re, S)$. Hence, Equation (6.34) yields:

$$J_e = \frac{f(Re, S)}{8}Fr^2 \qquad (6.35)$$

where Fr, S, Re and $J_e = J_b$ are correlated. Therefore, the Froude number can be expressed as:

$$Fr = \sqrt{\frac{8J_b}{f(Re, S)}} = Fr(J_b, Re, S) \qquad (6.36)$$

and Equation (6.29a) yields:

$$Sh = f_3[Sc, We, Re, J_b, S] \tag{6.29b}$$

Third, as a first approximation, the influence of *We* could be discarded. Thus, Equation (6.29b) yields:

$$Sh = f_4[Sc, Re, J_b, S] \tag{6.29c}$$

where Sherwood number is affected by only *Sc*, *Re*, J_b and *S*. Note that f_4 function in Equation (6.29c) must be defined using experimental data.

6.4.2 Conceptual models of gas-transfer process at an unsheared interface

In Section 1.3 it was already pointed out that the concentration boundary sublayer interacts with turbulent motions close to the air-water interface. Hence, CBL characteristics are expected to usually change with the space and the time depending on turbulence parameters. However, key point is to relate gas-transfer rate with hydrodynamics parameters representing the turbulence conditions at the interface.

The earliest and simplest model for K_L is the Lewis-Whitman model. It states that a stagnant film exists very near the interface. The gas moves across the film only by molecular diffusion. From the discussion in Section 1.3 it may be derived that the concept of the stagnant film implies that the concentration boundary sublayer exhibits a kind of time and space-averaged thickness δ_{CBL}, that may be considered as having a constant value. Due to the steady uniform laminar flow in the film region, there is a linear concentration profile within the CBL and the gas flux J_{g-t} is:

$$J_{g-t} = -D_m \cdot \left. \frac{dC}{dz} \right|_{z=0} = -D_m \cdot \frac{C_{sat} - C_w}{\delta_{CBL}} \tag{6.37}$$

where C_w is gas concentration at $z = \delta_{CBL}$. Equations (6.19) and (6.37) yield:

$$K_L = \frac{D_m}{\delta_{CBL}} \tag{6.38}$$

Therefore, in this model, K_L is linearly proportional to D_m, as compared to the square-root dependence obtained in the stagnant case. This is due to the different concentration distribution holding in the two cases. However, Lewis-Whitman model does not provide any physical insight about δ_{CBL} prediction. Furthermore, since CBL characteristics are changing reflecting system hydrodynamics, the basic assumption of this model cannot mostly properly capture the physical mechanism that controls the concentration boundary layer thickness. Nevertheless, some models were recently proposed to provide an estimation of the thickness δ_{CBL} of the CBL to introduce in Equation (6.38). First, Atkinson *et al.* (1995) has considered two approaches; the first one compares molecular and turbulent diffusivities, while the second one relates δ_{CBL} to the smallest eddies in the flow according to the Kolmogorov microlength scale η (Atkinson *et al.*, 1995). This is the smallest scale of turbulent flow, at which turbulent kinetic energy is converted to heat. From dimensional analysis Kolmogorov microlength η can be defined as:

$$\eta \propto \frac{\nu^{3/4}}{\varepsilon^{1/4}} \tag{6.39}$$

The first approach considers that the vertical profile of turbulent diffusivity can be estimated using Elder's analysis (Elder, 1959). As the shear stress τ is a linear function of depth in open channel flow $\tau = \tau_b \cdot (z/h)$, a velocity profile must be assumed to estimate the gradient velocity. Using a logarithmic profile and assuming that the thickness of diffusive layer δ_{CBL} is the depth below the air-water interface where molecular viscosity ν is comparable with eddy viscosity ν_t, after some simplifications, it can be shown that:

$$\delta_{CBL} \approx c_1 \cdot \frac{\nu}{u^*} \cdot Sc^{-1/3} \tag{6.40}$$

where c_1 is a constant that can be set equal to $c_1 = 10$. From Equations (6.39) and (6.40), it yields:

$$K_L = \frac{D_m}{10 \cdot \nu} \cdot u^* \cdot Sc^{1/3} \tag{6.41}$$

The second approach proposed by Atkinson relates δ_{CBL} with the smallest eddies in the flow. Starting from Kolmogorov microlength scale η, as defined by Equation (6.39), after some algebra, δ_{CBL} can be estimated as (Atkinson et al., 1995):

$$\delta_{CBL} \cong c_2 \left(\frac{\nu^3 \cdot h}{u^3} \right)^{1/4} \tag{6.42}$$

where c_3 is a numerical constant, that is $c_2 = 2$. From Equations (6.38) and (42), K_L can be obtained as:

$$K_L = \frac{D_m}{2} \left(\frac{u^3}{\nu^3 \cdot h} \right)^{1/4} \tag{6.43}$$

Gualtieri and Gualtieri, comparing the laminar boundary sublayer at the air-water interface with the bottom classic laminar sublayer, proposed another model to estimate the thickness δ_{CBL} of the concentration boundary layer (Gualtieri and Gualtieri, 2004). The bottom sublayer lies on a solid boundary, which has an infinite surface tension. On the other hand, the VBL is below the air-water interface which can be considered, due to its surface tension, as a semi-solid boundary. To follow this analogy a first velocity distribution in the VBL can be defined starting from the velocity distribution in the laminar sublayer near the bottom, which is known. Furthermore, in the laminar sublayer near the bottom, introducing the expression of bottom shear stress $\tau_b = \rho \cdot u^{*2}$ into the Newtonian expression for τ_b, a linear velocity distribution can be derived for the bottom sublayer. Applying the analogy between the air-water interface and the bottom, a second velocity distribution can be derived for the VBL below the water surface. Comparing these velocity distributions in the VBL at the air-water interface, its thickness δ_{VBL} can be derived. Finally, from Equations (6.27) and (6.38), gas-transfer coefficient K_L can be obtained as:

$$K_L = (D_m)^{2/3} \left(\frac{g \cdot J_b}{2 \cdot v \cdot Re_{g-t}} \right)^{1/3} \tag{6.44}$$

where Re_{g-t} is a specific gas-transfer Reynolds number, which from the proposed approach should be $<< 25$ and should be calibrated from experimental data. Analysis of a large amount experimental field data collected in stream and rives allowed to calibrate Re_{g-t} as $Re_{g-t} = 0.750$ (Gualtieri and Gualtieri, 2004). Equation (6.44) can be modified to derive an equation comparable with *LE* and *SE* model, which is:

$$K_L^* = c_3 Sc^{-2/3} Re^{*-1/3} \tag{6.45}$$

where the exponent of Re^* is intermediate between those from *LE* and *SE* models. Also, it should be noted that the exponent of Sc in Equation (6.45), that is $-2/3$, is that expected for a solid boundary or for a film-covered or highly contaminated water surface (Jähne and Haußecker, 1998; Banerjee and MacIntyre, 2004; Hasegawa and Kasagi, 2008).

The models based upon the concept of *surface-renewal* assume that the fluid elements inside the CBL are periodically *refreshed* by turbulent eddies acknowledging the central role played by turbulence. The mechanism of *surface-renewal* is related to turbulent eddies that periodically bring liquid parcels from the bulk liquid to the air-water interface. During the short period of time spent at the interface, the liquid elements are exposed to the atmosphere and subjected to the gas-transfer process by molecular diffusion. After that, turbulent motions move again the water parcels down to the bulk liquid. The described cycle is a *surface-renewal event* and its frequency is a function of the turbulent characteristics of the flow. In this case, the concentration boundary sublayer is allowed to grow from zero depth until at some point the turbulence suddenly replaces the water parcel in the CBL, that is a renewal event occurs, and the sublayer growth starts over from the beginning. The CBL thickness is assumed larger than the depth that can be penetrated by molecular diffusion during the time of exposure to the atmosphere.

Key-point of *surface-renewal* models is the definition of the time between two renewal events. The first model based upon the outlined concepts is the *penetration model* by Higbie. The Higbie model assumes that all the liquid elements have the same time t_r of exposure at the air-water interface. The time t_r is often called *contact time*, *surface age* or *renewal time*. The governing transport equation and the initial condition and boundary conditions are the same as in the stagnant case, that is Equations (6.14), (6.15) and (6.16). Hence, the solution for the gas-transfer flux is Equation (6.18b), but it is valid during the time between two renewal events. The average flux of gas during one cycle is (Thibodeaux, 1997):

$$J_{g-t} = -(C_{sat} - C_w)\sqrt{\frac{4D_m}{\pi t_r}} \tag{6.46}$$

and the gas-transfer coefficient is:

$$K_L = \sqrt{\frac{4D_m}{\pi t_r}} \tag{6.47}$$

The basic assumption proposed by Higbie about the same exposure time of the water parcels was improved by Danckwerts, who introduced a random replacement time function, termed *surface-age distribution function*, which was more typical of what might be expected from a turbulent fluid. This function represents the probability that a parcel is exposed for a time t before being replaced by a new water element from the bulk fluid (Thibodeaux, 1997); this function if t_{r-avg} is the average renewal time is:

$$\phi = r \exp(-rt) \tag{6.48}$$

where $r = 1/t_{r-avg}$ is the fractional *renewal rate*; thus, water parcels can remain at the surface for variable times that may be any value from zero to infinity. Averaging $\pi/4$ term disappears and the average gas flux is:

$$J_{g-t} = -(C_{sat} - C_w)\sqrt{D_m r} \tag{6.49}$$

and the gas-transfer coefficient is:

$$K_L = \sqrt{D_m r} \tag{6.50}$$

which indicates that the gas-transfer rate is proportional to the frequency at which a renewal event occurs. Both Higbie and Danckwerts models have the weakness that their key parameter, that is t_r and r, respectively, is neither known nor immediately related to the turbulence near the air-water interface. As previously outlined, Komori et al. (1989) suggested that surface renewal eddies were originated in bursting phenomena occurring in the buffer region of the wall. Low speed fluid was ejected toward the interface from a wall burst, the fluid moved up to the surface to form a *surface-renewal patch*, and a downdraft developed after the interaction. They successfully correlated the frequency of both surface renewal and bursting and obtained that gas-transfer rate was proportional to the square-root of the surface renewal frequency confirming Equation (6.50) (Komori et al., 1989). However, further studies pointed out a more complex interaction between the free surface and the ejections from sheared region near the channel bed. Hence, the measurements of *surface-renewal eddies* are difficult to correlate with K_L, because the investigators themselves have to define what constitutes a *surface-renewal eddy* (Tamburrino and Gulliver, 2002).

Despite these difficulties, the *renewal rate r* may expected to be a characteristics of turbulent eddies and further research efforts were addressed to relate it with turbulence. We have already recalled that a characteristic feature of turbulent flow is the presence of a wide range of eddy sizes, ranging from the flow domain, i.e. integral scale eddies, to smaller sizes, i.e. Kolmogorov scale eddies (Pope, 2000). It is a common statement that the large eddies transfer their energy to the smaller ones. First, this transfer is efficient and very little kinetic energy is lost (Pope, 2000). When the eddies become small enough, in the order of Kolmogorov scale in size, viscosity takes over and the energy is damped out and converted into heat. This process is usually described as a turbulence cascade. Turbulent energy production and dissipation are almost in equilibrium in the intermediate region of a stream, whereas near the free surface dissipation is predominant (Nezu and Nakagawa, 1993; Nakayama, 2000). The dissipation rate of turbulent kinetic energy ε can be measured directly or calculated. Experimental data demonstrated that ε can be scaled as (Moog and Jirka, 1999):

$$\varepsilon \propto \frac{u^{*3}}{h} \tag{6.51}$$

As previously outlined, the scale at which turbulent kinetic energy is converted to heat is the Kolmogorov microlength scale η, which estimates the smallest turbulent eddies. This stage is characterized by an eddy Reynolds number approximately equal to 1, if the eddy Reynolds number is defined using the characteristic length and the velocity of smallest eddies. This reflects the idea that at these smallest scales of motion, the inertial strength of the eddy is approximately equal to its viscous transport strength, i.e. eddy viscosity $v_t = v$ (Pope, 2000).

At this point, two extreme estimates for r can be applied: one for the case that the concentration boundary layer is renewed by integral-scale eddies, that is called the *large-eddy* estimate, and another one for the case that the concentration boundary layer is renewed by Kolmogorov-scale eddies, that is called the *small-eddy* estimate (Moog and Jirka, 1999). In both cases, turbulence is assumed to be homogeneous and isotropic and this hypothesis is of course critical in the interfacial region.

In the first estimate, it could be assumed that the surface layer could be divided into a series of rotational cells having diameter and velocity proportional to h and u_{rms}, which is the root-mean-square value of turbulent velocity fluctuations, respectively. Also, the velocity of cells could be scaled by u^*. Thus, r parameter can be considered as $r \propto u^*/h$. Inserting this into Equation (6.50) and non-dimensionalizing, the *large-eddy model (LE)* by Fortescue and Pearson (1967) states that (Moog and Jirka, 1999):

$$K_L^* = \frac{K_L}{u_{rms}} \approx \frac{K_L}{u^*} \propto Sc^{-1/2} Re^{*-1/2} \tag{6.52}$$

where K_L^* is the dimensionless gas-transfer rate and Re^* is the shear Reynolds number, which is defined as $Re^* = u^* \cdot h/v$.

In the second estimate, considering the attenuation of vertical fluctuations due to the free surface, it could be assumed that smaller eddies may contribute to surface renewal (Moog & Jirka, 1999). They are dissipated by viscosity. Integrating a roll cell model over a wave number spectrum containing an inertial sub-range, it follows:

$$K_L \propto Sc^{-1/2} \cdot (\varepsilon \cdot v)^{1/4} \tag{6.53}$$

This is the *small-eddy model (SE)* by Banerjee *et al.* (1968), where energy dissipation may be also enhanced by many factors, such as wind shear, wave breaking, natural convection, rain (Banerjee and McIntyre, 2004). This model, considering Equation (6.43), gives (Moog & Jirka, 1999):

$$K_L^* \approx \frac{K_L}{u^*} \propto Sc^{-1/2} Re^{*-1/4} \tag{6.54}$$

Comparison between Equations (6.52) and (6.54) shows that the large-eddy and small-eddy models differ only by the Reynolds number exponent, so that these models have the general form:

$$K_L^* \approx \frac{K_L}{u^*} \propto Sc^{-1/2} Re^{*n} \tag{6.55}$$

where $n = -1/2$ holds for the *large-eddy* model and $n = -1/4$ holds for the *small-eddy* model.

A different expression for both *large-eddy* and *small-eddy* models can be derived considering that:

$$u^* = u \left(\frac{f}{8}\right)^{1/2} \tag{6.56}$$

where f is the Darcy-Weisbach friction coefficient. If gas-transfer rate K_L is non-dimensionalized using the mean streamflow velocity u, those models become:

$$\frac{K_L}{u} \propto Sc^{-1/2} Re^{-1/2} \left(\frac{f}{8}\right)^{1/4} \tag{6.57}$$

$$\frac{K_L}{u} \propto Sc^{-1/2} Re^{-1/4} \left(\frac{f}{8}\right)^{3/8} \tag{6.58}$$

for *large-eddy* and *small-eddy* models, respectively. Notably, in Equations (6.57) and (6.58) the boundary type of the flow are directly taken into account by the friction coefficient. Interestingly, Theofanous *et al.* (1976) suggested that there is a smooth transition between low Re values, where large-scale eddies control gas transfer, and high Re values, where small-scale eddies dominate. The transition occurs at $Re^* = 500$. In order to compare *large-eddy* and *small-eddy* models, Moog and Jirka (1999) carried out experimental works in open channel flow with shear Reynolds number Re^* from 350 to 4200. First, they observed that measurements in stirred tanks supported *small-eddy* model. Second, from their experimental data, they obtained $n = -0.29$, which supported *small-eddy* model and yielded (Moog and Jirka, 1999):

$$K_L^* = 0.161 Sc^{-1/2} Re^{*-1/4} \tag{6.59}$$

Further experimental works supported the *small-eddy* model (Chao *et al.*, 2007). However, observations at low Reynolds number suggested that large coherent structures such as bursts and upwellings are responsible for interfacial transport. Hence, to solve this conflict they argued that both scales would be involved in gas-transfer process in a framework termed *chain saw* model (Moog and Jirka, 1999). Large scale motions transport turbulent energy to the interface, creating active zones or *patches* for the gas-transfer. Within these zones, the transfer is controlled by small eddies at a rate which is related to near-surface turbulent dissipation rate. Moreover, the variation in active area decreases with the increasing Re^*, leading to the successful scaling of *small-eddy* model at higher Re^* and confirming Theofanous suggestion (Moog and Jirka, 1999).

Since both *large-eddy* and *small-eddy* models are based on a *global* property of turbulence, the next theoretical step in the literature was to relate gas-transfer process directly to a *local* property of turbulence, that is the turbulence characteristics near the air-water interface. Hanratty (1991) argued that, since the CBL is very thin, the derivative in z-direction is much larger that in the other directions. Hence, using a coordinate system embedded on the interface, the advection-diffusion equation for the gas in a turbulent flow near a free surface may be simplified as:

$$\frac{\partial C}{\partial t} + w' \frac{\partial \bar{C}}{\partial z} = D_m \frac{\partial^2 C}{\partial z^2} \tag{6.60}$$

where C and $\overline{C}$ are instantaneous concentration and its temporal mean, respectively, and w' is the fluctuating velocity normal to the interface. A series-expansion and order-of-magnitude analysis near the interface yielded the following relation for w' (McCready et al., 1986):

$$w' \approx \frac{\partial w'}{\partial z} z \tag{6.61}$$

where the vertical velocity gradient very near to the air-water interface is also called β parameter. This gradient is changing with the time and the distance parallel to the interface and is function of flow turbulence. Equation (6.60) highlights the importance of β parameter for the gas-transfer process. Note that the vertical velocity gradient $\partial w' / \partial z$ is unequal to zero when at the water surface 2D continuity equation in a control volume that moves vertically is not satisfied, that is:

$$\frac{\partial u'}{\partial x} + \frac{\partial v'}{\partial y} \neq 0 \tag{6.62}$$

where u' and v' are the fluctuating velocity in the streamwise direction x and in the spanwise direction y, respectively, which are both tangential to the interface. Indeed, on a free water surface tangential velocity fluctuations are possible. Hence, from 3D continuity equation, the vertical velocity gradient may be derived as:

$$\beta = \frac{\partial w'}{\partial z} = -\left(\frac{\partial u'}{\partial x} + \frac{\partial v'}{\partial y} \right) \tag{6.63}$$

where the term in brackets is termed *surface divergence*. The physical meaning of Equation (6.62) or (6.63) is that there are convergence or divergence zone at the water surface, that is surface fluid elements are dilated or contracted due to turbulent motions that bring bulk fluids to the interface (Jähne and Haußecker, 1998; Banerjee and McIntyre, 2004). Thus, if free-surface turbulence can be measured or estimated, Equation (6.63) provides the value of β parameter which is related to gas-transfer rate.

The surface divergence cannot be predicted without a theory. Hence the *blocking theory* by Hunt and Graham (1978) was used by Banerjee (1990) to relate surface divergence to the far-field turbulence characteristics when they are homogeneous and isotropic. Using this approach, gas-transfer coefficient for an unsheared interface and high Sc was derived as (Banerjee, 1990):

$$K_L = c_4 u Sc^{-1/2} Re^{-1/2} [0.3(2.83 Re^{3/4} - 2.14 Re^{2/3})]^{1/4} \tag{6.64}$$

which is also termed as *surface divergence (SD)* model. The quantity within the brackets is the square of the nondimensional surface divergence. This model applies also to a rigid slip surface because Hunt and Graham theory holds for this case. However, since the air-water interface is mobile and can deform following the motions on the liquid side, the surface divergence for a deformable interface may be expected to be less and the constant c_4 using experimental data was equated to 0.20 (Banerjee and MacIntyre, 2004). Equation (6.64) is asymptotic to $Re^{-1/2}$ at small Re and almost to $Re^{-1/4}$ at large Re, which is in line with *LE* and *SE* models and confirms Theofanous suggestion. Also, in Equation (6.64) the friction factor is not present.

Following surface divergence approach, other researchers proposed predictive equations different from Equation (6.64). Tamburrino and Gulliver (2002) measured free surface turbulence in a fully developed, open channel flow and estimated β parameter. They argued that high values of β were not a primary result of large upwellings moving to the air-water interface. The spatial scales of β were more closely related to the high velocity gradients of surface vorticity, which can originated by large upwelling, but were not previously identified as source of surface renewal (Tamburrino and Gulliver, 2002). Using previous experimental data, they finally proposed that (Tamburrino and Gulliver, 2002):

$$K_L^* \approx \frac{K_L}{u^*} \propto 0.24 Sc^{-1/2} S_{\beta max}^{*-1/2} \tag{6.65}$$

where $S_{\beta max}^*$ is dimensionless maximum value of the β spectrum.

Law and Khoo (2002) measured β parameter for the two different near-surface turbulence conditions: one where turbulence was generated from beneath the interface; the other was wind-shear induced turbulence. They found that a general correlation can be obtained relating the dimensionless β to the scalar transport rate (Law and Khoo, 2002):

$$K_L^+ = \frac{K_L}{w_{\beta-rms}} = 0.22 Sc^{-1/2} \beta_{rms}^+ \tag{6.66}$$

where w_β is the vertical velocity at which the vertical velocity with respect to the interface w_r departs from the linear behavior and the subscript *rms* means its root mean square. Finally, if β_{rms} is the root mean square of the vertical velocity gradient at the interface, the dimensionless β is:

$$\beta_{rms}^+ = \frac{\beta_{rms} \nu}{w_{\beta-rms}^2} \tag{6.67}$$

Later, Sugihara and Tsumori (2005) carried out experiments in oscillating-grid turbulent flows to investigate the relation between gas-transfer rate and turbulence characteristics at the air-water interface. They obtained the following relation (Suhihara and Tsumori, 2005):

$$K_L = 0.30 (D_m \beta_{rms})^{1/2} \tag{6.68}$$

which may be rewritten as (Sugihara and Tsumori, 2005):

$$\frac{K_L}{k^{1/2}} = 0.18 Sc^{-1/2} Re_{k\varepsilon}^{-1/4} \tag{6.69}$$

where k is turbulent kinetic energy and $Re_{k\varepsilon}$ is a turbulent Reynolds number defined with k and ε parameters, that is $Re_{k\varepsilon} = k^2/\varepsilon\,\nu$. Note that the exponent $-1/4$ of Reynolds number would support the *small-eddy* model, Equation (6.55).

Xu et al. used an innovative particle image velocimetry-based measurement method to investigate interfacial turbulence and to assess β parameter (Xu et al., 2006). Several distinctly different flow conditions, including turbulence induced by wind shear from above, turbulence generated from the bottom and a combination of simultaneously contributing conditions from above and beneath the interface, were investigated. They suggested a general predictive equation to correlate the gas-transfer rate with the surface divergence (Xu *et al.*, 2006):

$$K_L = 0.20Sc^{-1/2} \, (\beta_{rms} v)^{1/2} \tag{6.70}$$

Recently, Janzen *et al.* (2010) used simultaneous measurements of velocity and concentration fields at the air-water interface to investigate the interaction between gas-transfer and turbulence generated in an oscillating-grid tank. They demonstrated that the application of the surface divergence concept provided good predictions for the gas-transfer rate also for low values of β. Furthermore, they also tested different conceptual models, such as the Lewis-Whitman model, the surface renewal model and the large-eddy model pointing out that the Lewis-Whitman model provided reasonable results.

6.4.3   Numerical simulation of gas-transfer process at an unsheared interface

Despite the rapid evolution of investigation techniques, especially in the last two decades, experimental methods cannot yet provide all the data required to a complete knowledge of gas-transfer process. In fact, to understand gas-transfer process, it is needed to perform simultaneous analyses of concentration and velocity fluctuations in the interfacial region in terms of both statistics and turbulent structures. If laser-induced fluorescence (LIF) technique allows to reveal the concentration distribution within the CBL (Münsterer and Jähne, 1998; Herlina and Jirka, 2004; Janzen *et al.* 2010; Jirka *et al.*, 2010) and microprobes are capable to follow concentration fluctuations (Chu and Jirka, 1992), these techniques still have difficulties in resolving the uppermost layer of the flow. On the other hand, particle image velocimetry provides an adequate picture of turbulence characteristics near the air-water interface (Kumar *et al.*, 1998; Xu *et al.*, 2006). Thus, the application of high-resolution numerical simulations has been increasingly proposed to provide a detailed and precise determination of velocity and concentration fields very near to the air-water interface.

The first numerical method applied was the Direct Numerical Simulation (DNS) of time-dependent three-dimensional Navier-Stokes equations, which have often been used in the field of physics and engineering. Several studies on gas-transfer process based on DNS are available in literature (Komori *et al.*, 1993; Nagaosa, 1999; Handler *et al.*, 1999; Shen *et al.*, 2001; Kermani and Shen, 2009; Kermani *et al.*, 2011). These studies sometimes confirmed findings of previous experimental or theoretical works but often provided detailed, novel insights on three-dimensional structures responsible for surface renewal and its net contribution to the dynamics of free-surface turbulence. Numerical results pointed out that vertical motions were restrained in the interfacial region by the damping effect, and the turbulent kinetic energy associated with them was redistributed mainly to the spanwise motions through the pressure fluctuation (Komori *et al.*, 1993). Also, they confirmed that large-eddies generated by bursts in the wall region were advected up to the free-surface producing *surface-renewal events*. (Komori *et al.*, 1993; Nagaosa, 1999). To be more in details, two types of vortex tubes were observed below the free surface (Nagaosa, 1999). The first type were elongated, near-horizontal, quasi-streamwise vortices, parallel to the main stream and the free surface. The interactions between these vortices and the air-water interfaces produced splats and antisplats at the free surface. The balance of intercomponent energy transfer between the spanwise and surface–normal direction via the pressure-strain effect was determined by the splats and antisplats, which furthermore are responsible for *surface renewal events* at the free surface. The second type were the surface-attached vortices, which were perpendicular to the interface and were established by connections of quasi-streamwise vortices to the free surface. The surface-attached vortices did not produce splats and antisplats at the interface. Hence, the direct contribution of the surface-attached vortices to the dynamics of the free surface turbulence, for example, the intercomponent energy transfer or turbulent gas-transfer across the free surface, is believed to be very small (Nagaosa, 1999). Overall, DNS results confirmed the close link between bursting phenomena from the bottom region, on one hand, and interfacial turbulence and gas-transfer process, on the other. Also, studies based on DNS provided detailed statistics of the dynamics and concentration fields and of the structure interfacial turbulence which are still beyond the capabilities of laboratory experiments.

However, the main limitation of these DNS studies arises from their low Reynolds number. In fact, the aforementioned studies were carried out for a shear Reynolds number $Re*$ and Schmidt or Prandtl number Sc or Pr ranging from 150 to 180 and 1 to 5, respectively (Komori et al., 1993; Nagaosa, 1999; Handler et al., 1999; Shen et al., 2001). Hence, Schmidt number were very far from the typical values of substances being involved in the gas-transfer process at the free-surface of streams and rivers. Also, in that range of $Re*$ the free surface lies within the logarithmic layer of the mean velocity profile. Therefore, the turbulence *seen* by the free surface is strongly anisotropic and interacts directly with the dynamics of the bed region (Magnaudet and Calmet, 2006). This makes the near-surface velocity and concentration fields observed in these DNS studies quite specific for low Reynolds number wall-bounded shear flows and do not allows to extend some of their conclusions to the environmental flows. Therefore, it should be very useful to perform numerical simulation with higher Reynolds number to investigate instantaneous and statistical structure of velocity and concentration fields in free-surface flows where turbulence is closer to isotropy and almost independent from the way it is generated in the bed region (Magnaudet and Calmet, 2006). However, it is well known that in the DNS the number of grid points required to capture the smallest scales grows with the Reynolds number and the Schmidt number like $Sc^3 Re^{9/4}$ (Pope, 2000). To overcome this limitation, Large Eddy Simulation (LES) has been recently applied to study the structure of interfacial velocity and concentration fields and the mechanism of gas-transfer process assuming that the small-scales do not prevail in this process (Calmet and Magnaudet, 1998; Magnaudet and Calmet, 2006).

Calmet and Magnaudet (1998) first applied LES to investigate gas-transfer across a flat, shear-free interface for a shear Reynolds number $Re*$ of 1280 and two values of Schmidt number, that is $Sc = 1$ and $Sc = 200$. They demonstrated that the concentration boundary sublayer is the related to the viscous boundary sublayer as:

$$\delta_{CBL} = \frac{\delta_{VBL}^{1/2}}{Sc^{1/2}} \tag{6.71}$$

which confirms results later obtained by Lorke et al. for the sediment-water interface (Lorke et al., 2003). Moreover, the analysis of vertical velocity and concentration fluctuation w' and C' revealed that the dynamics of the concentration field was closely correlated with large-scale structures present near the air-water interface confirming that the driving mechanism of gas-transfer is the *surface-renewal* by the structures coming from the bottom wall (Calmet and Magnaudet, 1998). In fact, large-structures that reach the interface ($w' > 0$), that is upwellings, or move downward from it ($w' < 0$), that is downwellings, carried low ($C' < 0$) and high ($C' > 0$) concentration, respectively. Also, the analysis of horizontal motions confirmed the role of surface divergence β, that is $\beta > 0$ corresponded to upwellings motions and compression of the interface and $\beta < 0$ corresponded to downwellings motions and dilatation of the interface (Calmet and Magnaudet, 1998). Finally, LES results were used to estimate gas-transfer rate. Comparison with both *large-eddy* and *small-eddy* models, which assume homogeneous and isotropic turbulence, could not allow to prefer one model on the other one (Calmet and Magnaudet, 1998).

Later, Large Eddy Simulation was again applied by Magnaudet and Calmet (2006) investigate gas-transfer across a flat unsheared interface of a turbulent channel flow for a shear Reynolds number $Re*$ of 1280 over a wide range of Schmidt number, that is from 1 to 200. LES results provided a detailed picture of the structure of the uppermost layers below the air-water interface. In fact, they identified an inner concentration sublayer, where mean concentration profile was linear, and an outer concentration sublayer, where root-mean-square concentration fluctuation grew up from zero at the surface to a maximum at the outer edge (Magnaudet and Calmet, 2006). This difference is analogous to that previously underlined between diffusive and concentration sublayers. Also, the thickness of the inner and the outer CBL was proportional to $Sc^{-1/2} Re^{-3/4}$ and $Sc^{-1/2} Re^{-1/4}$, respectively. Hence, the former corresponds to the Batchelor microscale. Notably, Lorke *et al.* (2003) demonstrated that the scaling of the diffusive sublayer height with the Batchelor microscale provided an adequate description of the sediment-water exchange of oxygen observed in the field (Lorke *et al.*, 2003). When plotted against the dimensionless distance to $Sc^{1/2}zu/v$, the near-surface profiles of the normalized concentration variance and of all terms contributing to its budget were shown to be independent of the Schmidt number (Magnaudet and Calmet, 2006). Moreover, LES results pointed out that the region where w' grows linearly with the distance from the interface, that is the Kolmogorov sublayer, evolved with $Re^{-3/4}$, whereas the viscous boundary sublayer thickness scaled with $Re^{-1/2}$. The simultaneous analysis of near-surface velocity and concentration fluctuations confirmed the central role of upwellings and downwellings and the typical horizontal size of these structures is found to be about $2L_I$, corresponding to the turbulence macroscale. Also, the thickness of the diffusive boundary sublayer was seen to undulate slightly, owing to the alternate compression and dilation induced by the upwellings and downwellings (Magnaudet and Calmet, 2006). The high-concentration structures driven by the downwellings mostly took the form of *needles* penetrating the bulk flow and represented the main way by which a gas could be transferred from the interface down to the bulk fluid. Obviously, due to LES characteristics, this picture could describe only horizontal large-size structures, whereas other methods could capture concentration smaller size structures. Finally, by a frequency analysis of the concentration equation, Magnaudet and Calmet demonstrated that the $Re^{-3/4}$ scaling of the inner CBL resulted directly in the scaling of K_L with the variance of the surface divergence β elevated to $1/4$:

$$K_L \approx D_m^{1/2}(\overline{\beta^2})^{1/4} \qquad (6.72)$$

which in turns, since $\overline{\beta^2} \approx \varepsilon/v$, implies that (Magnaudet and Calmet, 2006):

$$\frac{K_L}{u} \approx Sc^{-1/2} Re^{-1/4} \qquad (6.73)$$

which is identical to the *small-eddy* model. However, Equation (6.73) was derived only by the $Re^{-3/4}$ scaling of the inner CBL and did not mean that gas-transfer process is controlled by the small scale eddies, but, on the contrary, the role of large-scale structures such as upwellings and downwellings, remained, according to Magnaudet and Calmet (2006), dominant, as also highlighted by experimental observations (Rashidi and Banerjee, 1988; Komori *et al.*, 1989; Kumar *et al.*, 1998). Interestingly, Equation (6.73) supports the recent experimental results by Sugihara and Tsumoto (2005), that is Equation (6.69).

Recently Kermani and co-workers used DNS to study characteristics of interfacial transfer of gas and heat in free-surface turbulence (Kermani and Shen, 2009; Kermani *et al.*, 2011).

First, they argued that the *surface-renewal* models did not account for effects of vertical turbulent advection after any *surface-renewal event*. Second they highlighted that to study the surface renewal process, it is essential to obtain the surface age information. Using Lagrangian tracing, they directly quantified the *surface age*. Their results demonstrated that at the early stage of surface renewal, vertical advection associated with upwellings greatly enhanced the gas-transfer flux. After a fluid particle left the upwelling region, it may enter a nearby downwelling region immediately, where the gas flux was sharply reduced but the variation in surface temperature was small. Alternatively, the fluid particle could travel along the surface for some time before it was absorbed by a downwelling, where the surface temperature has changed significantly due to long duration of diffusion and the gas flux was also reduced. Hence they concluded that classical *surface-renewal* models, which are considering only diffusion, are not able to properly predict gas-transfer flux.

6.5 CONCLUSIVE REMARKS

Gas-transfer across the turbulent air-water interface of a surface water body is a relevant process in the Environmental Fluid Mechanics area. The movement through this interface of oxygen, carbon dioxide, nitrogen and toxic chemicals can greatly affect water quality levels.

In this chapter the gas-transfer of sparingly soluble gas, such as oxygen, carbon dioxide and many environmental contaminants, across the free surface of rivers and streams was discussed in details in terms of experimental measurements and observations, predictive models and numerical simulations. The transfer of these substances across the air-water interface is controlled by the processes occurring in a thin region below the interface. Also, in open channel flows turbulence is mostly generated at the channel bottom wall and is then self-transported towards the free surface. Hence, this condition leads to define the air-water interface as *unsheared* or *shear-free*. Both experimental and numerical studies as well as theoretical analysis have pointed out the role played by turbulence characteristics into the gas-transfer process. Turbulent structures produced in the bed region move upward to the free surface and interact with it producing a renewal of the near-surface layers of flow, which controls the gas-transfer process. Although the classic analysis leading to the *surface renewal* theory by Higbie and Danckwerts can be considered as an adequate general *picture* of the process, considerable efforts are currently produced to understand how turbulent coherent structures affect *surface-renewal*. Hence, conceptual models proposed to describe this process and to predict its rate K_L have tried to relate it to both global and local properties of turbulence. The models based on *global* properties, such as *large-eddy* and *small-eddy* models, relate K_L to the Schmidt number and the turbulent Reynolds number of the flow, which is defined with the aid of the integral length scale and some velocity scale. They basically differ on the range of scales which is assumed to control the gas-transfer process, that is large scale or small scale. However, to solve this conflict, it was proposed that both scales would be involved in the process and their relative importance would depend on the value of turbulent Reynolds number. More recently, models based on *local* properties of turbulence, that is interfacial turbulence characteristics, were proposed. Basic concept of these models is the *surface divergence*, that is β parameter, which is the vertical velocity gradient. This parameter is related to the horizontal velocity fluctuations. Recent numerical simulations were able to resolve both velocity and concentration fields near the air-water interface. Numerical results have pointed out that positive and negative values of β correspond to large-structures reaching the interface or moving downward from it, carrying low and high concentration.

Despite these important advances in gas-transfer understanding and modelling, many efforts are still needed to achieve a complete knowledge of this process. First, even if recent developments in experimental techniques are encouraging, they still require improvements to made measurements very close to the air-water interface, in the uppermost layers of the flow, which control the transfer of the gas across the free surface. Also, it is very important that detailed measurements of concentration field would be linked with simultaneous measurements of near-surface velocity field. Second, even if numerical methods have provided detailed and precise determination of velocity and concentration fields very near to the air-water interface pointing out relevant features of the interaction between turbulence and gas-transfer process, these methods should be extended to higher both Schmidt and Reynolds numbers to encompass typical conditions existing in streams and rivers. Also, the influence of the turbulent anisotropy on the relationship between gas-transfer rate and Reynolds number should be further investigated. Finally, future modelling efforts should be addressed to take into account the role of all the turbulent scales in the gas-transfer process.

APPENDIX A – LIST OF SYMBOLS

List of Symbols		
Symbol	Definition	Dimensions or Units
C	concentration	$[\text{M} \cdot \text{L}^{-3}]$
$\overline{C}$	temporal mean concentration	$[\text{M} \cdot \text{L}^{-3}]$
C_g	gas concentration in the bulk gas	$[\text{M} \cdot \text{L}^{-3}]$
C_i	gas concentration at the air-water interface	$[\text{M} \cdot \text{L}^{-3}]$
C_{sat}	gas concentration at saturation	$[\text{M} \cdot \text{L}^{-3}]$
C_w	gas concentration in the bulk water	$[\text{M} \cdot \text{L}^{-3}]$
D_m	molecular diffusion coefficient	$[\text{L}^2 \, \text{T}^{-1}]$
D_t	turbulent diffusion coefficient	$[\text{L}^2 \, \text{T}^{-1}]$
Fr	Froude number	
H_e	dimensional Henry constant	$[\text{M} \cdot \text{L}^2 \cdot \text{T}^{-2} \cdot \text{mole}^{-1}]$
H	dimensionless Henry constant	
J_b	channel bed slope	
J_e	energy line slope	
J_{g-t}	gas-transfer flux	$[\text{M} \, \text{L}^{-2} \, \text{T}^{-1}]$
$J_{g-t-gas}$	gas-transfer flux across the CBL on the air-side	$[\text{M} \, \text{L}^{-2} \, \text{T}^{-1}]$
J_{g-t-z}	gas-transfer flux in the vertical direction	$[\text{M} \, \text{L}^{-2} \, \text{T}^{-1}]$
$J_{g-t-water}$	gas-transfer flux across the CBL on the water-side	$[\text{M} \, \text{L}^{-2} \, \text{T}^{-1}]$
K_L	gas-transfer coefficient	$[\text{L} \cdot \text{T}^{-1}]$
$K_L{}^*$	dimensionless gas-transfer coefficient	
L_I	turbulent integral scale	$[\text{L}]$
Q	water discharge	$[\text{L}^3 \cdot \text{T}^{-1}]$
R	universal gas constant	$[\text{M} \cdot \text{L}^2 \cdot \text{T}^{-2} \cdot \text{K}^{-1} \cdot \text{mole}^{-1}]$
R_g	gas-transfer resistance in the CBL on the air-side	$[\text{T} \cdot \text{L}^{-1}]$
R_{tot}	total resistance to the gas-transfer	$[\text{T} \cdot \text{L}^{-1}]$
R_w	gas-transfer resistance in the CBL on the water-side	$[\text{T} \cdot \text{L}^{-1}]$
R_h	channel hydraulic radius	$[\text{L}]$
Re	Reynolds number	
Re^*	shear Reynolds number	
Re_{g-t}	gas-transfer Reynolds number	
$Re_{k\varepsilon}$	turbulent k-ε Reynolds number	

(Continued)

List of Symbols

Symbol	Definition	Dimensions or Units
S	relative roughness	
Sc	Schmidt number	
Sc_t	turbulent Schmidt number	
Sh	Sherwood number	
$S_{\beta max}{}^*$	dimensionless maximum value of the β spectrum	
T_a	absolute temperature	[K]
T_s	water surface tension	$[\text{M T}^{-2}]$
Vol	gas volume	$[\text{L}^3]$
We	Weber number	
c_1, c_2, c_3, c_4	numerical constants	
f	Darcy-Weisbach friction factor	
g	gravitational acceleration constant	$[\text{L T}^{-2}]$
h	channel water mean depth	[L]
k	turbulent kinetic energy per unit mass	$[\text{L}^2\,\text{T}^{-2}]$
k_g	gas-transfer velocity in the CBL on the air-side	$[\text{L}\cdot\text{T}^{-1}]$
k_w	gas-transfer velocity in the CBL on the water-side	$[\text{L}\cdot\text{T}^{-1}]$
n_m	number of moles	
p	gas partial pressure	$[\text{M}\cdot\text{L}^{-1}\cdot\text{T}^{-2}]$
p_g	gas pressure in the bulk gas	$[\text{M}\cdot\text{L}^{-1}\cdot\text{T}^{-2}]$
p_i	gas pressure at the air-water interface	$[\text{M}\cdot\text{L}^{-1}\cdot\text{T}^{-2}]$
r	renewal rate	$[\text{T}^{-1}]$
t	time	[T]
t_r	renewal time or surface age	[T]
t_{r-avg}	average renewal time	[T]
u	mean streamflow velocity	$[\text{L}\cdot\text{T}^{-1}]$
u_{rms}	root-mean-square of turbulent streamflow velocity	$[\text{L}\cdot\text{T}^{-1}]$
u^*	shear velocity	$[\text{L}\cdot\text{T}^{-1}]$
u'	fluctuating velocity in the streamwise direction x	$[\text{L}\cdot\text{T}^{-1}]$
v'	fluctuating velocity in the spanwise direction y	$[\text{L}\cdot\text{T}^{-1}]$
w_r	vertical velocity with respect to the interface	$[\text{L}\cdot\text{T}^{-1}]$
w_β	vertical velocity where w_r departs from the linearity	$[\text{L}\cdot\text{T}^{-1}]$
$w_{\beta-rms}$	root-mean-square of w_β	$[\text{L}\cdot\text{T}^{-1}]$
w'	fluctuating velocity in the vertical direction z	$[\text{L}\cdot\text{T}^{-1}]$
z	vertical coordinate	[L]
β	surface divergence	$[\text{T}^{-1}]$
β_{rms}	root mean square of β	$[\text{T}^{-1}]$
$\overline{\beta^2}$	variance of surface divergence	$[\text{T}^{-2}]$
δ_{CBL}	thickness of the concentration boundary layer	[L]
δ_{VBL}	thickness of the velocity boundary layer	[L]
ε	dissipation rate of turbulent kinetic energy per unit mass	$[\text{L}^2\,\text{T}^{-3}]$
ε_{cw}	Colebrook-White roughness coefficient	[L]
η	Kolmogorov microlenght scale	[L]
θ	temperature correction factor for K_L	
κ	Von Kármán constant	

(Continued)

List of Symbols

Symbol	Definition	Dimensions or Units
μ	water dynamic viscosity	$[M\,L^{-1}\,T^{-1}]$
ν	water kinematic viscosity	$[L^2\,T^{-1}]$
ν_t	water turbulent kinematic viscosity	$[L^2\,T^{-1}]$
V_t	threshold wind velocity	$[ms^{-1}]$
ρ	water density	$[M\,L^{-3}]$
τ	shear stress	$[M\,L^{-1}\,T^{-2}]$
τ_b	bed shear stress	$[M\,L^{-1}\,T^{-2}]$
$\tau_s = T_s/\rho$	ratio between water surface tension and water density	$[L^3\,T^{-2}]$
ϕ	surface-age distribution function	
ψ	shape factor of stream transverse section	

APPENDIX B – SYNOPSIS

Several chemicals are transferred upward to the air or downward to the water depending on the substances involved and departure from equilibrium (Henry's Law). This process is termed gas-transfer. Hence, gas transfer of a volatile or semi-volatile chemical is a two-way process involving both dissolution by the water and volatilization into the air across an air-water interface. Gas-transfer is controlled both by the characteristics of the substance being transferred and by the interaction between turbulence in the bulk fluid and the air-water interface. Several conceptual models have been proposed to predict the gas-transfer rate. An adequate general picture of the gas-transfer process is provided by the surface renewal approach, which assumes that the fluid elements at the air-water interface are periodically refreshed by turbulent eddies that bring liquid parcels from the bulk liquid to the air-water interface and vice versa. If the models based on surface renewal approach, such as large-eddy model and small-eddy model, estimate the gas-transfer rate as function of a global property of turbulence, further developments, such as the surface divergence model, relates gas-transfer to local property of turbulence. Finally, numerical simulations of gas-transfer process provided a detailed and precise determination of velocity and concentration fields very near to the air-water interface confirming previous theoretical efforts and experimental observations.

APPENDIX C – KEYWORDS

By the end of the chapter you should have encountered the following terms. Ensure that you are familiar with them!

Concentration boundary layer
Gas-transfer
Gas-transfer coefficient
Henry constant

Large-eddy model
Reaeration
Schmidt number
Small-eddy model

Surface divergence
Surface renewal
Velocity boundary layer
Volatilization

APPENDIX D – QUESTIONS

What is the gas-transfer process?
How the physicochemical characteristics of the substance being transferred affect gas-transfer process?

How turbulence interacts with the air-water interface for a substance being controlled by the liquid phase?

Which are the main conceptual models for the prediction of gas-transfer rate?

Which are the main findings from the numerical simulation of the gas-transfer process?

APPENDIX E – PROBLEMS

E1. Describe the hierarchal structure of layers at the air-water interface and discuss the parameters which control momentum and mass transport from the bulk fluid to the air-water interface in a turbulent flow. Compare this case to that of the gas-transfer into a stagnant water body.

E2. Describe in their basic assumptions the two-film model by Lewis and Whitman and of the surface renewal model by Higbie and Danckwerts. Compare Eqs. (6.38) and (6.50) and explain how turbulence characteristics could be used to derive the renewal rate according the large-eddy model and the small-eddy model.

E3. Describe the concept of surface divergence and the gas-transfer models derived from this concept.

E4. Compare values for δ_{CBL} calculated from Eqs. (6.40) and (6.42) for oxygen and water at 17°C. Note that at this temperature, $v = 1.1 \times 10^{-6}$ m^2/s, $D_m = 1.7 \times 10^{-9}$ m^2/s and $S_c = 653$. Use $u^* = 0.0334$ m/s, $u = 0.717$ m/s and h $= 0.05$ m. The values for the constants c_1 and c_2 are 10 and 2, respectively. Compare the values for δ_{CBL}.

Then, calculate gas-transfer rate K_L from Eqs. (6.41) and (6.43) and compare them with the experimental data from Moog and Jirka (1999), that is $K_L = 3.19$ m/day.

E5. Using the equations for the large-eddy model (Eq. 6.52) and the small-eddy model (Eq. 6.54) calculate the dimensionless gas-transfer coefficient K_L^* for oxygen for the experimental conditions listed in Table 6.3 (Lau, 1975). Note that water temperature was 20°C. at 20°C. At 20°C, $v = 1.0 \times 10^{-6}$ m^2/s, $D_m = 1.8 \times 10^{-9}$ m^2/s and $Sc = 548$. Compare the calculated coefficient to the experimental results. Comment on the comparison.

Table 6.3. Experimental data from Lau, 1975.

$u^* -$ m/s	Re^*	$K_{L-mis} -$ m/s
0.01087	295	4.6E−06
0.01469	399	6.4E−06
0.01782	484	6.8E−06
0.02473	672	9.2E−06
0.01600	435	7.4E−06
0.01345	365	5.8E−06
0.00939	401	4.3E−06
0.01359	580	5.9E−06
0.01382	590	7.1E−06
0.01508	644	6.7E−06
0.01890	807	1.2E−05
0.00282	77	1.5E−06
0.00351	95	1.5E−06

REFERENCES

Atkinson, J.F., Blair, S., Taylor, S. and, Ghosh, U., 1995, Surface aeration, *J. Environmental Engineering Division, ASCE*, **121**, 1, January–February 1995, pp. 113–118

Banerjee, S., 1990, Turbulence structure and transport mechanisms at interface, *Proceedings of IX Heat Transfer Conference*, Keynote Lectures, **1**, pp. 395–418

Banerjee, S., Rhodes, E., and Scott, D. S., 1968, Mass transfer through falling wavy liquid films in turbulent flow, *Ind. Eng. ChE. Fundamentals*, **7**, pp. 22–28

Banerjee, S., and MacIntyre, S., 2004, The air-water interface: turbulence and scalar exchange, in *Advances in Coastal and Ocean Engineering* (P.L.F.Liu ed.), 9, World Scientific, Hackensack, N. J., USA, pp. 181–237

Calmet, I., and Magnaudet, J., 1998, High-Schmidt number mass transfer through turbulent gas-liquid interfaces, *Int. J. Heat and Fluid Flow*. **19**, pp. 522–532

Chao, X., Jia, Y., and Wang, S.S.Y., 2007 Atmospheric reaeration in open channel flow 2007 *Restoring Our Natural Habitat – Proceedings of the 2007 World Environmental and Water Resources Congress*, Tampa, Florida, USA, May 15–19, 2007

Chapra, S.C., 1997, *Surface water quality modeling*, McGraw-Hill, New-York, USA

Chu, C. R., and Jirka, G. H., 1992, Turbulent gas flux measurements below the air-water interface of a grid-stirred tank, *Int. J. Heat and Mass Transfer*. **35**, pp. 1957–1968

Elder, J. W., 1959, The dispersion of marked fluid in turbulent shear flow, *J. Fluid Mech.*, **5**, pp. 544–560

Fortescue, G. E., and Pearson, J. R. A., 1967, On gas absorption into a turbulent liquid, *Chem. Engr. Science*, **22**, pp. 1163–1176

Gualtieri, C., Gualtieri P., and Pulci Doria, G., 2002, Dimensional analysis of reaeration rate in streams. *Journal of Environmental Engineering, ASCE*, **128**, n.1, January 2002, pp. 12–18

Gualtieri, C. and Gualtieri P., 2004, Turbulence-based models for gas transfer analysis with channel shape factor. *Environmental Fluid Mechanics*, **4**, n.3, September 2004, pp. 249–271

Gualtieri C., Pulci Doria G., and D'Avino, A., 2006, Gas-transfer coefficient in a smooth channel. An equation based on dimensional analysis. *Proceedings of XXX Hydraulics and Waterworks Conference*, Rome, Italy, September 10/15, 2006

Gualtieri C., 2005a, Discussion on Higashino, M., Stefan, H.G., and Gantzer, C.J.: Periodic diffusional mass transfer near sediment/water interface: *Theory. J. Env. Eng.*, ASCE, vol. 129, n. 5, May 2003, pp.447–455. *Journal of Environmental Engineering, ASCE*, **131**, n.1, January 2005, pp. 171–172

Gualtieri C., 2005b, Discussion on Chu, C.R., and Jirka, G.H.: Wind and stream induced reaeration. J. Env. Eng., ASCE, vol. 129, n. 12, December 2003, *Journal of Environmental Engineering, ASCE*, **131**, 8, August 2005, pp. 1236–1238

Gualtieri C., 2006, Verification of wind-driven volatilization models. *Environmental Fluid Mechanics*, **6**, n.1, February 2006, pp. 1–24

Handler, R. A., Saylor, J. R., Leighton, R. I., and Rovelstad, A. L., 1999, Transport of a passive scalar at a shear-free boundary in fully developed turbulent open channel flow, *Physics of Fluids*, **11**, 9, September 1999, pp. 2607–2625

Hanratty, T. J., 1991, Effect of gas flow on physical absorption, In *Gas-transfer at Water Surfaces*, edited by Wilhelm, S.C., and Gulliver, J.S., New York, ASCE, pp. 10–33

Hasegawa, Y., and Kasagi, N., 2008, Systematic analysis of high Schmidt number turbulent mass transfer across clean, contaminated and solid interfaces, *International Journal of Heat and Fluid Flow*, 29, pp. 765–773

Herlina, and Jirka, G. H., 2004, Application of LIF to investigate gas transfer near the air-water interface in a grid-stirred tank, *Experiments in Fluids*, **37**, pp. 341–349

Hunt, J. C. R., and Graham, J. M. R., 1978, Free stream turbulence near plane boundaries, *J. Fluid Mechanics*. **84**, Part 2, pp. 209–235

Jähne, B., and Haußecker, H., 1998, Air-water gas exchange, *Annual Review of Fluid Mechanics*, **30**, pp. 443–468

Janzen, J.G., Herlina, H., Jirka, G.H., Schulz, H.E., and Gulliver, J.S., 2010, Estimation of mass transfer velocity based on measured turbulence parameters, *AIChE Journal*, **56**, n.8, August 2010, pp. 2005–2017

Jirka, G.H., Herlina, H., and Niepelt, A., 2010, Gas transfer at the air-water interface: experiments with different turbulence forcing mechanisms, *Experiments in Fluids*, **49**, pp. 319–327

Kermani, A., and Shen, L., 2009, Surface age of surface renewal in turbulent interfacial transport, *Geophysical Research Letters*, **36**, L10605

Kermani, A., Khakpour, H.R., Shen, L., and Igusa, T., 2011, Statistics of surface renewal of passive scalars in free-surface turbulence, *J. Fluid Mechanics*, **678**, pp. 379–416

Komori, S., Murakami, Y., and Ueda, H., 1989, The relationship between surface-renewal and bursting motions in an open channel flow, *J. Fluid Mechanics*. **203**, pp. 103–123

Komori, S., Nagaosa, R., Murakami, Y., Chiba, S., Ishii, K., and Kuwahara. K., 1993, Direct numerical simulation of three-dimensional open-channel flow with zero-shear gas-liquid interface, *Physics of Fluids*, **5**, 1, January 1993, pp. 115–125

Kumar, S., Gupta, R., and Banerjee, S., 1998, An experimental investigation of the characteristic of free-surface turbulence in channel flow, *Physics of Fluids*, **10**, 2, February 1998, pp. 437–456

Lau, Y.L., 1975, An experimental investigation of reaeration in open channel flow. *Prog. in Water Tech.*, **7**, n.3/4, pp. 519–530

Law, C.N.S, and Khoo, B.C., 2002, Transport across a turbulent air-water interface, *AIChE Journal*, **48**, n.9, September 2002, pp. 1856–1868

Lorke, A., Müller, B., Maerki, M., and Wüest, A., 2003, Breathing sediments: The control of diffusive transport across the sediment-water interface by periodic boundary-layer turbulence. *Limnology and Oceanography*, **48**, n.6, pp. 2077–2085

Lorke, A., and Peeters, F., (2006), Toward a unified scaling relation for interfacial fluxes, *Journal of Physical Oceanography*, **36**, May 2006, pp. 955–961

Magnaudet, J., and Calmet, I., 2006, Turbulent mass transfer through a flat shear-free surface, *J. Fluid Mechanics*, **553**, pp. 155–185

McCready, M. A., Vassiliadou, E., and Hanratty, T. J., 1986, Computer simulation of turbulent mass transfer at a mobile interface, *AIChE Journal*, **32**, 7, July 1986, pp. 1108–1115

Melching, C. S., and Flores, H. E., 1999, Reaeration equations derived from U.S. Geological Survey database. *J. Environmental Engineering*, ASCE, **125**, 5, May 1999, pp. 407–414

Moog, D. B., and Jirka, G. H., 1999, Air-water gas transfer in uniform channel flow. *J. Hydraulic Engineering*, ASCE, **125**, 1, January 1999, pp. 3–10

Münsterer, T., and Jähne, B., 1998, LIF measurements of concentration profiles in the aqueous mass boundary layer, *Experiments in Fluids*, **25**, pp. 190–196

Nagaosa, R., 1999, Direct numerical simulation of vortex structures and turbulent scalar transfer across a free surface in a fully developed turbulence, *Physics of Fluids*, **11**, 6, June 1999, pp. 1581–1595

Nakayama, T., 2000, *Turbulence and coherent structures across air-water interface and relationship with gas-transfer*, PhD Thesis, Kyoto University, Japan

Nezu, I., and Nakagawa, H., 1993, *Turbulence in open-channel flows*, IAHR Monograph Series, Balkema, Rotterdam, The Netherlands

NIST, 2000, Standard Reference Database Number 69. February 2000

Pope, S. B., 2000, *Turbulent flows*, Cambridge University Press, Cambridge, U.K.

Rashidi, M., and Banerjee, S., 1988, Turbulence structure in free-surface channel flows, *Physics of Fluids*, **31**, 9, September 1998, pp. 2491–2503

Rathbun, R. E., and Tai, D. Y., 1982, Volatilization of organic compounds from streams, *Journal of Environmental Engineering Division, ASCE*, **108**, 5, October 1982, pp. 973–989

Rathbun, R. E., 1998, Transport, behavior and fate of volatile organic compounds in streams. *U.S. Geological Survey Professional Paper 1589*, Washington, DC

Schwarzenbach, R. P., Gschwend, P. M., and Imboden, D. M., 1993. *Environmental organic chemistry*. Wiley-Interscience, New York, USA

Shen, L., Triantafyllou, G. S., and Yue, D. K. P., 2001, Mixing of a passive scalar near a free surface, *Physics of Fluids*, **13**, 4, April 2001, pp. 913–926

Socolofsky, S. A., and Jirka G. H., (2002), *Environmental Fluid Mechanics. Part I: Mixing, Transport and Transformation*, Engineering Lectures, Institut für Hydromechanik, University of Kalsruhe, Germany

Sugihara, Y., and Tsumori, H., 2005, Surface-renewal eddies at the air-water interface in oscillating-grid turbulence, in *Proceedings of Environmental Hydraulics and Sustainable Management* (Lee and Lam eds.), Taylor & Francis Group, London, U.K., pp. 199–205

Tamburrino, A., and Gulliver, J. S., 2002, Free-surface turbulence and mass transfer in a channel flow, *AIChE Journal*, **48**, 12, December 2002, pp. 2732–2743

Theofanous, T. G., Houze, R. N., and Brumfield, L. K., 1976, Turbulent mass transfer at free, gas-liquid interfaces, with applications to open-channel, bubble and jet flows, *Int. J. Heat Mass Transfer*. **19**, pp. 613–624

Thibodeaux, L. J., 1996, *Environmental chemodynamics*. John Wiley & Sons, Chichester, U.K.

USEPA, 1985, *Rates, constants and kinetics formulation in surface water quality modeling*. U.S. EPA, Office of Research and Development, Environmental Research Laboratory, Athens, GA, USA, 1985

USEPA, 1997, *Deposition of air pollutants to the Great Waters*, U.S. EPA, Office of Air Qualiy Planning and Standards, Research Triangle Park, NC, USA

Weber, W. J., and DiGiano, F. A., 1996, *Process dynamics in environmental systems*. John Wiley & Sons, Chichester, U.K.

Wüest, A., and Lorke, A., 2003, Small-scale hydrodynamics in lakes. *Annual Review in Fluid Mechanics*, **35**, pp. 373–412

Xu, Z. F., Khoo, B. C., and Carpenter K., 2006, Mass transfer across the turbulent gas-water interface, *AIChE Journal*, **52**, 10, October 2006, pp. 3363–3374

CHAPTER SEVEN

Advective diffusion of air bubbles in turbulent water flows

Hubert Chanson

Professor in Civil Engineering, The University of Queensland, Brisbane, Australia

ABSTRACT

Air bubble entrainment is defined as the exchange of air between the atmosphere and flowing water. Also called self-aeration, the continuous exchange between air and water is most important for the biological and chemical equilibrium on our planet. Air bubble entrainment is observed in chemical, coastal, hydraulic, mechanical and nuclear engineering applications as well as in the natural environment such as waterfalls, mountain streams and river rapids, and breaking waves on the ocean surface. The resulting "white waters" provide some spectacular effects. The entrainment of air bubbles may be localised at a flow discontinuity or continuous along an air-water free-surface: i.e., singular and interfacial aeration respectively. At a flow singularity, the air bubbles are entrained locally at the impinging perimeter and advected in a region of high turbulent shear stresses. The interfacial aeration is the air bubble entrainment process along an air-water interface which is parallel to the flow direction. The onset of air bubble entrainment may be expressed in terms of the tangential Reynolds stress and the fluid properties. Once self-aeration takes place, the distributions of void fraction may be modelled by some analytical solutions of the advective diffusion equation for air bubbles. The microscopic structure of turbulent bubbly flows is complex and a number of examples are discussed. The results reveal the turbulent nature of the complex two-phase flows and the complicated interactions between entrained air bubbles and turbulence.

7.1 INTRODUCTION

The exchange of air between the atmosphere and flowing water is usually called air entrainment, air bubble entrainment or self-aeration. The continuous exchange between air and water is most important for the biological and chemical equilibrium on our planet. For example, the air-water mass transfer at the surface of the oceans regulates the composition of the atmosphere. The aeration process drives the exchange of nitrogen, oxygen and carbon dioxide between the atmosphere and the sea, in particular the dissolution of carbon dioxide into the oceans and the release of supersaturated oxygen to the atmosphere. Another form of flow aeration is the entrainment of un-dissolved air bubbles at the air-water free-surface. Air bubble entrainment is observed in chemical, coastal, hydraulic, mechanical and nuclear engineering applications. In Nature, air bubble entrainment is observed at waterfalls, in mountain streams and river rapids, and in breaking waves on the ocean surface. The resulting "white waters" provide some spectacular effects (Fig. 7.1 to 7.4). Figures 7.1 to 7.3 illustrates the air bubble entrainment in hydraulic structures during river floods, and

Figure 7.1. Air bubble entrainment on the Wivenhoe Dam spillway (Qld, Australia) on 17 January 2011.

Figure 7.2. Air bubble entrainment downstream of Burdekin Falls Dam (Qld, Australia) in February 2009 (Courtesy of Queensland Department of Environment and Mineral Resources (DERM, Dam Safety) and David Li) – Looking upstream at the chute flow and aerated jet formation at spillway toe – Note the "brownish" dark colour of the flow caused by the suspended load and the "white" waters downstream of the spillway toe highlighting the air bubble entrainment.

Figure 7.3. Free-surface aeration during the overtopping of Mount Crosby weir and bridge (Qld, Australia) on 17 Jan. 2011 – Flow direction from left to right.

Figure 7.4. Air entrainment at wave breaking – Honeymoon Bay, Moreton Island (Qld, Australia) on 7 July 2011 (Shutter speed 1/500 s).

Figure 7.4 presents some air bubble entrainment at a plunging breaking wave. Note that the free-surface aeration in large systems may be seen from space (Chanson 2008).

Herein we define air bubble entrainment as the entrainment or entrapment of un-dissolved air bubbles and air pockets that are advected within the flowing waters. The term air bubble is used broadly to describe a volume of air surrounded continuously or not by some liquid and encompassed within some air-water interface(s). The resulting air-water mixture consists of both air packets within water and water droplets surrounded by air, and the flow structure may be quite complicated.

Further the entrainment of air bubbles may be localised at a flow discontinuity or continuous along an air-water free-surface: i.e., singular or interfacial aeration respectively.
Examples of singular aeration include the air bubble entrainment by a vertical plunging
jet. Air bubbles are entrained locally at the intersection of the impinging water jet with
the receiving body of water. The impingement perimeter is a source of both vorticity and
air bubbles. Interfacial aeration is defined as the air bubble entrainment process along an
air-water interface, usually parallel to the flow direction. It is observed in spillway chute
flows and in high-velocity water jets discharging into air.

After a review of the basic mechanisms of air bubble entrainment in turbulent water flows,
it will be shown that the void fraction distributions may be modelled by some analytical
solutions of the advective diffusion equation for air bubbles. Later the micro-structure of the
air-water flow will be discussed and it will be argued that the interactions between entrained
air bubbles and turbulence remain a key challenge.

7.2 FUNDAMENTAL PROCESSES

7.2.1 Inception of air bubble entrainment

The inception of air bubble entrainment characterises the flow conditions at which some
bubble entrainment starts. Historically the inception conditions were expressed in terms of
a time-averaged velocity. It was often assumed that air entrainment occurs when the flow
velocity exceeds an onset velocity V_e of about 1 m/s. The approach is approximate and it
does not account for the complexity of the flow nor the turbulence properties. More detailed
studies linked the onset of air entrainment with a characteristic level of normal Reynolds
stress(es) next to the free-surface. For example, Ervine an Falvey (1987) and Chanson
(1993) for interfacial aeration, Cummings and Chanson (1999) for plunging jet aeration,
Brocchini and Peregrine (2001). Although present knowledge remains empirical and often
superficial, it is thought that the inception of air entrainment may be better described in
terms of tangential Reynolds stresses.

In turbulent shear flows, the air bubble entrainment is caused by the turbulence acting
next to the air-water interface. Through this interface, air is continuously being trapped
and released, and the resulting air-water mixture may extend to the entire flow. Air bubble
entrainment occurs when the turbulent shear stress is large enough to overcome both surface
tension and buoyancy effects (if any). Experimental evidences showed that the free-surface of
turbulent flows exhibits some surface "undulations" with a fine-grained turbulent structure
and larger underlying eddies. Since the turbulent energy is high in small eddy lengths close
to the free surface, air bubble entrainment may result from the action of high intensity
turbulent shear close to the air-water interface.

Free-surface breakup and bubble entrainment will take place when the turbulent shear
stress is greater than the surface tension force per unit area resisting the surface breakup.
That is:

$$\left| \rho_w \times v_i \times v_j \right| > \sigma \times \frac{\pi \times (r_1 + r_2)}{A} \qquad\qquad \text{inception of air entrainment (7.1)}$$

where ρ_w is the water density, v is the turbulent velocity fluctuation, (i, j) is the directional
tensor $(i, j = x, y, z)$, σ is the surface tension between air and water, $\pi \times (r_1 + r_2)$ is the
perimeter along which surface tension acts, r_1 and r_2 are the two principal radii of curvature
of the free surface deformation, and A is surface deformation area. Equation (7.1) gives a
criterion for the onset of free-surface aeration in terms of the magnitude of the instantaneous
tangential Reynolds stress, the air/water physical properties and the free-surface deformation

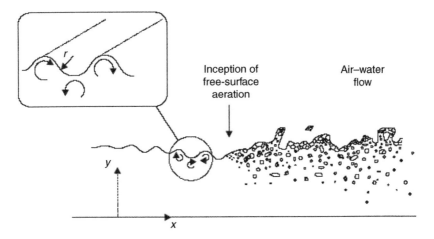

y

x

Figure 7.5. Inception of free-surface aeration in a two-dimensional flow.

properties. Simply air bubbles cannot be entrained across the free-surface until there is sufficient tangential shear relative to the surface tension force per unit area.

Considering a two-dimensional flow for which the vortical structures next to the free-surface have axes predominantly perpendicular to the flow direction, the entrained bubbles may be schematised by cylinders of radius r (Fig. 7.5). Equation (7.1) may be simplified into:

$$\left|\rho_w \times v_i \times v_j\right| > \frac{\sigma}{\pi \times r} \qquad \text{cylindrical bubbles \quad (7.2a)}$$

where x and y are the streamwise and normal directions respectively. For a three-dimensional flow with quasi-isotropic turbulence, the smallest interfacial area per unit volume of air is the sphere (radius r), and Equation (1) gives:

$$\left|\rho_w \times v_i \times v_j\right| > \frac{\sigma}{2 \times \pi \times r} \qquad \text{spherical bubbles \quad (7.2b)}$$

Equation (7.2) shows that the inception of air bubble entrainment takes place in the form of relatively large bubbles. But the largest bubbles will be detrained by buoyancy and this yields some preferential sizes of entrained bubbles, observed to be about 1 to 100 mm in prototype turbulent flows (e.g. Cain 1978, Chanson 1993, 1997).

7.2.2 Bubble breakup

The size of entrained air bubbles in turbulent shear flows is an important parameter affecting the interactions between turbulence and air bubbles. Next to the entrainment point, a region of strong mixing and momentum losses exists in which the entrained air is broken into small bubbles while being diffused within the air-water flow.

At equilibrium, the maximum bubble size in shear flows may be estimated by the balance between the surface tension force and the inertial force caused by the velocity changes over distances of the order of the bubble size. Some simple dimensional analysis yielded a criterion for bubble breakup (Hinze 1955). The result is however limited to some equilibrium situations and it is often not applicable (Chanson 1997, pp. 224–229).

In air-water flows, experimental observations of air bubbles showed that the bubble sizes are larger than the Kolmogorov microscale and smaller than the turbulent macroscale. These observations suggested that the length scale of eddies responsible for breaking up the bubbles is close to the bubble size. Larger eddies advect the bubbles while eddies with length-scales substantially smaller than the bubble size do not have the necessary energy to break up air bubbles.

In turbulent flows, the bubble break-up occurs when the tangential shear stress is greater than the capillary force per unit area. For a spherical bubble, it yields a condition for bubble breakup:

$$\left| \rho_w \times v_i \times v_j \right| > \frac{\sigma}{\pi \times d_{ab}} \qquad \text{spherical bubble} \quad (7.3a)$$

where d_{ab} is the bubble diameter. Equation (7.3a) holds for a spherical bubble and the left handside term is the magnitude of the instantaneous tangential Reynolds stress. More generally, for an elongated spheroid, bubble breakup takes place for:

$$\left| \rho_w \times v_i \times v_j \right| > \sigma \times \frac{\pi \times (r_1 + r_2)}{2 \times \pi \times r_1 \times \left(r_1 + r_2 \times \dfrac{\text{Arcsin}\left(\sqrt{1 - \dfrac{r_1^2}{r_2^2}} \right)}{\sqrt{1 - \dfrac{r_1^2}{r_2^2}}} \right)}$$

$$\text{elongated spheroid} \quad (7.3b)$$

where r_1 and r_2 are the equatorial and polar radii of the ellipsoid respectively with $r_2 > r_1$. Equation (7.3b) implies that some turbulence anisotropy (e.g. v_x, $v_y \gg v_z$) must induce some preferential bubble shapes.

7.3 ADVECTIVE DIFFUSION OF AIR BUBBLES. BASIC EQUATIONS

7.3.1 Presentation

Turbulent flows are characterised by a substantial amount of air-water mixing at the interfaces. Once entrained, the air bubbles are diffused through the flow while they are advected downstream. Herein their transport by advection and diffusion are assumed two separate additive processes; and the theory of superposition is applicable.

In the bubbly flow region, the air bubble diffusion transfer rate in the direction normal to the advective direction varies directly as the negative gradient of concentration. The scalar is the entrained air and its concentration is called the void fraction C defined as the volume of air per unit volume of air and water. Assuming a steady, quasi-one-dimensional flow, and for a small control volume, the continuity equation for air in the air-water flow is:

$$div(C \times \overrightarrow{V}) = div(D_t \times \overrightarrow{grad}\, C - C \times \overrightarrow{u_r}) \qquad (7.4)$$

where C is the void fraction, $\overrightarrow{V}$ is the advective velocity vector, D_t is the air bubble turbulent diffusivity and $\overrightarrow{u_r}$ is the bubble rise velocity vector that takes into account the effects of buoyancy. Equation (7.4) implies a constant air density, neglects compressibility effects, and is valid for a steady flow situation.

Equation (7.4) is called the advective diffusion equation. It characterises the air volume flux from a region of high void fraction to one of smaller air concentration. The first term $(C \times V)$ is the advective flux while the right handside term is the diffusive flux. The latter includes the combined effects of transverse diffusion and buoyancy. Equation (7.4) may be solved analytically for a number of basic boundary conditions. Mathematical solutions of the diffusion equation were addressed in two classical references (Carslaw and Jaeger 1959, Crank 1956). Since Equation (7.4) is linear, the theory of superposition may be used to build up solutions with more complex problems and boundary conditions. Its application to air-water flows was discussed by Wood (1984, 1991) and Chanson (1988, 1997).

7.3.2 Buoyancy effects on submerged air bubbles

When air bubbles are submerged in a liquid, a net upward force is exerted on each bubble. That is, the buoyancy force which is the vertical resultant of the pressure forces acting on the bubble. The buoyant force equals the weight of displaced liquid.

The effects of buoyancy on a submerged air bubble may be expressed in terms of the bubble rise velocity u_r. For a single bubble rising in a fluid at rest and in a steady state, the motion equation of the rising bubble yields an exact balance between the buoyant force (upwards), the drag force (downwards) and the weight force (downwards). The expression of the buoyant force may be derived from the integration of the pressure field around the bubble and it is directly proportional to minus the pressure gradient $\partial P / \partial z$ where P is the pressure and z is the vertical axis positive upwards. In a non-hydrostatic pressure gradient, the rise velocity may be estimated to a first approximation as:

$$u_r = \pm (u_r)_{Hyd} \times \sqrt{\frac{\left| \frac{\partial P}{\partial z} \right|}{\rho_w \times g}} \qquad (7.5)$$

where $(u_r)_{Hyd}$ is the bubble rise velocity in a hydrostatic pressure gradient (Fig. 7.6), ρ_w is the liquid density, herein water, and z is the vertical direction positive upwards. The sign of the rise velocity u_r depends on the sign of $\partial P / \partial z$. For $\partial P / \partial z < 0$, u_r is positive. Experimental results of bubble rise velocity in still water are reported in Figure 7.6. Relevant references include Haberman and Morton (1954) and Comolet (1979a,b).

7.3.3 A simple application

Let us consider a two-dimensional steady open channel flow down a steep chute (Fig. 7.7). The advective diffusion equation becomes:

$$\frac{\partial}{\partial x}(V_x \times C) + \frac{\partial}{\partial y}(V_y \times C) = \frac{\partial}{\partial x}\left(D_t \times \frac{\partial C}{\partial x} \right) + \frac{\partial}{\partial y}\left(D_t * \frac{\partial C}{\partial y} \right)$$

$$- \frac{\partial}{\partial x}(-u_r \times \sin \theta \times C) - \frac{\partial}{\partial y}(u_r \times \cos \theta * C) \qquad (7.6)$$

where θ is the angle between the horizontal and the channel invert, x is the streamwise direction and y is the transverse direction (Fig. 7.7). In the uniform equilibrium flow region,

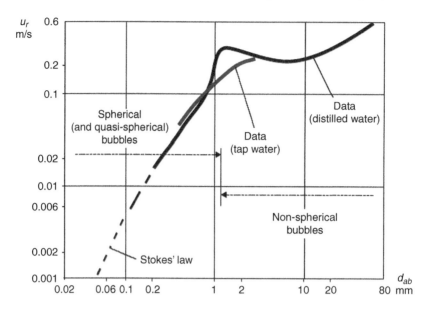

Figure 7.6. Bubble rise velocity in still water.

the gravity force component in the flow direction is counterbalanced exactly by the friction and drag force resultant. Hence $\partial/\partial x = 0$ and $V_y = 0$. Equation (7.6) yields:

$$0 = \frac{\partial}{\partial y}\left(D_t \times \frac{\partial C}{\partial y}\right) - \cos\theta \times \frac{\partial}{\partial y}(u_r \times C) \qquad (7.7)$$

where D_t is basically the diffusivity in the direction normal to the flow direction.

At a distance y from the invert, the fluid density is $\rho = \rho_w \times (1 - C)$ where C is the local void fraction. Hence the expression of the bubble rise velocity (Eq. (7.5)) becomes:

$$u_r = (u_r)_{Hyd} \times \sqrt{1 - C} \qquad (7.8)$$

Equation (7.8) gives the rise velocity in a two-phase flow mixture of void fraction C as a function of the rise velocity in hydrostatic pressure gradient. The buoyant force is smaller in aerated waters than in clear-water. For example, a heavy object might sink faster in "white waters" because of the lesser buoyancy.

The advective diffusion equation for air bubbles may be rewritten in dimensionless terms:

$$\frac{\partial}{\partial y'}\left(D' \times \frac{\partial C}{\partial y'}\right) = \frac{\partial}{\partial y'}(C \times \sqrt{1 - C}) \qquad (7.7b)$$

where $y' = y/Y_{90}$, Y_{90} is the characteristic distance where $C = 0.90$, $D' = D_t/((u_r)_{Hyd} \times \cos\theta \times Y_{90})$ is a dimensionless turbulent diffusivity and the rise velocity in hydrostatic pressure gradient $(u_r)_{Hyd}$ is assumed a constant. D' is the ratio of the air bubble diffusion coefficient to the rise velocity component normal to the flow direction time the characteristic transverse dimension of the shear flow.

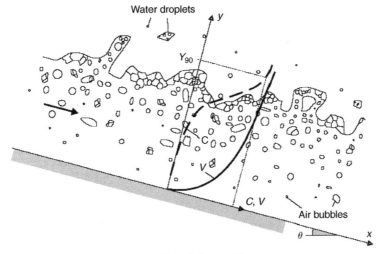

(A) Definition sketch.

(B) Self-aeration down Jordan weir on Lockyer Creek (Qld, Australia) on March 2011 (shutter speed: 1/8,000 s).

Figure 7.7. Self-aeration in a high-velocity open channel flow.

A first integration of Equation (7.7) leads to:

$$\frac{\partial C}{\partial y'} = \frac{1}{D'} \times C \times \sqrt{1 - C} \qquad (7.9)$$

Assuming a homogeneous turbulence across the flow ($D' = $ constant), a further integration yields:

$$C = 1 - \tanh^2\left(K' - \frac{y'}{2 \times D'}\right) \tag{7.10}$$

where K' is an integration constant and $\tanh(x)$ is the hyperbolic tangent function. The void fraction distribution (Eq. (7.10)) is a function of two constant parameters: the dimensionless diffusivity D' and the dimensionless constant K'. A relationship between D' and K' is deduced at the boundary condition $C = 0.90$ at $y' = 1$:

$$K' = K^* + \frac{1}{2 \times D'} \tag{7.11}$$

where $K^* = \tanh^{-1}(\sqrt{0.1}) = 0.32745015\dots$ If the diffusivity is unknown, it can deduced from the depth averaged void fraction C_{mean} defined as:

$$C_{mean} = \int\limits_0^1 C \times dy' \tag{7.12}$$

It yields:

$$C_{mean} = 2 \times D' \times \left(\tanh\left(K^* + \frac{1}{2 \times D'}\right) - \tanh(K^*)\right) \tag{7.13}$$

7.4 ADVECTIVE DIFFUSION OF AIR BUBBLES. ANALYTICAL SOLUTIONS

In turbulent shear flows, the air bubble entrainment processes differ substantially between singular aeration and interfacial aeration. Singular (local) air entrainment is localised at a flow discontinuity: e.g., the intersection of the impinging water jet with the receiving body of water. The air bubbles are entrained locally at the flow singularity: e.g., the toe of a hydraulic jump or at the impact of a plunging breaking wave (Fig. 7.4). The impingement perimeter is a source of air bubbles as well as a source of vorticity. Interfacial (continuous) aeration takes place along an air-water free-surface, usually parallel to the flow direction: e.g., spillway chute flow (Fig. 7.7). Across the free-surface, air is continuously entrapped and detrained, and the entrained air bubbles are advected in regions of relatively low shear.

In the following paragraphs, some analytical solutions of Equation (7.4) are developed for both singular and interfacial air entrainment processes.

7.4.1 Singular aeration

7.4.1.1 Air bubble entrainment at vertical plunging jets

Considering a vertical plunging jet, air bubbles may be entrained at impingement and carried downwards below the pool free surface (Fig. 7.8). This process is called plunging jet entrainment. In chemical engineering, plunging jets are used to stir chemicals as well as to increase gas-liquid mass transfer. Plunging jet devices are used also in industrial processes (e.g. bubble flotation of minerals) while planar plunging jets are observed at dam spillways

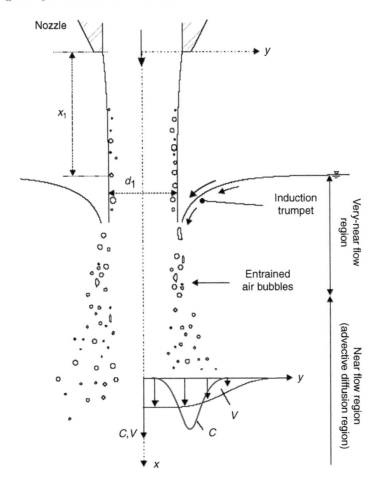

Figure 7.8. Advection of air bubbles downstream of the impingement of a vertical plunging jet.

and overfall drop structures. A related flow situation is the plunging breaking wave in the ocean (Fig. 7.3).

The air bubble diffusion at a plunging liquid jet is a form of advective diffusion. For a small control volume and neglecting the buoyancy effects, the continuity equation for air bubbles becomes:

$$div(C \times \overrightarrow{V}) = div(D_t \times \overrightarrow{grad}\, C) \tag{7.14}$$

In Equation (7.14), the bubble rise velocity term may be neglected because the jet velocity is much larger than the rise velocity.

For a circular plunging jet, assuming an uniform velocity distribution, for a constant diffusivity (in the radial direction) independent of the longitudinal location and for a small control volume delimited by streamlines (i.e. stream tube), Equation (7.14) becomes a simple advective diffusion equation:

$$\frac{V_1}{D_t} \times \frac{\partial C}{\partial x} = \frac{1}{r} \times \frac{\partial}{\partial y}\left(y \times \frac{\partial C}{\partial y}\right) \tag{7.15}$$

where x is the longitudinal direction, y is the radial distance from the jet centreline, V_1 is the jet impact velocity and the diffusivity term D_t averages the effects of the turbulent diffusion and of the longitudinal velocity gradient.

The boundary conditions are: $C(x < x_1, y \leq d_1/2) = 0$ and a circular source of total strength Q_{air} at $(x - x_1 = 0, y = d_1/2)$ where d_1 is the jet diameter at impact (Fig. 7.8). Equation (7.15) can be solved analytically by applying a superposition method. The general solution of the advective diffusion equation is:

$$C = \frac{Q_{air}}{Q_w} \times \frac{1}{4 \times D^{\#} \times \dfrac{x - x_1}{d_1/2}} \times \exp\left(-\frac{1}{4 \times D^{\#}} \times \frac{\left(\dfrac{y}{d_1/2}\right)^2 + 1}{\dfrac{x - x_1}{d_1/2}}\right)$$

$$\times I_o \left(\frac{1}{2 \times D^{\#}} \times \frac{\dfrac{y}{d_1/2}}{\dfrac{x - x_1}{d_1/2}}\right) \qquad \text{Circular plunging jet} \quad (7.16)$$

where I_o is the modified Bessel function of the first kind of order zero and $D^{\#} = D_t/(V_1 \times d_1/2)$.

For a two-dimensional free-falling jet, the air bubbles are entrapped at the point sources $(x = x_1, y = +d_1/2)$ and $(x = x_1, y = -d_1/2)$. Assuming an uniform velocity distribution, for a diffusion coefficient independent of the transverse location and for a small control volume (dx, dy) limited between two streamlines, the continuity equation (Eq. (7.14)) becomes a two-dimensional diffusion equation:

$$\frac{V_1}{D_t} \times \frac{\partial C}{\partial x} = \frac{\partial^2 C}{\partial y^2} \tag{7.17}$$

where y is the distance normal to the jet centreline (Fig. 7.8). The problem can be solved by superposing the contribution of each point source. The solution of the diffusion equation is:

$$C = \frac{1}{2} \times \frac{Q_{air}}{Q_w} \times \frac{1}{\sqrt{4 * \pi * D^{\#} \times \dfrac{x - x_1}{d_1}}}$$

$$\times \left(\exp\left(-\frac{1}{4 \times D^{\#}} \times \frac{\left(\dfrac{y}{d_1} - 1\right)^2}{\dfrac{x - x_1}{d_1}} \right) + \exp\left(-\frac{1}{4 \times D^{\#}} \times \frac{\left(\dfrac{y}{d_1} + 1\right)^2}{\dfrac{x - x_1}{d_1}} \right) \right)$$

$$\text{Two-dimensional plunging jet} \quad (7.18)$$

where Q_{air} is the entrained air flow rate, Q_w is the water flow rate, d_1 is the jet thickness at impact, and $D^{\#}$ is a dimensionless diffusivity: $D^{\#} = D_t/(V_1 \times d_1)$.

Discussion
Equations (7.16) and (7.18) are the exact analytical solutions of the advective diffusion of air bubbles (Eq. (7.18)). The two-dimensional and axi-symmetrical solutions differ because of the boundary conditions and of the integration method. Both solutions are three-dimensional

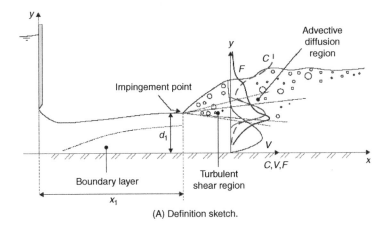

(A) Definition sketch.

(B) Hydraulic jump in a rectangular channel ($V_1/\sqrt{g \times d_1} = 6.5$, $\rho_w \times V_1 \times d_1/\mu_w = 1.5\ E + 5$) - Flow from left to right.

(C) Hydraulic jump downstream of Moree weir (NSW, Australia) on 16 December 1997 during some runoff – Flow from top left to bottom right.

Figure 7.9. Advection of air bubbles in hydraulic jumps.

solutions valid in the developing bubbly region and in the fully-aerated flow region. They were successfully compared with a range of experimental data.

7.4.1.2   Air bubble entrainment in a horizontal hydraulic jump

A hydraulic jump is the sudden transition from a supercritical flow into a slower, subcritical motion (Fig. 7.9). It is characterised by strong energy dissipation, spray and splashing and air bubble entrainment. The hydraulic jump is sometimes described as the limiting case of a horizontal supported plunging jet.

Assuming an uniform velocity distribution, for a constant diffusivity independent of the longitudinal and transverse location, Equation (7.14) becomes:

$$V_1 \times \frac{\partial C}{\partial x} + u_r \times \frac{\partial C}{\partial y} = D_t \times \frac{\partial^2 C}{\partial y^2} \tag{7.19}$$

where V_1 is the inflow velocity and the rise velocity is assumed constant. With a change of variable ($X = x - x_1 + u_r/V_1 \times y$) and assuming $u_r/V_1 \ll 1$, Equation (7.19) becomes a two-dimensional diffusion equation:

$$\frac{V_1}{D_t} \times \frac{\partial C}{\partial X} = \frac{\partial^2 C}{\partial y^2} \tag{7.20}$$

In a hydraulic jump, the air bubbles are supplied by a point source located at ($X = u_r/V_1 \times d_1, y = +d_1$) and the strength of the source is Q_{air}/W where W is the channel width.

The diffusion equation can be solved by applying the method of images and assuming an infinitesimally long channel bed. It yields:

$$C = \frac{Q_{air}}{Q_w} \times \frac{1}{\sqrt{4 \times \pi \times D^\# * X'}}$$

$$\times \left(\exp\left(-\frac{1}{4 \times D^\#} \times \frac{\left(\frac{y}{d_1} - 1\right)^2}{X'} \right) + \exp\left(-\frac{1}{4 \times D^\#} * \frac{\left(\frac{y}{d_1} + 1\right)^2}{X'} \right) \right) \tag{7.21}$$

where d_1 is the inflow depth, $D^\#$ is a dimensionless diffusivity: $D^\# = D_t/(V_1 \times d_1)$ and:

$$X' = \frac{X}{d_1} = \frac{x - x_1}{d_1} \times \left(1 + \frac{u_r}{V_1} \times \frac{y}{x - x_1} \right)$$

Equation (7.21) is close to Equation (7.18) but the distribution of void fraction is shifted upwards as a consequence of some buoyancy effect. Further the definition of d_1 differs (Fig. 7.9). In practice, Equation (7.21) provides a good agreement with experimental data in the advective diffusion region of hydraulic jumps with partially-developed inflow conditions.

7.4.2   Interfacial aeration

7.4.2.1   Interfacial aeration in a water jet discharging into the atmosphere

High velocity turbulent water jets discharging into the atmosphere are often used in hydraulic structures to dissipate energy. Typical examples include jet flows downstream of a ski jump at the toe of a spillway, water jets issued from bottom outlets, flows above a bottom aeration device along a spillway and water jets in fountains (Fig. 7.10). Other applications include mixing devices in chemical plants and spray devices. High-velocity water jets are used also for fire-fighting jet cutting (e.g. coal mining), with Pelton turbines and for irrigation.

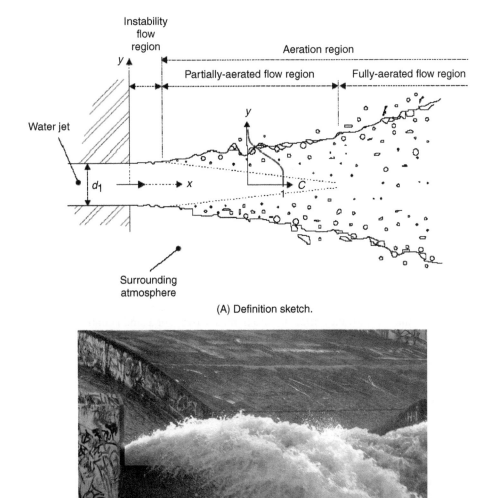

(A) Definition sketch.

(B) High-velocity water jet taking off the flip bucket of Lake Kurongbah Dam spillway (Qld, Australia) on 22 May 2009 during a low overflow event.

Figure 7.10. Advective dispersion of air bubbles in a turbulent water jet discharging into air.

(C) Vertical water jet (Jet d'eau) at the Bassin de l'Obélisque, Château de Versailles (France) on 20 June 1998 – Designed between 1704 and 1705 by Jules Hardouin-Mansart.

Figure 7.10. Continued

Considering a water jet discharging into air, the pressure distribution is quasi-uniform across the jet and the buoyancy effect is zero in most cases. For a small control volume, the advective diffusion equation for air bubbles in a steady flow is:

$$div(C \times \overrightarrow{V}) = div(D_t \times \overrightarrow{grad}\ C) \qquad (7.14)$$

For a circular water jet, the continuity equation for air becomes:

$$\frac{\partial}{\partial x}(C \times V_1) = \frac{1}{y} \times \frac{\partial}{\partial y}\left(D_t \times y \times \frac{\partial C}{\partial y}\right) \qquad (7.22)$$

where x is the longitudinal direction, y is the radial direction, V_1 is the jet velocity and D_t is the turbulent diffusivity in the radial direction.

Assuming a constant diffusivity D_t in the radial direction, and after separating the variables, the void fraction:

$$C = \mathbf{u} \times \exp\left(-\frac{D_t}{V_1} \times \alpha_n^2 \times x\right)$$

is a solution of the continuity equation provided that u is a function of y only satisfying the Bessel's equation of order zero:

$$\frac{\partial^2 \mathbf{u}}{\partial y^2} + \frac{1}{y} \times \frac{\partial \mathbf{u}}{\partial y} + a_n^2 \times \mathbf{u} = 0 \qquad (7.23)$$

At each position x, the diffusivity D_t is assumed a constant independent of the transverse location y. The boundary conditions are $C = 0.9$ at $y = Y_{90}$ for $x > 0$ and for $C = 0$ for $x < 0$. An analytical solution is a series of Bessel functions:

$$C = 0.9 - \frac{1.8}{Y_{90}} \times \sum_{n=1}^{+\infty} \frac{J_o(y \times \alpha_n)}{\alpha_n \times J_1(Y_{90} \times \alpha_n)} \times \exp\left(-\frac{D_t}{V_1} \times \alpha_n^2 \times x\right) \tag{7.24}$$

where J_o is the Bessel function of the first kind of order zero, α_n is the positive root of: $J_o(Y_{90} \times \alpha_n) = 0$, and J_1 is the Bessel function of the first kind of order one. Equation (7.24) was numerically computed by Carslaw and Jaeger (1959) for several values of the dimensionless diffusivity $D'' = D_t \times x/(V_1 \times Y_{90}^2)$.

Equation (7.24) is valid close to and away from the jet nozzle. It is a three-dimensional solution of the diffusion equation that it is valid when the clear water core of the jet disappears and the jet becomes fully-aerated.

For a two-dimensional water jet, assuming an uniform velocity distribution, and for a constant diffusivity independent of the longitudinal and transverse location, Equation (7.14) becomes:

$$V_1 \times \frac{\partial C}{\partial x} = D_t \times \frac{\partial^2 C}{\partial y^2} \tag{7.25}$$

where V_1 is the inflow depth. Equation (7.25) is a basic diffusion equation (Crank 1956, Carslaw and Jaeger 1959).

The boundary conditions are: $\lim(C(x > 0, y \to +\infty)) = 1$ and $\lim(C(x > 0, y \to -\infty)) = 1$ where the positive direction for the x- and y-axes is shown on Figure 7.10A. Note that, at the edge of the free-shear layer, the rapid change of shear stress is dominant. The effect of the removal of the bottom shear stress is to allow the fluid to accelerate. Further downstream the acceleration decreases rapidly down to zero.

The analytical solution of Equation (7.25) is:

$$C = \frac{1}{2} \times \left(2 + erf\left(\frac{\frac{y}{d_1} - \frac{1}{2}}{2 \times \sqrt{\frac{D_t}{V_1 \times d_1}} \times \frac{x}{d_1}}\right) + erf\left(\frac{\frac{y}{d_1} + \frac{1}{2}}{2 \times \sqrt{\frac{D_t}{V_1 \times d_1}} \times \frac{x}{d_1}}\right)\right) \tag{7.26}$$

where d_1 is the jet thickness at nozzle, *erf* is the Gaussian error function, and the diffusivity D_t averages the effect of the turbulence on the transverse dispersion and of the longitudinal velocity gradient. The boundary conditions imply the existence of a clear-water region between the air-bubble diffusion layers in the initial jet flow region as sketched in Figure 7.10A.

The two-dimensional case may be simplified for a two-dimensional free-shear layer: e.g. an open channel flow taking off a spillway aeration device or a ski jump. The analytical solution for a free shear layer is:

$$C = \frac{1}{2} \times \left(1 + erf\left(\frac{\frac{y}{d_1}}{2 \times \sqrt{\frac{D_t}{V_1 \times d_1}} \times \frac{x}{d_1}}\right)\right) \tag{7.27}$$

where $y = 0$ at the flow singularity (i.e. nozzle edge) and $y > 0$ towards the atmosphere.

7.4.2.2 Interfacial aeration in a high-velocity open channel flow

For a two-dimensional steady open channel flow, a complete solution was developed in section 7.3.3. Assuming a homogeneous turbulence across the flow ($D' = \text{constant}$), the integration of the advective diffusion equation yields:

$$C = 1 - \tanh^2\left(K' - \frac{y'}{2 \times D'}\right) \tag{7.10}$$

where K' is an integration constant (Eq. (7.11)) and $\tanh(x)$ is the hyperbolic tangent function. The void fraction distribution (Eq. (7.10)) is a function of the dimensionless diffusivity $D' = D_t/((u_r)_{Hyd} \times \cos\theta \times Y_{90})$ assuming that both the turbulent diffusivity D_t and bubble rise velocity in hydrostatic pressure gradient $(u)_{Hyd}$ are constant.

Equation (7.10) was successfully tested against prototype and laboratory data, and a pertinent discussion is developed in Chanson (1997).

7.4.3 Discussion

The above expressions (Sections 7.4.1 & 7.4.2) were developed assuming a constant, uniform air bubble diffusivity. While the analytical solutions are in close agreement with experimental data (e.g. Chanson 1997, Toombes 2002, Gonzalez 2005, Murzyn et al. 2005), the distributions of turbulent diffusivity are unlikely to be uniform in complex flow situations. Two well-documented examples are the skimming flow on a stepped spillway and the flow downstream of a drop structure (Fig. 7.11).

For a two-dimensional open channel flow, the advective diffusion equation for air bubbles yields:

$$\frac{\partial}{\partial y'}\left(D' \times \frac{\partial C}{\partial y'}\right) = \frac{\partial}{\partial y'}(C \times \sqrt{1 - C}) \tag{7.7b}$$

where $y' = y/Y_{90}$, Y_{90} is the characteristic distance where $C = 0.90$, and $D' = D_t/((u_r)_{Hyd} \times \cos\theta \times Y_{90})$ is a dimensionless turbulent diffusivity that is the ratio of the air bubble diffusion coefficient to the rise velocity component normal to the flow direction time the characteristic transverse dimension of the shear flow. In a skimming flow on a stepped chute (Fig. 7.11A), the flow is extremely turbulent and the air bubble diffusivity distribution may be approximated by:

$$D' = \frac{D_o}{1 - 2 \times \left(y' - \frac{1}{3}\right)^2} \tag{7.28}$$

The integration of the air bubble diffusion equation yields a S-shape void fraction profile:

$$C = 1 - \tanh^2\left(K' - \frac{y'}{2 \times D_o} + \frac{\left(y' - \frac{1}{3}\right)^3}{3 \times D_o}\right) \tag{7.29}$$

where K' is an integration constant and D_o is a function of the mean void fraction only:

$$K' = K^* + \frac{1}{2 \times D_o} - \frac{8}{81 \times D_o} \quad \text{with } K^* = 0.32745015\ldots \tag{7.30}$$

$$C_{mean} = 0.7622 \times (1.0434 - \exp(-3.614 \times D_o)) \tag{7.31}$$

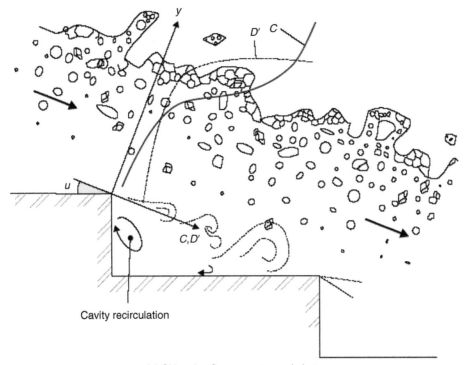

(a) Skimming flow on a stepped chute.

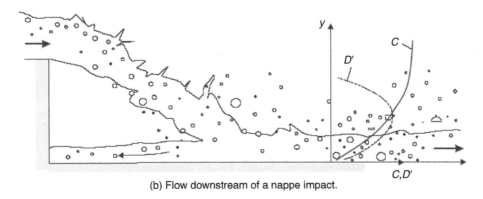

(b) Flow downstream of a nappe impact.

Figure 7.11. Advective dispersion of air bubbles in highly-turbulent open channel flows.

Equations (7.31) and (7.29) are sketched in Figure 7.11A. They were found to agree well with experimental measurements at step edges.

Downstream of a drop structure (Fig. 7.11B), the flow is fragmented, highly aerated and extremely turbulent. A realistic void fraction distribution model may be developed assuming a quasi-parabolic bubble diffusivity distribution:

$$D' = \frac{C \times \sqrt{1 - C}}{\lambda \times (K' - C)} \tag{7.32}$$

The integration of Equation (7.7b) yields:

$$C = K' \times (1 - \exp(-\lambda \times y')) \qquad (7.33)$$

where K' and λ are some dimensionless functions of the mean air content only:

$$K' = \frac{0.9}{1 - \exp(-\lambda)} \qquad (7.34)$$

$$C_{mean} = K' - \frac{0.9}{\lambda} \qquad (7.35)$$

Equations (7.32) and (7.33) are sketched in Figure 7.11B. In practice, Equation (7.33) applies to highly-aerated, fragmented flows like the steady flows downstream of drop structures and spillway bottom aeration devices, and the transition flows on stepped chutes, as well as the leading edge of unsteady surges. Note that the depth-averaged air content must satisfy $C_{mean} > 0.45$.

7.5   STRUCTURE OF THE BUBBLY FLOW

In Sections 7.3 and 7.4, the advective diffusion equation for air bubbles is developed and solved in terms of the void fraction. The void fraction is a gross parameter that does not describe the air-water structures, the bubbly flow turbulence nor the interactions between entrained bubbles and turbulent shear. Herein recent experimental developments are discussed in terms of the longitudinal flow structure and the air-water time and length scales following Chanson and Carosi (2007).

7.5.1   Streamwise particle grouping

With modern phase-detection intrusive probes, the probe output signals provide a complete characterisation of the streamwise air-water structure at one point. Figure 7.12 illustrates the operation of such a probe. Figure 7.12B shows two probes in a bubbly flow, while Figure 7.12A presents the piercing of air bubbles by the probe sensor. Some simple signal processing yields the basic statistical moments of air and water chords as well as the probability distribution functions of the chord sizes.

In turbulent shear flows, the experimental results demonstrated a broad spectrum of bubble chords. The range of bubble chord lengths extended over several orders of magnitude including at low void fractions. The distributions of bubble chords were skewed with a preponderance of small bubbles relative to the mean. The probability distribution functions of bubble chords tended to follow a log–normal and gamma distributions. Similar findings were observed in a variety of flows encompassing hydraulic jumps, plunging jets, dropshaft flows and high-velocity open channel flows.

In addition of void fraction and bubble chord distributions, some further signal processing may provide some information on the streamwise structure of the air-water flow including bubble clustering. A concentration of bubbles within some relatively short intervals of time may indicate some clustering while it may be instead the consequence of a random occurrence. The study of particle clustering events is relevant to infer whether the formation frequency responds to some particular frequencies of the flow. Figure 7.13 illustrates some occurrence of bubble pairing in the shear layer of a hydraulic jump. The binary pairing indicator is unity if the water chord time between adjacent bubbles is less than 10% of the

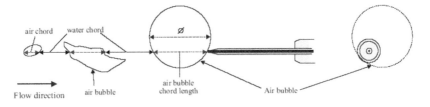

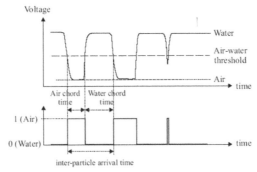

(A) Sketch of a phase-detection intrusive probe and its signal output.

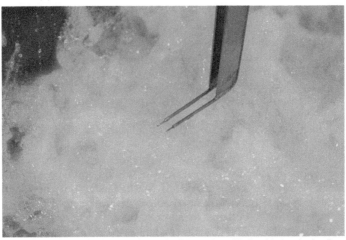

(B) Photograph of two single-tip conductivity probes side-by-side in a hydraulic jump ($Fr_1 = 6.6$, $\rho_w \times V_1 \times d_1/\mu_w = 8.6\ E+4$) – Flow from bottom left to top right.

Figure 7.12. Phase-detection intrusive probe in turbulent air-water flows.

median water chord time. The pattern of vertical lines seen in Figure 7.13 is an indication of patterns in which bubbles tend to form bubble groups.

One method is based upon the analysis of the water chord between two adjacent air bubbles (Fig. 7.12A). If two bubbles are closer than a particular length scale, they can be considered a group of bubbles. The characteristic water length scale may be related to the water chord statistics: e.g., a bubble cluster may be defined when the water chord was less than a given percentage of the mean water chord. Another criterion may be related to the leading bubble size itself, since bubbles within that distance are in the near-wake of and may be influenced by the leading particle.

Binary pairing indicator

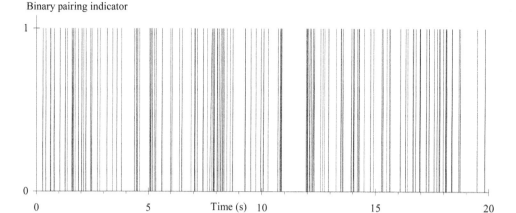

Figure 7.13. Closely spaced bubble pairs in the developing shear layer of a hydraulic jump $- Fr_1 = 8.5$, $\rho_\omega \times V_1 \times d_1/\mu_\omega = 9.8\,E + 4$, $x - x_1 = 0.4$ m, $d_1 = 0.024$ m, $y/d_1 = 1.33$, $C = 0.20$, $F = 158$ Hz.

Typical results may include the percentage of bubbles in clusters, the number of clusters per second, and the average number of bubbles per cluster. Extensive experiments in open channels, hydraulic jumps and plunging jets suggested that the outcomes were little affected by the cluster criterion selection. Most results indicated that the streamwise structure of turbulent flows was characterised by about 10 to 30% of bubbles travelling as parts of a group/cluster, with a very large majority of clusters comprising of 2 bubbles only. The experimental experience suggested further that a proper cluster analysis requires a high-frequency scan rate for a relatively long scan duration. However the analysis is restricted to the longitudinal distribution of bubbles and does not take into account particles travelling side by side.

Some typical result is presented in Figure 7.14. Figure 7.14 shows the vertical distribution of the percentage of bubbles in clusters (lower horizontal axis) and average number of bubbles per cluster (upper horizontal axis) in the advective diffusion region of a hydraulic jump. The void fraction distribution is also shown for completeness. The criterion for cluster existence is a water chord less than 10% of the median water chord. For this example, about 5 to 15% of all bubbles were part of a cluster structure and the average number of bubbles per cluster was about 2.1.

For a dispersed phase, a complementary approach is based upon an inter-particle arrival time analysis. The inter-particle arrival time is defined as the time between the arrival of two consecutive bubbles recorded by a probe sensor fixed in space (Fig. 7.12A). The distribution of inter-particle arrival times provides some information on the randomness of the structure. Random dispersed flows are those whose inter-particle arrival time distributions follow inhomogeneous Poisson statistics assuming non-interacting point particles (Edwards and Marx 1995a). In other words, an ideal dispersed flow is driven by a superposition of Poisson processes of bubble sizes, and any deviation from a Poisson process indicates some unsteadiness and particle clustering.

In practice, the analysis is conducted by breaking down the air-water flow data into narrow classes of particles of comparable sizes that are expected to have the same behaviour (Edwards and Marx 1995b). A simple means consists in dividing the bubble/droplet population in terms of the air/water chord time. The inter-particle arrival time analysis may provide some information on preferential clustering for particular classes of particle sizes.

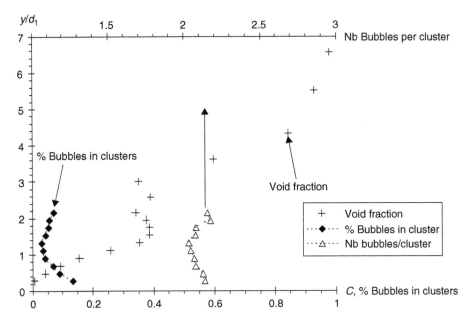

Figure 7.14. Bubble clustering in the bubbly flow region of a hydraulic jump: percentage of bubbles in clusters, average number of bubbles per cluster and void fraction – Cluster criterion: water chord time $<10\%$ median water chord time – $Fr_1 = 8.5$, $\rho_w \times V_1 \times d_1/\mu_w = 9.8\,E + 4$, $x - x_1 = 0.3$ m, $d_1 = 0.024$ m.

Some results in terms of inter-particle arrival time distributions are shown in Figure 7.15 for the same flow conditions and at the same cross-section as the data presented in Figure 7.14. Chi-square values are given in the Figure 7.7 captions. Figure 7.15 presents some inter-particle arrival time results for two chord time classes of the same sample (0 to 0.5 ms and 3 to 5 ms). For each class of bubble sizes, a comparison between data and Poisson distribution gives some information on its randomness. For example, Figure 7.15A shows that the data for bubble chord times below 0.5 m did not experience a random behaviour because the experimental and theoretical distributions differed substantially in shape. The second smallest inter-particle time class (0.5–1 m) had a population that was 2.5 times the expected value or about 11 standard deviations too large. This indicates that there was a higher probability of having bubbles with shorter inter-particle arrival times, hence some bubble clustering occurred. Simply the smallest class of bubble chord times did not exhibit the characteristics of a random process.

Altogether both approaches are complementary, although the inter-particle arrival time analysis may give some greater insight on the range of particle sizes affected by clustering.

7.5.2 Correlation analyses

When two or more phase detection probe sensors are simultaneously sampled, some correlation analyses may provide additional information on the bubbly flow structure. A well-known application is the use of dual tip probe to measure the interfacial velocity (Fig. 7.16). With large void fractions ($C > 0.10$), a cross-correlation analysis between the two probe sensors yields the time averaged velocity:

$$V = \frac{\Delta x}{T} \tag{7.36}$$

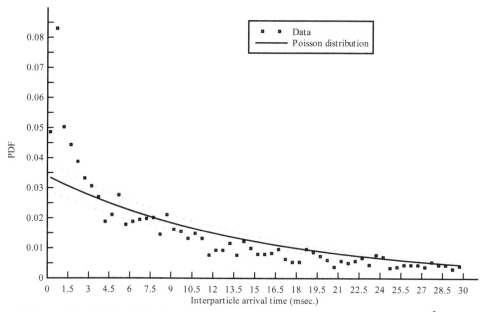

(A) Inter-particle arrival time distributions for bubble chord times between 0 and 0.5 ms, 3055 bubbles, $\chi^2 = 461$.

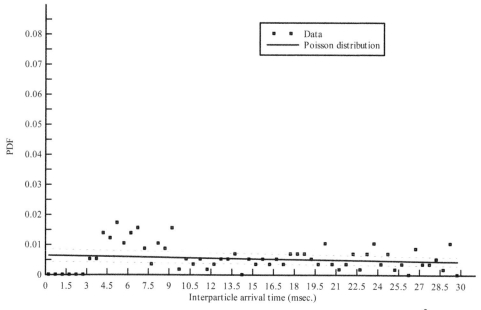

(B) Inter-particle arrival time distributions for bubble chord times between 3 and 5 ms, 581 bubbles, $\chi^2 = 110$.

Figure 7.15. Inter-particle arrival time distributions in the bubbly flow region of a hydraulic jump for different classes of air chord times – Comparison between data and Poisson distribution – Expected deviations from the Poisson distribution for each sample are shown in dashed lines – $Fr_1 = 8.5$, $\rho_\omega \times V_1 \times d_1 / \mu_\omega = 9.8\,E+4$, $x - x_1 = 0.3$ m, $d_1 = 0.024$ m.

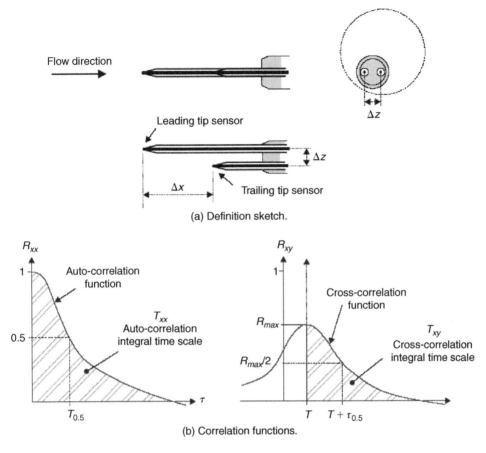

Figure 7.16. Dual sensor phase detection probe.

where T is the air-water interfacial travel time for which the cross-correlation function is maximum and Δx is the longitudinal distance between probe sensors (Fig. 7.16). Turbulence levels may be further derived from the relative width of the cross-correlation function:

$$Tu = 0.851 \times \frac{\sqrt{\tau_{0.5}^2 - T_{0.5}^2}}{T} \qquad (7.37)$$

where $\tau_{0.5}$ is the time scale for which the cross-correlation function is half of its maximum value such as: $R_{xy}(T + \tau_{0.5}) = 0.5 \times R_{xy}(T)$, R_{xy} is the normalised cross-correlation function, and $T_{0.5}$ is the characteristic time for which the normalised auto-correlation function equals: $R_{xx}(T_{0.5}) = 0.5$ (Fig. 7.16). Physically, a thin narrow cross-correlation function $((\tau_{0.5} - T_{0.5})/T \ll 1)$ must correspond to little fluctuations in the interfacial velocity, hence a small turbulence level Tu. While Equation (7.37) is not the true turbulence intensity u'/V, it is an expression of some turbulence level and average velocity fluctuations.

More generally, when two probe sensors are separated by a transverse or streamwise distance, their signals may be analysed in terms of the auto-correlation and cross-correlation functions R_{xx} and R_{xy} respectively. Figure 7.12B shows two probe sensors separated by a transverse distance Δz, while Figure 7.16 presents two probe sensors separated by a streamwise distance Δx. Practically the original data set may be segmented because the periodogram resolution is inversely proportional to the number of samples and it could be biased with large data sets (Hayes 1996).

Basic correlation analysis results include the maximum cross-correlation coefficient $(R_{xy})_{max}$, and the integral time scales T_{xx} and T_{xy} where:

$$T_{xx} = \int_{\tau=0}^{\tau=\tau(R_{xx}=0)} R_{xx}(\tau) \times d\tau \qquad (7.38)$$

$$T_{xy} = \int_{\tau=\tau(R_{xy}=(R_{xy})_{max})}^{\tau=\tau(R_{xy}=0)} R_{xy}(\tau) \times d\tau \qquad (7.39)$$

where R_{xx} is the normalised auto-correlation function, τ is the time lag, and R_{xy} is the normalised cross-correlation function between the two probe output signals (Fig. 7.16). The auto-correlation integral time scale T_{xx} represents the integral time scale of the longitudinal bubbly flow structure. It is a characteristic time of the eddies advecting the air-water interfaces in the streamwise direction. The cross-correlation time scale T_{xy} is a characteristic time scale of the vortices with a length scale y advecting the air-water flow structures. The length scale y may be a transverse separation distance Δz or a streamwise separation Δx.

When identical experiments are repeated with different separation distances y ($y = \Delta z$ or Δx), an integral turbulent length scale may be calculated as:

$$L_{xy} = \int_{y=0}^{y=y((R_{xy})_{max}=0)} (R_{xy})_{max} \times dy \qquad (7.40)$$

The length scale L_{xy} represents a measure of the transverse/streamwise length scale of the large vortical structures advecting air bubbles and air-water packets.

A turbulence integral time scale is:

$$\mathbf{T} = \frac{\int_{y=0}^{y=y((R_{xy})_{max}=0)} (R_{xy})_{max} \times T_{xy} \times dy}{L_{xy}}$$

The turbulence integral time scale $\mathbf{T}$ represents the transverse/streamwise integral time scale of the large eddies advecting air bubbles.

Figures 7.17 to 7.19 present some experimental results obtained in a hydraulic jump on a horizontal channel and in a skimming flow on a stepped channel. In both flow situations, the distributions of integral time scales showed a marked peak for $0.4 \le C \le 0.6$ (Fig. 7.17 and 7.18). Note that Figure 7.17 presents some transverse time scales T_{xy} while Figure 7.18 shows some longitudinal time scales T_{xy}. The distributions of transverse integral length scales exhibited some marked differences that may reflect the differences in turbulent mixing and air bubble advection processes between hydraulic jump and skimming flows. In Figure 7.19, the integral turbulent length scale L_{xy} represents a measure of the transverse

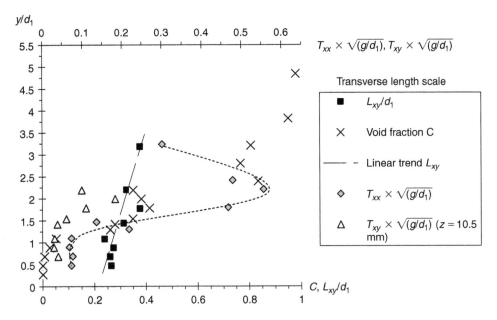

Figure 7.17. Dimensionless distributions of auto- and cross-correlation time scales $T_{xx} \times \sqrt{g/d_1}$ and $T_{xy} \times \sqrt{g/d_1}$ (transverse time scale, $y = \Delta z = 10.5$ mm), and transverse integral turbulent length scale L_{xy}/d_1 of Expression in a hydraulic jump – $Fr_1 = 7.9$, $\rho_\omega \times V_1 \times d_1/\mu_\omega = 9.4\ E+4$, $x - x_1 = 0.1$ m, $d_1 = 0.0245$ m.

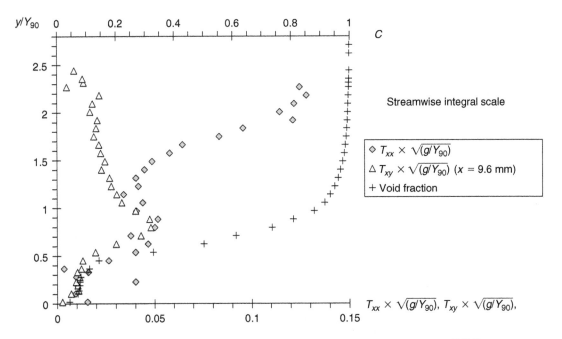

Figure 7.18. Dimensionless distributions of auto- and cross-correlation time scales $T_{xx} \times \sqrt{g/Y_{90}}$ and $T_{xy} \times \sqrt{g/Y_{90}}$ (longitudinal time scale, $y = \Delta x = 9.6$ mm) in a skimming flow on a stepped chute – $d_c/h = 1.15$, $\rho_\omega \times V_1 \times d_1/\mu_\omega = 1.2\ E+5$, Step 10, $Y_{90} = 0.0574$ m, $h = 0.1$ m, $\theta = 22°$.

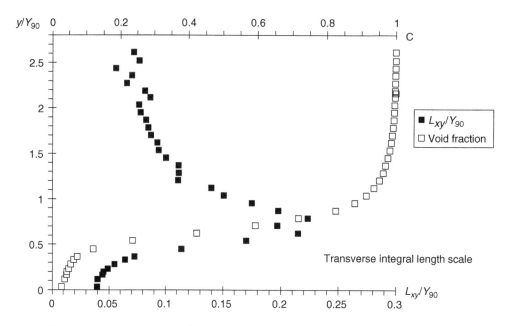

Figure 7.19. Dimensionless distributions of transverse integral turbulent length scale L_{xy}/Y_{90} in a skimming flow on a stepped chute – $d_c/h = 1.15$, $\rho_\omega \times V_1 \times d_1/\mu_\omega = 1.2\,E + 5$, Step 10, $Y_{90} = 0.0598$ m, $h = 0.1$ m, $\theta = 22°$.

size of large vortical structures advecting air bubbles in the skimming flow regime. The air-water turbulent length scale is closely related to the characteristic air-water depth Y_{90}: i.e., $0.05 \leq L_{xy}/Y_{90} \leq 0.2$ (Fig. 7.19). Note that both the integral turbulent length and time scales were maximum for about C = 0.5 to 0.7 (Fig. 7.18 & 7.19). The finding emphasises the existence of large-scale turbulent structures in the intermediate zone ($0.3 < C < 0.7$) of the flow, and it is hypothesised that these large vortices may play a preponderant role in terms of turbulent dissipation.

7.6 CONCLUSION

In turbulent free-surface flows, the strong interactions between turbulent waters and surrounding atmosphere may lead to some self-aeration, air entrainment, spray and splashing. This is the entrainment/entrapment of air bubbles which are advected within the bulk of the flow and the light diffraction on the entrained bubble interfaces gives a whitish appearance to the waters, called commonly white waters. In Nature, free-surface aeration may be encountered at waterfalls, in steep mountain streams and river rapids, as well as at breaking. The 'white waters' provide always some spectacular effect as illustrated in Figures 7.20 and 7.21. Although classical examples include the tidal bore of the Qiantang river in China, the Zambesi rapids in Africa, and the 980 m high Angel Falls in South America, 'white waters' are observed also in smaller streams, torrents and rivers. The rushing waters may become gravitationless in waterfalls, impacting downstream on rocks and water pools where their impact is surrounded by splashing, spray and fog as at Niagara Falls and Iguazu Falls. Self-aeration in man-made structures is also common, ranging from artistic fountains, attraction parks to engineering and industrial applications (Fig. 7.20 & 7.21).

Figure 7.20. Bassin de Latone,Château de Versailles (France) on 27 July 2008 (shutter speed: 1/800 s) – Built between 1668 et 1670 by André Le Nôtre, the fountain was inspired by the Metamorphosis by Ovid.

The entrainment of air bubbles may be localised at a flow discontinuity or continuous along an air-water free-surface: i.e., singular and interfacial aeration respectively. At a flow singularity, the air bubbles are entrained locally at the impinging perimeter and they are advected in a region of high shear. Interfacial aeration is the air bubble entrainment process

Figure 7.21. Splashing ahead of a water slide ride at Window of China (Taoyuan, Taiwan) on 15 November 2010
(shutter speed: 1/200 s).

along an air-water interface that is parallel to the flow direction. A condition for the onset of air bubble entrainment may be expressed in terms of the tangential Reynolds stress and the fluid properties. With both singular and interfacial aeration, the void fraction distributions may be modelled by some analytical solutions of the advective diffusion equation for air bubbles. Examples are illustrated and some comparison between physical data ad analytical models is presented.

The microscopic structure of turbulent bubbly flows is discussed based upon some developments in metrology and signal processing. The findings may provide new information on the air-water flow structure and the turbulent eddies advecting the bubbles.

The results bring new information on the fluid dynamics of air-water flows. They revealed the turbulent nature of the complex two-phase flows. Further developments are needed. For example, physical studies at prototype scale could be undertaken, while numerical modelling of air-water flows may be a future research topic. The computing approach will not be easy because the turbulent free-surface flows encompass many challenges including two-phase flow, turbulence, free surface fluctuations … It is believed that the interactions between entrained air bubbles and turbulence will remain a key challenge for the 21st century researchers.

7.7 MATHEMATICAL AIDS

Definition	Expression	Remarks
Surface area of a spheroid radii r_1, r_2	$A = 2 \times \pi \times r_1^2 + \pi \times \dfrac{r_2^2}{\sqrt{1 - \dfrac{r_2^2}{r_1^2}}} \times Ln\left(\dfrac{1 + \sqrt{1 - \dfrac{r_2^2}{r_1^2}}}{1 - \sqrt{1 - \dfrac{r_2^2}{r_1^2}}}\right)$	r_1: equatorial radius, r_2: polar radius. Oblate spheroid $(r_1 > r_2)$.
	$A = 2 \times \pi \times r_1 \times \left(r_1 + r_2 \times \dfrac{Arcsin\left(\sqrt{1 - \dfrac{r_1^2}{r_2^2}}\right)}{\sqrt{1 - \dfrac{r_1^1}{r_2^2}}}\right)$	Prolate spheroid $(r_1 < r_2)$.
Bessel function of the first kind of order zero	$J_o(u) = 1 - \dfrac{u^2}{2^2} + \dfrac{u^4}{2^2 \times 4^2} - \dfrac{u^6}{2^2 \times 4^2 * 6^2} + \dots$	also called modified Bessel function of the first kind of order zero
Bessel function of the first kind of order one	$J_1(u) = \dfrac{u}{2} - \dfrac{u^3}{2^2 \times 4} + \dfrac{u^5}{2^2 \times 4^2 \times 6} - \dfrac{u^7}{2^2 \times 4^2 * 6^2 \times 8} + \dots$	
Gaussian error function	$erf(u) = \dfrac{2}{\sqrt{p}} \times \int_0^u \exp(-t^2) \times dt$	also called error function.

APPENDIX A – LIST OF SYMBOLS

	List of Symbols	
Symbol	Definition	Dimensions or Units
A	bubble surface area	$[L^2]$
C	void fraction defined as the volume of air per unit volume of air and water	
C_{mean}	depth-averaged void fraction	
D'	ratio of air bubble diffusion coefficient to rise velocity component normal to the flow direction time the characteristic transverse dimension of the shear flow	
D_t	air bubble turbulent diffusion coefficient	$[L^2 \, T^{-1}]$
D_0	dimensionless function of the void fraction	
$D^{\#}$	dimensionless air bubble turbulent diffusion coefficient	

(Continued)

List of Symbols

Symbol	Definition	Dimensions or Units
F	air bubble count rate defined as the number of bubbles impacting the probe sensor per second	[Hz]
Fr_1	inflow Froude number of a hydraulic jump	
J_0	Bessel function of the first kind of order zero	
J_1	Bessel function of the first kind of order one	
K'	dimensionless integration constant	
L_{xy}	integral turbulent length scale	[L]
P	pressure	$[\text{N L}^{-2}]$
Q_{air}	entrained air flow rate	$[\text{L}^3 \cdot \text{T}^{-1}]$
Q_{water}	water discharge	$[\text{L}^3 \cdot \text{T}^{-1}]$
R_{xx}	normalized auto-correlation function	
R_{xy}	normalized cross-correlation function	
T	air-water interfacial travel time for which R_{xy} is maximum	[T]
T	transverse/streamwise turbulent integral time scale	[T]
$T_{0.5}$	characteristic time for which $R_{xx} = 0.5$	[T]
T_{xx}	auto-correlation integral time scale	[T]
T_{xy}	cross-correlation integral time scale	[T]
Tu	turbulence intensity	
V_e	onset velocity for air entrainment	[m s^{-1}]
V_x	streamwise velocity	[m s^{-1}]
V_y	transverse velocity	[m s^{-1}]
V_1	jet impact velocity or inflow velocity in the hydraulic jump	[m s^{-1}]
$\vec{V}$	advective velocity vector	[m s^{-1}]
Y_{90}	characteristic distance where $C = 0.90$	[L]
d_{ab}	air bubble diameter	[L]
d_1	jet thickness at impact or inflow depth in hydraulic jump	[L]
erf	Gaussian error function	
g	gravitational acceleration constant	$[\text{L T}^{-2}]$
r	radius of sphere	[L]
r_1	radius of curvature of the free surface deformation	[L]
r_2	radius of curvature of the free surface deformation	[L]
r_1	equatorial radius of the ellipsoid	[L]
r_2	polar radius of the ellipsoid	[L]
t	time	[T]
$\vec{u_r}$	bubble rise velocity vector	$[\text{m} \cdot \text{s}^{-1}]$
u_r	bubble rise velocity	$[\text{m} \cdot \text{s}^{-1}]$
u_r	bubble rise velocity in a hydrostatic pressure gradient	$[\text{m} \cdot \text{s}^{-1}]$
v_i	turbulent velocity fluctuation in the streamwise direction	$[\text{m} \cdot \text{s}^{-1}]$
v_j	turbulent velocity fluctuation in the normal direction	$[\text{m} \cdot \text{s}^{-1}]$
x	longitudinal/streamwise direction	[L]
x_1	distance between the gate and the jump toe	[L]
y	transverse or radial direction	[L]
y'	dimensionless transverse or radial direction: $y' = y/Y_{90}$	
z	vertical direction positive upward	[L]

(Continued)

List of Symbols		
Symbol	Definition	Dimensions or Units
Δx	longitudinal distance between probe sensors	[L]
Δy	transverse distance between probe sensors	[L]
α_n	positive root for $J = (Y_{90} * \alpha_n) = 0$	
θ	angle between the horizontal and the channel invert	
λ	dimensionless function of the mean air content	
μ_w	water dynamic viscosity	$[\text{M L}^{-1}\,\text{T}^{-1}]$
ρ_w	water density	$[\text{kg m}^{-3}]$
σ	surface tension between air and water	$[\text{N m}^{-1}]$
τ	time lag	[T]
$\tau_{0.5}$	time scale for which $R_{xy} = 0.5 \times R_{xy}(\text{T})$	[T]

APPENDIX B – SYNOPSIS

Self-seration is the entrainment/entrapment of air bubbles which are advected in the bulk of the turbulent flow. Tthe light diffraction on the entrained bubble interfaces gives a whitish appearance to the waters, called commonly white waters. Self-aeration or free-surface aeration may be encountered at Nature as well as in man-made engineering applications. There are two dominant types of self-aeration: singular and interfacial aeration. Singular aeration is the entrainment of air bubbles localised at a flow discontinuity. Interfacial aeration is the continuous air bubble entrainment along an air-water free-surface. The onset of air bubble entrainment may be expressed in terms of the tangential Reynolds stress and the fluid properties. With both singular and interfacial aeration, the void fraction distributions may be modelled by some analytical solutions of the advective diffusion equation for air bubbles. While the advective diffusion equation is identical, differences in boundary conditions lead to different analytical solutions. The microscopic structure of turbulent bubbly flows is complex and its analysis requires some advanced metrology and signal processing. The results highlight the turbulent nature of the complex two-phase flows, while the interactions between entrained air bubbles and turbulence will remain a key challenge for the 21st century researchers.

APPENDIX C – KEYWORDS

Air bubble entrainment
Self-aeration
Interfacial aeration
Singular aeration
Plunging jet
Hydraulic jumps
Advective diffusion equation
Interactions between turbulence and entrained air
Bubble size distributions
Bubble clusters

APPENDIX D – QUESTIONS

What is self-aeration?
Where can we find self-aerated flow?
What is the colour of self-aerated flows?
How can we define the onset of self-aeration?
Can you give several examples of singular aeration?
Can you give at least two examples of interfacial aeration?
What is a bubble cluster?

APPENDIX E - PROBLEMS

E1. For a three-dimensional flow, plot the relationship between turbulent stress and radius for air and water, and air and glycerine (viscosity $1.4 \, \text{Pa} \cdot \text{s}$ and surface tension $0.06 \, \text{N/m}$ at $20°\text{C}$).

E2. A circular water jet discharges into the atmosphere. The nozzle diameter is 5 mm (ID). The flow rate is $0.95 \, \text{l/s}$. Calculate the void fraction distribution at two sections located respectively at 5 and 25 diameters?

E3. A smooth spillway discharges $25 \, \text{m}^2/\text{s}$. At a sampling location, the air-water depth Y_{90} is estimated to 1.85 m and the depth-averaged void fraction is 0.21. Calculate the flow velocity and plot the void fraction distribution.

REFERENCES

Brocchini, M., and Peregrine, D.H. (2001). "The Dynamics of Strong Turbulence at Free Surfaces. Part 1. Description." *Jl Fluid Mech.*, Vol. 449, pp. 225–254.

Cain, P. (1978). "Measurements within Self-Aerated Flow on a Large Spillway." *Ph.D. Thesis*, Ref. 78–18, Dept. of Civil Engrg., Univ. of Canterbury, Christchurch, New Zealand.

Carslaw, H.S., and Jaeger, J.C. (1959). "Conduction of Heat in Solids." *Oxford University Press*, London, UK, 2nd ed., 510 pages.

Chanson, H. (1988). "A Study of Air Entrainment and Aeration Devices on a Spillway Model." *Ph.D. thesis*, Ref. 88-8, Dept. of Civil Engrg., University of Canterbury, New Zealand.

Chanson, H. (1993). "Self-Aerated Flows on Chutes and Spillways." *Jl of Hyd. Engrg.*, ASCE, Vol. 119, No. 2, pp. 220–243. Discussion: Vol. 120, No. 6, pp. 778–782.

Chanson, H. (1997). "Air Bubble Entrainment in Free-Surface Turbulent Shear Flows." *Academic Press*, London, UK, 401 pages.

Chanson, H. (2008). "Advective Diffusion of Air Bubbles in Turbulent Water Flows." in "Fluid Mechanics of Environmental Interfaces", *Taylor & Francis*, Leiden, The Netherlands, C. Gualtieri and D.T. Mihailovic Editors, Chapter 7, pp. 163–196.

Chanson, H., and Carosi, G. (2007). "Advanced Post-Processing and Correlation Analyses in High-Velocity Air-Water Flows." *Environmental Fluid Mechanics*, Vol. 7, No. 6, pp. 495–508 (DOI 10.1007/s10652-007-9038-3).

Comolet, R. (1979). "Vitesse d'Ascension d'une Bulle de Gaz Isolée dans un Liquide Peu Visqueux." ('The Terminal Velocity of a Gas Bubble in a Liquid of Very Low Viscosity.') *Jl de Mécanique Appliquée*, Vol. 3, No. 2, pp. 145–171 (in French).

Comolet, R. (1979). "Sur le Mouvement d'une bulle de gaz dans un liquide." ('Gas Bubble Motion in a Liquid Medium.') *Jl La Houille Blanche*, No. 1, pp. 31–42 (in French).

Crank, J. (1956). "The Mathematics of Diffusion." *Oxford University Press*, London, UK.

Cummings, P.D., and Chanson, H. (1999). "An Experimental Study of Individual Air Bubble Entrainment at a Planar Plunging Jet." *Chem. Eng. Research and Design*, Trans. IChemE, Part A, Vol. 77, No. A2, pp. 159–164.

Edwards, C.F., and Marx, K.D. (1995a). "Multipoint Statistical Structure of the Ideal Spray, Part I: Fundamental Concepts and the Realization Density." *Atomizati & Sprays*, Vol. 5, pp. 435–455.

Edwards, C.F., and Marx, K.D. (1995b). "Multipoint Statistical Structure of the Ideal Spray, Part II: Evaluating Steadiness using the Inter-particle Time Distribution." *Atomizati & Sprays*, Vol. 5, pp. 435–455.

Ervine, D.A., and Falvey, H.T. (1987). "Behaviour of Turbulent Water Jets in the Atmosphere and in Plunge Pools." *Proc. Instn Civ. Engrs., London*, Part 2, Mar. 1987, 83, pp. 295–314. Discussion: Part 2, Mar.-June 1988, 85, pp. 359–363.

Gonzalez, C.A. (2005). "An Experimental Study of Free-Surface Aeration on Embankment Stepped Chutes." *Ph.D. thesis*, Department of Civil Engineering, The University of Queensland, Brisbane, Australia, 240 pages.

Haberman, W.L., and Morton, R.K. (1954). "An Experimental Study of Bubbles Moving in Liquids." *Proceedings*, ASCE, 387, pp. 227–252.

Hayes, M.H. (1996). "Statistical, Digital Signal Processing and Modeling." *John Wiley*, New York, USA.

Hinze, J.O. (1955). "Fundamentals of the Hydrodynamic Mechanism of Splitting in Dispersion Processes." *Jl of AIChE*, Vol. 1, No. 3, pp. 289–295.

Murzyn, F., Mouaze, D., and Chaplin, J.R. (2005). "Optical Fibre Probe Measurements of Bubbly Flow in Hydraulic Jumps" *Intl Jl of Multiphase Flow*, Vol. 31, No. 1, pp. 141–154.

Toombes, L. (2002). "Experimental Study of Air-Water Flow Properties on Low-Gradient Stepped Cascades." *Ph.D. thesis*, Dept of Civil Engineering, The University of Queensland.

Wood, I.R. (1984). "Air Entrainment in High Speed Flows." *Proc. Intl. Symp. on Scale Effects in Modelling Hydraulic Structures*, IAHR, Esslingen, Germany, H. Kobus editor, paper 4.1.

Wood, I.R. (1991). "Air Entrainment in Free-Surface Flows." *IAHR Hydraulic Structures Design Manual No. 4*, Hydraulic Design Considerations, Balkema Publ., Rotterdam, The Netherlands, 149 pages.

Bibliography

Bombardelli, F., and Chanson, H. (2009). "Environmental Fluid Mechanics: Special Issue: Recent Advances on Multi-Phase Flows of Environmental Importance." *Environmental Fluid Mechanics*, Vol. 9, No. 2, pp. 121–266.

Brattberg, T., and Chanson, H. (1998). "Air Entrapment and Air Bubble Dispersion at Two-Dimensional Plunging Water Jets." *Chemical Engineering Science*, Vol. 53, No. 24, Dec., pp. 4113–4127. Errata: 1999, Vol. 54, No. 12, p. 1925.

Brattberg, T., Chanson, H., and Toombes, L. (1998). "Experimental Investigations of Free-Surface Aeration in the Developing Flow of Two-Dimensional Water Jets." *Jl of Fluids Eng.*, Trans. ASME, Vol. 120, No. 4, pp. 738–744.

Brocchini, M., and Peregrine, D.H. (2001b). "The Dynamics of Strong Turbulence at Free Surfaces. Part 2. Free-surface Boundary Conditions." *Jl Fluid Mech.*, Vol. 449, pp. 255–290.

Carosi, G., and Chanson, H. (2006). "Air-Water Time and Length Scales in Skimming Flows on a Stepped Spillway. Application to the Spray Characterisation." *Report No. CH59/06*, Div. of Civil Engineering, The University of Queensland, Brisbane, Australia, July, 142 pages.

Cartellier, A., and Achard, J.L. (1991). "Local Phase Detection Probes in Fluid/Fluid Two-Phase Flows." *Rev. Sci. Instrum.*, Vol. 62, No. 2, pp. 279–303.

Chachereau, Y., and Chanson, H. (2011a). "Free-Surface Fluctuations and Turbulence in Hydraulic Jumps." *Experimental Thermal and Fluid Science*, Vol. 35, No. 6, pp. 896–909 (DOI: 10.1016/j.expthermflusci.2011.01.009).

Chachereau, Y., and Chanson, H. (2011b). "Bubbly Flow Measurements in Hydraulic Jumps with Small Inflow Froude Numbers." *International Journal of Multiphase Flow*, Vol. 37, No. 6, pp. 555–564 (DOI: 10.1016/j.ijmultiphaseflow.2011.03.012).

Chang, K.A., Lim, H.J., and Su, C.B. (2003). "Fiber Optic Reflectometer for Velocity and Fraction Ratio Measurements in Multiphase Flows." *Rev. Scientific Inst.*, Vol. 74, No. 7, pp. 3559–3565. Discussion & Closure: 2004, Vol. 75, No. 1, pp. 284–286.

Chanson, H. (1989). "Study of Air Entrainment and Aeration Devices." *Jl of Hyd. Res.*, IAHR, Vol. 27, No. 3, pp. 301–319.

Chanson, H. (2002a). "Air-Water Flow Measurements with Intrusive Phase-Detection Probes. Can we Improve their Interpretation ?." *Jl of Hyd. Engrg.*, ASCE, Vol. 128, No. 3, pp. 252–255.

Chanson, H,. (2002b). "An Experimental Study of Roman Dropshaft Operation: Hydraulics, Two-Phase Flow, Acoustics." *Report CH50/02*, Dept of Civil Eng., Univ. of Queensland, Brisbane, Australia, 99 pages.

Chanson, H. (2004a). "Environmental Hydraulics of Open Channel Flows." *Elsevier Butterworth-Heinemann*, Oxford, UK, 483 pages.

Chanson, H. (2004b). "Unsteady Air-Water Flow Measurements in Sudden Open Channel Flows." *Experiments in Fluids*, Vol. 37, No. 6, pp. 899–909.

Chanson, H. (2004). "Fiber Optic Reflectometer for Velocity and Fraction Ratio Measurements in Multiphase Flows. Letter to the Editor" *Rev. Scientific Inst.*, Vo. 75, No. 1, pp. 284–285.

Chanson, H. (2006). "Air Bubble Entrainment in Hydraulic Jumps. Similitude and Scale Effects." *Report No. CH57/05*, Dept. of Civil Engineering, The University of Queensland, Brisbane, Australia, Jan., 119 pages.

Chanson, H. (2007a). "Bubbly Flow Structure in Hydraulic Jump." *European Journal of Mechanics B/Fluids*, Vol. 26, No. 3, pp. 367–384 (DOI:10.1016/j.euromechflu.2006.08.001).

Chanson, H. (2007b). "Air Entrainment Processes in Rectangular Dropshafts at Large Flows." *Journal of Hydraulic Research*, IAHR, Vol. 45, No. 1, pp. 42–53.

Chanson, H. (2008a). "Physical Modelling, Scale Effects and Self-Similarity of Stepped Spillway Flows." Proc. World Environmental and Water Resources Congress 2008 Ahupua'a, ASCEEWRI, 13–16 May, Hawaii, R.W. Badcock Jr and R. Walton Editors, Paper 658, 10 pages (CD-ROM).

Chanson, H. (2008b). "Métrologie des Ecoulements Cisaillés à Surface Libre en présence de Mélange Gaz-Liquide." ('Measurements in Free-Surface Turbulent Shear Flows in the Gas-Liquid Mixture.') *Proceedings of 11ème Congrès Francophone de Techniques Laser CFTL 2008*, Poitiers Futuroscope, France, 16–19 Sept., J.M. Most, L. David, F. Penot and L.E. Brizzi Editors., Invited plenary lecture, pp. 13–28.

Chanson, H. (2009a). "Turbulent Air-water Flows in Hydraulic Structures: Dynamic Similarity and Scale Effects." *Environmental Fluid Mechanics*, Vol. 9, No. 2, pp. 125–142 (DOI: 10.1007/s10652-008-9078-3).

Chanson, H. (2009b). "Current Knowledge In Hydraulic Jumps And Related Phenomena. A Survey of Experimental Results." European Journal of Mechanics B/Fluids, Vol. 28, No. 2, pp. 191–210 (DOI: 10.1016/j.euromechflu.2008.06.004).

Chanson, H. (2010). "Convective Transport of Air Bubbles in Strong Hydraulic Jumps." *International Journal of Multiphase Flow*, Vol. 36, No. 10, pp. 798–814 (DOI: 10.1016/j.ijmultiphaseflow.2010.05.006).

Chanson, H., Aoki, S., and Hoque, A. (2006). "Bubble Entrainment and Dispersion in Plunging Jet Flows: Freshwater versus Seawater." *Jl of Coastal Research*, Vol. 22, No. 3, May, pp. 664–677.

Chanson, H., and Carosi, G. (2007). "Turbulent Time and Length Scale Measurements in High-Velocity Open Channel Flows." *Experiments in Fluids*, Vol. 42, No. 3, pp. 385–401 (DOI 10.1007/s00348-006-0246-2).

Chanson, H., and Gualtieri, C. (2008). "Similitude and Scale Effects of Air Entrainment in Hydraulic Jumps." *Journal of Hydraulic Research*, IAHR, Vol. 46, No. 1, pp. 35–44.

Chanson, H., and Manasseh, R. (2003). "Air Entrainment Processes in a Circular Plunging Jet. Void Fraction and Acoustic Measurements." *Jl of Fluids Eng.*, Trans. ASME, Vol. 125, No. 5, Sept., pp. 910–921.

Chanson, H., and Toombes, L. (2002). "Air-Water Flows down Stepped chutes: Turbulence and Flow Structure Observations." *Intl Jl of Multiphase Flow*, Vol. 28, No. 11, pp. 1737–1761.

Crowe, C., Sommerfield, M., and Tsuji, Y. (1998). "Multiphase Flows with Droplets and Particles." *CRC Press*, Boca Raton, USA, 471 pages.

Cummings, P.D. (1996). "Aeration due to Breaking Waves." *Ph.D. thesis*, Dept. of Civil Engrg., University of Queensland, Australia.

Cummings, P.D., and Chanson, H. (1997a). "Air Entrainment in the Developing Flow Region of Plunging Jets. Part 1 Theoretical Development." *Jl of Fluids Eng.*, Trans. ASME, Vol. 119, No. 3, pp. 597–602.

Cummings, P.D., and Chanson, H. (1997b). "Air Entrainment in the Developing Flow Region of Plunging Jets. Part 2: Experimental." *Jl of Fluids Eng.*, Trans. ASME, Vol. 119, No. 3, pp. 603–608.

Cummings, P.D., and Chanson, H. (1999). "An Experimental Study of Individual Air Bubble Entrainment at a Planar Plunging Jet." *Chem. Eng. Research and Design*, Trans. IChemE, Part A, Vol. 77, No. A2, pp. 159–164.

Felder, S., and Chanson, H. (2009a). "Turbulence, Dynamic Similarity and Scale Effects in High-Velocity Free-Surface Flows above a Stepped Chute." Experiments in Fluids, Vol. 47, No. 1, pp. 1–18 (DOI: 10.1007/s00348-009-0628-3).

Felder, S., and Chanson, H. (2009b). "Energy Dissipation, Flow Resistance and Gas-Liquid Interfacial Area in Skimming Flows on Moderate-Slope Stepped Spillways." Environmental Fluid Mechanics, Vol. 9, No. 4, pp. 427–441 (DOI: 10.1007/s10652-009-9130-y).

Felder, S., and Chanson, H. (2011). "Air-Water Flow Properties in Step Cavity down a Stepped Chute." International Journal of Multiphase Flow, Vol. 37, No. 7, pp. 732–745 (DOI: 10.1016/j.ijmultiphaseflow.2011.02.009).

Gonzalez, C.A., and Chanson, H. (2004). "Interactions between Cavity Flow and Main Stream Skimming Flows: an Experimental Study." *Can Jl of Civ. Eng.*, Vol. 31.

Gonzalez, C.A., Takahashi, M., and Chanson, H. (2005). "Effects of Step Roughness in Skimming Flows: an Experimental Study." *Research Report No. CE160*, Dept. of Civil Engineering, The University of Queensland, Brisbane, Australia, July, 149 pages.

Gualtieri, C., and Chanson, H. (2007). "Clustering Process Analysis in a Large-Size Dropshaft and in a Hydraulic Jump." *Proceedings of the 32nd IAHR Biennial Congress*, Venice, Italy, G. Di Silvio and S. Lanzoni Editors, Topic C1.b, 11 pages (CD-ROM).

Gualtieri, C., and Chanson, H. (2010). "Effect of Froude Number on Bubble Clustering in a Hydraulic Jump." *Journal of Hydraulic Research, IAHR*, Vol. 48, No. 4, pp. 504–508 (DOI: 10.1080/00221686.2010.491688).

Heinlein, J., and Frtisching, U. (2006). "Droplet Clustering in Sprays." *Experiments in Fluids*, Vol. 40, No. 3, pp. 464–472.

Jones, O.C., and Delhaye, J.M. (1976). "Transient and Statistical Measurement Techniques for two-Phase Flows: a Critical Review." *Intl Jl of Multiphase Flow*, Vol. 3, pp. 89–116.

Kobus, H. (1984). "Scale Effects in Modelling Hydraulic Structures." *Proc. Intl Symp. on Scale Effects in Modelling Hydraulic Structures*, IAHR, Esslingen, Germany, H. Kobus Editor.

Kucukali, S., and Chanson, H. (2008). "Turbulence Measurements in Hydraulic Jumps with Partially-Developed Inflow Conditions." *Experimental Thermal and Fluid Science*, Vol. 33, No. 1, pp. 41–53 (DOI: 10.1016/j.expthermflusci.2008.06.012).

Lubin, P. Glockner, S., and Chanson, H. (2009). "Numerical Simulation of Air Entrainment and Turbulence in a Hydraulic Jump." *Proc. Colloque SHF Modèles Physiques Hydrauliques: Outils Indispensables du XXIe Siècle?*, Société Hydrotechnique de France, Lyon, France, 24–25 Nov., pp. 109–114.

Luong, J.T.K., and Sojka, P.E. (1999). "Unsteadiness in Effervescent Sprays." *Atomization & Sprays*, Vol. 9, pp. 87–109.

Murzyn, F., and Chanson, H. (2008). "Experimental Assessment of Scale Effects Affecting Two-Phase Flow Properties in Hydraulic Jumps." *Experiments in Fluids*, Vol. 45, No. 3, pp. 513–521 (DOI: 10.1007/s00348-008-0494-4).

Murzyn, F., and Chanson, H. (2009a). "Free-Surface Fluctuations in Hydraulic Jumps: Experimental Observations." *Experimental Thermal and Fluid Science*, Vol. 33, No. 7, pp. 1055–1064 (DOI: 10.1016/j.expthermflusci.2009.06.003).

Murzyn, F., and Chanson, H. (2009b). "Two-Phase Flow Measurements in Turbulent Hydraulic Jumps." *Chemical Engineering Research and Design*, Trans. IChemE, Part A, Vol. 87, No. 6, pp. 789–797 (DOI: 10.1016/j.cherd.2008.12.2003).

Noymer, P.D. (2000). "The Use of Single-Point Measurements to Characterise Dynamic Behaviours in Spray." *Experiments in Fluids*, Vol. 29, pp. 228–237.

Rao, N.S.L., and Konus, H.E. (1971). "Characteristics of Self-Aerated Free-Surface Flows." *Water and Waste Water/Current Research and Practice*, Vol. 10, Eric Schmidt Verlag, Berlin, Germany.

Straub, L.G., and Anderson, A.G. (1958). "Experiments on Self-Aerated Flow in Open Channels." *Jl of Hyd. Div.*, Proc. ASCE, Vol. 84, No. HY7, paper 1890, pp. 1890-1 to 1890-35.

Toombes, L., and Chanson, H. (2007a). "Free-Surface Aeration and Momentum Exchange at a Bottom Outlet." Journal of Hydraulic Research, IAHR, Vol. 45, No. 1, pp. 100–110.

Toombes, L., and Chanson, H. (2007b). "Surface Waves and Roughness in Self-Aerated Supercritical Flow." *Environmental Fluid Mechanics*, Vol. 5, No. 3, pp. 259–270 (DOI 10.1007/s10652-007-9022-y).

Toombes, L., and Chanson, H. (2008). "Interfacial Aeration and Bubble Count Rate Distributions in a Supercritical Flow Past a Backward-Facing Step." *International Journal of Multiphase Flow*, Vol. 34, No. 5, pp. 427–436 (doi.org/10.1016/j.ijmultiphaseflow.2008.01.005).

Wood, I.R. (1983). "Uniform Region of Self-Aerated Flow." *Jl Hyd. Eng.*, ASCE, Vol. 109, No. 3, pp. 447–461.

Internet resources

Chanson, H. (2000). "Self-aeration on chute and stepped spillways – Air entrainment and flow aeration in open channel flows." *Internet resource*.
(Internet address: http://www.uq.edu.au/~e2hchans/self_aer.html)

Chanson, H., and Manasseh, R. (2000). "Air Entrainment at a Circular Plunging Jet. Physical and Acoustic Characteristics – Internet Database." *Internet resource*.
(Internet address: http://www.uq.edu.au/~e2hchans/bubble/)

Cummings, P.D., and Chanson, H. (1997). "Air Entrainment in the Developing Flow Region of Plunging Jets. Extended Electronic Manuscript." *Jl of Fluids Engineering – Data Bank*, ASME (Electronic Files: 6,904 kBytes). (Internet address: http://www.uq.edu.au/~e2hchans/data/jfe97.html)

Open access research reprints in air-water flows. (Internet address: http://espace.library.uq.edu.au/list/author_id/193/)

Open access database in air-water flows. (Internet address: http://www.ands.org.au/)

CHAPTER EIGHT

Exchanges at the bed sediments-water column interface

Fabián A. Bombardelli

Department of Civil and Environmental Engineering,
University of California, Davis, United States of America

Patricio A. Moreno

Department of Civil and Environmental Engineering,
University of California, Davis, United States of America
Departmento de Ingeniería en Obras Civiles, Universidad Diego Portales,
Santiago, Chile

ABSTRACT

Interactions at the interface between bed sediments and the overlying waters have a tremendous importance for diverse natural and man-made processes such as fining and armouring in rivers, erosion/sedimentation in bays, and the cycling of different contaminants in water bodies at large. This chapter benefits from a focus on the interface of bed sediments and water to present and discuss the vast area of phenomena related to the transport of sediment particles and contaminants in water resources. We start by presenting the "modes" of sediment transport and follow with the discussion of the concept of incipient motion and the mass balance of solids at the interface – the Exner Equation. We then turn to predictors of different variables needed for the mass balance, such as bed load flow rates, entrainment functions, and the settling velocity. We continue with the theory of suspended sediment and of bed load, exposing separate and consistent treatments. Next section addresses the problem of sediment-laden transport of contaminants in water bodies. The chapter closes with an evaluation of future work in this important subject.

8.1 FOREWORD

The study of the exchange processes occurring at the interface bed sediments-water column, in addition to the theory of the general transport of solid particles in water bodies constitute a fascinating chapter of basic and applied science. This fascinating chapter combines ideas and concepts coming from branches of the physical and chemical fields of inquiry (Pope, 2000) including, among others, fluid mechanics, sediment transport, as well as biogeochemistry (García, 2008). Besides the interest from a purely scientific point of view there is a notable *practical* interest on the topic, associated with the water-quality implications of processes in which the interface participates, and the consequences of phenomena of erosion and deposition on the overall sediment distribution in the water body. It is thus no surprise that numerous efforts have been, and currently directed to the observation, interpretation and prediction of sediment transport in general, and of the exchanges at the interface of bed sediments with the water column in particular.

The motion of solid particles in open channels has been traditionally investigated via experiments. The need for prediction of erosion and deposition in streams, rivers, harbors and coastal areas led to the development of regressions and rules of thumb conducive to the provision of solutions in practical cases, in the old fashion of hydraulics. With the advent of the use of fluid-mechanics principles to understand problems of hydraulics, sediment transport started to benefit from a more mechanistic approach to the subject (Yen, 1992a). In this new century, more systematic analyses of the transport of solid particles in diverse water bodies have led to an improved understanding of the phenomenon. Within the above framework, more comprehensive models, rooted on the two-phase flow theory, have been developed to shed light into the interactions between the water column and the bed sediments (see Bombardelli and Jha, 2009, as example).

These interactions are extremely complicated given a host of factors. In first place, the location of the interface varies in space and time as a consequence of the differences in volume flow rates of the sediment particles in both the water column and in the neighborhood of the bed. Second, the interface plays a notably active role in the transport of contaminants in water bodies in general (Chung et al., 2009a; b; Massoudieh et al., 2010). This role is two-fold. On one hand, bed sediments act as the final repository for pollutants via continuous deposition throughout the years; on another hand, pollutant-rich sediment layers take part on the exchange of mass with the overlying waters. In this way, the contaminants which have been depositing for decades in river beds and in the bottom of lakes and estuaries can strongly influence the cycling of certain contaminants in the years to come, in spite of the eventual elimination of all discharges of pollutants to the water body.

The purpose of this chapter is to present a global, mechanistic view of the processes associated with the water column-bed sediments interface, mainly for non-cohesive sediments. We start by presenting the "modes" of sediment transport in water bodies in general, and follow with the discussion of the concept of incipient motion, and the mass balance of solids at the interface. We then turn to predictors of different variables needed for the mass balance, such as bed load flow rates, entrainment functions, and the settling velocity. We continue with the theory of suspended sediment and of bed load, exposing separate and consistent treatments. Next section addresses the problem of sediment-laden transport of contaminants in water bodies.

We aim at bridging the gap between sediment-transport theories available in open-channel flows and coastal environments, which are usually covered in different portions of the literature, and at eliminating the divide that seems to exist among findings of some research groups on sediment transport around the world.

8.2 SEDIMENT MASS CONSERVATION AT THE INTERFACE – THE EXNER EQUATION, AND RELATED TOPICS

8.2.1 Sediment transport modes

Rivers and streams transport sediment particles of diverse size, shape and density, a fact already identified and observed during Antiquity in China, Mesopotamia, and Egypt (van Rijn, 1993). Intuitively, this transport is possible because the intensity of the water current is strong enough to move the finer particles away from the bed, and to carry them along the stream. Heavier particles, on the other hand, are harder to be moved and either remain on the bed, move close to the bed, or deposit once moved. Some of these concepts are summarized in the beautiful plot by Hjulström, developed in 1935 (Fig. 8.1). (The notions conveyed in this diagram are slightly different than those of the current understanding on the sediment-transport processes, as discussed in next sections; see also Miedema, 2008.) According to Hjulström's diagram, as the velocity increases for a given particle size (of diameter d_p), the sediment may either undergo deposition, transport, or erosion. Notice that there is a limiting particle size for which no deposition occurs because sediment is very fine (smaller than about 15 microns). The plot underscores the complex nature of the processes associated with sediment motion, and emphasizes that those processes are highly non-linear. In other words, there is no simple proportionality between the intensity of the current and the amount of sediment transported. In addition to this non-linearity, geological processes affect the interaction of sediments with water (García, 2008). The diagram is presented herein for conceptual information purposes only, since it is not much used in current engineering practice.

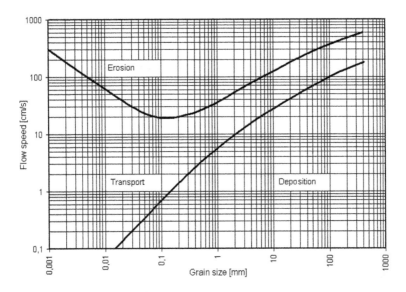

Figure 8.1. Hjulström's diagram, developed in 1935 (taken under the GNU project).

Although the phenomenon of sediment transport ultimately involves the motion of solid particles in a carrier fluid (i.e., a two-phase flow) regardless of the zone of the water column, engineers and scientists usually approach the problem of sediment transport postulating the existence of "modes," as follows: the *bed load*, the *suspended load*, and the *wash-load* (MacArthur et al., 2008). Fig. 8.2 shows a schematic of the different modes for particle motion. For convenience, a rather arbitrary distance *b* from the bed is established as a separation of bed load and suspended load layers (Parker, 2004; García, 2008).

Many observations on multiple rivers, streams and coastal areas in the world have shown that larger particles move close to the bottom as *bed load*, by either *sliding, rolling* and *saltating* (Francis, 1973; Abbott and Francis, 1977). The percentages of the total number of particles moving in each mode of bed load depend upon the particle size, d_p, and the flow intensity, expressed through the parameter $\tau_* = \tau_0/(\rho_s - \rho)gd_p$, where τ_0 is the bed shear stress, ρ_s and ρ denote the density of the sediment and water, respectively, and g is the acceleration of gravity. The fraction of rolling drastically decreases as τ_* increases, while the fraction of saltation in turn increases (Hu and Hui, 1996). Saltation is considered to be the main form of bed load motion in most natural conditions, where τ_* is relatively large (Einstein, 1950; Sekine and Kikkawa, 1992; Lee et al., 2000). Regarding *suspended load*, this is the mode of most of the fine sand grains in rivers, which are light enough to be lifted up. The *wash-load* is defined as the transport of very fine material in suspension with little interaction with the bed (MacArthur et al., 2008). The diverse modes can be determined with the help of the Rouse number, defined as the ratio of the fall (settling) velocity, w_s, and the product of the shear velocity, u_*, and von Kármán constant, κ. When the Rouse number is smaller than 0.8, the transport occurs as wash-load; when it varies between 0.8 and 2.5, the transport is within the suspended mode; when it goes beyond 2.5, the transport is as bed load (Whipple, 2004). Note that the shear velocity is a surrogate of the shear stress and can be related to the cross-sectional average velocity (U), the water depth (H) and the bottom roughness height (k) through Keulegan's resistance relation (García, 1999), as follows:

$$u_* = \frac{U\kappa}{\ln\left(11H/k\right)} \tag{8.1}$$

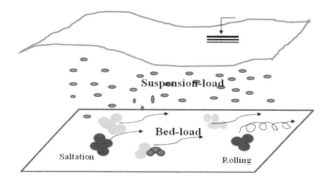

Figure 8.2. Schematic showing the classification of "modes" in sediment transport.

Based on knowledge coming from diverse experimental observations, it is usually assumed that the bed load layer is about a few particle diameters thick (Julien, 2010; page 195), and that thickness is customarily expressed as a fraction of the water depth.

8.2.2 Incipient sediment motion

The notion of incipient motion is associated with a threshold condition which basically separates erosion from deposition (Julien, 2010). Most treatises on sediment transport address in first place the meaning of incipient motion under non-submerged conditions, leading to the concept of angle of repose (Julien, 2010; Parker, 2004).

Under submerged conditions, particles are subjected to forces which tend to destabilize them, and also to stabilizing forces (associated basically to the weight). The movement of a particle (detachment) will occur when the resultant of those instantaneous forces acting on the particle points in a direction away from the other particles of the bed (Yalin, 1977).

The problem of incipient motion can be attacked from a dimensional analysis, as described below. Consider the following variables in the analysis (similar to those discussed by Yalin, 1977): ρ, ρ_s, μ (dynamic viscosity of the carrier fluid), d_p, u_*, and g, which can be applied to characterize any aspect of sediment transport (Yalin, 1977). Using the Buckingham's Pi Theorem (Kundu and Cohen, 2008), any given variable P of interest can be expressed as:

$$P = function\left(\frac{u_* d_p}{\nu}, \frac{\rho_s}{\rho}, \frac{\rho u_*^2}{(\rho_s - \rho)g d_p}\right) \tag{8.2}$$

where ν is the kinematic viscosity. The third term in Equation (8.2) is the ratio of the "tractive" fluid dynamics force acting on a particle, divided by the submerged weight of the particle (Yalin, 1977), used before to identify the different modes of bed load transport (Section 8.2.1). In turn, the first number is called the *boundary* Reynolds number (García, 1999).

The incipient-motion problem was tackled by Shields (1936) in his doctoral Dissertation (Kennedy, 1995) through experiments. Shields employed notions of dimensional analysis to formulate a relationship between the dimensionless shear stress discussed above, now called the *Shields number*, and the boundary Reynolds number, as follows:

$$\tau_{*,c} = \frac{\tau_{0,c}}{(\rho_s - \rho)g d_p} = f\left(\frac{u_{*,c}\, d_p}{\nu}\right) \tag{8.3}$$

Shields' experiments yielded a cloud of points which have become legendary (Kennedy, 1995; García, 2008). Assuming a binary behavior of the bed particles (i.e., motion/no motion), and although the original points show a considerable scatter, a curve was added later (see Fig. 8.3). the curve on Fig. 8.3 (expressed in terms of the *explicit* particle Reynolds number, $Re_p = (gRd_p)^{1/2}d_p/\nu$, with $R = (\rho_s - \rho)/\rho$ reflects the condition under which the sediment grains are "at the verge" of being moved (García, 1999; 2008). Rigorously speaking, there is no such a thing as a "threshold" of motion (Parker, 2004). Paintal (1971) developed experiments for large periods of time, noticing that particles were in motion for much lower values of any measure of threshold (Parker, 2004).

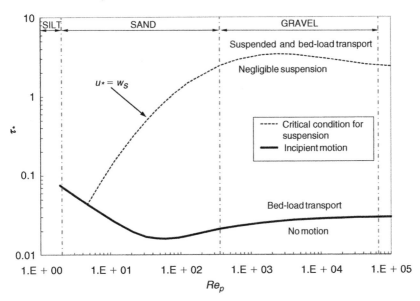

Figure 8.3. Shields-Parker diagram, with criterion for significant suspension (adapted from Parker, 2004; García, 2008).

Three portions can be identified in the Shields' diagram. On the left side, i.e., for small values of the particle Reynolds number, the conditions correspond to relatively small particles for a given flow intensity, u_*; the granular bed degenerates into a "muddy" substance (Yalin, 1977), the individuality of grains disappears and, then, the relation becomes:

$$\tau_{*,c} = \frac{\tau_{0,c}}{(\rho_s - \rho)g d_p} = \frac{const.}{\dfrac{u_{*,c}\, d_p}{\nu}} \tag{8.4}$$

where there is no dependence of the shear stress on the particle diameter. For large particle diameters, the viscosity does not play any role in the incipient motion and, thus, the equation transforms to:

$$\tau_{*,c} = \frac{\tau_{0,c}}{(\rho_s - \rho)gd_p} = const. \tag{8.5}$$

for which the curve is parallel to abscissas. In the middle portion of the diagram, there is a transitional behavior regarding the influence of the particle diameter. The features of all these portions of the diagram have been predicted very satisfactorily by the Ikeda-Iwagaki-Coleman model (see García, 1999; 2008; and Problems E1 and E2 in this chapter). It is worth mentioning that Parker et al. (2003) provided a regression for the Shields curve which allows for a quick computation of the critical shear stress as a function of the explicit particle Reynolds number:

$$\tau_{*,c} = 0.5\,[0.22Re_p^{-0.6} + 0.06 \cdot 10^{(-7.7\,Re_p^{-0.6})}] \tag{8.6}$$

leading to a value of 0.03 at high values of Re_p. An alternative piece-wise function is offered on page 149 of the book by Julien (2010).

Fig. 8.3 also shows the separation between the motion as bed load and sediment in suspension or, in other words, "the onset of significant suspension" (Parker, 2004). That curve is dictated by the equation $u_* = w_s$. This criterion points to a simple balance between the action of gravity in the vertical direction through the settling velocity, and a measure of the action of turbulence keeping the particles in suspension, which is a pervasive notion in sediment transport, as discussed below (see Parker, 2004).

Parker added data on sand and gravel rivers to the diagram, making it amenable for analysis of real streams. García dubs the resulting diagram the *Shields-regime diagram* (García, 1999) or the *Shields-Parker river sedimentation diagram* (García, 2008).

8.2.3 The Exner Equation

Consider a portion of sediment interface (Fig. 8.4a and 8.4b). In Fig. 8.4, x, y and z refer to the quasi stream-wise, quasi-transverse, and quasi-vertical coordinates, η indicates the position of the bed with respect to a certain datum, and q_{bx} and q_{by} denote the components of the bed load volume flow rate vector per unit width (Parker, 2004). Assume that the layer of sediment possesses a constant porosity λ_p. The mass conservation for sediment can be expressed via a book-keeping of every sediment particle in a control volume, as follows (Parker, 2004; García, 2008):

$$\frac{\partial}{\partial t}[mass\ within\ bed\ layer] = Mass\ flow\ rate\ in\ as\ bedload$$
$$- Mass\ flow\ rate\ out\ as\ bedload \tag{8.7}$$
$$+ Net\ vertical\ mass\ flux\ to\ the\ bed$$

In Equation (8.7), the last term includes the contribution of the sediment in suspension on the condition of the bed. Replacing intuitive formulations for the above terms, it is possible to write:

$$\frac{\partial}{\partial t}[\eta\Delta x\Delta y(1 - \lambda_p)\rho_s] = -\rho_s[q_{bx}(x + dx) - q_{bx}(x)]\Delta y$$
$$- \rho_s[q_{by}(y + dy) - q_{by}(y)]\Delta x \tag{8.8}$$
$$+ Net\ vertical\ mass\ flux\ to\ the\ bed$$

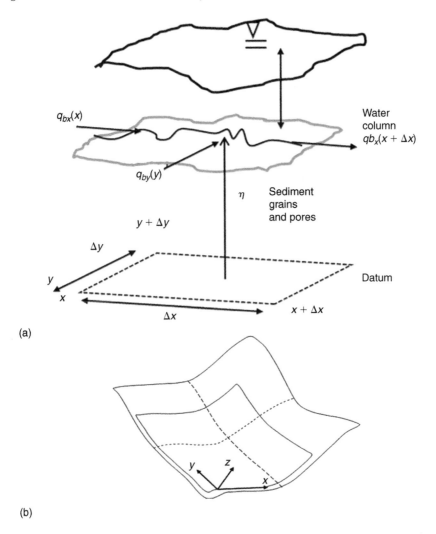

Figure 8.4. Schematic indicating: (a) the water column and bed sediments, and (b) a detail of the interface water column-bed sediments (adapted from Parker, 2004).

Whereas the negative sign in (8.8) refers to the fact that an increase in the mass flow rate with space leads to a decrease in the elevation of the interface, some considerations are needed for the vertical flow rate towards the bed, which are provided below. In the limit when the differentials approach 0, the Exner Equation is obtained, as follows:

$$(1 - \lambda_p)\frac{\partial \eta}{\partial t} = -\frac{\partial q_{bx}}{\partial x} - \frac{\partial q_{by}}{\partial y} + \textit{Vertical contribution} \tag{8.9}$$

In (8.9), the last term represents the difference between the number of particles being pulled down by gravity (called deposition), and the particles which are re-suspended by the action of the shear stress on the bed (called entrainment), expressed with units of velocity. The ability of a water current to destabilize sediment particles at the bed of a given size and to put them in suspension, can be defined through a non-dimensional number, E_s, the *entrainment function*, or dimensionless "sediment entrainment rate" (García and Parker, 1991; Parker, 2004; García, 2008):

$$E_r = w_s E_s \tag{8.10}$$

where E_r indicates the dimensional entrainment rate (in units of velocity). In turn, deposition (also called sedimentation) can be modeled in different ways, using for instance the rate of accumulation of solids at the bed, or the rate at which the location of the interface changes (DiToro, 2001). In the most common approach, deposition is modeled directly as the product of the fall (settling) velocity and a time-averaged sediment concentration near the bed, $\bar{c}_b$. (Concentrations are expressed herein as volume of sediment divided by total volume.) Replacing the models for entrainment and deposition in Equation (8.9) yields:

$$(1 - \lambda_p)\frac{\partial \eta}{\partial t} = -\frac{\partial q_{bx}}{\partial x} - \frac{\partial q_{by}}{\partial y} + w_s(\bar{c}_b - E_s) \tag{8.11}$$

$$(1 - \lambda_p)\frac{\partial \eta}{\partial t} = -\nabla_h \cdot \underline{q_b} + w_s (\bar{c}_b - E_s) \tag{8.12}$$

where the underline indicates vector; the Nabla operator with subindex h followed by a dot refers to the horizontal divergence of a vector; in this case, the vector is the volume flow rate vector per unit width, q_b.

Exner proposed this equation in 1925, for a one-dimensional (1-D) case (García, 2008). The Exner Equation is valid in general terms, regardless of the particle size, and whether the sediments are cohesive or non-cohesive. Parker et al. (2000) developed a probabilistic form of the Exner Equation.

8.2.4 Predictors for the volume flow rates as bed load

The volume flow rates embedded in Equation (8.12) are obtained mostly through experimental observations in laboratory flumes. As such, their use should be restricted to conditions similar to those under which they were obtained. All equations can be cast as a function of the flow intensity (shear stress or velocity), and representations of the particle density and size, using the following Einstein number,

$$q_* = \frac{q_b}{\sqrt{Rgd_p}\, d_p} \tag{8.13}$$

where q_b indicates in this case the stream-wise component of the volume flow rate.

Several formulation developed in the last century are summarized in Table 8.1, where $\tau_{*,c}$ refers to the dimensionless critical shear stress discussed in Section 8.2.2. The quoted formulas constitute a small number of the numerous available formulas.

8.2.5 Predictors for the entrainment of sediment into suspension

The entrainment is computed in practical applications via a series of predictors which have been also obtained essentially from laboratory experiments. A distinction needs to be made for cohesive and non-cohesive sediment. Most expressions to quantify sediment

Table 8.1. Summary of the most common formulations for bed load transport rates (García, 1999; Parker, 2004; Julien, 2010; García, 2008).

Author (year)	Formula	Observations
Ashida and Michihue (1972)	$q_* = 17(\tau_* - \tau_{*,c}) \times [(\tau_*)^{1/2} - (\tau_{*,c})^{1/2}]$	Verified with uniform sediment in the range of 0.3 to 7 mm. $\tau_{*,c} = 0.05$
Meyer-Peter and Muller (1948)	$q_* = 8(\tau_* - \tau_{*,c})^{3/2}$	Based on the median of the sediment distribution. $\tau_{*,c} = 0.047$. Verified with uniform gravel
Engelund and Fredsoe (1976)	$q_* = 18.74(\tau_* - \tau_{*,c}) \times [(\tau_*)^{1/2} - 0.7(\tau_{*,c})^{1/2}]$	$\tau_{*,c} = 0.05$
Fernandez Luque and van Beek (1976)	$q_* = 5.7(\tau_* - \tau_{*,c})^{3/2}$	$\tau_{*,c}$ ranges from 0.05 for 0.9 mm to 0.058 for 3.3 mm material
Wilson (1966)	$q_* = 12(\tau_* - \tau_{*,c})^{3/2}$	High rates of bed load transport
Yalin (1963)	$q_* = 0.635s(\tau_*)^{1/2} \times \left(1 - \dfrac{\ln(1 + a_2 s)}{a_2 s}\right)$	$a_2 = 2.45(R+1)^{0.4}(\tau_{*,c})^{1/2}$ $s = \dfrac{\tau_* - \tau_{*,c}}{\tau_{*,c}}$
Einstein (1950)	$1 - \dfrac{1}{\sqrt{\pi}} \displaystyle\int_{-(0.413/\tau_*)-2}^{(0.413/\tau_*)-2} e^{-t^2} dt = \dfrac{43.5q_*}{1 + 43.5q_*}$	Used for uniform sand and gravel
Parker (1979)	$q_* = 11.2 \dfrac{(\tau_* - 0.03)^{4.5}}{\tau_*^3}$	Gravel beds and bed load

resuspension for *cohesive* sediments are of the type (Mehta et al., 1982; Raudkivi, 1998; Sanford and Maa, 2001):

$$E_{r, cohesive} = \alpha \left[\frac{\tau_0 - \tau_{0,c}}{\tau_{0,c}}\right]^m \quad \text{for } \tau_0 \geq \tau_{0,c} \tag{8.14}$$

$$E_{r, cohesive} = 0 \quad \text{for } \tau_0 < \tau_{0,c} \tag{8.15}$$

where $E_{r,cohesive}$ is the erosion rate of cohesive sediment in mass area^{-1} time^{-1}; α is a coefficient; m is an exponent (set to 1 when a linear relationship is assumed); Chung et al., (2009a).

For non-cohesive sediments, the expressions proposed by Einstein (1950), Engelund and Fredsøe (1976; 1982), Smith and McLean (1977), van Rijn (1984), García and Parker (1991), and Zyserman and Fredsøe (1994), summarized in Table 8.2, are the most used ones for open-channel flows. These expressions rely on empirical parameters, making them unsafe to use outside of the calibration ranges (Zhong et al., 2011). (A more recent expression based on

Table 8.2. Summary of common expressions for the prediction of sediment entrainment rates under equilibrium conditions for non-cohesive sediment. Modified from García (2008).

Author (year)	Formula	Parameters	Reference height
Einstein (1950)	$\bar{c}_b = \dfrac{q_*}{23.2(\tau_{*,s})^{0.5}}$		$b = 2d_p$
Engelund and Fredsøe (1976; 1982)	$\bar{c}_b = \dfrac{0.65}{\left(1 + \lambda_b^{-1}\right)^3}$	$\lambda_b = \left[\dfrac{\tau_{*,s} - 0.06 - \dfrac{\beta p \pi}{6}}{0.027(R+1)\tau_{*,s}}\right]^{0.5}$ $p = \left[1 + \left(\dfrac{\frac{\beta \pi}{6}}{\tau_{*,s} - 0.06}\right)^4\right]^{-0.25}$ $\beta = 1$	$b = 2d_p$
Smith and McLean (1977)	$\bar{c}_b = \dfrac{0.65 \gamma_o T}{1 + \gamma_o T}$	$T = \dfrac{\tau_{*,s} - \tau_{*,c}}{\tau_{*,c}}$ $\gamma_o = 2.4 \times 10^{-3}$	$b = \alpha_i \Delta \tau \, d_p + k$ $\Delta \tau = \tau_{*,s} - \tau_{*,c}$ $\alpha_i = 26.3$
van Rijn (1984)	$\bar{c}_b = 0.015 \dfrac{d_p}{b} \dfrac{T^{1.5}}{d_*^{0.3}}$	$d_* = d_p \left(\dfrac{gR}{\nu^2}\right)^{1/3}$ $\Delta_b: \; mean \; dune \; heigh$	$b = \dfrac{\Delta_b}{2}$ or $b = 0.01H$
García and Parker (1991)	$E_S = \dfrac{A Z_u^5}{1 + \frac{A}{0.3} Z_u^5}$	$n = 0.6; \; A = 1.3 \times 10^{-7}$ $Z_u^5 = \dfrac{u_{*,s}}{w_s} R_p^n$	$b = 0.05H$
Zyserman and Fredsøe (1994)	$\bar{c}_b = \dfrac{0.331 \left(\tau_{*,s} - 0.045\right)^{1.75}}{1 + \frac{0.331}{0.46}\left(\tau_{*,s} - 0.045\right)^{1.75}}$		$b = 2d_p$

Note: the subscript s indicates "skin" variables (shear stress and shear velocity) as opposed to "form (pressure)" variables. Please see García (2008) for more information regarding this.

the kinetic theory was derived by Zhong et al. (2011); in that approach, several assumptions were made to obtain a simplified expression.) Equations in Table 8.2 correspond to the so-called "equilibrium conditions," where the mass flow rate of particles moving away from the bed equals the mass flow rate of particles moving towards the bed.

8.2.6 Predictors for the settling (fall) velocity

The remaining parameter needed to apply the Exner Equation is the settling velocity. The settling velocity of a particle is established by considering a sediment grain falling in a stagnant (quiescent) volume of fluid, which has reached a constant, equilibrium velocity (Julien, 2010). This condition is achieved when the two main forces acting on the particle, namely the drag and submerged weight, are equal. Fig. 8.5 shows the balance of forces on a spherical particle; F_D, F_{SW}, F_G, and F_{Bu} denote drag, submerged weight, weight, and buoyancy forces, respectively; A depicts the front area of the sphere; C_D is the drag coefficient; and V represents the relative particle velocity in the stagnant fluid, equal to w_S. In this case, the settling velocity of a spherical particle can be calculated as:

$$w_s = \left[\frac{4}{3} \frac{gRd_p}{C_D} \right]^{1/2} \tag{8.16}$$

This expression is not very practical since the drag coefficient is a function of the particle Reynolds number, defined as $R_p = w_s d_p/v$, which in turn is a function of the settling velocity. (Please note that this particle Reynolds number differs from the *explicit* particle Reynolds number, Re_p.) Thus, the calculation of the settling velocity is not explicit in the above expression, and it must be computed by iterative methods.

Drag Force

$$F_D = \frac{1}{2} \rho \, C_D \, A \, V^2 = \frac{1}{2} \rho \, C_D \left(\frac{\pi d_p^2}{4} \right) w_s^2$$

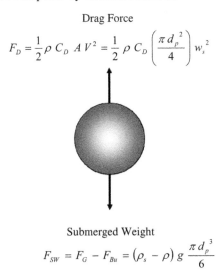

Submerged Weight

$$F_{SW} = F_G - F_{Bu} = (\rho_s - \rho) \, g \, \frac{\pi d_p^3}{6}$$

Figure 8.5. Forces acting on a settling sphere in a stagnant volume of fluid.

Several expressions are available on the literature for C_D. For creeping flow, i.e., for particle Reynolds numbers smaller than 1 (Kundu and Cohen, 2008), $C_D = 24/R_p$. Unfortunately, for flows of more practical interest (Reynolds numbers larger than 1 and non-spherical particles) the behavior of C_D is not completely known with accuracy (Yen, 1992b). Many authors have proposed different expressions to extend the use of C_D for spherical particles and larger particle Reynolds numbers. Rubey (1933) proposed a simple approximation to the drag coefficient:

$$C_D = \frac{24}{R_p} + 2 \tag{8.17}$$

Appropriate for the particular case of natural sands and gravels, the expression proposed by Engelund and Hansen (1967) is very close to the one derived by Rubey (1933):

$$C_D = \frac{24}{R_p} + 1.5 \tag{8.18}$$

Karamanev (2001) suggested that one of the best correlations for a freely settling sphere to experimental results was proposed by Turton and Levenspiel (1986):

$$C_D = \frac{24}{R_p}\left(1 + 0.173R_p^{0.6257}\right) + \frac{0.413}{1 + 16300^4 R_p^{-1.09}} \tag{8.19}$$

Yen (1992b) proposed another approximation to the drag coefficient that has been used in numerical simulations of saltating particles near the bed (Niño and García, 1998; González et al., 2008; Bombardelli et al., 2012; Moreno and Bombardelli, in press) with good results:

$$C_D = \frac{24}{R_p}\left(1 + 0.15\sqrt{R_p} + 0.017R_p\right) - \frac{0.208}{1 + 10^4 R_p^{-0.5}} \tag{8.20}$$

A comparison of results coming from all four expressions for the drag coefficient is shown in González (2008), and adapted here as Fig. 8.6. Relatively large differences start to be apparent at about $R_p = 10$. The formulas by Turton and Levenspiel (1986) and Yen (1992b) achieve similar results throughout the entire range of R_p.

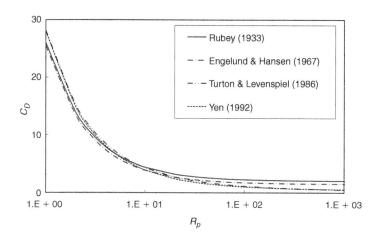

Figure 8.6. Comparison of different expressions for the drag coefficient (adapted from González, 2008)

When calculating the fall velocity of very small sediment particles, like silts, the particle Reynolds number drops below 1, yielding the following expression:

$$w_s = \frac{gRd_p^2}{18v} \tag{8.21}$$

For larger particles, the expressions presented in Table 8.3 have been proposed by several authors in the last decades.

Expressions of settling velocity for cohesive sediment can be found in Mehta and McAnally (2008). In this case, the fall velocity can be affected by both the size of suspended "flocs," and the sediment concentration.

Table 8.3. Summary of the common formulations for the settling (fall) velocity (García, 2008)

Author (year)	Formula	Observations
Dietrich (1982)	$w_s = \exp(-b_1 + b_2 S_l - b_3 S_l^2 - b_4 S_l^3 + b_5 S_l^4)\sqrt{g R d_p}$ $S_l = \ln(Re_p)$	$b_1 = 2.891394;$ $b_2 = 0.95296;$ $b_3 = 0.056835;$ $b_4 = 0.002892;$ $b_5 = 0.000245;$ d_p: mean sieve particle diameter. Applicable to non-spherical natural particles
Jimenez and Madsen (2003)	$w_s = \dfrac{\sqrt{g R d_N}}{\left(A_C + \frac{B_C}{S_*}\right)}$ $S_* = \dfrac{d_N}{4v}\sqrt{g R d_N}$	d_N: nominal particle diameter; A_C and B_C are function of shape factorr and particle roundnessr (Jimenez and Madsen, 2003). Suggested for natural quartz sediments of d_p range [0.063–2] mm
Soulsby (1997)	$w_s = \dfrac{v}{d_p}\left(\sqrt{10.36^2 + 1.049d_*^3} - 10.36\right)$ $d_* = \left[\dfrac{gR}{v^2}\right]^{1/3} d_p$	Aplicable to natural sand particles in marine environments

8.3 SEDIMENT TRANSPORT IN SUSPENSION: CONCEPTS AND THEORY

Relatively fine particles travel in suspension in open channels when turbulent diffusion in the vertical direction is strong enough to counterbalance the effect of particle settling, which pulls the sediment grains towards the bed (Parker, 2004). This allows particles to move along the channel without coming in contact with the bottom for long distances.

The mass conservation equation for particles in suspension (Parker, 2004; García, 2008) can be expressed through an advection-diffusion equation, as follows:

$$\frac{\partial \bar{c}}{\partial t} + \bar{u}\frac{\partial \bar{c}}{\partial x} + \bar{v}\frac{\partial \bar{c}}{\partial y} + (\bar{w} - w_s)\frac{\partial \bar{c}}{\partial z} = -\frac{\partial \left(\overline{u'c'}\right)}{\partial x} - \frac{\partial \left(\overline{v'c'}\right)}{\partial y} - \frac{\partial \left(\overline{w'c'}\right)}{\partial z} \tag{8.22}$$

where $\bar{c}$ denotes the volume concentration of suspended sediment averaged over turbulence; $\bar{u}$, $\bar{v}$ and $\bar{w}$ represent the turbulence-averaged flow velocity components in the x, y and z directions, respectively; c' refers to the fluctuation of the sediment concentration; u', v' and w' are the fluctuations of flow velocity components in the three abovementioned directions; and $\overline{u'c'}$, $\overline{v'c'}$, and $\overline{w'c'}$ indicate the sediment fluxes, or Reynolds fluxes of sediment.

The above equation is valid only for dilute mixtures of suspended sediment (volume fractions of sediment no greater than 2–4%; see Jha and Bombardelli, 2009) and small particle size, namely $d_p < 0.1\,\text{mm}$, since it implicitly assumes that the sediment moves horizontally at the velocity of the water, while in the z direction the particle travels at a velocity different from the fluid – the fall velocity w_s. This does not hold for particles of larger size, or for concentrations of sediment which are in the non-dilute category (Muste et al., 2005; Bombardelli and Jha, 2009; Jha and Bombardelli, 2010). To close the advection-diffusion equation for suspended sediment, the sediment fluxes need to be approximated in terms of the turbulence-averaged concentration. This is attained by assuming a direct proportionality between the sediment fluxes and the gradients of sediment concentration:

$$\overline{u'c'} = -D_{xx}\frac{\partial \overline{c}}{\partial x} \tag{8.23a}$$

$$\overline{v'c'} = -D_{yy}\frac{\partial \overline{c}}{\partial y} \tag{8.23b}$$

$$\overline{w'c'} = -D_{zz}\frac{\partial \overline{c}}{\partial z} \tag{8.23c}$$

In Equations (8.23), D_{xx}, D_{yy} and D_{zz} denote the eddy diffusivity of sediment in each spatial coordinate, which has dimensions of L^2/T. The formulation embedded on Equations (8.23) rests on the gradient-diffusion hypothesis, mathematically similar to Fick's law of molecular diffusion and Fourier's law of heat conduction (Pope, 2000). This concept has been put forward "without justification or criticism" (Pope, 2000). In his book, Pope (2000; page 94) offers several reasons why the gradient-diffusion hypothesis has conceptual limitations "which should be borne in mind," in spite of its wide use. Another important concept worth mentioning here consists in that the eddy diffusivity is a second-order tensor; further, the definitions of Equations (8.23) refer to the components in the diagonal of the tensor.

If a steady state is assumed (i.e., the concentration does not vary in time), a null vertical water velocity is enforced, and it is assumed that the horizontal gradients of concentration are negligible in comparison with the vertical gradients (see Fig. 8.7), Equation (8.22) can be simplified to the following expression (Parker, 2004; García, 2008):

$$-D_{zz}\frac{d\overline{c}}{dz} - w_s\overline{c} = 0 \tag{8.24}$$

The first term in Equation (8.24) denotes the diffusive flux of particles generated by turbulence; the second term represents the rate of sediment deposition from suspension to the bed. Both effects are in equilibrium (Parker, 2004). In order for the upward flux to cause sediment particles to maintain their suspended state, the mean concentration gradient in the water column should be negative, i.e., $d\overline{c}/dz < 0$. This means that turbulent fluxes diffuse sediment particles away from areas of higher concentration (near the bed) to areas of lower concentration (away from the bed).

Next step is to specify an expression of the eddy diffusivity of sediment as a function of the vertical coordinate. Hunter Rouse, a pioneer in introducing turbulence concepts to the theory of sediment transport, analyzed two versions of the eddy diffusivity. In the first version, he wanted to analyze the experimental results he himself obtained with the "jar" tests (Rouse, 1937; García, 2008), where the turbulence was quasi-homogeneous and quasi-isotropic. Under those conditions, Rouse assumed that D_{zz} was constant, and integrated Equation (8.24) to obtain an exponential decay of the concentration, which very nicely matched the experimental points. In the second case, it was an open-channel flow, the case portrayed by Equation (8.24) and Fig. 8.7. In this opportunity, he assumed that the eddy diffusivity of sediment was related to the eddy diffusivity of momentum (i.e., the eddy viscosity; v_T) through the Schmidt number, as follows:

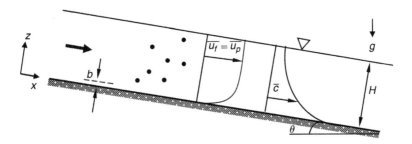

Figure 8.7. Schematic of sediment transport velocity and concentration in a wide rectangular open channel. Adapted from Bombardelli and Jha (2009). The overbar denotes average over turbulence.

$$Sch = \frac{v_T}{D_{zz}} \qquad (8.25)$$

and adopted a parabolic distribution for the eddy viscosity, which is congruent with the law of the wall (Parker, 2004; García, 2008). The integration of Equation (8.24) was developed by Rouse between $z = b$ and the free surface, yielding:

$$\frac{\bar{c}}{\bar{c}_b} = \left[\frac{(H - z)/z}{(H - b)/b} \right]^{Z_R} \qquad (8.26)$$

where:

$$Z_R = \frac{w_s}{\beta_s \kappa u_*} \qquad (8.27)$$

$\bar{c}_b$ is the near-bed, reference sediment concentration, and β_S is the inverse of the Schmidt number (Julien, 2010). The above equation is known as the Rousean distribution of suspended sediment concentration in the water column, and Z_R is the Rouse number. Some sample concentration profiles provided by Equation (8.26) are illustrated in Fig. 8.8, following the values of Z_R suggested by Vanoni (1975). Notice that the smaller the Rouse number is, i.e., the smaller the settling velocity is and thus the particle size is, the more uniform the sediment distribution in the vertical is, which follows intuition. According to Vanoni (1975) "the Z_R values used are those that give the best fit of Equation (8.26) [in this work], and thus the graph indicates only how well data fit the form of the equation" (brackets added by the authors of the chapter). Through this statement, it is possible to infer that the Rousean distribution gives very good approximations to the real concentration profiles when tuning or manipulating the values of the Rouse number. Thus, the statement suggests that the theory does not predict the distribution of relative concentration with accuracy, particularly because of the inherent relative inaccuracy in the prediction of β_s and w_s. In turn, $\kappa \approx 0.41$ can be used in these computations (García, 2008).

The "standard" Rousean solution is found in some books as Equation (8.26) where $\beta_s = 1$. However, assuming $\beta_s = 1$ is only reasonable at high Reynolds numbers when particles are small enough relative to the surrounding flow turbulent structures (Pope, 2000); in other words, the Kolmogorov length scales (η_k) are larger than the diameter of suspended particles (Lyn, 2008). In practice, β_s is used in numerical simulations as an empirical fitting parameter (Greimann et al., 1999). Through numerical simulations using the theory of multi-component fluids (Drew and Passman, 1999) and comparison with experimental data, Jha and Bombardelli (2009) found that for different turbulent closures the Schmidt number is smaller than one for dilute mixtures, and larger than one for non-dilute mixtures. In other words, the eddy diffusivity of sediment is smaller for non-dilute conditions, a result which agrees with the reduction of the diffusion coefficient in gases when the density of the gas increases (Jha and Bombardelli, 2009). More work is needed in this area to obtain more conclusive evidence regarding the appropriate values for β_s.

In the Rouse number, the value of the settling velocity is imposed rather than calculated, using, for example, one of the expressions of Table 8.3. As mentioned in the previous sub-section, this velocity is empirically calculated for quiescent fluids and assumed constant for the entire simulation. Additionally, for higher concentration of particles, the sediment velocity will differ from that of a single settling particle (Vanoni, 1975); hence, different results should be expected according to the diverse arrangements of particles throughout the open channel.

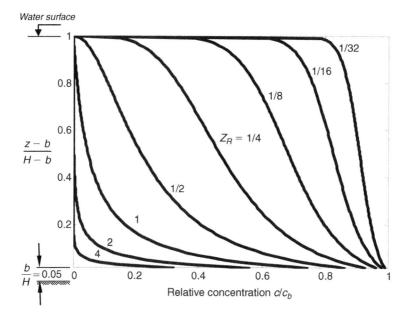

Figure 8.8. Rousean distribution of suspended sediment concentration for selected z_r values. The vertical axis shows the relative distance from the reference level $z = b$, for $b/H = 0.05$.

Interestingly, variants to Equation (8.26) have been put forward for oscillatory, sediment-laden flows. This is particularly useful in the case of coastal flows, whether in seas or lakes. Horikawa (1978) presented on pages 265 and 266 of his book a very interesting analysis in which he derives a more involved balance equation for sediment in suspension than Equation (8.24), but an equation which possesses nonetheless the same ingredients of Equation (8.24).

This procedure leads to an exponential decay of the sediment concentration. On page 269, Horikawa uses precisely Equation (8.24) to backcalculate the eddy diffusivity of sediment from field data. Chung et al. (2009a) in turn used signals of the instrument AWAC (Acoustic Wave and Current profiler) to infer Rousean-like distributions of sediment concentrations in the Salton Sea (California), when averaged over several wave periods. This is another area in which more research is needed.

The advection-diffusion Equation (8.22) represents the sediment particles as a scalar field with concentration $\bar{c}$. Consequently, it can be interpreted as a quasi single-phase flow approximation to the real phenomenon. However, relatively recent experimental results show a clear lag between the fluid and the sediment velocities (Rashidi et al., 1990; Kaftori et al., 1995; Muste and Patel, 1997; Greimann et al., 1999). In addition, several authors (Cellino and Graf, 2002; Greimann et al., 1999; Muste et al., 2005; Nezu and Azuma, 2004; Muste et al., 2005) have found that the distribution of concentration of sediments in the wall-normal direction differs appreciably from that obtained through the use of the "standard" Rousean equation (Bombardelli and Jha, 2009). All this calls for a more sophisticated approach to the sediment-transport problem.

A better description of the suspended sediment distribution can be obtained when applying the two-phase flow theory, where the carrier fluid is the liquid phase, and the sediment particles are the solid phase. The two-phase flow enables the calculation of the velocity of each phase in each coordinate axis through the application of mass and momentum balance equations for fluid and the sediment phases, separately (see Fig. 8.9). Additionally, the two-phase flow approach is more versatile and it could be used for dilute and non-dilute mixtures, cohesive and non-cohesive sediment.

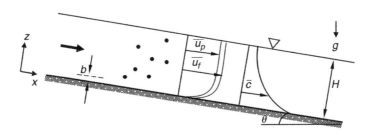

Figure 8.9. Schematic of sediment transport velocity and concentration in a wide rectangular open channel, using the two-phase flow approach.

8.4 SEDIMENT TRANSPORT AS BED LOAD: CONCEPTS AND THEORY

Sediment particles are considered to be transported in bed load mode when they maintain a quasi-permanent contact with the bed in a very narrow region called the bed layer (Niño et al., 1994a; García, 2008; Julien, 2010). As the ratio between the flow shear stress and the critical shear stress slowly exceeds one, the sediment particles usually slide or roll above the bed surface, as stated in Section 8.2. A slight increase from the previous transport stage causes particles to start a hopping motion, i.e., particles hop up a few diameters away from the bed to immediately collide with it (see Fig. 8.10), and/or eventually with other particles (García, 2008; Bombardelli et al., in review). In the bed load layer, the influence of turbulence in mixing is so small, that suspension is not possible (García, 2008).

In essence, particles subjected to saltation experience two major sets of processes. First, hydrodynamic forces lift the particles; then, particles fly over the bed, and are finally pulled

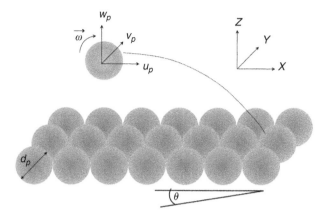

Figure 8.10. Sketch of a saltating particle, where u_p, v_p, and w_p denote the particle velocity components of the vector $\underline{u}_p$ in the three major axes; $\underline{\omega}$ represents the particle angular velocity vector; and θ describes the angle of the channel slope with respect to the horizontal plane. Adapted from Bombardelli et al. (in review).

down by gravity. Second, particles collide with the bed. The transfer of momentum in the vertical direction related to the collision of the particle with the bed causes the particle to hop up again.

A two-phase flow approach to model particle saltation unveils the different characteristics of both the fluid and grain motions. In an Eulerian-Lagrangian framework, i.e., an Eulerian description of the fluid flow and a Lagrangian description of the sediment particles, the following equations are used:

- Fluid (continuous, or carrier phase), using an Eulerian approach:
  - Mass balance
  - Linear momentum balance
- Sediment particles (dispersed phase), using a Lagrangian approach:
  - Linear momentum equation
  - Angular momentum equation.

(It is worth mentioning here that the mass conservation of the disperse phase is enforced by a "de facto" analysis of each particle. In addition, the energy equation for the fluid flow is not required in this case, assuming that no heat exchange between the particles and the surrounding fluid is present.) For the continuous phase, the ideal situation would be to use the Navier-Stokes Equations, valid for an incompressible, Newtonian fluid. Following this paradigm, a very detailed simulation of the flow in between particles is needed, which is extremely demanding from the computational point of view. Some authors have followed this route with interesting results for cases of isotropic and homogeneous turbulence, and gas-solid flows (see Squires and Eaton, 1990; Elghobashi and Truesdell, 1993; Ferrante and Elghobashi, 2003; Yamamoto et al., 2001; Portela and Oliemans, 2002; Vance et al., 2006; Dritselis and Vlachos, 2011). In open-channels, turbulence is non homogeneous and non isotropic; thus, those results cannot be extrapolated in direct form.

The simplest alternative is to use a known "solution" for the fluid flow, consisting in the law of the wall. The law of the wall represents the averaged velocity profile of the turbulent flow near the wall and, therefore, the velocity fluctuations are not included in the analysis. The law of the wall can be expressed as:

$$\frac{\overline{u_f}(z)}{u_*} = \frac{1}{\kappa} \ln\left(\frac{z}{z_0}\right) \tag{8.28a}$$

$$\overline{v_f}(z) = 0 \tag{8.28b}$$

$$\overline{w_f}(z) = 0 \tag{8.28c}$$

where $\overline{u_f}$ denotes the time-averaged flow velocity at a distance z above the bed, and z_0 indicates a reference length scale. That reference length scale is equal to $k/30$ for a rough boundary, and equal to $9v/u_*$ for a smooth wall. In this equation, only the velocity in the stream-wise direction is changing as a function of z, while the mean velocities in the transverse and wall normal directions are assumed to be nil.

For the dispersed phase, the two sets of processes mentioned before lead to two submodels (González, 2008; Bombardelli et al., in review): 1) The sub-model associated with the "free flight" of the particle throughout the flow field, away from the wall; 2) the submodel describing the collision of the particles with the bed. In the second sub-model, a combination of stochastic and geometrical expressions, in addition to considerations on the nature of collisions, are applied.

Particle-particle collisions could be assumed as another separate stage, or as an extension of the second one. These collisions may cause significant changes in sediment diffusion (also called "scatter") in the transverse direction to the flow in open channels. Sub-models for particle-particle collisions can use the conservation of linear and angular momentum of two colliding particles during the collision. In particular, the collision of the saltating particle with *the bed* could be interpreted in this light as the collision between two particles, one of which has a diameter tending to infinity. For dilute mixtures, the *hard-sphere model* (Crowe et al., 2012), applicable only to binary collisions, is an adequate choice for application to stage two (as well as for inter-particle collisions).

In what follows, we present the mathematical sub-models for each stage, for a three-dimensional case (see also González, 2008; Bombardelli et al., 2010; Moreno et al., 2011; Bombardelli et al., in review).

"Free flight" stage: Consider a sediment particle of mass m that moves with velocity $u_p \equiv (u_p, v_p, w_p)$ in a turbulent flow described by the velocity $\overline{u_f} \equiv (u_f, 0, 0)$, following Equations (8.28). The application of Newton's second law indicates that the acceleration of each sediment particle times its mass will be the result of all the forces acting on the particle. These forces are (Bombardelli et al., in review): Submerged weight, F_{SW}, which combines the action of gravity and the buoyancy force; drag, F_D; lift, F_L; Basset, F_B; Magnus, F_M; added-mass or virtual-mass, F_{VM}, and fluid acceleration, F_{FA} (if different from 0). Hence, the equation of motion of a saltating particle can be written as:

$$m\frac{du_p}{dt} = \underline{F_{SW}} + \underline{F_D} + \underline{F_L} + \underline{F_B} + \underline{F_M} + \underline{F_{VM}} + \underline{F_{FA}} \tag{8.29}$$

Different expressions have been proposed to model the forces of Equation (8.29). No absolute agreement exists among all researchers on the details of the forces for relatively large particle Reynolds numbers, i.e., for particles of finite size. Recently, Loth and Dorgan (2009) provided an interesting review of models. Therefore, there is no "correct" form of these forces (Lukerchenko, 2010; Lukerchenko et al., 2012). A cursory review of the literature (including papers from the mechanical, nuclear, chemical, and civil engineering fields) indicates that any belief on "correct" models is illusory. What researchers do seem to agree upon is on the general utility of these models, especially in order to predict global flow variables such as the total transport of particles as bed load. Introducing common models for the forces in (8.29), it is possible to obtain (Balachandar and Eaton, 2010):

$$
m \frac{du_p}{d\hat{t}} = \left(m - m_f \right) \underline{g} + 3\pi \mu d_p \left(\underline{u_f} - \underline{u_p} \right) \Phi(Re_p)
$$

$$
+ \frac{3}{2} d_p^2 \sqrt{\pi \rho \mu} \int_{-\infty}^{t} K(t, \tau) \frac{d}{d\tau} \left(\underline{u_f} - \underline{u_p} \right) d\tau
$$

$$
+ m_f C_m \left(\frac{D\underline{u_f}}{Dt} - \frac{d\underline{u_p}}{dt} \right) + m_f \frac{D\underline{u_f}}{Dt} \tag{8.30}
$$

where the forces of lift and Magnus have been omitted. In these equations, C_m is the virtual mass coefficient; m_f is the mass of fluid displaced by the particle; t refers to the time coordinate; τ is a dummy variable for integration; K is the kernel of the Basset force. The operator $d(\cdot)/dt$ indicates the material derivative using the particle velocity, and the operator $D(\cdot)/dt$ uses the flow velocity. Φ refers to the correction function in the drag force for large Reynolds numbers. Obviously, this model applies to any general time-dependent velocity field.

The non-dimensional particle rotation vector, $\underline{\omega}$ (made non-dimensional by using the shear velocity and the particle diameter), can be calculated at each time step of the simulation using the angular momentum equation. In this respect, several authors have included the expression proposed by Yamamoto et al. (2001) to their free-flight models (Harada and Gotoh, 2006; González, 2008; Lukerchenko et al., 2009; Bombardelli et al., 2012; Moreno and Bombardelli, in press), with good results. Yamamoto et al.'s expression is:

$$
\frac{d\underline{\omega_r}}{dt} = - C_t \frac{15}{16\pi} |\underline{\omega_r}| \underline{\omega_r} \tag{8.31}
$$

where $C_t = C_1/\sqrt{Re_R} + C_2/Re_R + C_3 Re_R$ denotes a non-dimensional coefficient which is a power-law function of the particle Reynolds number of the rotational motion, $Re_R = d_p^2 |\underline{\omega_r}|/4\nu$; the coefficients C_1, C_2, and C_3 are obtained from Table 8.4, originally published in Yamamoto et al. (2001); and $\underline{\omega_r}$ is the non-dimensional relative particle rotation vector with respect to the fluid vorticity.

Table 8.4. Values of coefficient C_1, C_2, and C_3 from Yamamoto et al. (2001).

Re_R	0–1	1–10	10–20	20–50	50+
C_1	0	0	5.32	6.44	6.45
C_2	50.27	50.27	37.2	32.2	32.1
C_3	0	0.0418	5.32	6.44	6.45

For simplicity, many authors have customarily used $C_m = 0.5$ for the virtual mass coefficient (Niño and García, 1994b and 1998; Schmeeckle and Nelson, 2003; González, 2008; Lukerchenko et al., 2009; Bombardelli et al., 2012; Moreno and Bombardelli, in press). This value of the coefficient corresponds to a translating particle without rotation; for a particle which is only rotating, Kendoush (2005) found that $C_m = 5$. No reliable coefficient for translating-rotating particles at large particle Reynolds numbers is available. For the lift coefficient, $C_L = 0.2$ was suggested by Wiberg and Smith (1985), and the coefficient has similar limitations to those of the coefficient of virtual mass. The drag coefficient can be computed employing any of the equations compared on Fig. 8.6, since the results are close in the range of particle Reynolds number found in bed load computations.

The numerical integration of the expression for the Basset force is particularly involved, because the integral becomes singular at the upper limit when usual formulations are used (see Bombardelli et al., 2008). In order to avoid this limitation, the calculation of the Basset term can be implemented in the computational model following the methodology proposed by Brush et al. (1964) or Bombardelli et al. (2008). Bombardelli et al. (2008) developed an efficient method which saves computational time by means of the use of fractional mathematics and the concept of "memory time." It is worth mentioning here that this methodology is first order accurate in time (van Hinsberg et al., 2011), and that it is valid as long as the kernel decays with the square root of time.

To obtain the particle velocity and rotation, Equation (30) can be integrated numerically by using the standard fourth-order Runge-Kutta method, although the final accuracy of the model will be limited by the order of accuracy of the integration of the Basset force.

Particle collision with the bed stage: The algorithm for this sub-model is extensively explained in González (2008), and Bombardelli et al. (in review), and it can be considered as an extension to three dimensions of the model by García and Niño (1992). The incident angles in the stream-wise and the span-wise directions (defining the three-dimensional behavior of the sediment motion) are calculated by using the particle velocity right before the collision with the wall. Then, a trigonometric relation is established through the angle between the tangent to the particle on the bed (considered to be a sphere for simulation purposes) at the impact point, and the channel surface. This trigonometric relation, for the vertical plane, can be written as follows (see Fig. 8.11):

$$\frac{r}{d_p} = \frac{1}{2}[\cos(\theta_b) - \tan(\theta_{in})\sin(\theta_b)] \tag{8.32}$$

where r/d_p is defined through a random number generator. In a similar way, a trigonometric relation for the span-wise direction can be developed (González, 2008; Bombardelli et al., in review).

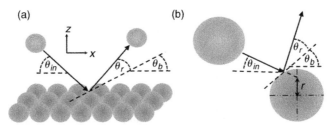

Figure 8.11. Schematic of the collision of a particle with the bed in a vertical plane. Left: View from a side. Right: θ_{in} is defined from the incident velocity of the particle, right before collision with the wall; θ_b denotes the angle between the plane tangent to the point of impact and the bed; and θ_r denotes the angle described for the particle after the collision with the bed, measured from θ_b (adapted from Bombardelli et al., in review).

Collisions of a particle with the wall produce variations of the position of particles in the transverse direction, often called "transverse diffusion," or "scatter." Niño and García (1994a, b, 1998) presented some particle trajectories where this effect is evident.

Particle-particle collisions stage: Not much is known regarding the collisions among particles in bed load transport. To the best of our knowledge, this problem has been addressed mostly through numerical simulations. Very recently, Bialik (2011a, b) and Moreno and Bombardelli (in press) presented an analysis of inter-particle collisions through a statistical approach. Most approaches coming from the mechanical engineering field are formulated via the use of the conservation principles of linear and angular momentum at the collision. To that end, the equations discussed by Crowe et al. (2012) are applied.

These models produce results in terms of particle jump length and height, allowing for the computation of the volume flow rate of particles moving as bed load. This has been done with good agreement with some of the expressions quoted on Table 8.1 (see González, 2008; Bombardelli et al., 2010; Moreno et al., 2011), and with experiments.

8.5 COMBINED MODELING OF BED LOAD TRANSPORT AND SUSPENDED LOAD MOTION

Recent contributions in the field are based on combined formulations of both bed load and sediment-load transport. In fact, the "scatter" produced by particle collisions with walls, as well as inter-particle collisions can be interpreted macroscopically as a diffusion process, with an associated diffusivity in the transverse direction. This, in addition to macroscopic diffusion and dispersion coefficients in the longitudinal direction lead to a generalized advection-diffusion equation which includes the suspended sediment load and the bed load (see Greimann et al., 2008).

8.6 CONTAMINANT EXCHANGES AT THE INTERFACE

Numerous water bodies in the world possess serious contamination with metals as a direct consequence of: a) anthropogenic activities such as mining, and b) dry and wet deposition. One typical case is that of mercury, which serves the purpose of illustrating the exchanges of contaminants at the interface in this chapter.

Many places in the state of California were subjected to mining activities during the Gold Rush era (1850s). In the so-called "hydraulic mining," high-pressure hoses were employed to dislodge the rocks containing gold. Those gravel particles formed "slurries" which were transported down-slope via sluices, facilitating the deposition of gold at the bottom of the sluice. Heavy metals such as mercury were used to ease the separation of gold from the gravel, and both sediment particles and metals were transported to the valley of the Sacramento River. Deposition in different portions of the river followed, thus increasing the damages of floods. Mercury remains to this day at the bottom sediments as a testament of those times.

Mercury is characterized by different chemical forms or oxidation states. Mercury's organic form is called Methylmercury (MeHg) and is the most bio-available. Mercury is believed to be methylated by bacteria in anaerobic waters and sediments (Morel et al., 1998). In the sediments, mercury then is transported usually upwards mainly through diffusive/dispersive processes in porous media (see Fig. 8.12). Once in the water column, mercury is transported by advection and turbulent diffusion (Massoudieh et al., 2009; 2010). Other species of mercury are: mercury divalent (HgII), Hg0, and cinnabar (HgS). The importance

List of Symbols

Symbol	Definition	Dimensions or Units
C_D	drag coefficient	
C_L	lift coefficient	
C_m	virtual mass coefficient	
C_t	coefficient in the expression for angular momentum of the particle	
d_p	particle diameter	$[L]$
D_{xx}, D_{yy}, D_{zz}	eddy diffusivity in the x, y and z directions	$[L^2 \cdot T^{-1}]$
E_r	erosion rate of sediment	$[L \cdot T^{-1}]$ or $[M \cdot L^{-2} \cdot T^{-1}]$
E_s	entrainment function of sediment into suspension	
F_B	Basset force	$[M \cdot L \cdot T^{-2}]$
F_{Bu}	buoyancy force	$[M \cdot L \cdot T^{-2}]$
F_D	drag force	$[M \cdot L \cdot T^{-2}]$
F_{FA}	fluid acceleration force	$[M \cdot L \cdot T^{-2}]$
F_G	gravity force	$[M \cdot L \cdot T^{-2}]$
F_L	lift force	$[M \cdot L \cdot T^{-2}]$
F_M	Magnus force	$[M \cdot L \cdot T^{-2}]$
F_{SW}	submerged weight	$[M \cdot L \cdot T^{-2}]$
F_{VM}	virtual (added) mass force	$[M \cdot L \cdot T^{-2}]$
g	acceleration of gravity	$[L \cdot T^{-2}]$
H	water depth	$[L]$
H_L	layer of constant thickness in the sediments	$[L]$
k	roughness height	$[L]$
k_d	decay constant	$[T]^{-1}$
m	mass of the particle	$[M]$
q_b	volumetric bed load transport rate	$[L^{-2} \cdot T^{-1}]$
q_*	Einstein number	
r	geometrical parameter linked to θ_{in} and θ_b	$[L]$
R	submerged specific gravity of sediment	
R_p	particle Reynolds number	
Re_p	explicit particle Reynolds number	
Re_R	rotational motion particle Reynolds number	
t	time	$[T]$
$\bar{u}$	mean flow velocity in the x direction	$[L \cdot T^{-1}]$
u'	velocity fluctuation in the x direction	$[L \cdot T^{-1}]$
u_*	shear velocity	$[L \cdot T^{-1}]$
u_{*s}	shear velocity due to skin friction	$[L \cdot T^{-1}]$
$\overline{u'c'}$	sediment (Reynolds) flux in the x direction	$[L \cdot T^{-1}]$
u_f	fluid velocity in the x direction	$[L \cdot T^{-1}]$
u_p	particle velocity in the x direction	$[L \cdot T^{-1}]$
U	cross-sectional average velocity	$[L \cdot T^{-1}]$
$\bar{v}$	mean flow velocity in the y direction	$[L \cdot T^{-1}]$
v'	velocity fluctuation in the y direction	$[L \cdot T^{-1}]$
v_p	particle velocity in the y direction	$[L \cdot T^{-1}]$

(Continued)

	List of Symbols	
Symbol	Definition	Dimensions or Units
$\overline{v'c'}$	sediment (Reynolds) flux in the y direction	$[L \cdot T^{-1}]$
w_b	burial velocity	$[L \cdot T^{-1}]$
$\overline{w}$	mean flow velocity in the z direction	$[L \cdot T^{-1}]$
w'	velocity fluctuation in the z direction	$[L \cdot T^{-1}]$
w_p	particle velocity in the z direction	$[L \cdot T^{-1}]$
w_s	settling (fall) velocity	$[L \cdot T^{-1}]$
$\overline{w'c'}$	sediment (Reynolds) flux in the z direction	$[L \cdot T^{-1}]$
Z_R	Rouse number	
β_s	ratio of sediment to momentum diffusivity	
∇	Nabla operator	$[L^{-1}]$
∇_h	horizontal divergence	$[L^{-1}]$
η	elevation of the bed with respect to datum	$[L]$
η_k	Kolmogorov length scale	$[L]$
θ	channel slope angle	$[rad]$
θ_b	angle formed by the tangent to the point of impact of the "flying" particle (with a particle on the bed) and the bed of the channel, defined in the x-z plane	$[rad]$
θ_{in}	angle formed by the "flying" particle right before colliding with the bed, defined in the x-z plane	$[rad]$
K	von Kármán constant	
λ_p	Porosity	
μ	dynamic viscosity	$[M \cdot L^{-1} \cdot T^{-1}]$
ν	kinematic viscosity	$[L^2 \cdot T^{-1}]$
ρ	water density	$[M \cdot L^{-3}]$
ρ_s	sediment density	$[M \cdot L^{-3}]$
Sch	Schmidt number	
$\tau_{*,s}$	Shields stress caused by skin friction	
τ_*	Shields parameter	
$\tau_{*,c}$	critical shear stress	
τ_0	bed shear stress	$[M \cdot L^{-1} \cdot T^{-2}]$
ϖ	dimensionless particle rotation	
ϖ_r	dimensionless particle relative rotation	
$\underline{\omega}$	dimensional particle rotation	$[rad \cdot T^{-1}]$

APPENDIX B – SYNOPSIS

The interface bed sediments-water column plays a tremendous role in the way the sediments and contaminants move in open channels. In fact, the location of that interface depends directly on the mass (volume) flow rates of sediment as bed load, and on the net vertical flow rate of particles towards the bed, associated with the sediment in suspension. (A third mode of sediment transport, called wash load exists as well, composed by very fine particles.) This mass balance of sediment at the interface leads to the well-known Exner Equation. For the application of that equation, predictors for the volume flow rates of sediment as bed load are needed. Several of those predictors were quoted in this chapter. Also, predictors for the settling velocity, and for the entrainment of sediment into suspension were provided, indicating the highly non-linear nature of the sediment resuspension process. Particles can be put into motion as bed load when the shear stress exerted by the flow on the bottom exceeds the critical shear stress. This critical shear stress can be obtained from regressions developed to experimental data. Sediment in suspension requires the turbulence to oppose (via diffusive processes) the tendency of particles to move towards the bed (due to gravity); when this condition is met and the horizontal gradients are small, an equilibrium condition takes place, leading to the Rousean Equation for sediment in suspension. Mechanistic models for bed load have provided successful results in terms of particle jump height and length. They have also been able to compute the volume flow rate as bed load, successfully. Finally, the chapter discussed the mass transfer of *contaminants* through the bed sediments-water column interface, signaling the similarities and differences with the case of sediment particles. The case of mercury was commented as a good example for the need of having comprehensive models for contaminants taking into account relevant processes at the interface.

APPENDIX C – KEYWORDS

The chapter includes the following terms.

Bed load transport	Mass conservation	Sediment transport
Drag coefficient	Particle saltation	Settling (fall) velocity
Exner Equation	Rousean distribution	Suspended load transport
Incipient sediment motion	Sediment entrainment	Two-phase flow

APPENDIX D – QUESTIONS

What is the meaning of "equilibrium" in the resuspension of sediment from the bottom in open channels? Under what conditions does this equilibrium state hold?

How do you quantify the flow rate of sediment towards the bed?

Which are the forces causing the motion of otherwise resting sediment particles in the bed under submerged conditions?

What is the conservation principle implied in the Exner Equation?

How can a two-phase flow approach lead to a better description of the sediment transport process?

What are the main assumptions made to obtain the advection-diffusion expression for suspended sediment?

What are the major processes involved in particle saltation?

Which processes affect contaminants at the bed sediments-water column interface?

APPENDIX E – PROBLEMS

E1. Derive the Ikeda-Coleman-Iwagaki model (see Equation (8.E1) below) assuming that the bed slope (angle α) is negligible, and that the particles are spherical. The flow is characterized by a velocity profile $\bar{u}(z)$, where z denotes the wall-normal coordinate, as shown in Fig. 8.E1. Consider the forces of drag, F_D, lift, F_L, the submerged weight, F_{SW} (i.e., particle weight minus buoyancy), normal force, F_N, exerted by the bed on the resting particle, and the resistive force, F_R.

$$\tau^* = \frac{4}{3} \frac{\mu}{(C_D + \mu C_L)} \frac{1}{F^2(u_{*,c} d_p/v)} \tag{8.E1}$$

where $\mu = \tan(\phi)$ denotes the coulomb friction coefficient, and ϕ indicates the angle of repose and $F(u_{*,c}\, d_p/v)$ is a function relating the flow velocity close to the bed and the shear velocity. Use the following expressions for the drag and lift forces:

$$F_D = \frac{1}{2}\rho C_D u^2 A \tag{8.E2}$$

$$F_D = \frac{1}{2}\rho\, C_L u^2 A \tag{8.E3}$$

where A denotes the transverse area of the spherical sediment particle.

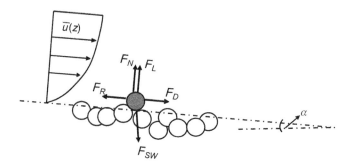

Figure 8.E.1. Sketch of incipient motion condition of a sediment particle lying on an open channel bed.

Hint: Use the known equilibrium condition $F_R = \mu\, F_N$, and the relation between drag and lift coefficient $C_L = 0.85 C_D$, where C_D can be calculated with the expression suggested by Yen (1992b).

E2. Use the Ikeda-Coleman-Iwagaki model (Eq. (8.E1)) for incipient motion to construct a "modified" Shields diagram that shows τ_* vs. Re_p.

For $F(u_{*,c}\, d_p/v)$ use the continuous function proposed by Swamee (1993).

$$F(u_{*,c}\, d_p/v) = \left\{ \left(\frac{2v}{u_*\, d_p}\right)^{10/3} + \left[\frac{1}{\kappa}\ln\left(1 + \frac{\frac{9}{2}\frac{u_*\, d_p}{v}}{1 + 0.3\frac{u_*\, d_p}{v}}\right) \right]^{-10/3} \right\}^{-0.3} \tag{8.E4}$$

Assume the following in the computations:

$\phi = 40°$ *(gravel)*

Hint: Use an iterative method to overcome the implicit nature of the equation (through u_*) assuming an initial guess value and then setting a tolerance value obtained from the above equations to check how close consecutive values of the guess are.

E3. Calculate the settling velocity for a sand particle of $d_{50} = 0.5$ mm, and a silt particle of $d_{50} = 25\,\mu$m using all expressions presented in Table 8.3. Consider a water temperature of 18°C. Compare results of different formulas with a plot.

E4. Plot the dimensionless bed load transport relations of Meyer-Peter and Muller (1948), Ashida and Michihue (1972), Fernandez-Luque and van Beek (1976), and Parker (1979) in a log-log scale, for values of τ_* ranging from 0.05 to 1, and a particle diameter of 0.9 mm.

E5. Plot in a unique figure the Rouse distribution for seven different sediment particle diameters: $4\,\mu m$ (coarse clay), $62\,\mu m$ (coarse silt), $0.25\,mm$ (fine sand), $0.5\,mm$ (medium sand), $2\,mm$ (very coarse sand), $16\,mm$ (medium gravel), and $64\,mm$ (very coarse gravel). Assume $b = 0.05H$ for an open channel with water depth of 3 m, and bed slope of $S = 0.008$. Calculate the settling velocity using the expression proposed by Soulsby (1997), adopting $R = 1.65$ and a water temperature of $20°C$. To calculate the Rouse number consider $\kappa = 0.41$, $\beta_S = 0.85$ and $\tau_b = \rho g H S$ (bed shear stress in a wide open channel). Discuss the differences between concentration profiles.

REFERENCES

Abbott, J. E., and Francis, J. R. D., 1977, Saltation and suspension of solid grains in a water stream, *Philosophical Transactions of the Royal Society of London Series A-Mathematical Physical and Engineering Sciences*, **284**, pp. 225–254

Ashida, K., and Michiue, M., 1972, Study on hydraulic resistance and bed load transport rate in alluvial streams, *Transactions Japan Society of Civil Engineers*, **206**, pp. 59–69

Balachandar, S., and Eaton, J. K., 2010, Turbulent dispersed multiphase flow, *Annual Reviews in Fluid Mechanics*, **42**, pp. 111–133

Bialik, R. J., 2011a, Numerical study of saltation of non-uniform grains, *Journal of Hydraulic Research*, **49(5)**, pp. 697–701

Bialik, R. J., 2011b, Particle-particle collision in Lagrangian modelling of saltating grains, *Journal of Hydraulic Research*, **49(1)**, pp. 23–31

Bombardelli, F. A., González, A. E., and Niño, Y. I., 2008, Computation of the particle Basset force with a fractional-derivative approach, *Journal of Hydraulic Engineering, ASCE*, **134(10)**, pp. 1513–1520

Bombardelli, F. A., and Jha, S. K., 2009, Hierarchical modeling of the dilute transport of suspended sediment in open channels, *Environmental Fluid Mechanics*, **9(2)**, pp. 207–235

Bombardelli, F. A., González, A. E., and Moreno, P. A., 2010, Numerical simulation of spheres moving and colliding close to bed streams, with a complete characterization of turbulence. In: *Proceedings of River Flow* 2010, Braunschweig, Germany, pp. 777–784

Bombardelli, F. A., González, A. E., Moreno, P. A., and Moniz, R., in review, Generalized algorithms for particle motion and collision with stream beds.

Brush, J. S., Ho, H. W., and Yen, B. C., 1964, Acceleration motion of sphere in a viscous fluid, *Journal of Hydraulic Division*, **90**, pp. 149—160.

Cellino, M., and Graf, W. H., 2002, Suspension flow in open channels; experimental study, *Journal of Hydraulic Research*, **15**, pp. 435–447

Chung, E. G., Bombardelli, F. A., and Schladow, S. G., 2009a, Sediment resuspension in a shallow lake, *Water Resources Research*, **45**, W05422

Chung, E. G., Bombardelli, F. A., and Schladow, S. G., 2009b, Modeling linkages between sediment resuspension and water quality in a shallow, eutrophic, wind-exposed lake, *Ecological Modelling*, **220**, pp. 1251–1265

Crowe, C. T., Schwarzkopf, J. D., Sommerfeld, M., and Tsuji, Y., 2012, *Multiphase flow with droplets and particles*, 2nd edition. CRC Press, Taylor & Francis Group, Florida, USA

Dietrich, W. E., 1982, Settling velocities of natural particles, *Water Resources Research*, **18(6)**, pp. 1615–1626

DiToro, D. M., 2001, *Sediment flux modelling*, Wiley-Interscience, New York, USA

Drew, D. A., and Passman, S. L., 1999, *Theory of multicomponent fluids*, Springer, Berlin, Germany

Dritselis, C., and Vlachos, N., 2011, Large Eddy simulation of gas-particle turbulent channel flow with momentum exchange between the phases, *International Journal of Multiphase Flow*, **37**, pp. 706–721

Einstein, H. A., 1950, *The bed load function for sediment transportation in open channels*, Technical Bulletin 1026, U.S. Dept. of Agric., Soil Conservation Service, Washington, D.C.

Elghobashi, S., and Truesdell, G., 1993, On the two-way interaction between turbulence and dispersed solid particles. 1. Turbulence modification, *Physics of Fluids A-Fluid Dynamics*, **5(7)**, pp. 1790–1801

Engelund, F., and Fredsøe, J., 1976, A sediment transport model for straight alluvial channels, *Nordic Hydrology*, **7**, pp. 293–306

Engelund, F., and Fredsøe, J., 1982, Sediment ripples and dunes, *Annual Reviews in Fluid Mechanics*, **14**, pp. 13–37

Fernandez Luque, R., and van Beek, R., 1976, Erosion and transport of bed sediment, *Journal of Hydraulic Research*, IAHR, **14(2)**, pp. 127–144

Ferrante, A., and Elghobashi, S., 2003, On the physical mechanism of two-way coupling in particle-laden isotropic turbulence, *Physics of Fluids*, **15(2)**, pp. 315–329

Francis, J. R. D., 1973, Experiments on the motion of solitary grains along the bed of a water-stream, *Proceedings of the Royal Society A.*, **332**, pp. 443–471.

García, M. H., and Parker, G., 1991, Entrainment of bed sediment into suspension, *Journal of Hydraulic Engineering*, ASCE, **117(4)**, pp. 414–435

García, M. H., and Niño, Y. I., 1992, Lagrangian description of beadload transport by saltating particles, In: *Proceedings of the 6th IAHR International Symposium on Stochastic Hydraulics*, Taipei, Taiwan, pp. 259–266

García, M. H., 1999, "Sedimentation and erosion hydraulics", *Hydraulic design handbook*, Mays, L. W., ed., McGraw-Hill Handbook, New York, USA, Chap. 6, pp. 6.1–6.113

García, M. H., 2008, "Sediment transport and morphodynamics," *Sedimentation engineering: Processes, measurements, modeling and practice*, ASCE manual of practice 110, M. H. García, ed., ASCE, Reston, Va., USA, Chap. 2, pp. 21–163

González, A. E., 2008, *Coupled numerical modeling of sediment transport near the bed using a two-phase flow approach*, Ph.D. Thesis, University of California, Davis, USA

Greimann, B. P., Muste, M., and Holly, F. M. Jr., 1999, Two-phase formulation of suspended sediment transport, *Journal of Hydraulic Research*, **37**, pp. 479–500

Greimann, B. P., Lai, Y., and Huang, J., 2008, Two-dimensional total sediment load model equations, *Journal of Hydraulic Engineering*, ASCE, **134(8)**, pp. 1142–1146

Harada, E., and Gotoh, H., 2006, Influence of sand shape to vertical sorting under uniform flow condition, In: *Proceedings of River Flow 2006*, Lisbon, Portugal, pp. 853–858

Hjulström, F., 1935, *Studies of the morphological activity of rivers, as illustrated by the river Fyris*, University of Uppsala, Geological Institute Bulletin, **25**, pp. 221–557

Horikawa, K., 1978, *Coastal engineering. An introduction to ocean engineering*, University of Tokyo Press, Japan

Hu, C., and Hui, Y., 1996, Bed load transport II. Stochastic characteristics, *Journal of Hydraulic Engineering*, ASCE, **122**, pp. 255–261

Jha, S. K., and Bombardelli, F. A., 2009, Two-phase modelling of turbulence in dilute sediment-laden, open-channel flows, *Environmental Fluid Mechanics*, **9**, pp. 237–266

Jha, S. K., and Bombardelli, F. A., 2010, Toward two-phase flow modelling of nondilute sediment transport in open channels, *Journal of Geophysical Research-Earth Surface*, **115**, F03015

Jimenez, J. A., and Madsen, O. S., 2003, A simple formula to estimate settling velocity of natural sediments, *Journal of Waterways, Port, Coastal, and Ocean Engineering*, ASCE, **129(2)**, pp. 70–78

Julien, P. Y., 2010, *Erosion and sedimentation,* second edition, Cambridge University Press, Cambridge, UK

Kaftori, D., Hetsroni, G., and Banerjee, S., 1995, Particle behaviour in turbulent, boundary layer II. Velocity and distribution profiles, *Physics of Fluids,* **7(5),** pp. 1107–1121

Karamanev, D. G., 2001, The study of free rise of buoyant spheres in gas reveals the universal behaviour of free rising rigid spheres in fluid in general, *International Journal of Multiphase Flow,* **27,** pp. 1479–1489

Kendoush, A., 2005, The virtual mass of a rotating sphere in fluids, *Journal of Applied Mechanics,* **72,** pp. 801–802

Kennedy, J. F., 1995, The Albert Shields story, *Journal of Hydraulic Engineering,* ASCE, **121,** pp. 766–722

Kundu, P. K., and Cohen, I. M., 2008, *Fluid Mechanics,* 4th edition, Academic, San Diego, USA

Lee, H. Y., Chen, Y., You, J., and Lin, Y., 2000, On three-dimensional continuous saltating process of sediment particles near the channel bed, *Journal of Hydraulic Engineering,* ASCE, **126,** pp. 691–700

Loth, E., and Dorgan, A. J., 2009, An equation of motion for particles of finite Reynolds number and size, *Environmental Fluid Mechanics,* **9(2),** pp. 187–206

Lukerchenko, N., Piatsevich, S., Chara, Z., and Vlasak, P., 2009, 3D numerical model of the spherical particle saltation in channel with a rough fixed bed, *Journal of Hydrology and Hydromechanics,* **57(2),** pp. 100–112

Lukerchenko, N., 2010, Discussion of "Computation of the particle Basset force with a fractional-derivative approach," by F. A. Bombardelli, A. E. González, and Y. I. Niño, *Journal of Hydraulic Engineering,* ASCE, **136(10),** 853–854

Lukerchenko, N., Kvurt, Y., Keita, I., Chara, Z., and Vlasak, P., 2012, Drag force, drag torque, and Magnus force coefficients of rotating spherical particle moving in fluid, *Particulate Science and Technology,* **30,** pp. 55–67

Lyn, D. A., 2008, "Turbulence models for sediment transport engineering," *Sedimentation engineering: Processes, measurements, modeling and practice,* ASCE manual of practice 110, M. H. García, ed., ASCE, Reston, Va., USA, Chap. 16, pp. 763–825

MacArthur, R. C., Neill, C. R., Hall, B. R., Galay, V. J., and Shvidchenko, A. B., 2008, "Overview of sedimentation engineering," *Sedimentation engineering: Processes, measurements, modeling and practice,* ASCE manual of practice 110, M. H. García, ed., ASCE, Reston, Va., USA, Chap. 1, pp. 1–20

Massoudieh, A., Zagar, D., Green, P. G., Cabrera-Toledo, C., Horvat, M., Ginn, T. R., Barkouki, T., Weathers, T., and Bombardelli, F. A., 2009, *Advances in Environmental Fluid Mechanics,* D. Mihailovich, and Carlo Gualtieri, eds., World Scientific, Chapter 13, pp. 275–308

Massoudieh, A., Bombardelli, F. A., and Ginn, T. R., 2010, A biogeochemical model of contaminant fate and transport in river waters and sediments, *Journal of Contaminant Hydrology,* **112(1–4),** pp. 103–117

Mehta, A. J., Parchure, T. M., Dixit, J. G., Ariathurai, R., 1982, *Resuspension potential of deposited cohesive sediment beds,* Academic Press, New York, USA

Mehta, A. J., and McAnally, W. H., 2008, "Fine-grained sediment transport," *Sedimentation engineering: Processes, measurements, modeling and practice,* ASCE manual of practice 110, M. H. García, ed., ASCE, Reston, Va., USA, Chap. 4, pp. 253–306

Meyer-Peter, E., and Muller, R., 1948, Formulas for bedload transport, In: *Proceedings of the 2nd Congress, IAHR,* Stockholm, pp. 39–64

Miedema, S. A., 2008, An analytical method to determine scour, *WEDA XXVII & Texas A&M 39,* St. Louis, USA .

Morel, F. M. M., Kraepiel, A. M. L., and Amyot, M., 1998, The chemical cycle and bioaccumulation of mercury, *Annual Reviews of Ecological Systems,* **29,** pp. 543–566

Moreno, P. M., Bombardelli, F. A., González, A. E., and Calo, V. M., 2011, Three dimensional model for particle saltation close to stream beds, including a detailed description of the particle interaction with turbulence and inter-particle collisions, In: *Proceedings of World Environmental and Water Resources Congress*, ASCE, Palm Springs, California, USA, pp. 2075–2084

Moreno, P. A., and Bombardelli, F. A., in press, Numerical simulation of particle-particle collisions in saltation mode near stream beds.

Muste, M., and Patel, V. C., 1997, Velocity profiles for particles and liquid in open-channel flow with suspended sediment, *Journal of Hydraulic Engineering*, ASCE, **123(9)**, pp. 742–751

Muste, M., Fujita, K., Yu, I., and Ettema, R., 2005, Two-phase versus mixed-flow perspective on suspended sediment transport in turbulent channel flows, *Water Resources Research*, **41**, W10402.

Nezu, I., and Azuma, R., 2004, Turbulence characteristics and interaction between particles and fluid in particle-laden open channel flows, *Journal of Hydraulic Engineering*, ASCE, **130**, pp. 988–1001

Niño, Y., García, M., and Ayala, L., 1994a, Gravel Saltation. 1. Experiments, *Water Resources Research*, **30(6)**, pp. 1907–1914

Niño, Y., and García, M., 1994b, Gravel Saltation. 2. Modeling, *Water Resources Research*, **30(6)**, pp. 1915–1924

Niño, Y., and García, M., 1998, Using Lagrangian particle saltation observations for beadload sediment transport modelling, *Hydrological Processes*, **12**, pp. 1197–1218

Paintal, A. S., 1971, Concept of critical shear stress in loose boundary open channels, *Journal of Hydraulic Research*, IAHR, **9**, pp. 91–113

Parker, G., 1979, Hydraulic geometry of active gravel river, *Journal of Hydraulic Engineering*, ASCE, **105(9)**, pp. 1185–1201

Parker, G., Paola, C., and Leclair, S., 2000, Probabilistic form of Exner equation of sediment continuity for mixtures with no active layer, *Journal of Hydraulic Engineering*, ASCE, **126(11)**, pp. 818–826

Parker, G., Toro-Escobar, C. M., Ramey, M., and Beck, S., 2003, The effect of floodwater extraction on the morphology of mountain streams, *Journal of Hydraulic Engineering*, ASCE, **129(11)**, pp. 885–895

Parker, G., 2004, *1D Sediment transport morphodynamics with applications to rivers and turbidity currents*, ebook, http://hydrolab.illinois.edu/people/parkerg//morphodynamics_e-book.htm?q=people/parkerg/morphodynamics_e-book.htm

Parker, G., 2008, "Transport of gravel and sediment mixtures," *Sedimentation engineering: Processes, measurements, modeling and practice*, ASCE manual of practice 110, M. H. García, ed., ASCE, Reston, Va., USA, Chap. 3, pp. 165—251

Pope, S. B., 2000, *Turbulent flows*, Cambridge University Press, New York, USA

Portela, L. M., and Oliemans, R. V. A., 2002, Eulerian-Lagrangian DNS/LES of particle-turbulence interaction in wall-bounded flows, *International Journal for Numerical Methods in* Fluids, **43(9)**, pp. 1045–1065

Rashidi, M., Hetsroni, G., and Banerjee, S., 1990, Particle-turbulence interaction in a boundary layer, *International Journal of Multiphase Flow*, **16(6)**, pp. 935–949

Raudkivi, A. J., 1998, *Loose boundary hydraulics*, Pergamon Press, Oxford, UK

Rouse, H., 1937, Modern conceptions of the mechanics of turbulence, *Transactions of the American Society of Civil Engineers*, **102**, Paper No. 1965, pp. 463–543

Rubey, W. W., 1933, Settling velocities of gravel, sand and silt particles, *American Journal of Science*, **25**, pp. 325–338

Sanford, L. P., and Maa, J. P. Y., 2001, A unified erosion formulation for fine sediments, *Mar. Geol.*, **179**, pp. 9–23

Schmeeckle, M., and Nelson, J., 2003, Direct numerical simulation of bed load transport using a local, dynamic boundary condition, *Sedimentology*, **50**, pp. 279–301

Sekine, M., and Kikkawa, H., 1992, Mechanics of saltating grains II, *Journal of Hydraulic Engineering*, ASCE, **118**, pp. 536–557

Shields, A., 1936, *"Anwendung der Aechichkeits-Mechanic und der Turbuleng Forschung auf dir Geschiebewegung' Mitt Preussische,"* Versuchsanstalt für Wasserbau and Schiffbau, Berlin, Germany (translated to English by W. P. Ott and J. C. van Uchelen, California Institute of Technology, Pasadena, California, USA

Smith, J. D., and McLean, S. R., 1977, Spatially averaged flow over a wavy surface, *Journal of Geophysical Research*, **83**, pp. 1735–1746

Soulsby, R. L., 1997, *Dynamics of marine sands*, Thomas Telford, London, UK

Squires, K., and Eaton, J., 1990, Particle response and turbulence modification in isotropic turbulence, *Physiscs of Fluids A-Fluid Dynamics*, **2(7)**, pp. 1191–1203

Swamee, P. K., 1993, Generalized inner region velocity distribution equation, *Journal of Hydraulic Engineering*, ASCE, **119(5)**, pp. 651–656

Turton, R., and Levenspiel, O., 1986, A short note on the drag correlation for spheres, *Powder Technology*, **47**, pp. 86–86

van Hinsberg, M. A. T., Boonkkamp, J. H. M. T., and Clercx, H. J. H., 2011, An efficient, second order method for the approximation of the Basset history force, *Journal of Computational Physics*, **230(4)**, pp. 1465–1478

van Rijn, L. C., 1984, Sediment transport, part II: Suspended load transport, *Journal of Hydraulic Engineering*, ASCE, **110(11)**, pp. 1613–1641

van Rijn, L. C., 1993, *Principles of sediment transport in rivers, estuaries and coastal seas*, Aqua Publications, Amsterdam, Netherlands

Vance, M. W., Squires, K. D., and Simonin, O., 2006, Properties of the particle velocity field in gas-solid turbulent channel flow, *Physics of Fluids*, **18**, 063302

Vanoni, V. A., 1975, *Sedimentation engineering*, ASCE manual and reports on engineering practice 54, New York, USA

Whipple, K. X., 2004, 12.163 Course Notes, MIT Open Courseware. http://ocw.mit.edu/NR/rdonlyres/Earth–Atmospheric–and-Planetary-Sciences/12-163Fall-2004/9D91F442-7769-4CAE-853E-213FCBD04081/0/3_flow_around_ bends.pdf

Wiberg, P., and Smith, J. D., 1985, A theoretical model for saltating grains in water, *Journal of Geophysical Research*, **90**, pp. 7341–7354

Wilson, K. C., 1966, Bedload transport at high shear stresses, *Journal of the Hydraulic Division*, ASCE, **89(HY3)**, pp. 221–250

Yalin, M. S., 1963, An expression fo bedload transportation, *Journal of the Hydraulic Division*, ASCE, **92(HY6)**, pp. 49–59

Yalin, M. S., 1977, *Mechanics of sediment transport*, Pergamon Press, 2nd Ed., New York, USA

Yamamoto, Y., Potthoff, M., Tanaka, T., Kajishima, T., and Tsuji, Y., 2001, Large-eddy simulation of turbulent gas-particle flow in a vertical channel: effect of considering inter-particle collisions, *Journal of Fluid Mechanics*, **442**, pp. 303–334.

Yen, B. C., 1992a, *Channel flow resistance: Centennial of Manning's formula*, Water Resources Publications, Littleton, CO

Yen, B. C., 1992b, *Sediment fall velocity in oscillating flow*, Water Resources Environmental Engineering Resources, Report 11. Dept. of Civil Eng., Univeristy of Virginia.

Zhong, D., Guangqian, W., and Ding, Y., 2011, Bed sediment entrainment function based on kinetic theory, *Journal of Hydraulic Engineering*, ASCE, **137(2)**, pp. 222–233

Zyserman, J. A., and Fredsøe, J., 1994, Data analysis of bed concentration of suspended sediment, *Journal of Hydraulic Engineering*, ASCE, **120(9)**, pp. 1021–1042

CHAPTER NINE

Surface water and streambed sediment interaction: The hyporheic exchange

Daniele Tonina

Center for Ecohydraulics Research, University of Idaho, Boise, USA

ABSTRACT

Stream and pore waters continuously interact and mix within streambeds due to spatial and temporal variations in channel characteristics (e.g., spatiotemporal variations of streambed pressure, of volume of alluvial material surrounding a river, of streambed hydraulic conductivity, and sediment transport). This mixing is typically referred to as hyporheic exchange and the hyporheic zone defines the interfacial zone between rivers and their surround aquifers and riparian zones. It is an important ecotone and a place where many biogeochemical reactions occur such as nitrification and denitrification. The significance of hyporheic exchange in affecting surface and subsurface water quality and linking fluvial geomorphology, ground-water, and riverine habitat for aquatic and terrestrial organisms has been emerging in recent decades as an important component of conserving, managing, and restoring river-ine ecosystems. In this chapter, we present the concepts, characteristics and environmental effects of hyporheic exchange, and we review the methods for measuring and predicting its characteristics, i.e. hyporheic flux and hyporheic residence time.

9.1 HYPORHEIC EXCHANGE AND ITS IMPORTANCE

Whereas loosing and recharging water fluxes between rivers and aquifers have been inves-tigated for a long time, only in the last few decades the exchange between rivers and their surrounding sediments has been increasingly recognized as an important integral part of streams (Elliott and Brooks, 1997b; Marion, *et al.*, 2002; Packman and Bencala, 2000; Tonina and Buffington, 2007; Triska, *et al.*, 1989a). This exchange is characterized by river waters entering the streambed sediment in downwelling areas, (i.e., downwelling fluxes) and then emerging into the stream in upwelling areas, (i.e., upwelling fluxes) (Elliott and Brooks, 1997b; Marion, *et al.*, 2002; Packman and Bencala, 2000; Tonina and Buffington, 2007; Triska, *et al.*, 1989a). Upwelling and downwelling fluxes occurring in permeable and porous sediments of the streambed (usually with higher hydraulic conductivity than the contiguous aquifer) stem from spatial variations of near-bed pressure, of alluvium depth, of alluvium lateral confinement, of streambed sediment hydraulic conductivity, and from flow turbu-lence, which causes pressure and velocity fluctuations at the channel bottom (Section 9.3) (Buffington and Tonina, 2009; Packman, *et al.*, 2004; Salehin, *et al.*, 2004; Savant, *et al.*, 1987; Shimizu, *et al.*, 1990; Tonina and Buffington, 2009a; Vaux, 1968; Vollmer, *et al.*, 2002). Because these mechanisms depend on stream geometry and discharge, their relative importance on hyporheic exchange may depend on channel type and vary along the stream network (Buffington and Tonina, 2009). Consequently, the vertical and horizontal extent of the hyporheic exchange vary spatially due to changes in stream size, morphology, alluvial bed and aquifer conditions and seasonally due to physicochemical fluctuations, e.g., stream discharge, water temperature and stream solute loads. Downwelling fluxes may extend vertically up to tens of meters and horizontally up to more than a kilometer (Stanford and Ward, 1988). For instance, the hyporheic vertical domain in sand-bed streams, with dune-like topography, has been shown to be a function of stream Reynolds number (the ratio between the mean flow velocity-depth product and water kinematic viscosity). The maximum depth of penetration is generally slightly less than the bed form wavelength, which could range from few centimeters to several meters (Cardenas and Wilson, 2007b; Wörman, *et al.*, 2002). Whereas, it may be comparable to the channel width in gravel-bed rivers with pool-riffle topography (Marzadri, *et al.*, 2010; Tonina and Buffington, 2011).

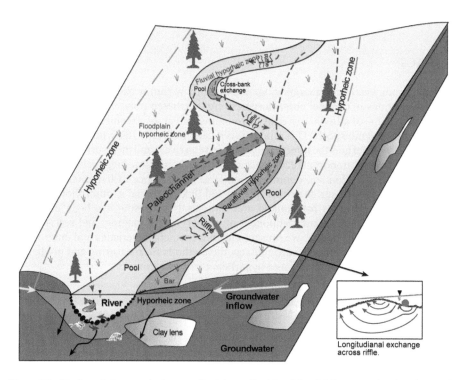

Figure 9.1. The hyporheic zones is shown as fluvial, parafluvial and floodplain zone and at the micro-scale around the log (close-up), at the unit-channel scale around the riffle in (close-up) and at the channel-reach scale as sequence of pool-riffle bed forms. Hyporheic domain is characterized by path lines exchanging between the river and the sediments (orange dashed lines), groundwater domain is characterized with dark blue with connection from the river to the groundwater (black solid lines) and from the groundwater to the river (yellow lines), modified from Tonina and Buffington (2009a).

Downwelling and upwelling fluxes have multiple consequences on stream and groundwater systems and form a constantly cycling connection between stream and surrounding saturated streambed sediments (Malard, *et al.*, 2002; Stanford and Ward, 1993). Downwelling fluxes transfer solutes, suspended particles and surface water into the sediment (e.g., Elliott and Brooks, 1997b; Ren and Packman, 2004b). They modify stream solute concentrations by mixing surface water with pore water (Bencala and Walters, 1983), influence groundwater habitat and ecosystems by delivering new solutes (Kim, *et al.*, 1992) and may impact hydraulic conductivity of the sediment (Nowinski, *et al.*, 2011; Packman and Brooks, 2001; Packman, *et al.*, 2000a; b; Packman and MacKay, 2003). Biofilms attached to streambed particles, and organisms dwelling within particle interstices uptake solutes and release transformed products, which hyporheic flow carries away (Bott, *et al.*, 1984; Triska, *et al.*, 1993a; Triska, *et al.*, 1993b). Because micro-organism population densities are elevated in the hyporheic zone relative to the water column, most microbially mediated transformations, such as nitrification and denitrification, occur in the hyporheic zone rather than in the water column (Master, *et al.*, 2005; Wuhrmann, 1972). These biogeochemical and transport processes generate concentration gradients, which sustain a reach ecotone (Edwards, 1998; Gibert, *et al.*, 1994; Stanford and Ward, 1993; Tonina and Buffington, 2009b). These gradients depend on reaction time, water temperature, solute concentrations, flow velocity, and length of the flow path (Findlay, *et al.*, 1993). In turn, upwelling fluxes carry stream water that has been exposed to the groundwater environment back into the river (Mulholland, *et al.*, 2008; Nagaoka and Ohgaki, 1990; Triska, *et al.*, 1993b; Triska, *et al.*, 1989b).

Field investigations on nutrient cycle, especially on nitrogen, have shown the importance of the hyporheic zone as a biogeoreaction zone (Duff and Triska, 2000; Fischer, *et al.*, 2005; Hill, *et al.*, 1998; Kjellin, *et al.*, 2007; Mulholland and DeAngelis, 2000; Mulholland, *et al.*, 2008; Mulholland, *et al.*, 2004; Storey, *et al.*, 2004; Triska, *et al.*, 1993b). Anthropogenic activities, primarily food and energy production, have altered the global nitrogen cycle, increasing reactive nitrogen availability in many aquatic ecosystems, which otherwise are nitrogen limited (Carpenter, *et al.*, 1998; Galloway, *et al.*, 2004; Vitousek, *et al.*, 1997). The hyporheic zone has been suggested to have an important role in processing reactive nitrogen (mainly in the form of ammonium) and nitrate in stream waters, and in returning it as nitrogen gases to the atmosphere (Duff and Triska, 2000; Hill, *et al.*, 1998; Kjellin, *et al.*, 2007; Mulholland and DeAngelis, 2000; Mulholland, *et al.*, 2008; Mulholland, *et al.*, 2004; Storey, *et al.*, 2004; Triska, *et al.*, 1993b). As nitrogen species enter streams via atmospheric deposition, overland flows, and groundwater infiltration, nitrification and denitrification within the hyporheic zone reduce their concentrations (Binley, 2005; Buss, *et al.*, 2005; Galloway, *et al.*, 2008; Master, *et al.*, 2005). Nitrification, which occurs in the aerobic zone of the hyporheic zone, transform reduced forms of nitrogen, chiefly ammonium, into nitrate and denitrification, which almost exclusively takes place in the anoxic zone of the hyporheic zone (Hill, *et al.*, 1998; Wagenschein and Rode, 2008), permanently removes between 30-70% of all the reactive nitrogen, mostly NO_3^-, entering streams. Other mechanisms such as assimilation by benthic algae and uptake by macrophytes may retain nitrogen within the ecosystem (Wagenschein and Rode, 2008). Additionally, downwelling water may carry colloids with pollutants, heavy metal and pathogens adhered on their surfaces within the sediment (Ren and Packman, 2004a; c). These particles may by trapped due to mechanical (straining) and adhesion (electrical forces) filtering within the sediments (Packman and Brooks, 1995; 2001; Packman, *et al.*, 1997; 2000b; Brunke, 1999; Ren and Packman, 2002; 2004a; b; c; 2007). Consequently, the hyporheic zone may be an important pathway or a temporary storage within the sediment for viruses, bacteria and other health-threatening substances (Maxwell, *et al.*, 2003; Redman, *et al.*, 1999). These substances may be successively released into the stream or migrate toward the riparian zone or floodplain.

Due to these potential effects on surface and subsurface water quality, stream restoration projects have recently started quantifying the hyporheic exchange (Fischer, *et al.*, 2005; Kasahara, *et al.*, 2009; Kasahara and Hill, 2006a; b; 2007). The common practice of adding boulders (Fischenich and Seal, 2000) or logjams (Sawyer, *et al.*, 2011) in a reach to restore ecological functions, or to restore previous morphologies (Kasahara, *et al.*, 2009; Kasahara and Hill, 2006a; b; 2007) have been analyzed for its hyporheic effects. Consequently, this chapter first defines the operational definition of hyporheic zone and then describes the main mechanisms driving hyporheic exchange; the spatial and temporal scales of hyporheic exchange and it concludes providing the state-of-the-art models for measuring and predicting hyporheic exchange.

9.2 DELINEATING THE HYPORHEIC ZONE

Presently there is not a unified definition of hyporheic zone whose operational delineation varies with applications and study goals (Bencala, 2000). However, three major definitions of the hyporheic zone have been historically proposed. They arise from the biologic, geochemical and hydraulic methods.

Biologists were the first to study hyporheic exchange because of its function in carrying oxygen-rich stream waters to the salmonid embryos incubating in their egg-nests called redds (Stuart, 1953). Salmonid bury their eggs within the hyporheic zone of gravel bed rivers at depths that range between 5 to 50cm below to original streambed surface depending on grain size, fish species and size (DeVries, 1997). Biologists also observed the presence of organisms, within the interstitial voids of the alluvial sediments, whose habitats require water properties similar to those of the surface water (Gibert, *et al.*, 1994; Orghidan, 1959; Stanford and Ward, 1988). The presence of these fauna indicates that streambed pore waters have chemistry that is more similar to the stream than subsurface waters. Consequently, the biological method defines the hyporheic zone in terms of the presence-absence of hyporheic fauna, called hyporheos (Orghidan, 1959). Because the hyporheic zone is a transitional zone between surface and subsurface environments (Edwards, 1998), its ecosystem is defined as an ecotone.

The geochemical method uses the different chemical signatures between the surface and subsurface waters. It defines the hyporheic zone as the volume of sediment containing an arbitrary amount of surface water, traditionally set at least 10% (Triska, *et al.*, 1989a). Consequently, the hyporheic zone is a transitional zone where surface and subsurface waters mix. The relative abundance of surface versus subsurface waters can be determined by measuring pH, electrical conductivity and temperature (Hendricks and White, 1991; White, *et al.*, 1987). Alternatively, conservative tracers (e.g., fluorescein, rhodamine, and various types of chlorides) could be added to the surface flow. Concentrations of natural-occurring or added tracers can be measured within the sediments with an array of sampling devices placed within the channel and the surrounding floodplain (Bencala and Walters, 1983; Castro and Hornberger, 1991; Jonsson, *et al.*, 2004; Kasahara and Hill, 2006a; b; Packman, *et al.*, 2000a; Packman, *et al.*, 2004; Triska, *et al.*, 1989a; b; Wondzell and Swanson, 1996). The extent of the hyporheic zone is then constructed from three-dimensional concentration maps interpolated from these measurements (e.g., Harvey, *et al.*, 1996).

Recently, the application of coupled surface-subsurface hydraulic models (Cardenas and Zlotnik, 2003; Elliott and Brooks, 1997a; b) and piezometric head (the sum of pressure and elevation heads) measurements in the field and flumes (Storey, *et al.*, 2003; Tonina and Buffington, 2007; Winter, *et al.*, 1998) started the use of the hydraulic approach. This method is based on the concept of hyporheic flow paths, which are the trajectories of stream water moving through the streambed sediment between its downwelling and upwelling points (Cardenas, *et al.*, 2004; Tonina and Buffington, 2007). This method defines the hyporheic zone as the volume of streambed sediment enveloped by all the flow paths of hyporheic exchange that begins and ends at the stream bed and banks. It does neglect those flow paths entrained within the groundwater system.

For all three methods, the hyporheic zone may include different biophysical zones (biotopes), representing different types, rates and magnitudes of physical, biological, and chemical processes (Boulton, *et al.*, 1992; Edwards, 1998; Stanford, *et al.*, 2005). Consequently, it can be subdivided into fluvial (river bed and banks), parafluvial (saturated sediments under exposed bars), and floodplain environments (Edwards, 1998; Stanford, 2006).

The three methods may lead to different extents and interpretations of the hyporheic zone because of the different operational and conceptual definitions. Whereas the geochemical and biological methods do not set any condition of stream water returning to the stream, the

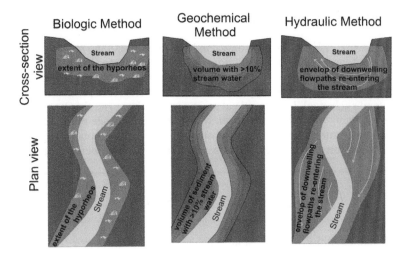

Figure 9.2. Hyporheic zone extension defined by the biological (left panel) geochemical (central panel) and hydraulic (right panel) methods (modified from Smith, 2005).

hydraulic approach does not consider as hyporheic fluxes those waters that are lost into the ground water or those which recharge the stream from the groundwater. This does not imply that the hydraulic method is more restrictive because some flow paths may have hundreds of days of residence time before re-entering the stream. Hence, their chemical signature could be more than the 90% different from that of the stream water and its fauna more similar to the groundwater than the hyporheos.

The biological method interprets the hyporheic zone as an ecotone, a transitional zone between river and groundwater ecosystems, and the geochemical method as a mixing zone between surface and subsurface waters. Instead, the hydraulic method considers hyporheic waters as an integral part of the river and treats them as stream waters flowing within the streambed interstices, matching the meaning of the term hyporheic from two Greek words 'hypo-' ($\upsilon\pi o$) under and 'rhe-' ($\rho\varepsilon$) flow: the flow underneath (Dahm, *et al.*, 2006; Orghidan, 1959).

The spatial extent of the hyporheic zone can also vary temporally because physicochemical properties, e.g., discharge, temperature, water table and solute concentrations of the surface and ground waters change over time (Storey, *et al.*, 2003). However, the hyporheic zone boundary defined with each method may change differently with these temporal physico-chemical changes. For instance, changes in water temperature may cause hyporheos to move closer to the riverbed or deeper into the sediment. It may also affect chemical reactions and thus the relative abundance of surface and subsurface constituents. This would alter the boundaries of the hyporheic zone as defined by the biological and probably geochemical methods but not by the hydraulic method with all else the same.

To overcome these limitations and to provide a unified method, a new definition has been recently proposed (Gooseff, 2010). It is based on the concept of residence time of stream water within the hyporheic. The hyporheic residence time may span several timescales (see Section 1.4 and 1.5) and this different time scale may help distinguish the zone of influence of the stream water; for instance 'the 24-h hyporheic zone' (Gooseff, 2010). The idea behind this approach is that because different processes have different rates and reaction times, this definition may adhere better to the processes taking place within the sediment. It could be used, for instance, to compare the zoning of the underflow of aerobic vs. anaerobic zones.

9.3   HYPORHEIC FLOW MECHANISMS

Following the work of Vaux (1968), Tonina and Buffington (2009a) summarized the major mechanisms that have been inferred to drive hyporheic exchange. These mechanisms can be shown from the mass balance of water within a fixed in space infinitesimal volume. Let's assume a saturated volume of the hyporheic zone with lateral sides parallel to the hyporheic flow (no-lateral exchange), bottom side at an impervious layer (no flow) and upper surface at the water-sediment interface. Consequently, the temporal change of the volume of water, V_w, within the volume depends on the subsurface inflow Q and outflow $Q + dQ/dx \cdot dl$, where dl is the infinitesimal length of the volume, and the hyporheic exchange e per unit length, such that

$$\frac{dV_W}{dt} = Q - \left(Q + \frac{dQ}{dx}dl\right) + e \cdot dl = \left(e - \frac{dQ}{dx}\right)dl \qquad (9.1)$$

For steady-state conditions, V_w does not change with time ($dV_w/dt = 0$), and $e = dQ/dx$. By assuming the Darcy (1856) equation:

$$Q = qA = -K_C \frac{dh}{dx} A \qquad (9.2)$$

where q is the subsurface flux ($q = n\,u$, where n is the sediment porosity and u the interstitial flow velocity), K_C is the hydraulic conductivity of the sediment, and dh/dx is the spatial gradient of the energy head, h, (Freeze and Cherry, 1979), the hyporheic exchange is

$$e = \frac{d}{dx}\left(-K_C\frac{dh}{dx}A\right) = \underbrace{-K_C A \frac{d^2h}{dx^2}}_{1} - \underbrace{K_C \frac{dA}{dx}\frac{dh}{dx}}_{2} - \underbrace{A \frac{dK_C}{dx}\frac{dh}{dx}}_{3} \qquad (9.3)$$

| Spatial changes in energy head | Spatial changes in alluvial area | Spatial changes in hydraulic conductivty |

This equation shows that e is driven by (1) spatial changes in the energy head, (d^2h/dx^2), (2) spatial changes in the cross-sectional area of alluvium (dA/dx), and (3) spatial changes in hydraulic conductivity (dK_C/dx).

    The energy head includes pressure, elevation and dynamic heads and their relative importance on driving hyporheic fluxes depends on bed form type and discharge (e.g., Buffington and Tonina, 2009; Tonina and Buffington, 2009a). For instance, dynamic heads are the main terms for fully submerged dune-like bed forms where flow detaches at the dune crest and attaches at the stoss side of the dune. This flow pattern results in areas of high pressure along the stoss side between the attachment point and the crest and of low pressure on the lee side between the crest and the attachment point (Elliott and Brooks, 1997a; b; Savant, et al., 1987; Thibodeaux and Boyle, 1987). Conversely, pressure and elevation heads are the dominant terms for pool-riffle morphology (Tonina and Buffington, 2007) and most likely for streambeds with step-pool morphology (Buffington and Tonina, 2009; Kasahara and Hill, 2006b) due to the large changes in water elevations induced by this macro bed forms. Consequently, water surface elevations could be a good proxy for the energy head for gravel-bed streams with pool-riffle and step-pool morphology (Anderson, et al., 2005). Energy heads may change temporally due to flood events (Boano, et al., 2007b) or wave motion (Qian, et al., 2008).

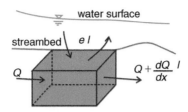

Figure 9.3. Hyporheic exchange, *e*, per unit length for an infinitesimal volume of length, *l*, of the hyporheic zone, modified from Tonina and Buffington (2009a).

Changes in cross-sectional area, which may include variations in alluvium depth and lateral constrain, may drive hyporheic exchange due to fluctuations of the subsurface volume available for water flow (Vaux, 1968). For instance, a reduction of alluvial area results in a smaller subsurface volume, which drives pore water out from the sediment to the river causing upwelling fluxes. Conversely, an expansion of the alluvial area increases the subsurface volume, which draws river water into the sediment and hence generates downwelling flux. Changes in alluvial depth may be caused by bedrock outcrop, layers of sediment with low hydraulic conductivity and bedrock (Baxter and Hauer, 2000). Changes in width may be due to local bedrock outcrop or changing from confined to unconfined reaches (Buffington and Tonina, 2009).

Conversely, heterogeneity of the hydraulic conductivity of the streambed material is expected to affect hyporheic fluxes in most streams because of the heterogeneity of the streambed material (Salehin, *et al.*, 2004). Results of the laboratory experiments of Salehin *et al.* (2004) in a flume with coarse material show that K_C heterogeneity of the stream substratum causes an increased spatial variability in hyporheic downwelling fluxes, development of preferential subsurface flow paths, greater average interfacial water flux, less vertical solute penetration, and a shorter mean hyporheic residence time than an equivalent homogenous substratum. However, others have reported that heterogeneity of the streambed material hydraulic conductivity may be of secondary importance respect to pressure head variations in meandering sand-bed rivers (Cardenas, *et al.*, 2004; Sophocleous, 1991). This highlights the stream type dependence of the hyporheic exchange and the variation of the relative importance of hyporheic mechanisms along the stream network (Buffington and Tonina, 2009).

Aside from these steady state mechanisms, turbulence and sediment transport may also drive hyporheic exchange (Tonina and Buffington, 2009a). Turbulence may present coherent flow structures at different scales. Flow–boundary interactions cause large-scale turbulent structure (wakes) (Buffin-Bélanger, *et al.*, 2000a). Large flow obstructions such as bed forms, large woody debris, boulders and large in-channel vegetation generate turbulent wakes at the meso-scale (Buffington, *et al.*, 2002; Middleton and Southard, 1984). Whereas small-scale wakes are shaded from local roughness such as protruding grains, particle clusters, and small in-channel vegetation (Brayshaw, *et al.*, 1983; Buffin-Bélanger, *et al.*, 2000b; Mendoza and Zhou, 1992; Ruff and Gelhar, 1972). Large, meso and small scale turbulence structures may cause pressure and velocity fluctuations at the water-sediment interface (Detert, *et al.*, 2007; Buffin-Bélanger, *et al.*, 2000a; Buffin-Bélanger, *et al.*, 2000b; Mendoza and Zhou, 1992; Shimizu, *et al.*, 1990; Middleton and Southard, 1984; Brayshaw, *et al.*, 1983; Ho and Gelhar, 1973; Ruff and Gelhar, 1972). These pressure and velocity fluctuations may cause exchange of mass between the river flow and near-surface pore water resulting in hyporheic exchange (Nagaoka and Ohgaki, 1990; Packman, *et al.*, 2004; Shimizu, *et al.*, 1990). Because turbulence decreases quickly within the sediment interstices, turbulence exchange is expected to be limited to a near-surface layer, whose thickness has been suggested to range between 2 and 10 times the mean grain size of the streambed sediment (Detert, *et al.*, 2007; Packman, *et al.*, 2004; Shimizu, *et al.*, 1990;

Tonina and Buffington, 2007). The turbulence dumping effect of the sediment increases with fine sediments such that sand-bed streams may have a thinner turbulence-induced hyporheic exchange than gravel-bed rivers.

Sediment transport in rivers can also cause hyporheic exchange due to the release of pore water in erosional areas and entrainment of stream water in depositional areas. Elliott and Brooks (1997a; 1997b) modeled this mechanism as a plug flow, with river water penetrating the sediment to the depth of scour. Consequently, the larger the bed form the deeper the exchange and the more active the sediment transport the faster the exchange. This mechanism is expected to be important in sand-bed streams where fine sediments are constantly transported by the flow and bed forms migrate (Buffington and Tonina, 2009). However, in gravel-bed rivers, sediment transport associated with bed form migration occurs only at very high flows, which are infrequent, and consequently this mechanism should be of secondary importance in these rivers (Hassan, 1990).

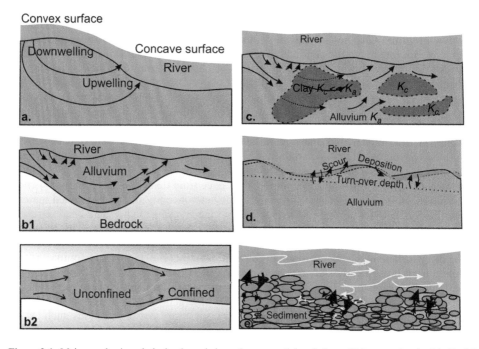

Figure 9.4. Major mechanisms inducing hyporheic exchange: spatial variations of (a) energy heads, (b) alluvial area, which depends on (b1) alluvial depth and (b2) valley constrain variations, (c) of hydraulic conductivity of the sediment and (d) sediment transport (turn-over) and (e) near-bed turbulence (modified from Tonina and Buffington (2009a)).

Recent research suggested another mechanism that has been overlooked in the paste: gravity-induced hyporheic fluxes (Boano, *et al.*, 2009). Flume experiments with flat featureless streambeds of fine sediment have shown that hyporheic flows could be driven by small differences in water density, as low as $\Delta\rho/\rho = 10^{-4}$, where ρ is the stream water density and $\Delta\rho$ is the density difference between surface and subsurface waters (Boano, *et al.*, 2009). These differences could stem from temperature or solute concentration gradients.

9.4 SPATIAL AND TEMPORAL SCALES OF HYPORHEIC EXCHANGE

The mechanisms driving hyporheic exchange described above occur over multiple, nested, spatial scales, resulting in nested scales of hyporheic circulation (Baxter and Hauer, 2000; Dent, *et al.*, 2001; Edwards, 1998; Malard, *et al.*, 2002; Stanford and Ward, 1988). The different scales of hyporheic exchange can be expressed in terms of channel width (W), typically at bankfull flow condition, to normalize rivers of different size and include: micro ($<10^{-1}W$), channel-unit (10^{-1}–$10W$), channel-reach (10–$10^3 W$), and valley-segment scales (10^3–$10^4 W$). Micro-scale circulation results from local, small-scale variations in channel characteristics (e.g., variation of head around a cluster of streambed particles, or variation of hydraulic conductivity around a sand lens). Typically, micro-scale includes ripples, dune, randomly distributed large boulders in plane-bed morphology (Buffington and Tonina, 2009), logs (Sawyer, *et al.*, 2011) or log-jams (Wondzell, 2006). Channel-unit circulation is associated with head variations around individual bed forms. The term channel-unit refers to individual morphologic elements, such as pools, bars, and steps (Bisson, *et al.*, 1982; Church, 1992). Bed forms result from interactions between stream flow and sediment transport, but can also be created by animal activity (White, 1990) such as salmonid redd (Tonina and Buffington, 2009b), or locally forced by wood debris (Buffington, *et al.*, 2002; Montgomery, *et al.*, 1995).

Channel reaches are collections of multiple, repeating sequences of channel units (e.g., pool-riffle or step-pool reaches; (Montgomery and Buffington, 1997; 1998)). Hyporheic circulation at this scale results from factors such as changes in reach slope, meso-scale changes in the volume of alluvium, cross-valley head differences between the main channel and secondary channels (Kasahara and Wondzell, 2003), flow through the floodplain (between meander bends (Boano, *et al.*, 2010a; Cardenas, 2009; Wroblicky, *et al.*, 1998), or within buried paleochannels (Stanford and Ward, 1993)). Reach-scale hyporheic circulation is also caused by irregularity amongst bed forms, with topographic low points driving larger-scale circulation and capturing hyporheic circulation of upstream channel units (Elliott and Brooks, 1997a).

Valley segments are collections of channel reaches that share similar characteristics. They exhibit similar degree of channel sinuosity, slope, and confinement (Montgomery and Buffington, 1997), with valley-segment circulation driven by changes in these characteristics (e.g., change from a confined, steep, straight valley segment to an unconfined, low gradient, meandering segment), by changes in geology (rock type, geologic history, underlying bedrock topography) or by changes in discharges after tributary confluences. For example, the frequency of the elevation changes in the underlying bedrock topography may structure broad-scale variations in the depth of alluvium that in turn generate deep, valley-scale circulation (Baxter and Hauer, 2000).

Hyporheic exchange at micro- and channel-unit scales has been studied primarily in flume experiments with sand beds organized into ripples and dune-like features (Elliott and Brooks, 1997a; b; Marion, *et al.*, 2002; Packman and Brooks, 2001; Savant, *et al.*, 1987) and individual obstruction such as log (Sawyer, *et al.*, 2011) or logjams (Endreny, *et al.*, 2011). Laboratory experiments have also been conducted with gravel beds shaped into two-dimensional dunes, or with stones placed on planar gravel beds (Cooper, 1965; Packman, *et al.*, 2004; Thibodeaux and Boyle, 1987). Recent laboratory studies have examined exchange across three-dimensional, pool-riffle topography typical of gravel-bed rivers (Tonina and Buffington, 2007). Field studies of hyporheic exchange through individual riffles and step-pool units have also been conducted in sand- and gravel-bed rivers (Harvey and Bencala, 1993; Hill, *et al.*, 1998; Hill and Lymburner, 1998; White, *et al.*, 1987), with recent interest in the effectiveness of channel units constructed for stream restoration projects (Hester and Doyle, 2006; Kasahara and Hill, 2006a; b).

Hyporheic exchange at reach and valley segment scales has been explored through field studies and numerical simulations in predominantly coarse-grained channels. These studies have shown the importance of bedrock knickpoints, side channels, buried paleochannels, changes in bed slope, sediment hydraulic conductivity heterogeneity, variation of channel units, and post-flood changes in channel morphology (Anderson, *et al.*, 2005; Baxter and Hauer, 2000; Bencala and Walters, 1983; Cardenas, *et al.*, 2004; Castro and Hornberger, 1991; Gooseff, *et al.*, 2006; Gooseff, *et al.*, 2007; Harvey and Bencala, 1993; Kasahara and Wondzell, 2003; Malard, *et al.*, 2002; Stanford and Ward, 1988; 1993; Wondzell, 2006; Wondzell and Swanson, 1996; 1999; Wörman, *et al.*, 2002; Wörman, *et al.*, 2006).

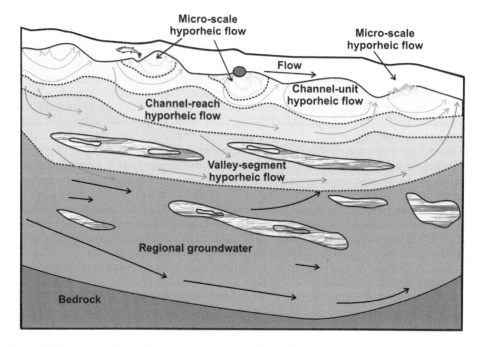

Figure 9.5. Four scales of nested hyporheic exchange: a) micro-scale (e.g. head variations induced by cluster of streambed particles, logs or biological formation such as salmon redds; fractions of a channel width (W) in length $<10^{-1}W$); b)channel-unit scale (head variations around individual bed forms (e.g. riffle); 10^{-1}–$10W$); c) channel-reach scale (e.g. sequence of pool-riffle; 10–$10^3 W$) and d) valley-segment scales (e.g., variations in valley confinement, alluvial depth, or underlying bedrock topography; 10^3–$10^4 W$) (Baxter and Hauer 2000; Dent *et al.* 2001; Edwards 1998; Malard *et al.* 2002). Regional groundwater envelops the hyporheic exchange. Modified from Alley *et al.* (1999) and Buffington and Tonina (2009).

Several studies have demonstrated the nested behavior of different scales of hyporheic flow. For example, Tonina and Buffington (2009b) showed that hyporheic circulation caused by salmon redds is superposed on that due to pool-riffle topography. Similarly, Baxter and Hauer (2000) demonstrated that bed form circulation is nested within valley-scale hyporheic flow.

Although bed topography generally determines where upwelling and downwelling fluxes occur within a channel, these fluxes may represent combined effects of nested, multi-scale processes (Stonedahl, *et al.*, 2010). Consequently, it is important to consider the different scales of hyporheic exchange beyond those of interest in a given study, both to understand the source of the fluxes and the larger- or smaller-scale patterns of flow, which may not otherwise be apparent. For example, at reach and valley-segment scales a river may be characterized as loosing or gaining water, but closer inspection may reveal that this is the result of multiple smaller-scale upwelling and downwelling fluxes, rather than uniform losses or gains (Woessner, 2000), recognition of which may be biologically important (Boulton, *et al.*, 1998).

Besides these spatial scales, temporal variations affect hyporheic exchange. Daily and seasonal water table variations due to rainfall, snowmelt, and vegetation activity on the floodplain can change the groundwater and hyporheic flow fields (Constantz, *et al.*, 1994; Wagner and Bretschko, 2003). For example, vegetation can seasonally pump subsurface water through its roots, altering head gradients and the magnitude and pattern of hyporheic exchange (Duke, *et al.*, 2007). Similarly, daily and seasonal changes in river discharge influence head gradients that drive hyporheic exchange and that, in turn, affect ground-water flow fields and the relative contributions of surface versus groundwater flow to the hyporheic zone (Harvey and Bencala, 1993; Malcolm, *et al.*, 2005; Sear, *et al.*, 1999; Soulsby, *et al.*, 2001; Storey, *et al.*, 2003; Wroblicky, *et al.*, 1998). Furthermore, chemical and biological changes may be associated with temporal variations in physical characteristics, such as discharge; runoff following rainfall may increase the abundance of fine sediment or chemicals carried by the river and exchanged with the hyporheic zone, enhancing the temporal and spatial heterogeneity of hyporheic zone biochemistry.

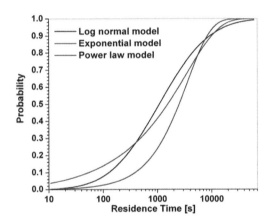

Figure 9.6. Cumulative residence time distribution (CDF) within the hyporheic zone. The residence time is modeled as a log normal, exponential or power law model. Data are from experiment 2 of Tonina and Buffington (2007).

9.5 RESIDENCE TIME OF HYPORHEIC EXCHANGE

The spatial and temporal variability of mechanisms driving hyporheic exchange produce a distribution of exchange path lengths and hyporheic residence times, which is the amount of time that river water spends traversing the subsurface sediment before re-emerging into the stream. The residence time is a key property of the hyporheic zone because biogeochemical reactions occurring within the sediment depend on the amount of time that river water flows within the hyporheic zone (Duff and Triska, 2000; Edwards, 1998; Hendricks and White, 1991; Mulholland and DeAngelis, 2000; Thomas, *et al.*, 2003). Hyporheic flow paths can be strongly three dimensional, both within the channel (Cardenas, *et al.*, 2004; Marzadri, *et al.*, 2010; Tonina and Buffington, 2007) and through the floodplain (Harvey and Bencala, 1993; Kasahara and Hill, 2007; Kasahara and Wondzell, 2003; Wroblicky, *et al.*, 1998), exhibiting a broad distribution of path lengths and exchange rates. Paleochannels buried within the floodplain can create preferential flow paths with fast hyporheic exchange

(Sophocleous, 1991; Stanford and Ward, 1993) and may be important biological hotspots and travel corridors for hyporheic fauna. The residence time distribution represents the combined interaction of all flow paths within the hyporheic zone. It is typically expressed as a probability or cumulative density function, the shape of which depends on site-specific factors. These factors include the spatial and temporal distribution of different types of hyporheic zones (e.g., fluvial, parafluvial, floodplain), sediment properties, and head distribution (Choi, *et al.*, 2000; Harvey, *et al.*, 1996). A variety of models has been proposed to describe residence time distributions. These include the exponential (Bencala and Walters, 1983; Runkel, 1998), log normal (Wörman, *et al.*, 2002; Marzadri, *et al.*, 2010; Tonina and Buffington, 2011), and power functions (Haggerty, *et al.*, 2002; Cardenas, *et al.*, 2008a), as well as one approach that fits separate exponential distributions to each component of the hyporheic zone (e.g., fluvial, parafluvial, floodplain) (Choi, *et al.*, 2000).

9.6 MEASURING HYPORHEIC EXCHANGE

A variety of methods has been used to study hyporheic exchange, ranging from direct observations to numerical predictions. In this section, we review some of the historic and emerging measurement techniques for measuring hyporheic exchange. In the next section, we will focus on the empirical, numerical and analytical solutions to quantify hyporheic exchange.

9.6.1 Biological methods

The biological methods focus mainly on defining the extent of the hyporheic exchange. They are primarily sampling methods, which may include series of sediment cores (Williams and Hynes, 1974), installation of piezometer (standpipes) from which subsurface water is pumped (Malard, *et al.*, 2002) or in which multi-level pit traps are installed (Danielopol, 1989; Danielopol and Niederreiter, 1987) and the Karaman-Chappuis pit method (Boulton, *et al.*, 2004). Organisms collected are then classified to define the extent of the hyporheic zone.

9.6.2 Geophysical methods

Groundwater penetrating radar (Naegeli, *et al.*, 1996) tomography (Acworth and Dasey, 2003) and recently, electrical resistivity imaging methods, coupled with electrically conductive stream tracers, such as chloride, have been proposed for mapping the hyporheic zone (Crook, *et al.*, 2008; Ward, *et al.*, 2010). Besides providing snapshots of the hyporheic volume boundaries, these techniques may be used to derive information on the hyporheic flow field by studying the evolution over time of the injected electrically conductive solute plume.

9.6.3 Tracers

Passive tracers, such as salt, fluorescein, rhodamine, and various chlorides are commonly used in field and laboratory experiments to quantify rates and magnitudes of the hyporheic exchange (e.g., Bencala and Walters, 1983; Castro and Hornberger, 1991; Gooseff, *et al.*, 2007; Gooseff, *et al.*, 2003; Harvey and Bencala, 1993; Tonina and Buffington, 2007; Triska, *et al.*, 1993a; Triska, *et al.*, 1993b; Triska, *et al.*, 1989a). A known amount of the tracer is released into the stream using either a pulsed or constant injection and its concentration is monitored at one or more locations downstream of where the tracer is believed to be completely mixed within the surface flow (allowing unbiased measurements at any location within the downstream channel) (Church, 1974; Day, 1977; Fischer, 1967). Characteristics of the hyporheic exchange are interpreted from the shape of the breakthrough curve (a plot of concentration versus time) (Bencala and Walters, 1983; Castro and Hornberger, 1991; Gooseff, *et al.*, 2007; Gooseff, *et al.*, 2003; Harvey, *et al.*, 1996; Packman, *et al.*, 2004; Triska, *et al.*, 1989a; b). Tracers are most commonly used to examine ensemble reach-average exchange rates, but can also be used to examine spatial variability of sub-reach exchange (Zarnetske, *et al.*, 2008). Field tracer tests are typically post-processed with numerical models, which may include OTIS (Runkel, 1998), STIR (Marion, *et al.*, 2008b) and Lagrangian-based models with transfer function (e.g., Haggerty and Reeves, 2002; Wörman, *et al.*, 2002), which are described in section 1.7. The applicability of tracer tests in streams to detect hyporheic exchange depends on the following Damköhler number, *DaI*, (Wagner and Harvey, 1997):

$$DaI = \frac{\alpha(1 + A_W/A_{sz})L}{U} \tag{9.4}$$

where U is the mean stream velocity, L is the length of the reach, α is the stream-storage exchange coefficient (see OTIS model in section 1.7) and A_W and A_{sz} are the reach-scale mean stream and storage zone cross-sectional areas, respectively. Values of *DaI* close to 1 minimize uncertainty on the values of α and A_{sz}, which express the importance of the hyporheic exchange.

Natural tracers, like organic carbon, nitrogen and isotopes, have been used to investigate geochemical and biochemical reactions within the hyporheic zone (Hendricks and White, 1991; Payn, *et al.*, 2009; Triska, *et al.*, 1993a; Triska, *et al.*, 1993b; Triska, *et al.*, 1989a; b; Zarnetske, *et al.*, 2011). These tracers are reactive, rather than passive, and thus mass balance analysis needs to account for their biogeochemical transformations. Water temperature is a particularly useful natural tracer, because it is relatively economical and easy to measure (Constantz and Thomas, 1997; Hendricks and White, 1991; Lewandowski, *et al.*, 2011; White, *et al.*, 1987). It also has biological relevance for aquatic organisms (Bjornn and Reiser, 1991; Hynes, 1970). Use of fiber optics to sense temperature is an emerging technology that is particularly attractive because of its broad spatial resolution (0.01–10,000 m) (Selker, *et al.*, 2006). Typically, sensors are buried into the ground at specific locations and the analysis of the time shift (time lag between the signals) and reduction of the amplitude of the dial temperature signal between stream and hyporheic waters is used to predict mean vertical hyporheic velocity. Temperature in the streambed sediments can be expressed using a 1-D advection-dispersion equation.

$$\frac{\partial T}{\partial t} = \kappa_e \frac{\partial^2 T}{\partial z^2} - \frac{nv}{\gamma} \frac{\partial T}{\partial z} \tag{9.5}$$

where T is temperature, z is vertical coordinate, κ_e is effective thermal diffusivity, γ is the ratio of heat capacity of the streambed to the fluid, n is porosity, and v is vertical interstitial fluid velocity (Hatch, *et al.*, 2006; Stallman, 1960). The thermal front velocity can be written as $v_T = v/\gamma$. The solution of equation (9.5) with periodic sinusoidal temperature variations as boundary conditions at the streambed and a thermal gradient equal to zero at an infinite streambed depth is (Hatch, *et al.*, 2006; Keery, *et al.*, 2007; Stallman, 1960):

$$T(z,t) = A_S \exp\left(\frac{v_T z}{2\kappa_e} - \frac{z}{2\kappa_e}\sqrt{\frac{\alpha_T + v_T^2}{2}}\right) \cos\left(\frac{2\pi t}{P} - \frac{z}{2\kappa_e}\sqrt{\frac{\alpha_T - v_T^2}{2}}\right) \tag{9.6}$$

where A_S is the amplitude of the surface temperature signal, P is the period of temperature variations (one day), and $\alpha_T = \sqrt{v_T^4 + (8\pi\kappa_e/P)^2}$. Hatch, *et al.* (2006) separated equation (9.6) into two components and solved for the seepage flux based on the observed amplitude ratio (A_r)

$$A_r = \frac{A_L}{A_S} = \exp\left[\frac{z_L - z_S}{2\kappa_e}\left(v - \sqrt{\frac{\alpha + v_T^2}{2}}\right)\right] \tag{9.7}$$

and phase shift (time lag $\Delta\phi$)

$$\Delta\phi = \phi_L - \phi_S = \frac{z_L - z_S}{2\kappa_e}\sqrt{\frac{\alpha - v_T^2}{2}} \tag{9.8}$$

of the temperature signal between two sensor locations separated by a distance ΔL (surface sensor and depth sensor). The L and S subscripts refer to the sensor at depth L and at the surface, respectively. They proposed the following two solutions for the vertical velocity:

$$v_T = \frac{2\kappa_e}{\Delta L}\ln(A_r) + \sqrt{\frac{\alpha_T + v_T^2}{2}} \tag{9.9a}$$

$$v_T = \sqrt{\alpha_T - 2\left(\frac{\Delta\phi 4\pi\kappa_e}{\Delta L P}\right)^2} \tag{9.9b}$$

The solutions contain thermal front velocities, v_T, on both sides of the equations either explicitly or embedded in the α_T term. Hence, an iterative solution scheme is necessary to solve for the seepage flux. Seepage flux magnitude and direction are estimated from equation (9.9a), while only the seepage magnitude is estimated from equation (9.9b). Although solution of both equations should match, sensitivities and limitations in both equations typically result in varying estimates of the seepage flux magnitude (Shanafield, *et al.*, 2011).

The calculated seepage flux depends partly on sediment thermal properties, which are typically estimated based on normal reported ranges. The sensitivities of the equations can result in estimate errors of upward or downward seepage flux, especially at low fluid velocities, depending on the estimated sediment properties. To minimize and prevent some limitations of the previous two solutions a third analysis has been suggested (Gariglio, *et al.*, 2011; Luce, *et al.*, in review). This is based on using information from both amplitude and phase components of equation (9.6). By relating the amplitude ratio (A_r) of the surface signal to the phase shift ($\Delta\phi$), the following metric can be calculated:

$$\eta = \frac{\ln\left(\frac{A_L}{A_S}\right)}{\phi_L - \phi_S} \tag{9.10}$$

Substituting equation (9.7) and (9.8) into (9.10) the vertical velocity can be defined directly from η

$$\eta = \frac{\sqrt{\alpha_T + v_T^2} - \sqrt{2}\,v_T}{\sqrt{\alpha_T - v_T^2}} \tag{9.11}$$

A diffusive thermal scaling velocity can be defined as $v_D = 2\sqrt{\omega \kappa_e}$ with $\omega = 2\pi/P$, which can be used to normalize the velocity v_T such that

$$\eta = \frac{\sqrt{\sqrt{v_*^4 + 1 + v_*^2} - \sqrt{2}v_*}}{\sqrt{\sqrt{v_*^4 + 1} - v_*^2}} \qquad (9.12)$$

This dimensionless expression shows that if $\eta = 1$ then the observed amplitude ratio and phase shift follow a purely diffusive heat transport regime. The seepage flux is downwelling (signal is less dumped and travels faster than the purely conductive case) if $\eta < 1$, whereas it is upwelling (signal is more dumped and travels slower than the purely conductive case) $\eta > 1$. Equation (9.12) can be rearranged to provide an explicit solution for the dimensionless hyporheic flux:

$$v_* = \frac{1}{2} \frac{\frac{1}{\eta} - \eta}{\sqrt{\frac{1}{\eta} + \eta}} \qquad (9.13)$$

The use of η can provide the information about upwelling and downwelling fluxes at the sensor location avoiding uncertainty due to sensor location and sediment thermal properties, which have the largest impact of equations (9.9) accuracy (Shanafield, et al., 2011). Additionally, once the distance between sensors is known, this approach can provide information about the sediment thermal properties by using one of the following two equations, which stem from combining equations (9.9) and (9.13) (Luce, et al., in review):

$$\kappa_e = \left[\frac{\Delta L}{-\ln(A_r)} \sqrt{\frac{\omega}{2}} \left(\sqrt{\sqrt{v_*^4 + 1} + v_*^2} - \sqrt{2}v_* \right) \right]^2$$

$$\kappa_e = \left(\frac{\Delta L}{-\Delta\phi} \right)^2 \frac{\omega}{2} \left(\sqrt{v_*^4 + 1} - v_*^2 \right) \qquad (9.14)$$

In case of steady state temperature, hyporheic flux could be quantified from vertical thermal profiles by solving the following implicit equation (Bredehoeft and Papadopulos, 1965):

$$\beta = \frac{\gamma q \Delta L}{\kappa_e}$$

$$\frac{T(z) - T_S}{T_L - T_S} = \frac{e^{(\beta z/\Delta L)} - 1}{e^\beta - 1} \qquad (9.15)$$

where T_S and T_L indicate the temperatures at the surface and at the lowest recorded sensor, respectively with ΔL the distance between the two sensors, and $T(z)$ is the temperature at a generic location z between S and L. The sign of β defines the direction of the fluxes, such as $\beta > 0$ flux is downwelling, $\beta < 0$ flux is upwelling and $\beta = 0$ the system is purely diffusive.

Recently, Haggerty, et al. (2006) proposed the use of smart tracers. These tracers change irreversibly into another detectable compound when exposed to aerobic conditions within the hyporheic zone. Consequently, they are potentially a new tool to study the extent and magnitude of hyporheic exchange.

Resins and degradable material have also been applied in measuring hyporheic flows at point scales (Clayton, *et al.*, 1996). For instance, Carling *et al.* (2006) used the uptake of rhodamine dye by carbon granules to map the subsurface velocities. Their approach, which can be used in either laboratory or field settings, provides an inexpensive method for creating detailed maps of hyporheic flow.

9.6.4 VHG

The most common field technique for assessing the local magnitude of hyporheic exchange is to measure the vertical head gradient (VHG) (Anderson, *et al.*, 2005; Baxter and Hauer, 2000; Geist, 2000; Kasahara and Hill, 2006b; Lee and Cherry, 1978; Terhune, 1958; Valett, *et al.*, 1994). For this method, a standpipe, called piezometer, is used to measure the pressure head at a given depth within the sediment. VHG is the difference in water elevation between the piezometer and the stream surface (Δh) divided by the depth of the subsurface measurement below the streambed (ΔL), VHG $= \Delta h \cdot \Delta L^{-1}$ (Lee and Cherry, 1978). Upwelling and downwelling fluxes are indicated by positive and negative values, respectively. VHG estimates can be improved by attaching a separate stilling well, which avoids uncertainty due to waves and water fluctuations in reading the water surface elevation on the outside of the piezometer (Baxter, *et al.*, 2003). Depending on local turbulence and dynamic head variations (those due to changes in velocity and momentum), the near-bed pressure may also differ from the static pressure (that determined from the water surface elevation) (Tonina and Buffington, 2007). Hence, it is better to measure the near-bed pressure with a stilling well than to estimate it from the water surface elevation. Alternatively, a manometer can be created for measuring Δh by attaching flexible tubing to the top of the piezometer and placing the other end of the tubing in the flow (Lee and Cherry, 1978). Additionally, paired pressure transducers with built in data loggers, one recording pressure at the channel bottom and a companion sensor buried within the streambed sediment at a prescribed depth, could be used to record pressure variation continuously.

VHG point measurements made at different locations within a stream can be contoured to determine reach-scale patterns of hyporheic exchange (Baxter and Hauer, 2000; Geist, 2000; Valett, *et al.*, 1994). However, depending on the site-specific, three-dimensional structure of hyporheic flow paths, VHG values may change with the depth of subsurface measurement, a factor that typically has not been accounted for in prior studies. Both the magnitude and sense of hyporheic exchange (upwelling vs. downwelling) can change with depth of measurement. For example, although the near-bed pressure is fixed at a given location, the subsurface pressure and consequent VHG can vary widely depending on the depth of measurement because of the typically complex three-dimensional structure of hyporheic flow (Buffington and Tonina, 2007; Cardenas and Zlotnik, 2003; Marzadri, *et al.*, 2010). Consequently, the VHG approach should be viewed as a first-order approximation of the actual structure of hyporheic exchange, because the hyporheic extent predicted depends on sampling effort, which is typically of limited spatial extent and resolution due to the cost of installing piezometers or other measurement devices (Edwards, 1998; Freeze and Cherry, 1979).

In addition to determining VHG, the vertical hydraulic conductivity of the sediment at each measurement location (K_{Cz}) can be quantified from falling head tests (Baxter, *et al.*, 2003; Freeze and Cherry, 1979; Geist, 2000). This allows calculation of vertical hyporheic flux (q_z) from Darcy's law ($q_z = -K_{Cz} \cdot \Delta h \cdot \Delta L^{-1}$) (Freeze and Cherry, 1979). Horizontal hyporheic velocities may also be estimated in some cases (Baxter, *et al.*, 2003).

9.7 PREDICTING HYPORHEIC EXCHANGE

A variety of numerical models have been developed for predicting hyporheic exchange, ranging from simple one dimensional models to complex three dimensional ones, at the micro, channel-unit, channel-reach and valley-segment scales.

At the micro and channel-unit scales ($<10W$), the most common approach is to treat hyporheic exchange as a groundwater flow governed by Darcy's law within porous media (Elliott and Brooks, 1997b; Nagaoka and Ohgaki, 1990; Vaux, 1968). The premise of this approach is that stream flow–bed form interactions create spatial variations in streambed pressure that drive subsurface flow (bed form-induced convection, or pumping exchange) (Elliott and Brooks, 1997b; Harvey and Bencala, 1993; Thibodeaux and Boyle, 1987; Vaux, 1962; 1968). Consequently, this method requires measuring or predicting the pressure distribution along the streambed, as well as knowing the hydraulic conductivity of the sediment, but it does not require tracer experiments or calibration. Near-bed pressures can be predicted from hydraulic models (e.g., McDonald, *et al.*, 2005), derived from analytical solutions (e.g., Colombini, *et al.*, 1987; Elliott and Brooks, 1997b), measured from arrays of near-bed pressure transducers or piezometers (e.g., Tonina and Buffington, 2007) or estimated from surveys of water-surface elevation and bed topography (e.g., Gooseff, *et al.*, 2006; Kasahara and Wondzell, 2003; Wondzell and Swanson, 1996). Measurements of the pressure-head distributions within the floodplain are also required. They are used to define the lateral boundary conditions and typically determined from piezometer arrays (e.g., Kasahara and Wondzell, 2003; Wondzell and Swanson, 1996). The hydraulic conductivity of the sediment can be estimated as a function of grain size and sorting, or determined from sediment cores, falling head tests, or calibration of the groundwater model with known water surface elevations (Freeze and Cherry, 1979; Niswonger, *et al.*, 2005). Ground-penetrating radar coupled with field measurements of saturated conductivity has been used in recent studies to map the three-dimensional structure of sedimentary deposits and corresponding variation in hydraulic conductivity (Cardenas and Zlotnik, 2003).

Elliott and Brooks (1997a; 1997b) developed a Darcy-type model for predicting hyporheic exchange in sand-bedded channels with two-dimensional dune-like bed forms. Following the laboratory measurements of Vittal *et al.* (1977), they modeled the near-surface streambed pressure, h, as a regular sinusoidal distribution reflecting periodic drops in pressure due to bed form drag and turbulent energy dissipation.

$$h = h_m \sin\left(\frac{2\pi}{\lambda}x\right)$$

$$h_m = 0.28 \frac{U^2}{2g} \begin{cases} \left(\frac{Y^*}{0.34}\right)^{3/8} & Y^* \le 0.34 \\ \left(\frac{Y^*}{0.34}\right)^{3/2} & Y^* > 0.34 \end{cases} \tag{9.16}$$

where h_m is the near-bed dynamic head amplitude, U is the mean flow velocity, g is the gravity, Y^* is the dimensionless water depth, defined as the ratio between Y_0 the hydraulic flow depth and Δ the bed form amplitude, and λ is the bed form wavelength. This information allows quantifying the spatially averaged hyporheic flux $\bar{q}$ over one bed form of wavelength λ. In this formulation, the governing hypothesis are: Darcian flow, homogeneous hydraulic conductivity (K_C) of the sediment, neutral surface-subsurface gradient and constant alluvial depth d_b, yielding the following expression:

$$\bar{q} = \frac{2K_C h_m \tanh\left(\frac{2\pi}{\lambda}d_b\right)}{\lambda} \tag{9.17}$$

The residence time distribution $\overline{R}$ weighted over the downwelling fluxes can be estimated by the following implicit equation:

$$\frac{4\pi^2 K_C h_m t}{\lambda^2 n} = \frac{2a \cos(\overline{R}) \tanh\left(\frac{2\pi}{\lambda} d_b\right)}{\overline{R}} \tag{9.18}$$

The mean hyporheic downwelling flux and the hyporheic residence time distribution are essential for estimating the benthic and hyporheic habitat quality and for quantifying the biogeochemical reactions occurring in the sediment such as nitrification and denitrification, because they quantify the exchange flow and the contact time (Marzadri, *et al.*, 2011; Tonina, *et al.*, 2011). This model has been successfully tested with stationary bed forms. However, sand-bed streams are typically characterized by migrating bed forms. In this case, the turn-over model has been developed to account for pore-water released by erosion and stream water entrainment into the sediment during deposition (Elliott and Brooks, 1997a; b). For dune-like triangular bed form the mean downwelling flux is:

$$\overline{q} = \frac{Y_0}{\lambda} n U_b \tag{9.19}$$

where U_b is the migrating velocity of bed forms. For a random bed forms generated from a Gaussian process downwelling flux is:

$$\overline{q} = \left(\frac{\pi}{2}\right)^{1/2} \frac{Y_0}{\lambda} n U_b \tag{9.20}$$

Subsequent studies have tested and modified the original Elliott and Brooks approach (Marion, *et al.*, 2002; Packman and Bencala, 2000; Packman, *et al.*, 2000b; Tonina and Buffington, 2007; Zaramella, *et al.*, 2003). Salehin,*et al.* (2004) demonstrated the importance of sediment hydraulic heterogeneities in controlling the rate and spatial extent of hyporheic fluxes. Boano *et al.* (2007b) extended the analytical solution for the case of unsteady surface discharges, which in turn affect the pressure field within the sediment. They then investigated the effect of successive flooding event on hyporheic residence time (Boano, *et al.*, 2010c). Cardenas and Wilson (2007a) explored the influence of groundwater upwelling and downwelling fluxes on hyporheic exchange. In another work (Cardenas and Wilson, 2007b), they presented the relationship between surface hydraulics and hyporheic exchange and they suggested the following relationship between the vertical extent of hyporheic vertical, d_H, and the Reynolds number of the flow, $Re = \rho \cdot U \cdot Y_0/\mu$, where U, is the mean stream flow velocity, Y_0, the hydraulic water depth, ρ, the water density, μ, the dynamic viscosity of the water:

$$\frac{d_H}{\lambda} = \frac{Re^{0.3429}}{4.5158 + Re^{0.3429}} \qquad R^2 = 0.99 \tag{9.21}$$

The hyporheic extension d_H is not influence by the alluvium depth d_b as long as $d_H/d_b < 0.7$ (Cardenas and Wilson, 2007b).

Other works added the effects of hyporheic fluxes on colloid deposition within the hyporheic zone (Packman, *et al.*, 2000a; b; Ren and Packman, 2002; 2004a; b; c; 2007) and Marion, *et al.* (2008a) added the effect of streambed stratigraphy.

The Elliott and Brooks model is an exact solution of the hyporheic flow field in two dimensions. Three dimensional solutions for any streambed pressure distribution can be obtained by coupling the known pressure field with numerical groundwater models, such as MOD-FLOW (Cardenas, *et al.*, 2004; Gooseff, *et al.*, 2006; Kasahara and Hill, 2006a; b; Kasahara and Wondzell, 2003; Storey, *et al.*, 2003; Wondzell and Swanson, 1996; Wroblicky, *et al.*, 1998) or FLUENT (Tonina and Buffington, 2007). For example, Tonina and Buffington (2007) extended the Elliott and Brooks model to gravel-bed rivers with pool-riffle bed forms, which is an ubiquitous feature in both natural and regulated gravel-bed rivers. This bed form is characterized by a wavelength λ, linear distance between two pools and an amplitude, Δ, the difference between pool bottom and bar top. In a successive work, they showed the importance of accounting for the three dimensional structure of streambed pressures, the effects of changing discharge and alluvial depth in these channels (Tonina and Buffington, 2011). In this work, they analyzed the main physical variables thought to drive hyporheic flows in pool-riffle sequence. They selected fluid density, ρ, mean flow velocity U, hydraulic depth Y_0, fluid dynamic viscosity, μ, bed form amplitude, Δ, and wavelength, λ, gravitational acceleration, g, alluvium depth, d_b, permeability of the alluvium, K, and streambed slope, s_i. They applied the Buckingham Pi theorem to reduce these variables to a set of dimensionless numbers to predict the vertical extend of hyporheic exchange, d_H, the mean, μ_t, and standard deviation, σ_t, of the hyporheic residence time distribution for partially submerged bed forms. They proposed a set of five dimensionless numbers for predicting the hyporheic depth:

$$\Pi_1' = \frac{\rho U Y_0}{\mu}; \quad \Pi_2' = \frac{U}{\sqrt{g Y_0}}; \quad \Pi_3' = \frac{\Delta}{Y_0};$$
$$\Pi_4' = \frac{\lambda}{Y_0}; \quad \Pi_5' = s_i \tag{9.22}$$

The first two numbers are the stream flow Reynolds (Π_1') (ratio between inertial and viscous forces) and Froude (Π_2') (ratio between inertial and gravitational forces) numbers, the third (Π_3') and fourth (Π_4') dimensionless numbers characterize the pool-riffle geometry and water hydraulic depth and the last one is the streambed slope. The Buckingham theorem applied for predicting the mean and standard deviation of the residence time resulted in the following six dimensionless numbers:

$$\Pi_1 = \frac{\rho U \sqrt{K}}{\mu}; \quad \Pi_2 = \frac{U^2}{g\sqrt{K}}; \quad \Pi_3 = \frac{Y_0}{\sqrt{K}};$$
$$\Pi_4 = s_i; \quad \Pi_5 = \frac{\Delta}{\lambda}; \quad \Pi_6 = \frac{d_b}{\sqrt{K}} \tag{9.23}$$

where the first two dimensionless numbers are a modified version of the Reynolds and Froude numbers, respectively. They characterize the surface, via the mean flow velocity (U), and subsurface, through the sediment permeability (K), hydraulics. Regression analysis with the data from the flume experiments reported in Tonina and Buffington (2007) was used to find the coefficients A_i, B_i and B_i' of equations (9.24), (9.25) and (9.26) for the mean hyporheic depth, d_H, and the mean, μ_t, and standard deviation, σ_t, of the hyporheic residence time, respectively. The values of those coefficients are reported in (Table 9.1).

$$\frac{d_H}{d} = \exp\left(\sum_{i=1}^{5} A_i \ln(\Pi_i')\right) \tag{9.24}$$

Table 9.1. Empirically determined parameters for equations (9.24), (9.25) and (9.26) respectively.

i		1	2	3	4	5	6
Hyporheic mean depth (9.24)	A_i	0.152	−0.058	−0.509	0.074	0.906	
Mean hyporheic residence time (9.25)	B_i	−0.682	0.387	1.619	0.314	−1.339	0.407
Hyporheic residence time standard deviation (9.26)	B'_i	−0.533	0.652	1.369	0.098	−1.066	0.456

$$\frac{\mu_t U}{\sqrt{K}} = \exp\left(\sum_{i=1}^{6} B_i \ln(\Pi_i)\right) \tag{9.25}$$

$$\frac{\sigma_t U}{\sqrt{K}} = \exp\left(\sum_{i=1}^{6} B'_i \ln(\Pi_i)\right) \tag{9.26}$$

The permeability of the alluvium, K, may be quantified with the following equation (Freeze and Cherry, 1979):

$$K = 5.6 \cdot 10^{-3} \frac{n^3}{(1-n)^2} d_g^2 \tag{9.27}$$

where d_g is the geometric mean of the streambed material.

Information about the mean and standard deviations can be used to define the hyporheic residence time distribution entirely for partially and totally submerged pool-riffle sequence, because the residence times are log normally distributed for this bed form (Marzadri, *et al.*, 2010; Tonina and Buffington, 2011). Most recently, Marzadri, *et al.* (2010) developed an analytical three-dimensional model for hyporheic exchange in channels with submerged pool-riffle morphology. They proposed the following predictors for the mean and variance of the residence time distribution:

$$\mu_t^* = 1.39 Y^{*0.6}; \quad R^2 = 0.96$$

$$\mu_t = \frac{\lambda \mu_t^*}{K_C S_i C_z} \tag{9.28}$$

$$\sigma_t^{2*} = 2.07 Y^{*0.89}; \quad R^2 = 0.91$$

$$\sigma_t^2 = \left(\frac{\lambda}{K_C S_i C_z}\right)^2 \sigma_t^{2*} \tag{9.29}$$

where μ_t^* and σ_t^{2*} are the dimensionless mean and variance of the residence time distribution and C_z is the dimensionless Chezy number,

$$C_z = 6 + 2.5 \ln\left(\frac{1}{2.5 d_s}\right) \tag{9.30}$$

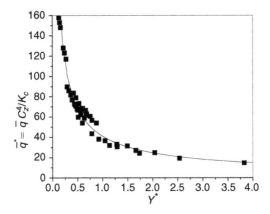

Figure 9.7. Averaged downwelling flux as a function of dimensionless water depth. Data from Marzadri *et al.* (2010).

d_s is the relative submergence, which is the ratio between the d_{50}, the median grain size of the streambed material and Y_0. Additionally, the following relationship for the downwelling flux averaged over one bed form unit, $\bar{q}$, can be interpolated with a regression analysis from the data of Marzadri, *et al.* (2010)

$$\bar{q}^* = 41.108 Y^{*-0.732}; \quad R^2 = 0.96$$

$$\bar{q} = \frac{\bar{q}^* K_C}{C_z^4} \tag{9.31}$$

where $\bar{q}^*$ is the dimensionless downwelling flux. Equations (9.28), (9.29) and (9.31) were derived for fully submerged pool-riffle in straight channels with limited lateral exchange.

Following previous works (Marion and Zaramella, 2005; Packman and Salehin, 2003) O'Connor and Harvey (2008) proposed an approached different from the pumping mechanism. They studied the hyporheic exchange as a diffusive process with an effective diffusivity D_e such that solute concentration, C, within the sediment is modeled with the following diffusion equation

$$\frac{\partial C}{\partial t} = D_e \frac{\partial^2 C}{\partial z^2} \tag{9.32}$$

whose solution for a fix concentration C_S in the stream and a 0 concentration flux deep in the sediment is:

$$C = C_S \left(1 - erf\left(\frac{z}{2\sqrt{D_e t}}\right)\right) \tag{9.33}$$

They suggested the following regression equation for the effective diffusivity D_e:

$$\frac{D_e}{D'_m} = \begin{cases} 5 \cdot 10^{-4} \, Re_* \, Pe_K^{6/5} & \text{for } Re_* \, Pe_K^{6/5} \geq 2000 \\ 1 & \text{otherwise} \end{cases}$$

$$Re_* = \frac{u_* k_s}{\nu}; \quad Pe_K = \frac{u_* \sqrt{K}}{D'_m}; \quad D'_m = \beta' D_m \tag{9.34}$$

where β' $(=(1+3(1+n))^{-1})$ is the tortuosity of the sediment voids (Iversen and Jørgensen, 1993), D_m the molecular diffusion coefficient, and k_s is the roughness height of the streambed. O'Connor and Harvey (2008) suggested the following expression for k_s (Van Rijn, 1984):

$$k_s = 3d_{90} + 1.1\Delta\left(1 - e^{-25\Delta/\lambda}\right) \tag{9.35}$$

where d_{90} is the particle diameter for which 90% of the particle sizes are finer. The shear velocity, u_*, is equal to the product $\rho \cdot R_H \cdot s_i$, where the hydraulic radius R_H is the ratio between the flow cross-sectional area, A_W, and the wetted perimeter of the cross-section A_W.

At the channel-reach scale (10^1–10^3 channel width), the transient storage model (Bencala, 2000; Harvey and Gorelick, 2000; Harvey, et al., 1996; Runkel, 1998), coded in the one-dimensional transport with inflow and storage program (OTIS) (Runkel, 1998), is presently the most applied model. It is a one dimensional advection-dispersion approach that treats hyporheic exchange in terms of fixed-volume storage elements and uses two parameters to characterize the hyporheic exchange: the storage area A_{sz}, and the exchange rate, α, along the river (Bencala, 2000; Harvey and Gorelick, 2000; Harvey, et al., 1996; Runkel, 1998). This model is a one-dimensional approach in two equations:

$$\frac{\partial C}{\partial t} + \frac{Q_R}{A_W}\frac{\partial C}{\partial x} = 7\frac{1}{A_W}\frac{\partial}{\partial x}\left(A_W D \frac{\partial C}{\partial x}\right) + \frac{q_L}{A_W}(C_L - C) + \alpha(C_{sz} - C)$$

$$\frac{\partial C_{sz}}{\partial t} = 7 - \alpha\frac{A_W}{A_{sz}}(C_{sz} - C) \tag{9.36}$$

where C is the stream solute concentration, Q_R the stream discharge, A_W the river cross sectional area, D the dispersion coefficient, q_L the lateral inflow, and C_L the solute concentration of the lateral inflow. The last term in the first equation and the second equation link the river flow with the storage volume with solute concentration within the storage zone C_{sz}. The parameters, A_{sz}, α, D and A_W are typically calibrated with measured breakthrough curves of tracer experiments and hence this method is typically coupled with tracer experiments. A drawback of this approach is that it does not separate transient storage caused by hyporheic exchange versus that due to "dead zones" within the river (e.g., backwater eddies, stagnant water at the bottom of pools, flow through vegetation) (Harvey and Wagner, 2000).

Although, it has been argued that transient storage in dead zones is small compared to that caused by hyporheic exchange especially in small streams (Mulholland, et al., 1997), their relative roles depend on discharge and channel type (Bencala, 2005; Gooseff, et al., 2003; Harvey, et al., 1996; Wondzell, 2006; Zaramella, et al., 2003). Both discharge and channel type influence the spatial extent of dead zones and the magnitude of head variations driving hyporheic exchange. Therefore, the transient storage approach is best applied to rivers with few dead zones. Those could be steep channels, which tend to have small shallow pools and limited backwater environments, as opposed to low-gradient channels that commonly have extensive pools and numerous backwaters, particularly in meandering floodplain rivers or forest channels with abundant wood debris (Buffington, et al., 2003; Buffington, et al., 2002; Montgomery and Buffington, 1997; 1998). The transient storage model tends to simplify the hyporheic exchange and may not correctly describe spatial and temporal rates of exchange, particularly the longer exchange paths (tail of the residence time distribution) (Zaramella, et al., 2003) that may have important biochemical implications. The model could be modified to account for multiple storage zones with different hyporheic areas and exchange coefficients (Choi, *et al.*, 2000). Nevertheless, the transient storage model is a useful first-order approximation that captures the bulk of the hyporheic exchange. Most of the exchange occurs in the shallow, near-surface portion of the hyporheic zone and is characterized by rapid exchange along short flow paths that are strongly coupled to surface hydraulics (Harvey and Wagner, 2000; Marzadri, *et al.*, 2010; Tonina and Buffington, 2007; Zaramella, *et al.*, 2003).

To overcome some limitations of the OTIS model, other researchers proposed different forms for the transfer function, which models the interaction between the stream water and the hyporheic zone. For instance, Wörman, *et al.* (2002) modeled the transfer function as a pumping mechanisms, whereas Boano, *et al.* (2007a) adopted the continuous time random walk (CTRW) theory. Most of these methods, which include the Solute Transport and Multirate Mass Transfer-Liner Coordinates (STAMMT-L) (Haggerty and Reeves, 2002) and the Solute Transport in River (STIR) (Marion, *et al.*, 2008b) primarily differ due to the selected form of the hyporheic residence time distribution. These methods have been successfully applied at both reach and valley-segment scales.

Whereas these methods are mostly surface water oriented, Stonedahl, *et al.* (2010) explored the role of different bed form scales, e.g., ripples, dune and planimetric features such as meanders, (see section 1.4) on hyporheic exchange as the domain increases from the micro to the channel-reach scales. They suggested a superposition of the effects for bed form induced hyporheic exchange at the micro, channel-unit and channel-reach scales. They applied their framework to show the effects of the local and large-scale topography on the hyporheic exchange. They assumed that dynamic heads dominate at the micro-scale, which they modeled as dune-like bed forms, and piezometric heads, due to water surface variations, dominate at the channel-reach scale.

Numerical models of hyporheic exchange have typically been applied at channel-unit and channel-reach scales, but Wörman *et al.* (2006) proposed a generalized three dimensional model that can be applied at landscape scales. Although they assumed a sinusoidal streambed pressure distribution throughout the entire river network conditions which is only suited to dune-ripple channels and an energy profile approximated by the land topography rather than that of the water surface elevation, their study is an important attempt to model hyporheic exchange at broad scales using digital elevation models. This approach could be used to provide a quantitative landscape perspective of hyporheic exchange that has been lacking.

9.8 CONCLUDING REMARKS

The major mechanisms that drive hyporheic flow have been identified. Yet fully developed models that characterize these drivers, for different stream types and morphologies, have not been presented. Predictors are available for dune-like (Elliott and Brooks, 1997a; b) and submerged (Marzadri, *et al.*, 2010) and partially submerged (Tonina and Buffington, 2011) alternate-bars under some restrictive conditions. Consequently, research is still needed to develop schemes for coupling surface and hyporheic hydraulics. New methods have been proposed to superpose the effects of different drivers at several scales (Stonedahl, *et al.*, 2010) and for large systems (Wörman, *et al.*, 2006). Only recently, new models have been emerging to couple hyporheic and stream hydraulics with non-conservative solutes (Boano, *et al.*, 2010b; Cardenas, *et al.*, 2008b; Marzadri, *et al.*, 2011; Rutherford, *et al.*, 1995; Tonina, *et al.*, 2011) and very fine sediments (Ren and Packman, 2002; 2004a; b; c; 2007).

Field observations have been studying the role of hyporheic fluxes on the fate of reactive solutes for several decades (e.g., Triska, *et al.*, 1993a; Triska, *et al.*, 1993b; Triska, *et al.*, 1994; Zarnetske, *et al.*, 2011). Recent modeling efforts have worked to understand the role of stream topography, stream flow and hyporheic fluxes and their relationship to the concentrations of reactive solute. These studies aimed to develop conceptual and numerical models of the effects of hyporheic fluxes on dissolved oxygen concentration (Rutherford, et al., 1995; Tonina, et al., 2011) and of reactive inorganic dissolved nitrogen (Boano, et al., 2010b; Cardenas, et al., 2008b; Marzadri, et al., 2011) to understand the role of streambed topography and stream flow. For instance, the work of Marzadri et al. (2011) presents a semi-analytical framework for nitrification-denitrification processes coupled with hyporheic hydraulics based on the hyporheic residence time approach. Their research suggests the ratio between the hyporheic median residence time, τ_{50}, and the oxygen time limit, τ_{lim} as an index of prevailing aerobic or anaerobic conditions within the hyporheic zone, (Marzadri, et al., 2012):

$$Da = \frac{\tau_{50}}{\tau_{lim}} \tag{9.37}$$

where τ_{lim} is the time within the hyporheic zone for which dissolved oxygen concentrations, C_O, reach a threshold value, C_{lim} below which aerobic reactions stop and anaerobic conditions take place. Marzadri, et al. (2011) suggest the following equation for τ_{lim}

$$\tau_{lim} = \frac{1}{K_{RN}} \ln\left(\frac{C_{O,stream}}{C_{lim}}\right) \tag{9.38}$$

where $C_{O,stream}$ is the dissolved oxygen concentration at the downwelling area and equals the stream concentration, and K_{RN} is the reaction rate of the dissolved oxygen consumption, which includes nitrification and respiration processes. Consequently, once the median residence time for a particular hyporheic zone is known then values of $Da > 1$ means that anaerobic conditions would dominate within the hyporheic zone such that denitrification has the potential to cause the hyporheic zone to be a sink of nitrates and a source of nitrogen gases. Conversely, when $Da < 1$, aerobic conditions would prevail and the hyporheic zone has the potential to be a source of nitrate because of nitrification processes. For alternate bar morphology, Marzadri, et al. (2012) show to following expression for τ_{50}

$$\tau_{50} = \frac{0.21\lambda}{K_C C_Z s_i} e^{1.22\,Y^*} \qquad R^2 = 0.99 \tag{9.39}$$

and for dune-like bed form with alluvium depth $z = d_b$:

$$\tau_{50} = \frac{4.189 \tanh(d_b/\lambda)}{4\pi^2 K_C h_m} \lambda \tag{9.40}$$

Understanding the role of surface and hyporheic hydraulics and biogeochemical reactions are key elements for up-scaling results from the micro- and channel-unit scales to the channel-reach and valley scales. Consequently, research advances should focus on quantifying the role of hyporheic processes in the fate of reactive solutes along streams and the role of the hyporheic zone as a buffer zone between streams and their surrounding riparian zone and aquifers at the valley and stream network scales (valley scale). At the same time, research is needed at the smallest scales to quantify the importance of surface water turbulence on the shallow and fast hyporheic exchange (micro-scale). Turbulence-induced hyporheic exchange may extend as deep as $10d_{50}$ in clear gravel (Salehin, et al., 2004) or $2d_{50}$ in sediment mixture of gravel and sand (Tonina and Buffington, 2007). These are considerable extents of the streambed sediment and may play an important role on stream and subsurface water quality.

APPENDIX A – LIST OF SYMBOLS

L: length; M: mass; T: time and Θ: temperature and the symbol –: dimensionless.

List of Symbols		
Symbol	Definition	Dimensions
A	infinitesimal cross-sectional area	$[L^2]$
A_L	amplitude of the temperature signal at the depth z_L within the streambed sediment	$[\Theta]$
A_S	amplitude of the temperature signal at the streambed surface	$[\Theta]$
A_W	stream flow cross-sectional area	$[L^2]$
A_i	regression coefficients for equation (9.24)	$[-]$
A_r	temperature signal amplitude ratio	$[-]$
A_{sz}	storage zone cross-sectional area	$[L^2]$
B_i'	regression coefficients for equation (9.26)	$[-]$
B_i	regression coefficients for equation (9.25)	$[-]$
C	solute concentration in the stream	$[M \cdot L^{-3}]$
C_H	solute concentration in the hyporheic sediments	$[M \cdot L^{-3}]$
C_O	dissolved oxygen concentration	$[M \cdot L^{-3}]$
$C_{O,stream}$	dissolved oxygen concentration at the downwelling area	$[M \cdot L^{-3}]$
C_S	solute concentration at the water-sediment interface	$[M \cdot L^{-3}]$
C_L	solute concentration of the lateral inflow	$[M \cdot L^{-3}]$
C_{lim}	threshold oxygen concentration between aerobic and anaerobic conditions	$[M \cdot L^{-3}]$
C_{sz}	solute concentration within the storage zone	$[M \cdot L^{-3}]$
C_z	dimensionless Chezy number	$[-]$
D	dispersion coefficient in the stream	$[L^2 \cdot T]$
Da	oxygen Damköhler number for hyporheic zone	$[-]$
DaI	Damköhler number	$[-]$
D_e	effective diffusion coefficient	$[L^2 \cdot T^{-1}]$
D_m	molecular diffusion coefficient	$[L^2 \cdot T^{-1}]$
D_m'	molecular diffusion coefficient corrected by sediment voids tortuosity	$[L^2 \cdot T^{-1}]$
K	permeability of the streambed sediment	$[L^2]$
K_C	hydraulic conductivity of the streambed sediment	$[L \cdot T^{-1}]$
K_{Cz}	hydraulic conductivity along the vertical direction	$[L \cdot T^{-1}]$
K_{RN}	reaction rate of the dissolved oxygen consumption	$[T^{-1}]$
L	length of the channel reach	$[L]$
P	period of the temperature fluctuation (1 day for daily fluctuations)	$[T]$
Q	subsurface discharge	$[L^3 \cdot T^{-1}]$
Q_R	river discharge	$[L^3 \cdot T^{-1}]$
$\overline{R}$	weighed hyporheic residence time	$[-]$
R_H	hydraulic radius	$[L]$
T	temperature	$[\Theta]$
T_S	constant temperature at the streambed surface	$[\Theta]$

(Continued)

List of Symbols

Symbol	Definition	Dimensions
T_L	constant temperature within the sediment at depth z_L	$[\Theta]$
U	stream velocity averaged at the reach scale	$[L \cdot T^{-1}]$
U_b	bed form migrating velocity	$[L \cdot T^{-1}]$
VHG	vertical hydraulic gradient	$[-]$
V_W	volume of water	$[L^3]$
W	bankfull channel width	$[L]$
Y^*	dimensionless stream hydraulic depth	$[-]$
Y_0	hydraulic depth of the stream	$[L]$
d_{90}	particle diameter for which 90% of the particle sizes are finer	$[L]$
d_b	alluvial depth	$[L]$
d_g	geometric mean of the streambed material	$[L]$
d_H	hyporheic depth	$[L]$
d_s	relative submergence	$[-]$
e	hyporheic exchange per unit length	$[L^2 \cdot T^{-1}]$
g	gravity acceleration	$[L \cdot T^{-1}]$
h	total energy head	$[L]$
h_m	amplitude of the dynamic head	$[L]$
k_s	roughness height of the streambed	$[L]$
l	infinitesimal length	$[L]$
n	porosity	$[-]$
$\bar{q}$	hyporheic downwelling flux averaged over one bed form	$[L \cdot T^{-1}]$
q	subsurface flux	$[L \cdot T^{-1}]$
q_l	lateral inflow per unit length of stream	$[L^2 \cdot T^{-1}]$
q_z	hyporheic vertical flux	$[L \cdot T^{-1}]$
s_i	bed slope	$[-]$
t	time variable	$[T]$
u	interstitial flow velocity	$[L \cdot T^{-1}]$
u_*	shear velocity	$[L \cdot T^{-1}]$
v	vertical interstitial fluid velocity	$[L \cdot T^{-1}]$
v^*	dimensionless front velocity	$[-]$
v_{Ar}	front thermal velocity based on the amplitude ratio method	$[L \cdot T^{-1}]$
v_D	diffusive thermal scaling velocity	$[L \cdot T^{-1}]$
v_T	front thermal velocity	$[L \cdot T^{-1}]$
$v_{\Delta\varphi}$	front thermal velocity based on the phase shift method	$[L \cdot T^{-1}]$
x	coordinate along the longitudinal direction	$[L]$
y	coordinate along the transversal direction	$[L]$
z	coordinate along the vertical direction	$[L]$
D	bed form amplitude	$[L]$
DL	distance between sensors	$[L]$
$\Delta\varphi$	Phase difference between the surface and the subsurface location	$[-]$
α	stream-storage zone exchange coefficient	$[T^{-1}]$
α_T	parameter for the temperature advecion-dispersion equation	$[L^2 \cdot T^{-2}]$

(Continued)

List of Symbols

Symbol	Definition	Dimensions
β	dimensionless parameter for the front thermal velocity for steady state conditions	$[-]$
β'	tortuosity of the sediment voids	$[-]$
γ	ratio of heat capacity of the streambed to the fluid	$[-]$
η	dimensionless number defining the direction of the hyporheic fluxes at the water-sediment interface and where the fluxes are diffusive or advective dominated	$[-]$
φ_L	phase of the temperature signal at depth z_L	$[-]$
φ_S	phase of the temperature signal at the water sediment interface	$[-]$
κ_e	effective thermal diffusivity	$[L^2 \cdot T^{-1}]$
λ	wavelength of bed form	$[L]$
μ	dynamic viscosity of the water	$[M \cdot T^{-1} \cdot L^{-1}]$
μ_τ	mean of the hyporheic residence time distribution	$[T]$
μ_τ^*	dimensionless mean hyporheic residence time distribution	$[-]$
Π	dimensional numbers for equations (9.21), (9.22) and (9.25)	$[-]$
Π'	dimensional numbers for equation (9.19) and (9.24)	$[-]$
π	a constant that is the ratio of circle's circumference to its diameter: 3.14159265	$[-]$
ρ	water density	$[M \cdot L^{-3}]$
σ_τ	standard deviation of the hyporheic residence time distribution	$[T]$
σ_τ^*	dimensionless variance of the hyporheic residence time distribution	$[-]$
$\tau_{\lim}$	hyporheic residence time for which the oxygen concentration reached the threshold value	$[T]$
τ_{50}	median hyporheic residence time	$[T]$

APPENDIX B – SYNOPSIS

Several methods have been proposed to define the hyporheic zone operatively, which include the biological, geophysical and hydrological models. Recently a fourth method based on the concept of residence time has been suggested. It defines the hyporheic zone based on its time distribution; for instance, the 24-hr hyporheic zone would encompass the streambed sediment with streamlines that have residence time shorter than 24 hours. This approach aims to recognize the multiple time scales that characterize the hyporheic zone and the important effects that those timescale have on the biogeochemical processes, which takes place within the streambed sediment. Recent analysis shows that hyporheic exchange stems from spatial and temporal variations of near-bed energy head variations, including turbulence, sediment transport, spatial changes of streambed sediment hydraulic properties and of the alluvial depth and width and density difference between stream and subsurface waters. Whereas progress has been made on defining these mechanisms and providing models to quantified the hyporheic exchange from streambed and flow interaction, their applicability is still limited to few bed forms under certain flow conditions and mostly at the micro-, channel-unit and channel-reach scales. As research improves the modeling of the hyporheic hydraulics at different scales and for different mechanisms, these models are coupled with transport of reactive solute to understand the role of hyporheic fluxes on the fate of nutrient, organic and inorganic compounds.

APPENDIX C – KEYWORDS

By the end of the chapter, you should have encountered the following terms. Ensure that you are familiar with them!

Hyporheic exchange	Hyporheic zone	Hyporheic exchange scales
Hyporheic residence time distribution	Downwelling fluxes	Upwelling fluxes
Energy heads	Hyporheic mechanisms	Pumping model
Diffusive model	Vertical hydraulic gradient	Turn-over model

APPENDIX D – QUESTIONS

Q1. What is hyporheic exchange?

Q2. What are the three major methods to define hyporheic zone?

Q3. What are the major mechanisms that drive hyporheic exchange under steady state conditions?

Q4. What type of mechanisms does the turn-over model simulate?

Q5. What type of energy heads are the key factor on controlling hyporheic hydraulics in gravel bed rivers with pool-riffle morphology and in sand-bed rivers with dune-like morphology?

APPENDIX E – PROBLEMS

E1. Describe the VHG field method.

E2. A stream with dune-like bedform has the following characteristics: $Y_0 = 0.3$ m, $U = 0.5$ m $\cdot$ s^{-1}, $s_i = 0.001$, $D_m = 10^{-6}$ m$^2 \cdot$ s^{-1}, $d_{90} = 0.001$ m, $d_g = 0.0005$ m, $n = 0.32$, $\Delta = 0.05$ m, $\lambda = 1$ m, $\rho = 1000$ kg $\cdot$ m^{-3}, $m = 1.3 \cdot 10^{-3}$ kg $\cdot$ m$^{-1} \cdot$ s^{-1}. Using the diffusion model quantify the vertical solute concentration profile at time $t = 10$ and 100 seconds.

E3. A stream with alternate-bar bedform has the following characteristics: $Y_0 = 0.3$ m, $U = 0.5$ m $\cdot$ s^{-1}, $s_i = 0.001$, $D_m = 10^{-6}$ m$^2 \cdot$ s^{-1}, $d_{90} = 0.03$ m, $d_g = 0.01$ m, $n = 0.34$, $\Delta = 0.3$ m, $\lambda = 3$ m, $\rho = 1000$ kg $\cdot$ m^{-3}, $\mu = 1.3 \cdot 10^{-3}$ kg $\cdot$ m$^{-1} \cdot$ s^{-1}. Using the diffusion model quantify the vertical solute concentration profile at time $t = 10$ and 100 seconds.

E4. A stream with fully submerged alternate-bar bedform has the following characteristics: $Y_0 = 0.3$ m, $s_i = 0.001$, $d_{50} = 0.01$ m, $K_C = 0.001$ m s^{-1}, $\Delta = 0.7$ m, $\lambda = 12$ m, $\rho = 1000$ kg $\cdot$ m^{-3}, $\mu = 1.3 \cdot 10^{-3}$ kg $\cdot$ m$^{-1} \cdot$ s^{-1}. Quantify the hyporheic residence time distribution and the downwelling averaged flux.

E5. Four temperature probes are buried within the streambed sediment at 0, 10, 20 and 50 cm below the streambed surface. During the winter period, they record the following temperature: 0.5°C, 0.7°C, 2°C and 4°C, respectively. During the summer season, they have the following amplitudes of 4.79°C, 2.01°C, 0.93°C and 0.12°C, and phases of 2.94, 2.35, 1.66 and 0.03 radiants, respectively for a daily period. The thermal properties of the sediment are: $\gamma = 0.655$, $\kappa_e = 6.5 \cdot 10^{-7}$ m$^2 \cdot$ s^{-1}. Calculate the hyporheic vertical fluxes at 10, 20 and 50 cm below the streambed surface.

REFERENCES

Acworth, R. I., and G. R. Dasey. 2003. Mapping of the hyporheic zone around a tidal creek using a combination of borehole logging, borehole electrical tomography and cross-creek electrical imaging, New South Wales, Australia. *Hydrogeology Journal*, 11: 368–377.

Anderson, J. K., S. M. Wondzell, M. N. Gooseff, and R. Haggerty. 2005. Patterns in stream longitudinal profiles and implications for hyporheic exchange flow at the H.J. Andrews Experimental Forest, Oregon, USA. *Hydrological Processes*, 19: 2931–2949. doi:10.1002/hyp.5791.

Baxter, C. V., F. R. Hauer, and W. W. Woessner. 2003. Measuring groundwater–stream water exchange: new techniques for installing minipiezometers and estimating hydraulic conductivity. *Transactions of the American Fisheries Society*, 132: 493–502.

Baxter, C. V., and R. F. Hauer. 2000. Geomorphology, hyporheic exchange, and selection of spawning habitat by bull trout (*Salvelinus confluentus*). *Canadian Journal of Fisheries and Aquatic Sciences*, 57: 1470–1481.

Bencala, K. E. 2000. Hyporheic zone hydrological processes. *Hydrological Processes*, 14: 2797–2798.

Bencala, K. E. 2005. Hyporheic exchange flows, in *Encyclopedia of Hydrological Sciences*. edited by M. G. Anderson.

Bencala, K. E., and R. A. Walters. 1983. Simulation of solute transport in a mountain pool-and-riffle stream: a transient storage model. *Water Resources Research*, 19 (3): 718–724.

Binley, A. 2005. *Groundwater – surface water interactions: a survey of UK field site infrastructure*. Science Report SC030155/SR5. Environmental Agency: Bristol, UK.

Bisson, P. A., J. L. Nielsen, R. A. Palmason, and L. E. Grove. 1982. A system of naming habitat types in small streams, with examples of habitat utilization by salmonids during low streamflow. Proceedings of Proceedings of a Symposium on Acquisition and Utilization of Aquatic Habitat Inventory Information. Western Division of the American Fisheries Society, Bethesda, MD. Portland, Or.

Bjornn, T. C., and D. W. Reiser. 1991. Habitat requirements of salmonids in streams, in *Influence of forest and rangeland management on salmonid fishes and their habitats*. edited by W. R. Meehan. Am. Fish. Soc. Spec. Publ. 19. Bethesda, Md. pp. 83–138.

Boano, F., C. Camporeale, and R. Revelli. 2010a. A linear model for the coupled surface-subsurface flow in a meandering stream. *Water Resources Research*, 46: W07535. doi:10.1029/2008WR007583.

Boano, F., A. Demaria, R. Revelli, and L. Ridolfi. 2010b. Biogeochemical zonation due to intrameander hyporheic flow. *Water Resources Research*, 46: W02511. doi:10.1029/2008WR007583.

Boano, F., A. I. Packman, A. Cortis, R. Revelli, and L. Ridolfi. 2007a. A continuous time random walk approach to the stream transport of solutes. *Water Resources Research*, 43 (W10425). doi:10.1029/2007WR006062.

Boano, F., D. Poggi, R. Revelli, and L. Ridolfi. 2009. Gravity-driven water exchange between streams and hyporheic zones. *Geophysical Research Letters*, 36: L20402. doi: 10.1029/2009GL040147.

Boano, F., R. Revelli, and L. Ridolfi. 2007b. Bedform-induced hyporheic exchange with unsteady flows. *Advances in Water Resources*, 30: 148–156.

Boano, F., R. Revelli, and L. Ridolfi. 2010c. Effect of streamflow stochasticity on bedform-driven hyporheic exchange. *Advances in Water Resources*, 33: 1367–1374. doi: 10.1016/j.advwatres.2010.03.005.

Bott, T. L., L. A. Kaplan, and F. T. Kuserk. 1984. Benthic bacterial biomass supported by stream water dissolved organica matter. *Microbial Ecology*, 10: 335–344.

Boulton, A. J., M.-J. Dole-Olivier, and P. Marmonier. 2004. Effects of sample volume and taxonomic resolution on assessment of hyporheic assemblage composition sampled using a Bou-Rouch pump. *Archiv für Hydrobiologie*, **159** (3): 327–355.

Boulton, A. J., S. Findlay, P. Marmonier, E. H. Stanley, and M. H. Valett. 1998. The functional significance of the hyporheic zone in streams and rivers. *Annual Review of Ecology and Systematics*, **29**: 59–81.

Boulton, A. J., H. M. Valett, and S. G. Fisher. 1992. Spatial distribution and taxonomic composition of the hyporheos of several Sonoran Desert streams. *Archiv für Hyrdobiologie*, **125** (1): 37–61.

Brayshaw, A. C., L. E. Frostick, and I. Reid. 1983. The hydrodynamics of particle clusters and sediment entrainment in coarse alluvial channels. *Sedimentology*, **30**: 137–143.

Bredehoeft, J. D., and I. S. Papadopulos. 1965. Rates of vertical groundwater movement estimated from the earth's thermal profile. *Water Resources Research*, **1** (2): 325–328. doi:10.1029/WR001i002p00325.

Brunke, M. 1999. Colmation and depth filtration within streambeds: retention of particles in hyporheic interstices. *International Review of Hydrobiology*, **84** (2): 99–117.

Buffin-Bélanger, T., A. G. Roy, and A. D. Kirkbride. 2000a. On large-scale flow structures in a gravel-bed river. *Geomorphology*, **32**: 417–435.

Buffin-Bélanger, T., A. G. Roy, and A. D. Kirkbride. 2000b. Vers l'integration des structures turbulentes de l'ecoulement dans la dynamique d'un cours d'eau a lit de graviers. *Geographie Physique et Quaternaire*, **54** (1): 105–117.

Buffington, J. M., D. J. Isaak, and R. F. Thurow. 2003. Hydraulic control on the spatial distribution of Chinook salmon spawning gravels in central Idaho: integrating reach-scale predictions via digital elevation models. Proceedings of AGU Fall Meeting. AGU. San Francisco.

Buffington, J. M., T. E. Lisle, R. D. Woodsmith, and S. Hilton. 2002. Controls on the size and occurrence of pools in coarse-grained forest rivers. *River Research and Applications*, **18**: 507–531.

Buffington, J. M., and D. Tonina. 2007. Discussion of "Evaluating vertical velocities between the stream and the hyporheic zone from temperature data" by I. Seydell, B.E. Wawra and U.C.E. Zanke, in *Gravel-Bed Rivers 6-From Process Understanding to River Restoration*. edited by H. Habersack, *et al.* Elsevier: Amsterdam, The Netherlands. pp. 128–131.

Buffington, J. M., and D. Tonina. 2009. Hyporheic exchange in mountain rivers II: Effects of channel morphology on mechanics, scales, and rates of exchange. *Geography Compass*, **3** (3): 1038–1062.

Buss, S. R., M. O. Rivett, P. Morgan, and C. D. Bemment. 2005. *Attenuation of nitrate in the sub-surface environment*. Science Report SC030155/SR2. Environmental Agency: Bristol, UK.

Cardenas, M. B. 2009. Stream-aquifer interactions and hyporheic exchange in gaining and losing sinuous streams. *Water Resources Research*, **45**: W06429. doi:10.1029/2008WR007651.

Cardenas, M. B., J. L. Wilson, and R. Haggerty. 2008a. Residence time of bedform-driven hyporheic exchange. *Advances in Water Resources*, **31** (10): 1382–1386. doi:10.1016/j.advwatres.2008.07.006.

Cardenas, M. B., P. L. M. Cook, H. Jiang, and P. Traykovski. 2008b. Constraining denitrification in permeable wave-influenced marine sediment using linked hydrodynamic and biogeochemical modeling. *Earth and Planetary Science Letters*, **275** (1–2): 127–137. doi:10.1016/j.epsl.2008.08.016.

Cardenas, M. B., J. F. Wilson, and V. A. Zlotnik. 2004. Impact of heterogeneity, bed forms, and stream curvature on subchannel hyporheic exchange. *Water Resources Research*, **40** (W08307). doi:10.1029/2004WR003008.

Cardenas, M. B., and J. L. Wilson. 2007a. Exchange across a sediment–water interface with ambient groundwater discharge. *Journal of Hydrology*, **346**: 69–80. doi:10.1016/j.jhydrol.2007.08.019.

Cardenas, M. B., and J. L. Wilson. 2007b. Hydrodynamics of coupled flow above and below a sediment–water interface with triangular bedforms. *Advances in Water Resources*, **30**: 301–313.

Cardenas, M. B., and V. A. Zlotnik. 2003. Three-dimensional model of modern channel bend deposits. *Water Resources Research*, **39** (6): 7. doi:10.1029/2002WR001383.

Carling, P. A., L. J. Whitcombe, I. A. Benson, B. G. Hankin, and A. M. Radecki-Pawlik. 2006. A new method to determine interstitial flow patterns in flume studies of sub-aqueous gravel bedforms such as fish nests. *River Research and Applications*, **22**: 691–701.

Carpenter, S. R., N. F. Caraco, D. L. Correll, R. W. Howarth, A. N. Sharpley, and V. H. Smith. 1998. Nonpoint pollution of surface waters with phosphorus and nitrogen. *Ecological Applications*, **8** (3): 559–568.

Castro, N. M., and G. M. Hornberger. 1991. Surface-subsurface water interaction in an alluviated mountain stream channel. *Water Resources Research*, **27** (7): 1613–1621.

Choi, J., J. W. Harvey, and M. H. Conklin. 2000. Characterizing multiple timescales of stream and storage zone interaction that affect solute fate and transport in streams. *Water Resources Research*, **36** (6): 1511–1518.

Church, M. 1974. *Electrochemical and fluorometric tracer techniques for streamflow measurements*. Technical Bulletin. British Geomorphological Research Group.

Church, M. 1992. Channel morphology and typology, in *The Rivers Handbook*. edited by P. Carlow and G. E. Petts. Blackwell Scientific Publications: Oxford. pp. 126–143.

Clayton, J. L., J. King, and R. F. Thurow. 1996. Evaluation of an ion adsorption method to estimate intragravel flow velocity in salmonid spawning gravel. *North American Journal of Fisheries Management*, **16**: 167–174.

Colombini, M., G. Seminara, and M. Tubino. 1987. Finite-amplitude alternate bars. *Journal Fluid Mechanics*, **181**: 213–232.

Constantz, J., and C. L. Thomas. 1997. Stream bed temperature profiles as indicators of percolation characteristics beneath arroyos in the middle Rio Grande basin, USA. *Hydrological Processes*, **11**: 1621–1634.

Constantz, J., C. L. Thomas, and G. Zellweger. 1994. Influence of diurnal variations in stream temperature on streamflow loss and groundwater recharge. *Water Resources Research*, **30** (12): 3253–3264.

Cooper, A. C. 1965. *The effect of transported stream sediments on the survival of sockeye and pink salmon eggs and alevin*. International Pacific Salmon Fisheries Commission Bulletin 18: New Westminster, B.C., Canada.

Crook, N., A. Binley, R. Knight, D. A. Robinson, J. P. Zarnetske, and R. Haggerty. 2008. Electrical resistivity imaging of the architecture of substream sediments. *Water Resources Research*, **44**: W00D13. doi:10.1029/2008WR006968.

Dahm, C. N., M. H. Valett, C. V. Baxter, and W. W. Woessner. 2006. Hyporheic Zones, in *Methods in Stream Ecology*. edited. Elsevier.

Danielopol, D. L. 1989. Groundwater fauna associated with riverine aquifers. *Journal of the North American Benthological Society*, **8** (1): 18–35.

Danielopol, D. L., and R. Niederreiter. 1987. A sampling device for groundwater organisms and oxygen measurements in multi-level monitoring wells. *Stygologia*, **3**: 252–263.

Darcy, H. 1856. *Les fontaines publiques de la ville de Dijon*. Paris, Libraire des Corps Imperiaux des Ponts et Chaussées et des Mines: Dalmont, Paris.

Day, T. J. 1977. Field procedures and evaluation of a slug dilution gauging method in mountain streams. *Journal of Hydrology, ASCE*, **16** (2): 113–133.

Dent, C. L., N. B. Grimm, and S. G. Fisher. 2001. Multiscale effects of surface-subsurface exchange on stream water nutrient concentrations. *Journal of the North American Benthological Society*, **20** (2): 162–181.

Detert, M., M. Klar, T. Wenka, and G. H. Jirka. 2007. Pressure- and velocity-measurements above and within a porous gravel bed at the threshold of stability, in *Gravel-Bed Rivers VI: From Process Understanding to River Restoration*. edited by H. Habersack, *et al.* Elsevier: Amsterdam, The Netherlands. pp. 85–105.

DeVries, P. 1997. Riverine salmonid egg burial depths: a review of published data and implications for scour studies. *Canadian Journal of Fisheries and Aquatic Sciences*, **54**: 1685–1698.

Duff, J. H., and F. J. Triska. 2000. Nitrogen biogeochemistry and surface-subsurface exchange in streams, in *Streams and ground waters*. edited by J. B. Jones and P. J. Mulholland. Academic Press: San Diego, Calif. pp. 197–220.

Duke, J. R., J. D. White, P. M. Allen, and R. S. Muttiah. 2007. Riparian influence on hyporheic-zone formation downstream of a small dam in the Blackland Prairie region of Texas. *Hydrological Processes*, **21**: 141–150. doi:10.1002/hyp.6228.

Edwards, R. T. 1998. The hyporheic zone, in *River ecology and management: Lessons from the Pacific coastal ecoregion*. edited by R. J. Naiman and R. E. Bilby. Springer-Verlag: New York. pp. 399–429.

Elliott, A., and N. H. Brooks. 1997a. Transfer of nonsorbing solutes to a streambed with bed forms: laboratory experiments. *Water Resources Research*, **33** (1): 137–151.

Elliott, A., and N. H. Brooks. 1997b. Transfer of nonsorbing solutes to a streambed with bed forms: theory. *Water Resources Research*, **33** (1): 123–136.

Endreny, T., L. Lautz, and D. I. Siegel. 2011. Hyporheic flow path response to hydraulic jumps at river steps: Flume and hydrodynamic models. *Water Resources Research*, **47**: W02517. doi:10.1029/2009WR008631.

Findlay, S., W. Strayer, C. Goumbala, and K. Gould. 1993. Metabolism of streamwater dissolved organic carbon in the shallow hyporheic zone. *Limnology and Oceanography*, **38**: 1493–1499.

Fischenich, C., and R. Seal. 2000. *Boulder Cluster*. USAE Research and Development Center, Environmental Laboratory: Vicksburg, MS 6p.

Fischer, H., F. Kloep, S. Wilzcek, and M. Pusch. 2005. A River's Liver: Microbial Processes within the Hyporheic Zone of a Large Lowland River. *Biogeochemestry*, **76** (2): 349–371.

Fischer, H. B. 1967. Flow measurements with fluorescent tracers. *Journal of the Hydraulics Division*, **93** (HY1): 139–140.

Freeze, R. A., and J. A. Cherry. 1979. *Groundwater*. Prentice Hall: Englewood Cliffs, New Jersey.

Galloway, J. N., F. J. Dentener, D. G. Capone, E. W. Boyer, R. W. Howarth, S. P. Seitzinger, G. P. Asner, C. C. Cleveland, P. A. Green, E. A. Holland, D. M. Karl, A. F. Michaels, J. H. Porter, A. R. Townsend, and Vörösmarty. 2004. Nitrogen cycles: past, present, and future. *Biogeochemestry*, **70**: 153–226.

Galloway, J. N., A. R. Townsend, J. W. Erisman, M. Bekunda, Z. Cai, J. R. Freney, L. A. Martinelli, S. P. Seitzinger, and M. A. Sutton. 2008. Transformation of the nitrogen cycle: recent trends, questions, and potential solutions. *Science*, **320** (5878): 889–892.

Gariglio, F., D. Tonina, and C. H. Luce. 2011. Quantifying Hyporheic Exchange Over a Long Time Scale Using Heat as a Tracer in Bear Valley Creek, Idaho. *EOS, Transactions of the American Geophysical Union*, **92**.

Geist, D. R. 2000. Hyporheic discharge of river water into fall chinook salmon (*Oncorhynchis tshawytscha*) spawning areas in the Hanford Reach, Columbia River. *Canadian Journal of Fisheries and Aquatic Sciences*, **57**: 1647–1656.

Gibert, J., J. A. Stanford, M.-J. Dole-Olivier, and J. V. Ward. 1994. Basic attributes of groundwater ecosystems and prospects for research, in *Groundwater ecology*. edited by J. Gilbert, et al. Academic Press: San Diego. pp. 8–40.

Gooseff, M. N. 2010. Defining hyporheic zones- Advancing our conceptual and operational definitions of where stream water and groundwater meet. *Geography Compass*, **4** (8): 945–955. doi: 10.1111/j.1749-8198.2010.00364.x.

Gooseff, M. N., J. K. Anderson, S. M. Wondzell, J. LaNier, and R. Haggerty. 2006. A modelling study of hyporheic exchange pattern and the sequence, size, and spacing of stream bedforms in mountain stream networks, Oregon, USA. *Hydrological Processes*, **20** (11): 2443–2457.

Gooseff, M. N., R. O. J. Hall, and J. L. Tank. 2007. Relating transient storage to channel complexity in streams of varying land use in Jackson Hole, Wyoming. *Water Resources Research*, **43** (W01417). doi:10.1029/2005WR004626.

Gooseff, M. N., S. M. Wondzell, R. Haggerty, and J. K. Anderson. 2003. Comparing transient storage modeling and residence time distribution (RTD) analysis in geomorphically varied reaches in the Lookout Creek basin, Oregon, USA. *Advances in Water Resources*, **26**: 925–937.

Haggerty, R., and P. Reeves. 2002. *STAMM-L version 1.0 user's manual*. Sandia National Laboratory: Albuquerque, NM. 76 pp.

Haggerty, R., S. M. Wondzell, and M. A. Johnson. 2002. Power-law residence time distribution in hyporheic zome of a 2nd-order mountain stream. *Geophysical Research Letters*, **29** (13): 4. doi:10.1029/2002GL014743.

Harvey, A. M., and B. J. Wagner. 2000. Quantifying hydrologic interactions between streams and their subsurface hyporheic zones, in *Streams and Ground Waters*. edited by J. B. Jones and P. J. Mulholland. Academic Press: San Diego. pp. 3–44.

Harvey, C., and S. M. Gorelick. 2000. Rate-limited mass transfer or macrodispersion: which dominates plume evolution at the Macrodispersion Experiment (MADE) site? *Water Resources Research*, **36** (3): 637–650.

Harvey, J. W., and K. E. Bencala. 1993. The effect of streambed topography on surface-subsurface water exchange in mountain catchments. *Water Resources Research*, **29** (1): 89–98.

Harvey, J. W., B. J. Wagner, and K. E. Bencala. 1996. Evaluating the reliability of the stream tracer approach to characterize stream-subsurface water exchange. *Water Resources Research*, **32** (8): 2441–2451.

Hassan, M. A. 1990. Scour, fill, and burial depth of coarse material in gravel bed streams. *Earth Surface Processes and Landforms*, **15**: 341–356.

Hatch, C., A. T. Fisher, J. S. Revenaugh, J. Constantz, and C. Ruehl. 2006. Quantifying surface water–groundwater interactions using time series analysis of streambed thermal records: Method development. *Water Resources Research*, **42**: W10410. doi:10.1029/2005WR004787.

Hendricks, S. P., and D. S. White. 1991. Physico-chemical patterns within a hyporheic zone of a northern Michigan river, with comments on surface water patterns. *Canadian Journal of Fisheries and Aquatic Sciences*, **48**: 1645–1654.

Hester, E. T., and M. W. Doyle. 2006. Impact of in-channel geomorphic structures on surface-subsurface exchange of water and heat in streams. *EOS, Transactions, American Geophysical Union*, **87** (52): Abstract B23A-1063.

Hill, A. R., C. F. Labadia, and K. Sanmugadas. 1998. Hyporheic zone hydrology and nitrogen dynamics in relation to the streambed topography of a N-rich stream. *Biogeochemistry*, **42**: 285–310.

Hill, A. R., and J. Lymburner. 1998. Hyporheic zone chemistry and stream-subsurface exchange in two groundwater-fed streams. *Canadian Journal of Fisheries and Aquatic Sciences*, **55**: 495–506.

Ho, R. T., and L. W. Gelhar. 1973. Turbulent flow with wavy permeable boundaries. *Journal Fluid Mechanics*, **58** (2): 403–414.

Hynes, H. B. N. 1970. *The Ecology of Running Waters*. University Press: Liverpool.

Iversen, N., and B. B. Jørgensen. 19993. Diffusion coefficients of sulfate and methane in marine sediments: Influence of porosity. *Geochimica et Cosmochimica Acta*, **57**: 571–578.

Jonsson, K., H. Johansson, and A. Wörman. 2004. Sorption behavior and long-term retention of reactive solutes in the hyporheic zone of streams. *Journal of Environmental Engineering*, **130** (5). doi:10.1061/(ASCE)0733–9372(2004)130:5(573).

Kasahara, T., T. Datry, M. Mutz, and A. J. Boulton. 2009. Treating causes not symptoms: restoration of surface-groundwater interactions in rivers. *Marine and Freshwater Research*, **60**: 976–981.

Kasahara, T., and A. R. Hill. 2006a. Effects of riffle–step restoration on hyporheic zone chemistry in N-rich lowland streams. *Canadian Journal of Fisheries and Aquatic Sciences*, **63**: 120–133. doi: 10.1139/F05-199.

Kasahara, T., and A. R. Hill. 2006b. Hyporheic exchange flows induced by constructed riffles and steps in lowland streams in southern Ontario, Canada. *Hydrological Processes*, **20**: 4287–4305. doi: 10.1002/hyp.6174.

Kasahara, T., and A. R. Hill. 2007. Lateral hyporheic zone chemistry in an artificially constructed gravel bar and a re-meandered stream channel, Southern Ontario, Canada. *Journal of the American Water Resources Association*, **43** (5).

Kasahara, T., and S. M. Wondzell. 2003. Geomorphic controls on hyporheic exchange flow in mountain streams. *Water Resources Research*, **39** (1): 1005. doi:10.1029/2002WR001386.

Keery, J., A. Binley, N. Crook, and J. W. N. Smith. 2007. Temporal and spatial variability of groundwater–surface water fluxes: Development and application of an analytical method using temperature time series. *Journal of Hydrology*, **336**: 1–16. doi:10.1016/j.jhydrol.2006.12.003.

Kim, B. K. A., A. P. Jackman, and F. J. Triska. 1992. Modeling biotic uptake by periphyton and transient hyporheic storage of nitrate in a natural stream. *Water Resources Research*, **28** (10): 2743–2752.

Kjellin, J., S. Hallin, and A. Wörman. 2007. Spatial variations in denitrification activity in wetland sediments explained by hydrology and denitrifying community structure. *Water Research*, **41**: 4710–4720.

Lee, D. R., and J. A. Cherry. 1978. A field exercise on groundwater flow using seepage meters and mini-piezometers. *Journal of Geological Education*, **27**: 6–10.

Lewandowski, J., L. Angermann, G. Nützmann, and J. H. Fleckenstein. 2011. A heat pulse technique for the determination of small-scale flow directions and flow velocities in the streambed of sand-bed streams. *Hydrological Processes*, **25** (20): 3244–3255. DOI: 10.1002/hyp.8062.

Luce, C. H., D. Tonina, and F. Gariglio. in review. Solutions for the diurnally forced advection-diffusion equation to estimate bulk fluid velocity and diffusivity in streambeds from temperature time series. *Water Resources Research*.

Malard, F., K. Tockner, M.-J. Dole-Olivier, and J. V. Ward. 2002. A landscape perspective of surface–subsurface hydrological exchanges in river corridors. *Freshwater Biology*, **47**: 621–640.

Malcolm, I. A., C. Soulsby, A. F. Youngson, and D. M. Hannah. 2005. Catchment-scale controls on groundwater-surface water interactions in the hyporheic zone: implications for salmon embryo survival. *River Research and Applications*, **21**: 977–989. DOI: 10.1002/rra.861.

Marion, A., M. Bellinello, I. Guymer, and A. I. Packman. 2002. Effect of bed form geometry on the penetration of nonreactive solutes into a streambed. *Water Resources Research*, **38** (10): 1209. doi:10.1029/2001WR000264.

Marion, A., A. I. Packman, M. Zaramella, and A. Bottacin-Busolin. 2008a. Hyporheic Flows in Stratified Beds. *Water Resources Research*, **44**: W09433. doi:10.1029/2007WR006079.

Marion, A., and M. Zaramella. 2005. Diffusive behavior of bedform-induced hyporheic exchage in rivers. *Journal of Environmental Engineering*, **131**: 1260–1266.

Marion, A., M. Zaramella, and A. Bottacin-Busolin. 2008b. Solute transport in rivers with multiple storage zones: The STIR model. *Water Resources Research*, **44**: W10406. doi:10.1029/2008WR007037.

Marzadri, A., D. Tonina, and A. Bellin. 2011. A semianalytical three-dimensional process-based model for hyporheic nitrogen dynamics in gravel bed rivers. *Water Resources Research*, **46**: W11518. doi:10.1029/2011WR010583.

Marzadri, A., D. Tonina, and A. Bellin. 2012. Morphodynamic controls on redox conditions and on nitrogen dynamics within the hyporheic zone: Application to gravel bed rivers with alternate-bar morphology. *Journal of Geophysical Research*.

Marzadri, A., D. Tonina, A. Bellin, and G. Vignoli. 2010. Effects of bar topography on hyporheic flow in gravel-bed rivers. *Water Resources Research*, **46**: W07531. doi:10.1029/2009WR008285,.

Master, Y., U. Shavit, and A. Shaviv. 2005. Modified isotope pairing technique to study N transformations in polluted aquatic systems: theory. *Environmental Science and Technology*, **39**: 1749–1756.

Maxwell, R. M., C. Welty, and A. F. B. Tompson. 2003. Streamline-based simulation of virus transport resulting from long term artificial recharge in a heterogeneous aquifer. *Advances in Water Resources*, **26**: 1075–1096.

McDonald, R. R., J. M. Nelson, and J. P. Bennett. 2005. *Multi-dimensional surface-water modeling system user's guide*. U.S. Geological Survey. pp. 136 pp.

Mendoza, C., and D. Zhou. 1992. Effects of porous bed on turbulent stream flow above bed. *Journal of Hydraulic Engineering*, **118** (9).

Middleton, G. V., and J. B. Southard. 1984. *Mechanics of sediment movement*. Short Course 3. Society for Economic Paleontologists and Mineralogists: Tulsa, Okla.

Montgomery, D. R., and J. M. Buffington. 1997. Channel-reach morphology in mountain drainage basins. *Geological Society of America Bulletin*, **109**: 596–611.

Montgomery, D. R., and J. M. Buffington. 1998. Channel processes, classification, and response, in *River Ecology and Management*. edited by R. Naiman and R. Bilby. Springer-Verlag: NY. pp. 13–42.

Montgomery, D. R., J. M. Buffington, R. D. Smith, K. M. Schmidt, and G. Pess. 1995. Pool spacing in forest channels. *Water Resources Research*, **31** (4): 1097–1105.

Mulholland, P. J., and D. L. DeAngelis. 2000. Surface-subsurface exchange and nutrient spiraling, in *Streams and ground waters*. edited by J. B. Jones and P. J. Mulholland. Academis Press: San Diego, Calif. pp. 149–168.

Mulholland, P. J., A. M. Helton, G. C. Poole, R. O. J. Hall, S. K. Hamilton, B. J. Peterson, J. L. Tank, L. R. Ashkenas, L. W. Cooper, C. N. Dahm, W. K. Dodds, S. E. G. Findlay, S. V. Gregory, N. B. Grimm, S. L. Johnson, W. H. McDowell, J. L. Meyer, M. H. Valett, J. R. Webster, C. P. Arango, J. J. Beaulieu, M. J. Bernot, A. J. Burgin, C. L. Crenshaw, L. T. Johnson, B. R. Niederlehner, J. M. O'Brien, J. D. Potter, R. W. Sheibley, D. J. Sobota, and S. M. Thomas. 2008. Stream denitrification across biomes and its response to anthropogenic nitrate loading. *Nature*, **452** (13).

Mulholland, P. J., E. R. Marzorf, J. R. Webster, D. D. Hart, and S. P. Hendricks. 1997. Evidence that hyporheic zones increase heterotrophic metabolism and phosphorus uptake in forest streams. *Limnology and Oceanography*, **42**: 443–451.

Mulholland, P. J., M. H. Valett, J. R. Webster, S. A. Thomas, L. W. Cooper, S. K. Hamilton, and B. J. Peterson. 2004. Stream denitrification and total nitrate uptake rates measured using a field ^{15}N tracer addition approach. *Limnology and Oceanography*, **49** (3): 809–820.

Naegeli, M. W., P. Huggenberger, and U. Uehlinger. 1996. Ground penetrating radar for assessing sediment structures in the hyporheic zone of a prealpine river. *Journal of the North American Benthological Society*, **15** (3): 353–366.

Nagaoka, H., and S. Ohgaki. 1990. Mass transfer mechanism in a porous riverbed. *Water Research*, **24** (4): 417–425.

Niswonger, R. G., D. E. Prudic, G. Pohll, and J. Constantz. 2005. Incorporating seepage losses into the unsteady streamflow equations for simulating intermittent flow along mountain front streams. *Water Resources Research*, **41**: W06006. doi:10.1029/2004WR003677.

Nowinski, J. D., B. M. Cardenas, and A. F. Lightbody. 2011. Evolution of hydraulic conductivity in the floodplain of a meandering river due to hyporheic transport of fine materials. *Geophysical Research Letters*, **38**: L01401. doi:10.1029/2010GL045819.

O'Connor, B. L., and J. W. Harvey. 2008. Scaling hyporheic exchange and its influence on biogeochemical reactions in aquatic ecosystems. *Water Resources Research*, **44**: W12423. doi:10.1029/2008WR007160.

Orghidan, T. 1959. Ein neuer lebensraum des unterirdischen wassers: Der hyporheische biotop. *Archiv für Hydrobiologie*, **55**: 392–414.

Packman, A. I., and K. E. Bencala. 2000. Modeling methods in study of surface-subsurface hydrological interactions, in *Streams and ground waters*. edited by J. B. Jones and P. J. Mulholland. Academic Press: San Diego, Calif. pp. 45–80.

Packman, A. I., and N. H. Brooks. 1995. Colloidal particle exchange between stream and stream bed in a laboratory flume. *Marine and Freshwater Research*, **46** (1).

Packman, A. I., and N. H. Brooks. 2001. Hyporheic exchange of solutes and colloids with moving bed forms. *Water Resources Research*, **37** (10): 2591–2605.

Packman, A. I., N. H. Brooks, and J. J. Morgan. 1997. Experiment techniques for laboratory investigation of clay colloid tranposrt and filtration in a stream with a sand bed. *Water, Air and Soil Pollution*, **99**: 113–122.

Packman, A. I., N. H. Brooks, and J. J. Morgan. 2000a. Kaolinite exchange between a stream and streambed: laboratory experiments and validation of a colloid transport model. *Water Resources Research*, **36** (8): 2363–2372.

Packman, A. I., N. H. Brooks, and J. J. Morgan. 2000b. A physicochemical model for colloid exchange between a stream and a sand streambed with bed forms. *Water Resources Research*, **36** (8): 2351–2361.

Packman, A. I., and J. MacKay. 2003. Interplay of stream-subsurface exchange, clay particle deposition, and streambed evolution. *Water Resources Research*, **39** (4). doi:10.1029/2002WR001432.

Packman, A. I., and M. Salehin. 2003. Relative roles of stream flow and sedimentary conditions in controlling hyporheic exchange. *Hydrobiologia*, **494**: 291–297.

Packman, A. I., M. Salehin, and M. Zaramella. 2004. Hyporheic exchange with gravel beds: basic hydrodynamic interactions and bedform-induced advective flows. *Journal of Hydraulic Engineering*, **130** (7).

Payn, R. A., M. N. Gooseff, B. L. McGlynn, K. E. Bencala, and S. M. Wondzell. 2009. Channel water balance and exchange with subsurface flow along a mountain headwater stream in Montana, United States. *Water Resources Research*, **45**: w11427. doi:10.1029/2008WR007644.

Qian, Q., V. R. Voller, and H. G. Stefan. 2008. A vertical dispersion model for solute exchange induced by underflow and periodic hyporheic flow in a stream gravel bed. *Water Resources Research*, **44**: W07422. doi:10.1029/2007WR006366.

Redman, J. A., S. B. Grant, T. M. Olson, J. M. Adkins, J. L. Jackson, M. S. Castillo, and W. A. Yanko. 1999. Physicochemical mechanisms responsable for the filtration and mobilization of a filamentous bacterophage in quartz sand. *Water Research*, **33** (1): 43–52.

Ren, J., and A. I. Packman. 2002. Effects of background water composition on stream–subsurface exchange of submicron colloids. *Journal of Environmental Engineering*, **128** (7): 624–634. DOI: 10.1061/(ASCE)0733-9372(2002)128:7(624).

Ren, J., and A. I. Packman. 2004a. Coupled stream-subsurface exchange of colloidal hematite and dissolved zinc, copper and phosphate. *Environmental Science and Technology*, **39** (17): 6387–6394.

Ren, J., and A. I. Packman. 2004b. Modeling of simultaneous exchange of colloids and sorbing contaminants between streams and streambeds. *Environmental Science and Technology*, **38** (10): 2901–2911. doi: 10.1021/es034852l.

Ren, J., and A. I. Packman. 2004c. Stream-subsurface exchange of zinc in the presence of silica and kaolinite colloids. *Environmental Science and Technology*, **38** (24): 6571–6581. doi: 10.1021/es035090x.

Ren, J., and A. I. Packman. 2007. Changes in fine sediment size distributions due to interactions with streambed sediments. *Sedimentary Geology*, **202**: 529–537. doi: 10.1016/j.sedgeo.2007.03.021.

Ruff, J. F., and L. W. Gelhar. 1972. Turbulent shear flow in porous boundary. *Journal of the Engineering Mechanics Division ASCE*, **98** (EM4).

Runkel, R. L. 1998. *One-dimensional transport with inflow and storage (OTIS): a solute transport model for streams and rivers*. Water-resources investigations. US Geological Survey: Denver, Colorado.

Rutherford, I. C., J. D. Boyle, A. H. Elliott, T. V. J. Hatherell, and T. W. Chiu. 1995. Modeling benthic oxygen uptake by pumping. *Journal of Environmental Engineering*, **121** (1). Paper No. 7216.

Salehin, M., A. I. Packman, and M. Paradis. 2004. Hyporheic exchange with heterogeneous streambeds: laboratory experiments and modeling. *Water Resources Research*, **40** (11). doi:10.1029/2003WR002567.

Savant, A. S., D. D. Reible, and L. J. Thibodeaux. 1987. Convective Transport Within Stable River Sediments. *Water Resources Research*, **23** (9): 1763–1768.

Sawyer, A. H., B. M. Cardenas, and J. Buttles. 2011. Hyporheic exchange due to channel-spanning logs.*Water Resources Research*, **47**: W08502. doi: 10.1029/2010WR010484.

Sear, D. A., P. D. Armitage, and F. H. Dawson. 1999. Groundwater dominated rivers. *Hydrological Processes*, **13**: 255–276.

Selker, J. S., L. Thévenaz, H. Huwald, A. Mallet, W. Luxemburg, N. van de Giesen, M. Stejskal, J. Zeman, M. Westhoff, and M. B. Parlange. 2006. Distributed fiber-optic temperature sensing for hydrologic systems. *Water Resources Research*, **42** (W12202).

Shanafield, M., C. Hatch, and G. Pohll. 2011. Uncertainty in thermal time series analysis estimates of streambed water flux. *Water Resources Research*, **47**: W03504. doi:10.1029/2010WR009574.

Shimizu, Y., T. Tsujimoto, and H. Nakagawa. 1990. Experiments and macroscopic modeling of flow in highly permeable porous medium under free-surface flow. *Journal of Hydraulic Engineering*, **8** (1): 69–78.

Smith, J. W. M. (2005), *Groundwater-surface water interaction in the hyporheic zone*, pp. 65, UK Environmental Agency, Bristol, UK.

Sophocleous, M. A. 1991. Stream-floodwave propagation through the Great Bend alluvial aquifer, Kansas: field meaurements and numerical simulations. *Journal of Hydrology*, **124**: 207–228.

Soulsby, C., I. A. Malcolm, and A. F. Youngson. 2001. Hydrochemistry of the hyporheic zone in salmon spawning gravels: a preliminary assessment in a degraded agricaltural stream. *Regulated Rivers: Research & Management*, **17** (6): 651–665.

Stallman, R. W. 1960. Notes on the use of temperature data for computing groundwater velocity. Proceedings of 6th Assembly on Hydraulics. Société de hydrotechnique de France. Nancy, France.

Stanford, J. A. 2006. Landscapes and riverscapes, in *Methods in Stream Ecology*. edited by F. R. Hauer and G. A. Lamberti. Academic Press: San Diego. pp. 3–21.

Stanford, J. A., M. S. Lorang, and R. F. Hauer. 2005. The shifting habitat mosaic of river ecosystems. *Verhandlungen Internationale Vereinigung für Theoretische und Angewandte Limnologie*, **29**. doi:0368–0770/05/1940–0013.

Stanford, J. A., and J. V. Ward. 1988. The hyporheic habitat of river ecosystems. *Nature*, **335**: 64–65.

Stanford, J. A., and J. V. Ward. 1993. An ecosystem perspective of alluvial rivers: connectivity and the hyporheic corridor. *Journal of the North American Benthological Society*, **12** (1): 48–60.

Stonedahl, S. H., J. W. Harvey, A. Wörman, M. Salehin, and A. I. Packman. 2010. A multi-scale model for integrating hyporheic exchange from ripples to meanders. *Water Resources Research*, **46**: W12539. DOI:10.1029/2009WR008865.

Storey, R. G., K. W. F. Howard, and D. D. Williams. 2003. Factor controlling riffle-scale hyporheic exchange flows and their seasonal changes in gaining stream: a three-dimensional groundwater model. *Water Resources Research*, **39** (2): 17, doi:10.1029/2002WR001367.

Storey, R. G., D. D. Williams, and R. R. Fulthorpe. 2004. Nitrogen processing in the hyporheic zone of a pastoral stream. *Biogeochemestry*, **69**: 285–313.

Stuart, T. A. 1953. Water currents through permeable gravel and their significance to spawning salmonids, etc. *Nature*, **172**: 407–408.

Terhune, L. D. B. 1958. The Mark VI Groundwater standpipe for measuring seepage through salmon spawning gravel. *J. Fish. Res. Board Canada*, **15** (5): 1027–1063.

Thibodeaux, L. J., and J. D. Boyle. 1987. Bedform-generated convective transport in bottom sediment. *Nature*, **325** (22): 341–343.

Thomas, S. A., M. H. Valett, J. R. Webster, and P. J. Mulholland. 2003. A regression approach to estimating reactive solute uptake in advective and transient storage zones of stream ecosystems. *Advances in Water Resources*, **26**: 965–976.

Tonina, D., and J. M. Buffington. 2007. Hyporheic exchange in gravel-bed rivers with pool-riffle morphology: laboratory experiments and three-dimensional modeling. *Water Resources Research*, **43** (W01421).

Tonina, D., and J. M. Buffington. 2009a. Hyporheic exchange in mountain rivers I: Mechanics and environmental effects *Geography Compass*, **3** (3): 1063–1086.

Tonina, D., and J. M. Buffington. 2009b. A three-dimensional model for analyzing the effects of salmon redds on hyporheic exchange and egg pocket habitat. *Canadian Journal of Fisheries and Aquatic Sciences*, **66**: 2157–2173. doi: 10.1139/F09–146.

Tonina, D., and J. M. Buffington. 2011. Effects of stream discharge, alluvial depth and bar amplitude on hyporheic flow in pool-riffle channels. *Water Resources Research*, **47**: W08508. doi:10.1029/2010WR009140.

Tonina, D., A. Marzadri, and A. Bellin. 2011. Effect of hyporheic flows induced by alternate bars on benthic oxygen uptake. Proceedings of 34th IAHR World Congress 2011. IAHR. Brisbane, Australia.

Triska, F. J., J. H. Duff, and R. J. Avanzino. 1993a. Patterns of hydrological exchange and nutrient transformation in the hyporheic zone of a gravel bottom stream: examining terrestrial-aquatic linkages. *Freshwater Biology*, **29**: 259–274.

Triska, F. J., J. H. Duff, and R. J. Avanzino. 1993b. The role of water exchange between a stream channel and its hyporheic zone in nitrogen cycling at the terrestrial-aquatic interface. *Hydrobiologia*, **251**: 167–184.

Triska, F. J., A. P. Jackman, J. H. Duff, and R. J. Avanzino. 1994. Ammonium sorption to channel and riparian sediments: A transient storage pool for dissolved inorganic nitrogen. *Biogeochemestry*, **26** (2): 67–83.

Triska, F. J., V. C. Kennedy, R. J. Avanzino, G. W. Zellweger, and K. E. Bencala. 1989a. Retention and transport of nutrients in a third-order stream in Northwestern California: Hyporheic processes. *Ecology*, **70** (6): 1893–1905.

Triska, F. J., V. C. Kennedy, R. J. Avanzino, G. W. Zellweger, and K. E. Bencala. 1989b. Retention and transport of nutrients in a third-order stream: channel processes. *Ecology*, **70**: 1894–1905.

Valett, M. H., S. G. Fisher, N. B. Grimm, and P. Camill. 1994. Vertical hydrologic exchange and ecologic stability of a desert stream ecosystem. *Ecology*, **75**: 548–560.

Van Rijn, L. C. 1984. Sediment transport, part III: Bed forms and alluvial roughness. *Journal of Hydraulic Engineering*, **110**: 1733–1754.

Vaux, W. G. 1962. *Interchange of stream and intragravel water in a salmon spawning riffle*. Special Scientific Report. US Fish and Wildlife Service, Bureau of Commercial Fisheries: Washington, DC. 11 pp.

Vaux, W. G. 1968. Intragravel flow and interchange of water in a streambed. *Fishery Bulletin*, **66** (3): 479–489.

Vitousek, P. M., J. D. Aber, R. W. Howarth, G. E. Likens, P. A. Matson, D. W. Schindler, W. H. Schlesinger, and D. G. Tilman. 1997. Human alteration of the global nitrogen cycle: sources and consequences. *Ecological Applications*, **7** (3): 737–750.

Vittal, N., K. G. Ranga Raju, and R. J. Garde. 1977. Resistance of two-dimensional triangular roughness. *Journal Hydraulic Research*, **15** (1): 19–36.

Vollmer, S., F. de los Santos Ramos, H. Daebel, and G. Kühn. 2002. Micro scale exchange processes between surface and subsurface water. *Journal of Hydrology*, **269**: 3–10.

Wagenschein, D., and M. Rode. 2008. Modelling the impact of river morphology on nitrogen retention—A case study of the Weisse Elster River (Germany). *Ecological Modelling*, **211** (1–2): 224–232. doi: 10.1016/j.ecolmodel.2007.09.009

Wagner, B. J., and J. W. Harvey. 1997. Experimental design for estimating parameters of rate-limited mass transfer: analysis of stream tracer studies. *Water Resources Research*, **33** (7): 1731–1741.

Wagner, F. H., and G. Bretschko. 2003. Riparian Trees and Flow Paths between the Hyporheic Zone
and Groundwater in the Oberer Seebach, Austria. *International Review of Hydrobiology*, **88** (2): 129–138.

Ward, A. S., M. N. Gooseff, and K. Singha. 2010. Imaging hyporheic zone solute transport using electrical resistivity. *Hydrological Processes*, **24**: 948–953. doi: 10.1002/hyp.7672.

White, D. F., C. H. Elzinga, and S. P. Hendricks. 1987. Temperature patterns within the hyporheic zone of a northern Michigan river. *Journal of the North American Benthological Society*, **62** (2): 85–91.

White, D. S. 1990. Biological relationships to convective flow patterns within stream beds. *Hydrobiologia*, **196**: 149–158.

Williams, D. D., and H. B. N. Hynes. 1974. The occurrence of benthos deep in the substratum of stream. *Freshwater Biology*, **4**: 233–256.

Winter, T. C., J. W. Harvey, O. L. Franke, and W. M. Alley. 1998. *Ground water and surface water a single resource*. Circular. U.S. Geological Survey.

Woessner, W. W. 2000. Stream and fluvial plain ground water interactions: rescaling hydrogeologic thought. *Ground Water*, **38** (3): 423–429.

Wondzell, S. M. 2006. Effect of morphology and discharge on hyporheic exchange flows in two small streams in the Cascade Mountains of Oregon, USA. *Hydrological Processes*, **20**: 267–287. doi: 10.1002/hyp.5902.

Wondzell, S. M., and F. J. Swanson. 1996. Seasonal and storm dynamics of the hyporheic zone of 4-th order mountain stream. 1: hydrologic processes. *North America Benthological Society*, **15**: 3–19.

Wondzell, S. M., and F. J. Swanson. 1999. Floods, channel change, and the hyporheic zone. *Water Resources Research*, **35** (2): 555–567.

Wörman, A., A. I. Packman, H. Johansson, and K. Jonsson. 2002. Effect of flow-induced exchange in hyporheic zones on longitudinal transport of solutes in streams and rivers. *Water Resources Research*, **38** (1): 15. doi:10.1029/2001WR000769.

Wörman, A., A. I. Packman, L. Marklund, and J. W. Harvey. 2006. Exact three-dimensional spectral solution to surface-groundwater interaction with arbitrary surface topography. *Geophysical Research Letters*, **33** (L07402). doi:10.1029/2006GL025747.

Wroblicky, G. J., M. E. Campana, M. H. Valett, and C. N. Dahm. 1998. Seasonal variation in surface-subsurface water exchange and lateral hyporheic area of two stream-aquifer systems. *Water Resources Research*, **34** (3): 317–328.

Wuhrmann, K. 1972. River purification, in *Water pollution microbiology*. edited by R. Mitchell. Wiley-Interscience: New York, NY. pp. 119–151.

Zaramella, M., A. I. Packman, and A. Marion. 2003. Application of the transient storage model to analyze advective hyporheic exchange with deep and shallow sediment beds. *Water Resources Research*, **39** (7): 1198. doi:10.1029/2002WR001344.

Zarnetske, J. P., M. N. Gooseff, B. W. G. Bowden, Morgan J., T. R. Brosten, J. H. Bradford, and J. P. McNamara. 2008. Influence of morphology and permafrost dynamics on hyporheic exchange in arctic headwater streams under warming climate conditions. *Geophysical Research Letters*, **35** (L02501). doi:10.1029/2007GL032049.

Zarnetske, J. P., R. Haggerty, S. M. Wondzell, and M. A. Baker. 2011. Dynamics of nitrate production and removal as a function of residence time in the hyporheic zone. *Journal of Geophysical Research*, **116**: G01025. doi:10.1029/2010JG001356.

CHAPTER TEN

Environmental fluid dynamics of tidal bores: Theoretical considerations and field observations

Hubert Chanson

Professor in Civil Engineering, The University of Queensland, Brisbane, Australia

ABSTRACT

A tidal bore is a series of waves propagating upstream as the tidal flow turns to rising in a river mouth during the early flood tide. The formation of the bore occurs is linked with a macro-tidal range exceeding 4.5 to 6 m, a funnel shape of the river mouth and estuarine zone to amplify the tidal range. After formation of the bore, there is an abrupt rise in water depth at the bore front associated with a flow singularity in terms of water elevation, and pressure and velocity fields. The application of continuity and momentum principles gives a complete solution of the ratio of the conjugate cross-section areas as a function of the upstream Froude number. The effects of the flow resistance are observed to decrease the ratio of conjugate depths for a given Froude number. The field observations show that the tidal bore passage is associated with large fluctuations in water depth and instantaneous velocity components associated with intense turbulent mixing. The interactions between tidal bores and human society are complex. A tidal bore impacts on a range of socio-economic resources, encompassing the sedimentation of the upper estuary, the impact on the reproduction and development of native fish species, and the sustainability of unique eco-systems. It can be a major tourism attraction like in North America, Far East Asia and Europe, and a number of bores are surfed with tidal bore surfing competitions and festivals. But a tidal bore is a massive hydrodynamic shock which might become dangerous and hinder the local traffic and economical development.

10.1 INTRODUCTION

A tidal bore is a series of waves propagating upstream as the tidal flow turns to rising. It is an unsteady flow motion generated by the rapid rise in free-surface elevation at the river mouth during the early flood tide. The formation of a bore occurs when the tidal range exceeds 4.5 to 6 m and the funnel shape of both river mouth and lower estuarine zone amplifies the tidal wave. The driving process is the large tidal amplitude and its amplification in the estuary. After formation of the bore, there is an abrupt rise in water depth at the bore front associated with a flow singularity in terms of water elevation, and pressure and velocity fields. The tidal bore is a positive surge also called hydraulic jump in translation. Figures 10.1 to 10.5 illustrate some tidal bores in China, France and Indonesia. Pertinent accounts include Moore (1888), Darwin (1897), Moule (1923), and Chanson (2011). The existence of the tidal bore is based upon a fragile hydrodynamic balance between the tidal amplitude, the freshwater river flow conditions and the river channel bathymetry, and this balance may be easily disturbed by changes in boundary conditions and freshwater inflow (Chanson 2011). A number of man-made interferences led to the disappearance of tidal bores in France, Canada, Mexico for example. While the fluvial navigation gained in safety, the ecology of the estuarine systems was affected adversely, e.g. with the disappearance of native fish species. Natural events do also affect tidal bores: e.g., the 1964 Alaska earthquake on the Turnagain and Knik Arms bores, the 2001 flood of Ord River (Australia), the combination of storm surge and spring tide in Bangladesh in November 1970.

A related process is the tsunami-induced bore. When a tsunami wave propagates in a river, its leading edge is led by a positive surge. The tsunami-induced bore may propagate far upstream. Some tsunami-induced river bores were observed in Hawaii in 1946, in Japan

(A) Breaking tidal bore downstream of St Pardon on 8 September 2010 (shutter speed: 1/320 s) – Looking downstream of the incoming bore.

(B) Tidal bore in front of St Pardon on 12 September 2010 at sunrise (shutter speed: 1/100 s) – Bore propagation from left to right.

Figure 10.1. Tidal bore of the Dordogne River (France).

(C) Undular tidal bore in front of St Sulpice de Faleyrens on 19 June 2011 evening (shutter speed: 1/500 s) (Courtesy of Mr and Mrs Chanson) – Bore propagation from right to left.

Figure 10.1. *Continued.*

in 1983, 2001, 2003 and 2011, and even in the River Yealm in United Kingdom on 27 June 2011. During the 11 March 2011 tsunami catastrophe in Japan, tsunami-induced bores were observed in several rivers in north-eastern Honshu and as far s North-America.

After a brief introduction on tidal bores, some basic theoretical developments are developed. Then some recent field observations are presented and discussed. The results are challenging since the propagation of tidal bores is associated with sediment scour, strong mixing and suspended sediment advection upstream. It will be shown that the hydrodynamics of tidal bores remains a challenge to engineers and scientists because of the unsteady nature and sharp discontinuity of the flow.

10.2 THEORETICAL CONSIDERATIONS

10.2.1 Presentation

A tidal bore is characterised by a sudden rise in free-surface elevation and a discontinuity of the pressure and velocity fields. In the system of reference following the bore front, the integral form of the continuity and momentum equations gives a series of relationships between the flow properties in front of and behind the bore (Rayleigh, 1914; Henderson, 1966; Liggett, 1994):

$$(V_1 + U) \times A_1 = (V_2 + U) \times A_2 \tag{10.1}$$

(A) Tidal bore at Podensac on 11 September 2010 morning (shutter speed: 1/320 s) – Looking downstream at the incoming tidal bore.

(B) Undular tidal bore upstream of Podensac on 16 August 2011 evening (shutter speed: 1/2,000 s) (Courtesy of Isabelle Borde) – Bore propagation from left to right.

Figure 10.2. Tidal bore of the Garonne River (France).

$$\rho \times (V_1 + U) \times A_1 \times (\beta_1 \times (V_1 + U) - \beta_2 \times (V_2 + U))$$

$$= \iint\limits_{A_2} P \times dA - \iint\limits_{A_1} P \times dA + F_{fric} - W \times \sin\theta \qquad (10.2)$$

where V is the flow velocity and U is the bore celerity for an observer standing on the bank (Fig. 10.6), ρ is the water density, g is the gravity acceleration, A is the channel

Figure 10.3. Tidal bore of the Kampar River (Indonesia) in September 2010 (Courtesy of Antony Colas) – Bore propagation from right to left.

Figure 10.4. Tidal bore of the Qiantang River(China) at Yanguan on 23 July 2009 (shutter speed: 1/500 s) (Courtesy of Jean-Pierre Girardot) – The tidal range was 4 m – Bore propagation from left to right.

(A) Breaking bore in front of Pointe du Grouin du Sud on 19 October 2008 morning (shutter speed: 1/640 s) – Bore propagation from right to left.

(B) Breaking tidal bore downstream of Pointe du Grouin du Sud on 24 September 2010 evening (shutter speed: 1/125 s) – Bore propagation from right to left.

Figure 10.5. Tidal bore of the Sélune River (France).

cross-sectional area measured perpendicular to the main flow direction, β is a momentum correction coefficient or Boussinesq coefficient, P is the pressure, F_{fric} is the flow resistance force, W is the weight force, θ is the angle between the bed slope and horizontal, and the subscripts 1 and 2 refer respectively to the initial flow conditions and the flow conditions immediately after the tidal bore. Note that U is positive for a tidal bore (Fig. 10.6).

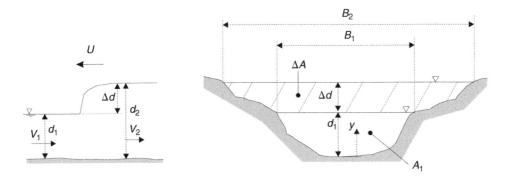

Figure 10.6. Definition sketch of a tidal bore propagating upstream.

10.2.2 Momentum considerations

The continuity and momentum equations provide some analytical solutions within some basic assumptions. First let us neglect the flow resistance and the effect of the velocity distribution ($\beta_1 = \beta_2 = 1$) and let us assume a flat horizontal channel ($\sin\theta \approx 0$). The momentum principle becomes:

$$\rho \times (V_1 + U) \times A_1 \times (V_1 - V_2) = \iint_{A_2} P \times dA - \iint_{A_1} P \times dA \tag{10.3}$$

In the system of reference in translation with the bore, the rate of change of momentum flux equals the difference in pressure forces. The latter may be expressed assuming a hydrostatic pressure distribution in front of and behind the tidal bore. The net pressure force resultant consists of the increase of pressure $\rho \times g \times (d_2 - d_1)$ applied to the initial flow cross-section area A_1 plus the pressure force on the area $\Delta A = A_2 - A_1$. This latter term equals:

$$\int_{A_1}^{A_2} \int \rho \times g \times (d_2 - y) \times dA = \frac{1}{2} \times \rho \times g \times (d_2 - d_1)^2 \times B' \tag{10.4}$$

where y is the distance normal to the bed, d_1 and d_2 are the flow depths in front of and behind the bore (Fig. 10.6), and B' is a characteristic free-surface width. It may be noted that $B_1 < B' < B_2$ where B_1 and B_2 are the upstream and downstream free-surface widths
  Another characteristic free-surface width B is defined as:

$$B = \frac{A_2 - A_1}{d_2 - d_1} \tag{10.5}$$

The equation of conservation of mass may be expressed as:

$$V_1 - V_2 = (V_1 + U) \times \frac{A_2 - A_1}{A_2} \tag{10.6}$$

The combination of the equations of conservation of mass and momentum (Eq. (10.3) and (10.6)) yields to the following expressions:

$$U + V_1 = \sqrt{\frac{1}{2} \times \frac{g \times A_2}{A_1 \times B} \times \left(\left(2 - \frac{B'}{B}\right) \times A_1 + \frac{B'}{B} \times A_2\right)} \tag{10.7}$$

$$V_1 - V_2 = \sqrt{\frac{1}{2} \times \frac{g \times (A_2 - A_1)^2}{B \times A_1 \times A_2} \left(\left(2 - \frac{B'}{B}\right) \times A_1 + \frac{B'}{B} \times A_2\right)} \tag{10.8}$$

After transformation, Equation (10.7) may be rewritten in the form:

$$Fr_1 = \sqrt{\frac{1}{2} \times \frac{A_2}{A_1} \times \frac{B_1}{B} \times \left(\left(2 - \frac{B'}{B}\right) + \frac{B'}{B} \times \frac{A_2}{A_1}\right)} \tag{10.9}$$

where Fr_1 is the tidal bore Froude number defined as:

$$Fr_1 = \frac{U + V_1}{\sqrt{g \times \dfrac{A_1}{B_1}}} \tag{10.10}$$

Equation (10.10) defines the Froude number for an irregular channel based upon momentum considerations. Interestingly the same expression may be derived from energy considerations (Henderson, 1966; Chanson, 2004).

Altogether Equation (10.9) provides an analytical solution of the tidal bore Froude number as a function of the ratios of cross-sectional areas A_2/A_1, and of characteristic widths B'/B and B_1/B. The effects of the celerity are linked implicitly with the initial flow conditions, including for a fluid initially at rest ($V_1 = 0$). Equation (10.9) may be rewritten to express the ratio of conjugate cross-section areas A_2/A_1 as a function of the upstream Froude number:

$$\frac{A_2}{A_1} = \frac{1}{2} \times \frac{\sqrt{\left(2 - \dfrac{B'}{B}\right)^2 + 8 \times \dfrac{\frac{B'}{B}}{\frac{B_1}{B}} \times Fr_1^2} - \left(2 - \dfrac{B'}{B}\right)}{\dfrac{B'}{B}} \tag{10.11}$$

which is valid for any tidal bore in an irregular flat channel (Fig. 10.6). The effects of channel cross-sectional shape are taken into account through the ratios B'/B and B_1/B.

Limiting cases

In some particular situations, the cross-sectional shape satisfies the approximation $B_2 \approx B \approx B' \approx B_1$: for example, a channel cross-sectional shape with parallel walls next to the waterline or a rectangular channel. In that case, Equations (10.7) and (10.8) may be simplified into:

$$U + V_1 = \sqrt{\frac{1}{2} \times \frac{g}{A_1} \times \frac{(A_1 + A_2) \times A_2}{B}} \qquad B_2 \approx B \approx B' \approx B_1 \qquad (10.12)$$

$$V_1 - V_2 = \sqrt{\frac{1}{2} \times \frac{g \times (A_1 + A_2) \times (A_2 - A_1)^2}{B \times A_1 \times A_2}} \qquad B_2 \approx B \approx B' \approx B_1 \qquad (10.13)$$

The above solution is close to the development of Lighthill (1978). Equation (10.12) may be expressed as the ratio of conjugate cross-section areas as a function of the upstream Froude number Fr_1:

$$\frac{A_2}{A_1} = \frac{1}{2} \times \left(\sqrt{1 + 8 \times Fr_1^2} - 1 \right) \qquad B_2 \approx B \approx B' \approx B_1 \qquad (10.14)$$

This last equation yields to the Bélanger equation for a rectangular horizontal channel in absence of friction:

$$\frac{d_2}{d_1} = \frac{1}{2} \times \left(\sqrt{1 + 8 \times Fr_1^2} - 1 \right) \qquad \text{Rectangular channel} \qquad (10.15)$$

Application

Several field observations of tidal bores were documented with detailed hydrodynamic and bathymetric conditions. The data are summarised in Table 10.1. Figure 10.5B shows the Sélune River channel during the field study on 24 September 2010 and the photograph highlights the wide and flat cross-sectional shape. Despite the range of channel cross-sectional shapes, the data indicated that the approximation $B_1 < B' < B < B_2$ held on average (Table 10.1).

The upstream Froude number was calculated using Equation (10.10) based upon the field measurements of velocity and bore celerity, and the data are summarised in Figure 10.7. Figure 10.7 presents the upstream Froude number as a function of the ratio of conjugate cross-section areas. The data are compared with Equation (10.11) for irregular channel cross-sections and with Equation (10.14) for channel cross-sectional shapes with parallel walls next to the waterline. The results highlight the effects of the irregular cross-section and illustrate that Equation (10.14) is not appropriate in an irregular channel like a natural estuarine system.

Further the definition of the Froude number Fr_1 (Eq. (10.10)) differs from the traditional approximation $V_1/(g \times d_1)^{0.5}$. For the data listed in Table 10.1 and shown in Figure 10.7, the difference varied between 12% and 74%!

10.2.3 Discussion: effects of flow resistance

In presence of some boundary friction and drag losses, the flow resistance force is non-zero and the equation of conservation of momentum may be solved analytically for a flat horizontal channel. The combination of the equations of conservation of mass and momentum yields:

$$U + V_1 = \sqrt{\frac{1}{2} \times \frac{g \times A_2}{A_1 \times B} \times \left(\left(2 - \frac{B'}{B} \right) \times A_1 + \frac{B'}{B} \times A_2 \right) + \frac{A_2}{A_2 - A_1} \times \frac{F_{fric}}{\rho \times A_1}}$$

$$(10.16)$$

Table 10.1. Hydrodynamics and bathymetric properties of tidal bores.

River	Bore	Date	Ref.	d_1 (m)	U (m/s)	Fr_1 (m)	$d_2 - d_1$	A_2/A_1
Dee	Breaking	2/07/03	[SFW04]	1.50	4.70	1.04	0.28	1.13
Daly	Undular	6/09/03	[WWSC04]	0.72	4.10	1.79	0.45	1.80
Garonne	Undular	10/08/10	[CLSR10]	1.77	4.49	1.30	0.50	1.37
Garonne	Undular	11/09/10	[CLSR10]	1.81	4.20	1.20	0.46	1.33
Sélune	Breaking	24/09/10	[MCS10]	0.38	2.00	2.35	0.34	6.19
Sélune	Breaking	25/09/10	[MCS10]	0.33	1.96	2.48	0.41	9.79

River	Bore	Date	B_2/B_1	B/B_1	B'/B_1	A_2/A_1 Eq. (10.11)
Dee	Breaking	2/07/03	1.013	1.007	1.001	1.052
Daly	Undular	6/09/03	1.066	1.030	1.085	2.09
Garonne	Undular	10/08/10	1.083	1.042	1.018	1.44
Garonne	Undular	11/09/10	1.076	1.032	1.021	1.30
Sélune	Breaking	24/09/10	3.37	2.33	1.92	4.92
Sélune	Breaking	25/09/10	3.53	2.33	1.98	5.18

Notes: d_1: initial water depth at sampling location; *Italic data*: incomplete data; [SFW04]: Simpson *et al.* (2004); [WWSC04]: Wolanski *et al.* (2004); [CLSR10]: Chanson *et al.* (2010); [MCS10]: Mouazé *et al.* (2010).

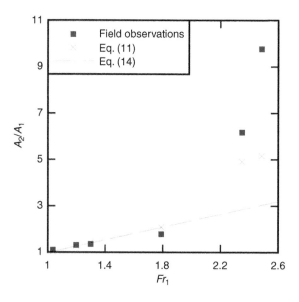

Figure 10.7. Momentum application to tidal bores in irregular cross-section channels – Comparison between field observations (Table 10.1) and Equations (10.11) and (10.14).

$$V_1 - V_2 = \sqrt{\frac{1}{2} \times \frac{g \times (A_2 - A_1)^2}{B \times A_1 \times A_2} \left(\left(2 - \frac{B'}{B} \right) \times A_1 + \frac{B'}{B} \times A_2 \right) + \frac{A_2}{A_2 - A_1} \frac{F_{fric}}{\rho \times g \times \frac{A_1^2}{B}}}$$

(10.17)

Equations (10.16) and (10.17) are the extension of Equations (10.7) and (10.8) in presence of flow resistance. The results may be transformed into:

$$Fr_1 = \sqrt{\frac{1}{2} \times \frac{A_2}{A_1} \times \frac{B_1}{B} \times \left(\left(2 - \frac{B'}{B} \right) + \frac{B'}{B} \times \frac{A_2}{A_1} \right) + \frac{A_2}{A_2 - A_1} \times \frac{F_{fric}}{\rho \times g \times \frac{A_1^2}{B}}}$$

(10.18)

The result (Eq. (10.18)) gives a relationship between the tidal bore Froude number, the ratio of the conjugate cross-section areas A_2/A_1 and the flow resistance force in flat channels of irregular cross-sectional shape. Figure 10.8 illustrates the effects of bed friction on the hydraulic jump properties for a irregular channel corresponding to the bathymetric conditions of the Sélune River listed in Table 10.1.

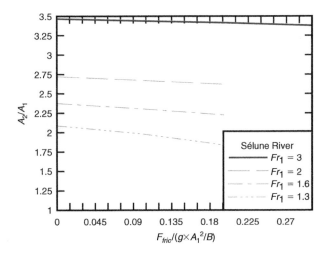

Figure 10.8. Effect of flow resistance on the tidal bore properties in an irregular channel.

The theoretical developments imply a smaller ratio of the conjugate depths d_2/d_1 with increasing flow resistance. The finding is consistent with laboratory data (for example Leutheusser and Schiller, 1975). It is more general and applicable to any irregular channel cross-sectional shape. Importantly the results highlighted that the effects of flow resistance decrease with increasing Froude number and become negligible for Froude numbers greater than 2 to 3 depending upon the cross-sectional properties (Fig. 10.8).

10.3 FIELD OBSERVATIONS

10.3.1 Presentation

When a tidal bore is formed, the flow properties immediately upstream and downstream of its front must satisfy the principles of continuity and momentum (section 10.2). Theoretical considerations demonstrate that a key dimensionless parameter is the tidal bore Froude number defined as:

$$Fr_1 = \frac{U + V_1}{\sqrt{g \times \frac{A_1}{B_1}}} \tag{10.10}$$

Fr_1 characterises the strength of the bore. If the Froude number Fr_1 is less than unity, the tidal bore cannot form. For a Froude number between 1 and 1.5 to 1.8, the bore is followed by a train of quasi-periodic secondary waves called whelps or undulations. This type of bore is the undular non-breaking bore illustrated in Figures 10.1, 10.2 and 10.9. The free-surface properties of undular tidal bores were investigated for more than a century (Boussinesq, 1877; Lemoine, 1948; Chanson, 2010). A recent review showed that the rate of energy dissipation is small to negligible, while the approximation of hydrostatic pressure is inaccurate. The free-surface profile present a pattern somehow comparable to the sinusoidal and cnoidal wave functions, although neither captures the fine details of the undulation shape and asymmetrical wave profile (Chanson, 2010). Field and laboratory data showed a maximum in wave amplitude and steepness for $Fr_1 = 1.3$ to 1.4 corresponding to the apparition of some breaking at the first wave crest.

For larger Froude numbers, the bore tidal has a breaking front with a marked roller. Examples are shown in Figures 10.3 to 10.5.

Figure 10.9. Undular tidal bore the Garonne River on 10 September 2010 at Arcins (shutter speed: 1/125 s) – Bore propagation from right to left – The surfer in the background rides the bore front.

Table 10.2. Detailed field measurements of turbulent velocity in tidal bores.

River	Country	Bore type	Date	Ref.	Instrument	Sampling rate (Hz)
Dee	UK	Breaking	2/07/03	[SFW04]	ADCP (1.2 MHz)	1
Daly	Australia	Undular	6/09/03	[WWSC04]	ADCP Nortek Aquadopp	2
Garonne	France	Undular	10/08/10	[CLSR10]	ADV Nortek Vector (6 MHz)	64
Garonne	France	Undular	11/09/10	[CLSR10]	ADV Nortek Vector (6 MHz)	64
Sélune	France	Breaking	24/09/10	[MCS10]	ADV Nortek Vector (6 MHz)	64
Sélune	France	Breaking	25/09/10	[MCS10]	ADV Nortek Vector (6 MHz)	64

River	Bore	Date	d_1 (m)	U (m/s)	Fr_1	z (m)	z/d_1
Daly	Undular	2/07/03	1.50	4.70	1.04	*0.75*	0.5
Dee	Breaking	6/09/03	0.72	4.10	1.79	0.55	0.76
Garonne	Undular	10/08/10	1.77	4.49	1.30	0.81 m below surface	0.54(*)
Garonne	Undular	11/09/10	1.81	4.20	1.20	0.81 m below surface	0.55(*)
Sélune	Breaking	24/09/10	0.38	2.00	2.35	0.225	0.59
Sélune	Breaking	25/09/10	0.33	1.96	2.48	0.10	0.30

Notes: d_1: initial water depth at sampling location; *Italic data*: incomplete data; z: sampling elevation above bed; [SFW04]: Simpson *et al.* (2004); [WWSC04]: Wolanski *et al.* (2004); [CLSR10]: Chanson *et al.* (2010); [MCS10]: Mouazé *et al.* (2010); (*): immediately prior to bore passage.

All the field observations highlighted the intense turbulence generated by the advancing bore (Fig. 10.1 to 10.5). Moule (1923) reported a description of the Qiantang River bore (China) from the 13th century: "*when the wave [or tide] comes it is steep as a mountain, roaring like thunder, a horizontal flying bank of water [or ice] and sidelong shooting precipice of snow plunging and leaping in a dreadful manner*". Bazin (1865) described the destructive power of the Hoogly River tidal bore in India: "*the tidal bore creates a 4 to 5 m high wall of water, and advances with a great noise announcing the flood tide; it entrains upstream all the floating debris and sinks the small boats on the shoals and in shallow waters*". La Condamine (1745) documented the impact of the passage of the Amazon River bore: "*One can see a wall of water of 4 to 5 m in height, then another, then a third one and sometimes a fourth one, that comes close together, and that occupies all the width of the channel; this bore advanced very rapidly, breaks and destroy everything*". A further illustration of intense turbulence is the number of field work incidents, encompassing studies in the Dee River, Rio Mearim, Daly River and Sélune River. In the Rio Mearim "*one sawhorse and instrument tumbled along the bottom for 1.4 km with currents exceeding 3 m·s^{-1}, was buried in a sand bank, and had to be abandoned*" (Kjerve and Ferreira, 1993). In the Sélune River, "*the field study experienced a number of problems and failures. About 40 s after the passage of the bore, the metallic frame started to move. The ADV support failed completely 10 minutes after the tidal bore*" (Mouazé *et al.*, 2010).

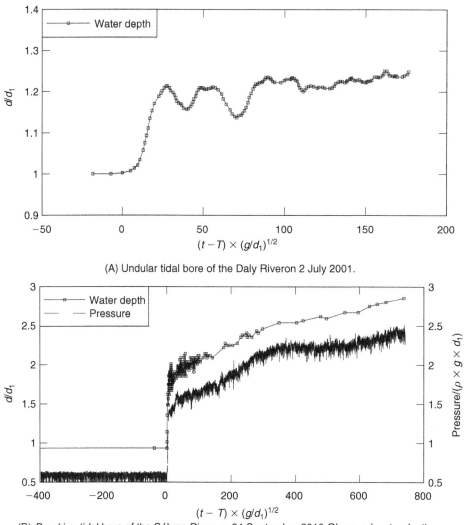

(A) Undular tidal bore of the Daly River on 2 July 2001.

(B) Breaking tidal bore of the Sélune River on 24 September 2010-Observed water depth and pressure sampled at 0.225 m above bed.

Figure 10.10. Time-variations of water depth and free-surface discontinuity across a tidal bore.

Some recent free-surface and turbulent velocity measurements were conducted in the field with detailed temporal and spatial resolutions (Table 10.2). The data provide an unique characterisation of the unsteady turbulent field and mixing processes. Table 10.2 summarises the basic flow conditions and includes details on the instrumentation. The basic outcomes are summarised in this section.

10.3.2   Field observations

The propagation of a tidal bore is associated with a sudden rise in free-surface elevation. Basically the passage of the tidal bore creates a sudden discontinuity in terms of the flow depth followed by large, long-lasting fluctuations of the free-surface behind the bore front.

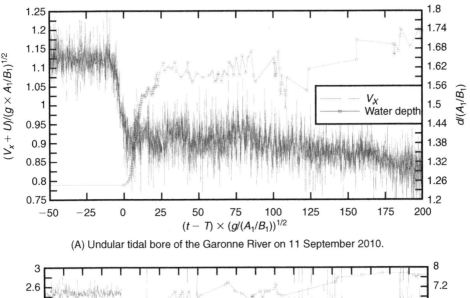

(A) Undular tidal bore of the Garonne River on 11 September 2010.

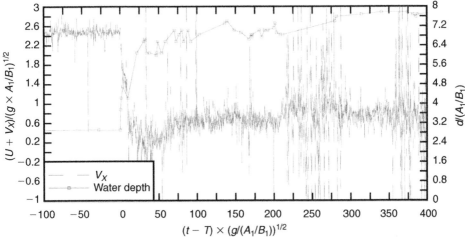

(B) Breaking tidal bore of the Sélune River on 25 September 2010-Observed water depth and press sampled at 0.10 m above bed.

Figure 10.11. Time-variations of the dimensionless water depth and longitudinal velocity during a tidal bore passage.

Typical field observations are presented in Figure 10.10 for an undular and breaking bore. Figure 10.10 shows the dimensionless water depth as a function of the dimensionless time $(t - T)$ where T is the passage time of the bore front. In Figure 10.10B, some pressure data recorded at 0.225 m above the bed are further included. Further details on each field study are reported in Table 10.2.

All the turbulent velocity data show a rapid deceleration of the flow associated with the passage of the bore as illustrated in Figure 10.11. Figure 10.11 shows some typical time variations of the longitudinal velocity component during the propagation of tidal bores. The data are presented in a dimensionless form based upon the momentum considerations developed above. In most natural systems, the bore passage is associated with a flow reversal $(V_x < 0)$ although this might not be always the case (Bazin, 1865; Kjerfve and Ferreira, 1993). Some large fluctuations of longitudinal velocities are observed during and shortly after the bore at all vertical elevations within the water column. The tidal bore acts as

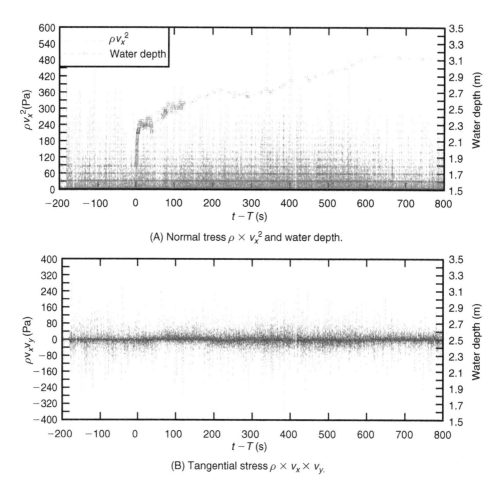

Figure 10.12. Time-variations of the turbulent Reynolds stresses during the passage of the undular tidal bore of
the Garonne River on 11 September 2010.

a hydrodynamic shock with a sudden change in velocity and pressure fields. The tidal
bore is always followed by some highly turbulent flow motion with long lasting effects.
Both the transverse and vertical velocity component data present some large and rapid
fluctuations with time immediately after the bore passage. The bore passage is further
associated with some relatively-long-period oscillations superposed to some high-frequency
turbulent fluctuations. The former may be linked with the formation, development and
advection of large-scale coherent structures behind the front, as hinted by some recent
numerical simulations (see for example Lubin *et al.*, 2010).

   The unsteady turbulent flow motion is characterised by large turbulent stresses, and
turbulent stress fluctuations, below the tidal bore and following whelp motion. The data
indicate that the turbulent stress magnitudes are larger than in the initial turbulent flow
shortly prior to the bore, and highlight the intense turbulent mixing beneath the tidal bore
(Fig. 10.12). Figure 10.12 presents some results in terms of normal and tangential Reynolds
stresses during an undular tidal bore; note that the results are presented in dimensional form.
The instantaneous turbulent shear stress magnitudes are larger than the critical threshold for
sediment motion and transport, although the comparison has some limits. In a turbulent
bore, the large scale vortices play an important role in terms of sediment material pickup

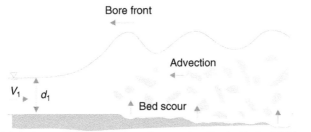

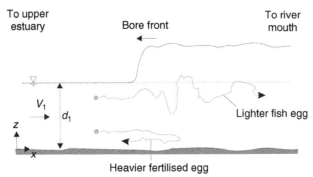

(A) Bed scour and upstream sediment advection behind a tidal bore.

(B) Selective dispersion of fish eggs beneath a tidal bore.

Figure 10.13. Conceptual sketches of the impact of tidal bores on estuarine systems.

and upward advection. Sediment motion occurs by convection since the turbulent length scale is much larger than the sediment characteristic size. Further the high levels of shear stresses revealed during the field measurements occur during very short transient times (turbulent bursts) rather at a continuous level like in a steady fluvial motion.

The field data illustrate that a tidal bore induces a very strong mixing in the natural channel, for which the classical mixing theories do not account for. During the tidal bore passage, and the eroded material and other scalars are advected upstream in the whelps and wave motion behind the bore front. The results are consistent with the very strong turbulent mixing observed in the tidal-bore affected estuaries, associated with the accretion and deposition of sediment materials in the upper estuarine zones.

10.3.3 Discussion

Both field measurements and laboratory studies (see bibliography) highlight some key features of the impact of tidal bores on the estuarine system.

The turbulent velocity measurements indicate the existence of energetic turbulent events during and behind the tidal bore (Fig. 10.11 & 10.12). These are highlighted by large and rapid fluctuations of turbulent velocities and Reynolds stresses. The duration of the turbulent events seem larger beneath undular bores, and shorter and more intense beneath breaking bores. This type of macro-turbulence can maintain its coherence as the eddies are advected behind the bore. Importantly the macro-turbulence contributes to significant sediment erosion from the bed and banks, and the upstream advection of the eroded material as illustrated in Figure 10.13A.

A recent study showed the preferential dispersion of fish eggs in a tidal bore affected estuary (Chanson and Tan, 2010). The fish eggs are typically advected downstream during the ebb tide. The arrival of the tidal bore induces a selective longitudinal dispersion of

(A) Tidal bore of the Sélune River on 7 April 2004 at sunrise - Bore propagation from left to right.

(B) Tidal bore of the Dordogne River on 12 September 2010 at sunrise - Bore propagation from left to right.

Figure 10.14. Tidal bore propagation at sunrise.

the eggs. The lightest, unfertilised fish eggs tend to flow downstream towards the river mouth, while the fertilised fish eggs are advected upstream behind the bore (Fig. 10.13B). The tidal bore induces a rapid longitudinal spread of the eggs with some preferential mixing depending upon their density and stages of development. The fertilised fish eggs are confined by the tidal bore to the upper estuary that is the known breeding grounds of juveniles. The unfertilised, neutrally buoyant eggs continue downstream possibly up to the river mouth, although the strong flood flow may bring them back into the upper estuary at a later stage of the tidal cycle. Figure 10.13B illustrates the selective dispersion process.

More generally the bore occurrence is essential to a number of ecological processes and the sustainability of unique eco-systems. The tidal bore propagation induces a massive

(A) Road sign to the benak (tidal bore) of the Batang Lupar, Sri Aman (Malaysia) (Courtesy of Antony Colas).
(B) Stamped postal envelop edited by the French Post Office in 2010 as part of a series on the tidal bores in
Gironde.

(C) Tidal bore of the Qiang Tang River (China) overtopping its banks on 31 August 2011 during a combination
of strong spring tide as the typhoon Nanmadol approached the coastline (Photo by ChinaFotoPress/Getty
Images).

Figure 10.15. Impact of tidal bores on the human society.

mixing of estuarine waters stirring the organic matter and creating some rich fishing grounds
(for example, the Rokan River in Indonesia).

10.4   CONCLUSION

A tidal bore is a hydrodynamic shock propagating upstream as the tidal flow turns to rising
and forming during the spring tides when the tidal range exceeds 5–6 m and the flood
tide is confined to a narrow funnelled estuary with low freshwater levels. The tidal bore
propagation induces a massive mixing of the natural system and its occurrence is critical to
the environmental balance of the estuarine zone.

The application of continuity and momentum principles gives a complete solution of
the ratio of the conjugate cross-section areas as a function of the upstream Froude number

$Fr_1 = (V_1 + U)/\sqrt{g \times A_1/B_1}$ for a range of channel cross-sections. The effects of the flow resistance are observed to decrease the ratio of conjugate depths for a given Froude number. The field observations show that the tidal bore passage is associated with large fluctuations in water depth and instantaneous velocity components associated with intense turbulent mixing. Some detailed turbulent velocity measurements at several vertical elevations during and shortly after the bore passage highlight some seminal features of tidal bores: namely some relatively-long-term oscillations in terms of flow depth and velocity superposed to some high-frequency turbulent fluctuations.

The interactions between tidal bores and human society are complex (Fig. 10.14 & 10.15). A tidal bore impacts on a range of socio-economic resources, encompassing the sedimentation of the upper estuary, the impact on the reproduction and development of native fish species, and the sustainability of unique eco-systems. A tidal bore can be a major tourism attraction like in North America, Far East Asia and Europe (Fig. 10.14 & 10.15A). A number of bores are surfed with tidal bore surfing competitions and festivals in South America, Europe and South-East Asia. But a tidal bore is a massive hydrodynamic shock which might become dangerous (Fig. 10.15C) and hinder the local traffic and development. A bore is an integral part of the environmental and socio-cultural heritage (Fig. 10.15B). It is a fascinating geophysical phenomenon in terms of geo-morphological and biological processes, as well as for the estuarine populations. Yet it remains a challenging research topic to the scientists, engineers and socio-environment experts.

ACKNOWLEDGEMENTS

The author thanks Dr Pierre Lubin, University of Bordeaux, for his valuable inputs and fruitful exchanges. He acknowledges the assistance of all the people (incl. colleagues, students, friends and relatives) who participated to the field works, and without whom this contribution could not have been possible. He thanks Dr Carlo Gualtieri for his encouragement to develop this contribution. The financial support of the University of Queensland, the Université de Bordeaux, the Australian Academy of Science and the Agence Nationale de la Recherche (Projet 10-BLAN-0911-01) is acknowledged.

APPENDIX A – LIST OF SYMBOLS

List of Symbols		
Symbol	Definition	Dimensions or Units
A	cross-section area	$[L^2]$
A_1	initial cross-section area	$[L^2]$
A_2	flow cross-section area behind the bore	$[L^2]$
B	characteristic free-surface width	$[L]$
B'	characteristic free-surface width	$[L]$
B_1	upstream free-surface width	$[L]$
B_2	free-surface width behind the bore	$[L]$
d_1	upstream flow depth	$[L]$
d_2	flow depth behind the bore	$[L]$
g	gravitational acceleration constant	$[L\,T^{-2}]$
Fr	tidal bore Froude number	–

(Continued)

List of Symbols

Symbol	Definition	Dimensions or Units
Fr_1	inflow Froude number of tidal bore	–
F_{fric}	friction force	[N]
g	gravity acceleration	$[L\,T^{-2}]$
P	pressure	$[N\,m^{-2}]$
Q	water discharge	$[L^3 \cdot T^{-1}]$
T	time of tidal bore passage	[T]
t	time	[T]
Tu	turbulence intensity	–
U	tidal bore celerity positive upstream	$[L\,T^{-1}]$
V_x	longitudinal velocity component	$[L\,T^{-1}]$
V_y	transverse velocity component	$[L\,T^{-1}]$
V_z	vertical velocity component	$[L\,T^{-1}]$
V_1	initial flow velocity positive downstream	$[L\,T^{-1}]$
V_2	flow velocity behind the bore positive downstream	$[L\,T^{-}]$
W	weight force	[N]
x	longitudinal/streamwise direction	[L]
y	transverse or radial direction	[L]
z	vertical direction positive upward	[L]
β	momentum correction coefficient	–
μ	water dynamic viscosity	$[M\,L^{-1}\,T^{-1}]$
θ	angle between bed slope and horizontal	–
ρ	water density	$[M\,L^{-3}]$
σ	surface tension between air and water	$[N\,m^{-1}]$

APPENDIX B – SYNOPSIS

A tidal bore is a hydrodynamic shock propagating upstream as the tidal flow turns to rising. A tidal bore forms during the spring tides when the tidal range exceeds 5–6 m and the flood tide is confined to a narrow funnelled estuary with low freshwater levels. The tidal bore propagation induces a massive mixing of the natural system. Its occurrence is critical to the environmental balance of the estuarine zone. The application of continuity and momentum principles gives a complete solution of the ratio of the conjugate cross-section areas as a function of the upstream Froude number $Fr_1 = (V_1 + U)/\sqrt{g \times A_1/B_1}$. The flow resistance is observed to decrease the ratio of conjugate depths for a given Froude number. The tidal bore passage is associated with large fluctuations in water depth and instantaneous velocity components. This is associated with intense turbulent mixing, and sediment scour and advection in a natural system. The interactions between tidal bores and human society are complex. Both positive and adverse impacts may be encountered. Tidal bore surfing is becoming a renown extreme sport.

APPENDIX C – KEYWORDS

Tidal bore	Turbulent mixing
Momentum considerations	Froude number
Undular bores	Turbulent stresses
Breaking bores	Sediment processes
Flow resistance	Hydrodynamic shock

APPENDIX D – QUESTIONS

What are the three basic requirements for the occurrence of tidal bores?
What is the main driving mechanism of a tidal bore?
How many tidal bores are observed worldwide?
What is the basic principle used to analyse a tidal bore flow motion?
Write the tidal bore Froude number and explain each term.
What is the effect of boundary friction on the tidal bore properties?
What are the potential impacts of a tidal bore in a natural estuarine system?
Where can we see tidal bore surfing?

APPENDIX E – PROBLEMS

E1. Using tide predictions for France and China, predict the likely dates of tidal bore occurrence in the Bay of Mont Michel and in the Qiantang River in September 2013.

This may require to surf the Internet to find the tide predictions for the Bay of Mont Michel and the Qiantang River.

E2. On the 27 Sept. 2000, the flow conditions of the tidal bore of the Dordogne River were: initial water depth $= 1.5$ m, initial flow velocity $= +0.22$ m/s, observed bore celerity: 4.8 m/s. Assuming a wide rectangular channel, calculate the flow velocity after the passage of the bore. (Use the downstream flow direction as positive axis.) Numerical solution: $V_2 = -1.26$ m/s (flow reversal), $d_2 = 2.13$ m.

E3. Plot the relationship between the ratio of conjugate cross-section areas and dimensionless flow resistance force for two Garonne River data sets listed in Table 10.1. Deduce the dimensionless flow resistance force from the observations.

REFERENCES

Bazin, H. (1865). "Recherches Expérimentales sur la Propagation des Ondes." ('Experimental Research on Wave Propagation.') Mémoires présentés par divers savants à l'Académie des Sciences, Paris, France, Vol. 19, pp. 495–644 (in French).
Boussinesq, J.V. (1877). "Essai sur la Théorie des Eaux Courantes." ('Essay on the Theory of Water Flow.') Mémoires présentés par divers savants à l'Académie des Sciences, Paris, France, Vol. 23, Série 3, No. 1, supplément 24, pp. 1–680 (in French).
Chanson, H. (2004). "The Hydraulics of Open Channel Flow: An Introduction." Butterworth-Heinemann, 2nd edition, Oxford, UK, 630 pages (ISBN 978 0 7506 5978 9).
Chanson, H. (2010). "Undular Tidal Bores: Basic Theory and Free-surface Characteristics." Journal of Hydraulic Engineering, ASCE, Vol. 136, No. 11, pp. 940–944 (DOI: 10.1061/(ASCE)HY.1943-7900.0000264).
Chanson, H. (2011). "Tidal Bores, Aegir, Eagre, Mascaret, Pororoca: Theory and Observations." World Scientific, Singapore, 220 pages (ISBN 9789814335416).
Chanson, H., Lubin, P., Simon, B., and Reungoat, D. (2010). "Turbulence and Sediment Processes in the Tidal Bore of the Garonne River: First Observations." Hydraulic Model Report No. CH79/10, School of Civil Engineering, The University of Queensland, Brisbane, Australia, 97 pages.
Chanson, H., and Tan, K.K. (2010). "Turbulent Mixing of Particles under Tidal Bores: an Experimental Analysis." Journal of Hydraulic Research, IAHR, Vol. 48, No. 5, pp. 641–649 (DOI: 10.1080/00221686.2010.512779 (ISSN 0022-1686).

Darwin, G.H. (1897). The Tides and Kindred Phenomena in the Solar System." Lectures delivered at the Lowell Institute, Boston, W.H. Freeman and Co. Publ., London, 1962.

Henderson, F.M. (1966). "Open Channel Flow." MacMillan Company, New York, USA.

Kjerfve, B., and Ferreira, H.O. (1993). "Tidal Bores: First Ever Measurements." Ciência e Cultura (Jl of the Brazilian Assoc. for the Advancement of Science), Vol. 45, No. 2, March/April, pp. 135–138.

La Condamine, C.H. de (1745). "Relation abrégée d'un voyage fait dans l'intérieur de l'Amérique méridionale, depuis la côte de la mer du sud jusqu'aux côtes du Brésil et de la Guyane, en descendant la rivière des Amazones." ('Diary of the Inland Travel to Southern America, from West to East up to Brazil and Guyana, following the Amazon River.') J.E.Dufour and P. Roux Libraires, Maestricht, 415 pages (in French).

Lemoine, R. (1948). "Sur les Ondes Positives de Translation dans les Canaux et sur le Ressaut Ondulé de Faible Amplitude." ('On the Positive Surges in Channels and on the Undular Jumps of Low Wave Height.') Jl La Houille Blanche, Mar–Apr., pp. 183–185 (in French).

Leutheusser, H.J., and Schiller, E.J. (1975). "Hydraulic Jump in a Rough Channel." Water Power & Dam Construction, Vol. 27, No. 5, pp. 186–191.

Liggett, J.A. (1994). "Fluid Mechanics." McGraw-Hill, New York, USA.

Lighthill, J. (1978). "Waves in Fluids." Cambridge University Press, Cambridge, UK, 504 pages.

Lubin, P., Glockner, S., and Chanson, H. (2010). "Numerical Simulation of a Weak Breaking Tidal Bore." Mechanics Research Communications, Vol. 37, No. 1, pp. 119–121 (DOI: 10.1016/j.mechrescom.2009.09.008).

Moore, R.N. (1888). "Report on the Bore of the Tsien-Tang Kiang." Hydrographic Office, London.

Mouazé, D., Chanson, H., and Simon, B. (2010). "Field Measurements in the Tidal Bore of the Sélune River in the Bay of Mont Saint Michel (September 2010)." Hydraulic Model Report No. CH81/10, School of Civil Engineering, The University of Queensland, Brisbane, Australia, 72 pages

Moule, A.C. (1923). "The Bore on the Ch'ien-T'ang River in China." T'oung Pao, Archives pour servir à l'étude de l'histoire, des langues, la geographie et l'ethnographie de l'Asie Orientale (Chine, Japon, Corée, Indo-Chine, Asie Centrale et Malaisie), Vol. 22, pp. 10–188.

Rayleigh, Lord (1908). "Note on Tidal Bores." Proc. Royal Soc. of London, Series A containing Papers of a Mathematical and Physical Character, Vol. 81, No. 541, pp. 448–449.

Simpson, J.H., Fisher, N.R., and Wiles, P. (2004). "Reynolds Stress and TKE Production in an Estuary with a Tidal Bore." Estuarine, Coastal and Shelf Science, Vol. 60, No. 4, pp. 619–627.

Wolanski, E., Williams, D., Spagnol, S., and Chanson, H. (2004). "Undular Tidal Bore Dynamics in the Daly Estuary, Northern Australia." Estuarine, Coastal and Shelf Science, Vol. 60, No. 4, pp. 629–636 (DOI: 10.1016/j.ecss.2004.03.001).

Bibliography: tidal bore research at the University of Queensland

Reviews

Chanson, H. (2004). "Mixing and Dispersion Role of Tidal Bores." in "Fluvial, Environmental & Coastal Developments in Hydraulic Engineering", Balkema, Leiden, The Netherlands, Proc. Intl Workshop on State-of-the-Art Hydraulic Engineering, 16–19 Feb. 2004, Bari, Italy, M. Mossa, Y. Yasuda and H. Chanson Ed., pp. 223–232.

Chanson, H. (2005). "Mascaret, Aegir, Pororoca, Tidal Bore. Quid ? Où? Quand? Comment? Pourquoi ?" Jl La Houille Blanche, No. 3, pp. 103–114 (in French).

Chanson, H. (2009). "Environmental, Ecological and Cultural Impacts of Tidal Bores, Benaks, Bonos and Burros." Proc. International Workshop on Environmental Hydraulics IWEH09, Theoretical, Experimental and Computational Solutions, Valencia, Spain, 29–30 Oct., P.A. Lopez-Jimenez, V.S. Fuertes-Miquel, P.L. Iglesias-Rey, G. Lopez-Patino, F.J. Martinez-Solano, and G. Palau-Selvador Eds., Invited keynote lecture, 20 pages (CD-ROM).

Chanson, H. (2010). "Tidal Bores, Aegir and Pororoca: the Geophysical Wonders." Proc. of 17th Congress of IAHR Asia and Pacific Division, IAHR-APD, Auckland, New Zealand, 21–24 Feb., B. Melville, G. De Costa, and T. Swann Eds., Invited keynote lecture, 18 pages.

Chanson, H. (2010). "Undular Bores." Proc. Second International Conference on Coastal Zone Engineering and Management (Arabian Coast 2010), November 1–3, 2010, Muscat, Oman, Invited plenary lecture, 12 pages.

Chanson, H. (2011). "Current Knowledge in Tidal bores and their Environmental, Ecological and Cultural Impacts." Environmental Fluid Mechanics, Vol. 11, No. 1, pp. 77–98 (DOI: 10.1007/s10652-009-9160-5)

Theoretical analyses

Chanson, H. (2010). "Undular Tidal Bores: Basic Theory and Free-surface Characteristics." Journal of Hydraulic Engineering, ASCE, Vol. 136, No. 11, pp. 940–944 (DOI: 10.1061/(ASCE)HY.1943-7900.0000264).

Chanson, H. (2012). "Momentum Considerations in Hydraulic Jumps and Bores." Journal of Irrigation and Drainage Engineering, ASCE, Vol. 138 (DOI: 10.1061/(ASCE)IR.1943-4774.0000409). [In Print].

Field observations

Chanson, H. (2001). "Flow Field in a Tidal Bore: a Physical Model." Proc. 29th IAHR Congress, Beijing, China, Theme E, Tsinghua University Press, Beijing, G. LI Ed., pp. 365–373. (CD-ROM, Tsinghua University Press, ISBN 7-900637-10-9.)

Chanson, H. (2004). "Coastal Observations: The Tidal Bore of the Sélune River, Mont Saint Michel Bay, France." Shore & Beach, Vol. 72, No. 4, pp. 14–16.

Chanson, H. (2005). "Tidal Bore Processes in the Baie du Mont Saint Michel (France): Field Observations and Discussion." Proc. 31st Biennial IAHR Congress, Seoul, Korea, B.H. Jun, S.I. Lee, I.W. Seo and G.W. Choi Editors, Theme E.4, Paper 0062, pp. 4037–4046.

Chanson, H. (2008). "Photographic Observations of Tidal Bores (Mascarets) in France." Hydraulic Model Report No. CH71/08, Div. of Civil Engineering, The University of Queensland, Brisbane, Australia, 104 pages, 1 movie and 2 audio files (ISBN 9781864999303).

Chanson, H. (2009). "The Rumble Sound Generated by a Tidal Bore Event in the Baie du Mont Saint Michel." Journal of Acoustical Society of America, Vol. 125, No. 6, pp. 3561–3568 (DOI: 10.1121/1.3124781).

Chanson, H., Lubin, P., Simon, B., and Reungoat, D. (2010). "Turbulence and Sediment Processes in the Tidal Bore of the Garonne River: First Observations." Hydraulic Model Report No. CH79/10, School of Civil Engineering, The University of Queensland, Brisbane, Australia, 97 pages.

Chanson, H. Reungoat, D., Simon, B., and Lubin, P. (2011). "High-Frequency Turbulence and Suspended Sediment Concentration Measurements in the Garonne River Tidal Bore." Estuarine Coastal and Shelf Science, Vol. 95, No. 2–3, pp. 298–306 (DOI 10.1016/j.ecss.2011.09.012)

Mouaze, D., Chanson, H., and Simon, B. (2010). "Field Measurements in the Tidal Bore of the Sélune River in the Bay of Mont Saint Michel (September 2010)." Hydraulic Model Report No. CH81/10, School of Civil Engineering, The University of Queensland, Brisbane, Australia, 72 pages (ISBN 9781742720210).

Simon, B., Lubin, P., Reungoat, D., and Chanson, H. (2011). "Turbulence Measurements in the Garonne River Tidal Bore: First Observations." Proc. 34th IAHR World Congress, Brisbane, Australia, 26 June–1 July, Engineers Australia Publication, Eric Valentine, Colin Apelt, James Ball, Hubert Chanson, Ron Cox, Rob Ettema, George Kuczera, Martin Lambert, Bruce Melville and Jane Sargison Editors, pp. 1141–1148.

Physical studies

Chanson, H. (2001). "Flow Field in a Tidal Bore : a Physical Model." Proc. 29th IAHR Congress, Beijing, China, Theme E, Tsinghua University Press, Beijing, G. LI Ed., pp. 365–373. (CD-ROM, Tsinghua University Press, ISBN 7-900637-10-9.)

Chanson, H. (2005). "Physical Modelling of the Flow Field in an Undular Tidal Bore." Jl of Hyd. Res., IAHR, Vol. 43, No. 3, pp. 234–244.

Chanson, H. (2008). "Turbulence in Positive Surges and Tidal Bores. Effects of Bed Roughness and Adverse Bed Slopes." Hydraulic Model Report No. CH68/08, Div. of Civil Engineering, The University of Queensland, Brisbane, Australia, 121 pages & 5 movie files (ISBN 9781864999198).

Chanson, H. (2009). "An Experimental Study of Tidal Bore Propagation: the Impact of Bridge Piers and Channel Constriction." Hydraulic Model Report No. CH74/08, School of Civil Engineering, The University of Queensland, Brisbane, Australia, 110 pages & 5 movie files (ISBN 9781864999600).

Chanson, H. (2010). "Undular Tidal Bores: Basic Theory and Free-surface Characteristics." Journal of Hydraulic Engineering, ASCE, Vol. 136, No. 11, pp. 940–944 (DOI: 10.1061/(ASCE)HY.1943-7900.0000264).

Chanson, H. (2010). "Unsteady Turbulence in Tidal Bores: Effects of Bed Roughness." Journal of Waterway, Port, Coastal, and Ocean Engineering, ASCE, Vol. 136, No. 5, pp. 247–256 (DOI: 10.1061/(ASCE)WW.1943-5460.0000048).

Chanson, H. (2011). "Undular Tidal Bores: Effect of Channel Constriction and Bridge Piers." Environmental Fluid Mechanics, Vol. 11, No. 4, pp. 385–404 & 4 videos (DOI: 10.1007/s10652-010-9189-5).

Chanson, H. (2011). "Turbulent Shear Stresses in Hydraulic Jumps and Decelerating Surges: An Experimental Study." Earth Surface Processes and Landforms, Vol. 36, No. 2, pp. 180–189 & 2 videos (DOI: 10.1002/esp.2031).

Chanson, H., and Docherty, N.J. (2010). "Unsteady Turbulence in Tidal Bores: Ensemble-Average or VITA?" Proc. 17th Australasian Fluid Mechanics Conference, Auckland, New Zealand, 5–9 Dec., 4 pages.

Chanson, H., and Docherty, N.J. (2012). "Turbulent Velocity Measurements in Open Channel Bores." European Journal of Mechanics B/Fluids, Vol. 31 (DOI 10.1016/j.euromechflu.2011.10.001). [In Print].

Chanson, H., and Tan, K.K. (2010). "Particle Dispersion under Tidal Bores: Application to Sediments and Fish Eggs." Proc. 7th International Conference on Multiphase Flow ICMF 2010, Tampa FL, USA, May 30–June 4, Paper No. 12.7.3, 9 pages (USB Memory Stick).

Chanson, H., and Tan, K.K. (2010). "Turbulent Mixing of Particles under Tidal Bores: an Experimental Analysis." Journal of Hydraulic Research, IAHR, Vol. 48, No. 5, pp. 641–649 (DOI: 10.1080/00221686.2010.512779 (ISSN 0022-1686).

Chanson, H., and Tan, K.K. (2011). "Dispersion of Fish Eggs under Undular and Breaking Tidal Bores." Fluid Dynamics & Materials Processing, Vol. [In Print].

Docherty, N.J., and Chanson, H. (2010). "Characterisation of Unsteady Turbulence in Breaking Tidal Bores including the Effects of Bed Roughness." Hydraulic Model Report No. CH76/10, School of Civil Engineering, The University of Queensland, Brisbane, Australia, 112 pages.

Docherty, N.J., and Chanson, H. (2011). "Unsteady Turbulence Measurements in Breaking Tidal Bores including the Effect of Bed Roughness." Proc. 34th IAHR World Congress, Brisbane, Australia, 26 June–1 July, Engineers Australia Publication, Eric Valentine, Colin Apelt, James Ball, Hubert Chanson, Ron Cox, Rob Ettema, George Kuczera, Martin Lambert, Bruce Melville and Jane Sargison Editors, pp. 1039–1046.

Donnelly, C., and Chanson, H. (2005). "Environmental Impact of Undular Tidal Bores in Tropical Rivers." Environmental Fluid Mechanics, Vol. 5, No. 5, pp. 481–494 (DOI: 10.1007/s10652-005-0711-0).

Gualtieri, C., and Chanson, H. (2011). "Hydrodynamics and Turbulence in Positive Surges. A Comparative Study." Proc. 34th IAHR World Congress, Brisbane, Australia, 26 June–1 July, Engineers Australia Publication, Eric Valentine, Colin Apelt, James Ball, Hubert Chanson, Ron Cox, Rob Ettema, George Kuczera, Martin Lambert, Bruce Melville and Jane Sargison Editors, pp. 1062–1069.

Khezri, N., and Chanson, H. (2011). "Inception of Gravel Bed Motion beneath Tidal Bores: an Experimental Study." Proc. 34th IAHR World Congress, Brisbane, Australia, 26 June–1 July, Engineers Australia Publication, Eric Valentine, Colin Apelt, James Ball, Hubert Chanson, Ron Cox, Rob Ettema, George Kuczera, Martin Lambert, Bruce Melville and Jane Sargison Editors, pp. 1077–1084.

Koch, C., and Chanson, H. (2005). "An Experimental Study of Tidal Bores and Positive Surges: Hydrodynamics and Turbulence of the Bore Front." Report No. CH56/05, Dept. of Civil Engineering, The University of Queensland, Brisbane, Australia, July, 170 pages (ISBN 978-1-86499-824-5).

Koch, C., and Chanson, H. (2008). "Turbulent Mixing beneath an Undular Bore Front." Journal of Coastal Research, Vol. 24, No. 4, pp. 999–1007 (DOI: 10.2112/06-0688.1).

Koch, C., and Chanson, H. (2009). "Turbulence Measurements in Positive Surges and Bores." Journal of Hydraulic Research, IAHR, Vol. 47, No. 1, pp. 29–40 (DOI: 10.3826/jhr.2009.2954).

Numerical modelling

Chanson, H., Lubin, P., and Glockner, S. (2011). "Unsteady Turbulence in a Shock: Physical and Numerical Modelling in Tidal Bores and Hydraulic Jumps." in "Turbulence: Theory, Types and Simulation", Nova Science Publishers, Hauppauge NY, USA. [In Print].

Furuyama, S., and Chanson, H. (2008). "A Numerical Study of Open Channel Flow Hydrodynamics and Turbulence of the Tidal Bore and Dam-Break Flows." Report No. CH66/08, Div. of Civil Engineering, The University of Queensland, Brisbane, Australia, May, 88 pages (ISBN 9781864999068).

Furuyama, S., and Chanson, H. (2010). "A Numerical Solution of a Tidal Bore Flow." Coastal Engineering Journal, Vol. 52, No. 3, pp. 215–234 (DOI: 10.1142/S057856341000218X).

Lubin, P., Glockner, S., and Chanson, H. (2010). "Numerical Simulation of a Weak Breaking Tidal Bore." Mechanics Research Communications, Vol. 37, No. 1, pp. 119–121 (DOI: 10.1016/j.mechrescom.2009.09.008).

Lubin, P., Chanson, H., and Glockner, S. (2010). "Large Eddy Simulation of Turbulence Generated by a Weak Breaking Tidal Bore." Environmental Fluid Mechanics, Vol. (DOI: 10.1007/s10652-009-9165-0).

Reichstetter, M., and Chanson, H. (2011). "Negative Surge in Open Channel: Physical, Numerical and Analytical Modelling." Proc. 34th IAHR World Congress, Brisbane,

Australia, 26 June–1 July, Engineers Australia Publication, Eric Valentine, Colin Apelt, James Ball, Hubert Chanson, Ron Cox, Rob Ettema, George Kuczera, Martin Lambert, Bruce Melville and Jane Sargison Editors, pp. 2306–2313.

Simon, B., Lubin, P., Glockner, S., and Chanson, H. (2011). "Three-Dimensional Numerical Simulation of the Hydrodynamics generated by a Weak Breaking Tidal Bore." Proc. 34th IAHR World Congress, Brisbane, Australia, 26 June–1 July, Engineers Australia Publication, Eric Valentine, Colin Apelt, James Ball, Hubert Chanson, Ron Cox, Rob Ettema, George Kuczera, Martin Lambert, Bruce Melville and Jane Sargison Editors, pp. 1133–1140.

Internet resources

Chanson, H. (2000). "The Tidal Bore of the Seine River, France." Internet resource (Internet address: http://www.uq.edu.au/~e2hchans/mascaret.html).

Chanson, H., (2003). "Free-surface undulations in open channel flows: Undular jumps, Undular bores, Standing waves." Internet resource (Internet address: http://www.uq.edu.au/~e2hchans/undular.html).

Chanson, H. (2004). "Tidal bores, Mascaret, Pororoca. Myths, Fables and Reality !!!." Internet resource (Internet address: http://www.uq.edu.au/~e2hchans/tid_bore.html).

Open access research reprints on tidal bores (Internet address: http://espace.library.uq.edu.au/list/author_id/193/).

Audio-visual resources

Chanson, H. (2005). "Tidal Bore of the Dordogne River (France) on 27 September 2000." Flowvis: the Art of Fluid Dynamics, Australian Institute of Physics' (AIP) "Physics for the Nation" Congress, Australian National University (ANU), School of Art, Canberra, Australia, 31 Jan. to 4 Feb. 2005.

Lespinasse, M. (2004). "La Tribu du Mascaret" ('Surfing the Dordogne') Grand Angle production, France, 30 minutes.

Lespinasse, M. (2005). "Les Fils de la Lune" ('The Children of the Moon') Grand Angle production, France, 50 minutes.

Lespinasse, M. (2008). "Rendez Vous avec le Dragon" ('Appointment with the Dragon') Thalassa, France, 30 minutes.

NHK Japan Broadcasting Corp (1989). "Pororoca: the Backward Flow of the Amazon." Videocassette VHS colour, NHK, Japan, 29 minutes.

Part four
Processes at interfaces of biotic systems

CHAPTER ELEVEN

Transport processes in the soil-vegetation-lower atmosphere system

Dragutin T. Mihailović

Department for Field and Vegetable Crops, Faculty of Agriculture, University of Novi Sad, Novi Sad, Serbia

ABSTRACT

The interaction of the land surface and the atmosphere may be summarised as follows: interaction of vegetation with radiation, evaporation from bare soil, evapotranspiration which includes transpiration and evaporation of intercepted precipitation and dew, conduction of soil water through the vegetation layer, vertical movement water in the soil, run-off, heat conduction in the soil, momentum transport, effects of snow presence and freezing or melting of soil moisture. Consequently, the processes parametrized in the land surface schemes can be divided into three parts: thermal and hydraulic processes, bare soil transfer processes and canopy transport processes. The chapter shortly describes these processes through a land surface scheme capturing the main processes in the soil-vegetation-lower atmosphere system. The biophysical processes in vegetation are elaborated using so-called "sandwich" representation where the vegetation is treated as a block of constant-density porous material "sandwiched" between two constant-stress layers with an upper boundary (the height of the canopy top) and a lower boundary (the height of the canopy bottom). For description of the transport processes in the soil, the three-soil layer approach is used. The chapter also includes a detailed description and explanation of governing equations, the representation of energy fluxes and radiation, the parameterization of aerodynamic characteristics, resistances and model hydrology. A special attention will be devoted to consideration of "K"-theory within and above canopy.

11.1 FOREWORD

The land surface is important in atmospheric modelling as it controls a number of key processes. The brightness of the surface (its albedo) determines how much of the incoming solar radiation is absorbed and how much is reflected. The total absorbed radiation is partitioned by the surface into land-atmosphere fluxes of heat and moisture, and a ground heat flux which may heat the soil or melt any lying snow. The nature of this partitioning affects the near surface conditions (for example, freely evaporating surfaces are cooler than dry surfaces) and also atmospheric processes such as cumulus convection. Surface flux partitioning is dependent on both the land cover and its hydrological state.

As experience with numerical modelling of atmospheric processes has progressed over the decades, the atmospheric modelling community has come to recognise that various aspects of the atmosphere–ecosystem–ocean system, which once were thought to play a relatively minor role, are actually very important in atmospheric circulations. Ecosystem, soil processes and their effect on the atmosphere are certainly in this category. Most mesoscale and global atmospheric models of 20 years ago either ignored or treated in an extremely simple manner interactions of the atmosphere with underlying soil and vegetated surfaces. Now, field and modelling studies have demonstrated that these interactions are extremely important in both long-term climate simulations and short-term weather forecasting applications (Dickinson 1995; Pielke *et al.,* 1998). Moreover, recent numerical studies strongly suggest that land-use change may cause significant weather, climate, and ecosystem change (Chase *et al.,* 1998; Baron *et al.,* 1998; Stohlgren *et al.,* 1998; Pielke *et al.,* 1999). Because the role of these interactions has become recognised, parameterizations of vegetation and soil processes have progressively become more sophisticated over the years in order to treat the complexities of the physical system. Soil–vegetation–atmosphere transfer (SVAT) schemes employed in general circulation, mesoscale, and small-scale atmospheric numerical models have become increasingly sophisticated (Deardorff 1978; Avissar *et al.,* 1985; Dickinson *et al.,* 1986; Sellers *et al.,* 1986; Noilhan and Planton 1989; Mihailović *et al.,* 1993; Acs 1994; Bosilovich and Sun 1995; Viterbo and Beljaars 1995; Pleim and Xiu 1995; Cox *et al.,* 1999; Walko *et al.,* 2000; Mihailović *et al.,* 2004). Also, our ability to sense characteristics of the land surface remotely has improved dramatically, enabling much better data to be used as inputs to the more sophisticated parameterizations (Loveland *et al.,* 1991; Lee *et al.,* 1995).

The Land Air Parameterization Scheme (LAPS) is one such SVAT scheme that has been developed at University of Novi Sad to be a component of any environmental model for agricultural purposes. The current version of LAPS is a representation of surface features that include vegetation, soil, lakes and oceans, and their influence on each other and on the atmosphere. LAPS includes prognostic equations for soil temperature and moisture for multiple layers, vegetation temperature, and surface water including dew and intercepted rainfall, and temperature and water vapour mixing ratio of canopy air. Exchange terms in these prognostic equations include turbulent exchange, heat conduction, and water diffusion and percolation in the soil, long-wave and short-wave radiative transfer, transpiration, and precipitation. This chapter provides a description of the current version of LAPS.

11.2   SCHEME STRUCTURE AND BASIC EQUATIONS

The net radiation absorbed by the canopy and soil is assumed to be partitioned into sensible heat, latent heat, and storage terms, as

$$R_{ng} = \lambda E_g + H_g + C_g \frac{\partial T_g}{\partial t} \tag{11.1}$$

$$R_{nf} = \lambda E_f + H_f + C_f \frac{\partial T_f}{\partial t} \tag{11.2}$$

where R_n is absorbed net radiation $[MT^{-3}]$, λ is latent heat of vaporisation $[L^2T^{-2}]$, E is evapotranspiration rate $[ML^{-2}T^{-1}]$, H is sensible heat flux $[MT^{-3}]$, C is heat capacity $[M\theta^{-1}T^{-2}]$, T is surface (canopy or soil) temperature $[\theta]$. The subscripts f, g refer to the canopy and soil respectively. The deep soil temperature $[\theta]$, T_d, is calculated from the equation (Mihailović *et al.,* 1999)

$$R_{ng} = \lambda E_g + H_g + \sqrt{365\pi} \frac{C_g}{2} \frac{\partial T_d}{\partial t} \tag{11.3}$$

The prognostic equations for the water stored on the canopy $[L]$, w_f, is

$$\frac{\partial w_f}{\partial t} = P_f - E_{wf}/\rho \tag{11.4}$$

where ρ is water density [ML^{-3}], P_f is water amount retained on the canopy [LT^{-1}], E_{wf} the evaporation rate of water from the wetted fraction of canopy [$ML^{-2}T^{-1}$]. When the conditions for dew formation are satisfied, the condensed moisture is added to the interception store, w_f. The parameterization of the soil content is based on the concept of the three-layer model (Mihailović, 1996). The governing equations take the form

$$\frac{\partial \vartheta_1}{\partial t} = \frac{1}{D_1}\left\{P_1 - F_{1,2} - \frac{E_g + E_{tf,1}}{\rho} - R_0 - R_1\right\} \tag{11.5}$$

$$\frac{\partial \vartheta_2}{\partial t} = \frac{1}{D_2}\left\{F_{1,2} - F_{2,3} - \frac{E_{tf,2}}{\rho} - R_2\right\} \tag{11.6}$$

$$\frac{\partial \vartheta_3}{\partial t} = \frac{1}{D_3}\left\{F_{2,3} - F_3 - R_3\right\} \tag{11.7}$$

where ϑ_i is volumetric soil water content [L^3L^{-3}] in the ith layer, P_1 is infiltration rate of precipitation into the upper soil moisture store [LT^{-1}]; D_i is thickness of the ith soil layer [L], $F_{i,i+1}$ is water flux between i and $i+1$ soil layer [LT^{-1}], F_3 is gravitational drainage flux from recharge soil water store [LT^{-1}], $E_{tf,1}$ and $E_{tf,2}$ are canopy extraction of soil moisture by transpiration from the rooted first and second soil layers [$ML^{-2}T^{-1}$] respectively; R_0 is surface run-off [LT^{-1}]; and R_i is subsurface run-off from the ith soil layer [LT^{-1}].

Eqs. (1)–(3) are solved using an implicit backward method, i.e.,

$$T_g^{n+1} = T_g^n + \frac{\Gamma_f^n\left(\frac{\partial \Gamma_g}{\partial T_f}\right)^n + \Gamma_g^n\left[\frac{C_f}{\Delta t} - \left(\frac{\partial \Gamma_f}{\partial T_f}\right)^n\right]}{\left(\frac{\partial \Gamma_f}{\partial T_g}\right)^n\left(\frac{\partial \Gamma_g}{\partial T_f}\right)^n - \left[\frac{C_f}{\Delta t} - \left(\frac{\partial \Gamma_f}{\partial T_f}\right)^n\right]\left[\frac{C_g}{\Delta t} - \left(\frac{\partial \Gamma_g}{\partial T_g}\right)^n\right]} \tag{11.8}$$

$$T_f^{n+1} = T_f^n + \frac{\Gamma_g^n\left(\frac{\partial \Gamma_g}{\partial T_f}\right)^n + \Gamma_f^n\left[\frac{C_g}{\Delta t} - \left(\frac{\partial \Gamma_g}{\partial T_g}\right)^n\right]}{\left(\frac{\partial \Gamma_f}{\partial T_g}\right)^n\left(\frac{\partial \Gamma_g}{\partial T_f}\right)^n - \left[\frac{C_f}{\Delta t} - \left(\frac{\partial \Gamma_f}{\partial T_f}\right)^n\right]\left[\frac{C_g}{\Delta t} - \left(\frac{\partial \Gamma_g}{\partial T_g}\right)^n\right]} \tag{11.9}$$

$$T_d^{n+1} = T_d^n + \frac{\Gamma_g}{\frac{\sqrt{365\pi}}{2\Delta t}C_g - \left(\frac{\partial \Gamma_g}{\partial T_d}\right)} \tag{11.10}$$

where: $\Gamma_f = R_{nf} - \lambda E_f - H_f$, $\Gamma_g = R_{ng} - \lambda E_g - H_g$, and Δt is time step. Eqs. (11.4)–(11.6) are solved using an explicit time scheme.

11.3 REPRESENTATION OF ENERGY FLUXES

Our treatment of the energy fluxes may be classified as the so-called "resistance" representation. Schematic diagram of the Land-Air Parameterization Scheme (LAPS) is shown in Fig. 11.1. The transfer pathways for latent sensible heat fluxes are shown on the left- and right-hand sides of the diagram respectively. The fluxes of sensible and latent heat from the soil and canopy are represented by electrical analogue models in which the fluxes are

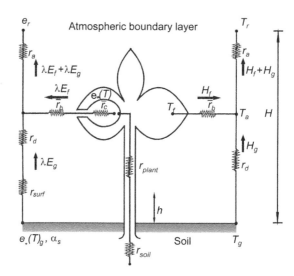

Figure 11.1. Schematic diagram of the Land-Air Parameterization Scheme (LAPS). The transfer pathways for latent sensible heat fluxes are shown on the left- and right-hand sides of the diagram respectively.

proportional to potential differences (in temperature or vapour pressure) and inversely proportional to resistances, which are equivalent to the inverse integrals of conductances over a specified length scale. The fluxes in Eqs. (11.1)–(11.3) are parametrized as follows.

The latent heat flux from canopy vegetation to canopy air space is given by

$$\lambda E_f = \frac{\rho_p c_p}{\gamma}\left[e_*(T_f) - e_a\right]\left(\frac{w_w}{\bar{r}_b} + \frac{1 - w_w}{\bar{r}_b + \bar{r}_c}\right),\tag{11.11}$$

where ρ_p, c_p are the density and specific heat of air [ML^{-3}, $L^2T^{-2}\theta^{-1}$], γ is the psychrometric constant $\times 10^2$[$ML^{-1}T^{-2}\theta^{-1}$], $e_*(T_f)$ is saturated vapour pressure at temperature $T_f \times 10^2$[$ML^{-1}T^{-2}$]; e_a is canopy air space vapour pressure [$ML^{-1}T^{-2}$], w_w is wetted fraction of canopy, $\bar{r}_b$ is bulk canopy boundary layer resistance [TL^{-1}] and $\bar{r}_c$ is bulk canopy stomatal resistance [TL^{-1}].

The evaporation rate E_{wf} from the wetted portion of canopy, with wetted fractions denoted by w_w according to Eq. (11.11) is

$$\lambda E_{wf} = \frac{\rho_p c_p}{\gamma}\left[e_*\left(T_f\right) - e_a\right]\frac{w_w}{\bar{r}_b}.\tag{11.12}$$

The fraction of the foliage that is wet, w_w, is parametrized according to Deardorff (1978). Transpiration occurs only from dry leaf and it is only outwards. This physiological process is parametrized with the equation

$$\lambda E_{tf} = \frac{\rho_p c_p}{\gamma}\left[e_*\left(T_f\right) - e_a\right]\frac{1 - w_w}{\bar{r}_b + \bar{r}_c}\tag{11.13}$$

where E_{tf} is the transpiration rate from foliage [$ML^{-2}T^{-1}$]. Dew formation occurs when $e_*(T_f) \leq e_a$. In that case the condensed moisture is added to the surface interception store, w_f. The transpiration rate is zero under this condition.

The latent heat flux from soil surface is parametrized as

$$\lambda E_g = \frac{\rho_p c_p}{\gamma} \frac{\alpha_s e_*(T_g) - e_a}{r_{surf} + r_d}(1 - \sigma_c) \tag{11.14}$$

where α_s is a factor to correct for soil dryness (Mihailović *et al.*, 1995), $e_*(T_g)$ is saturated vapour pressure at temperature T_g [ML^{-1}T^{-2}]; r_{surf} is soil surface resistance [TL^{-1}], r_d is aerodynamic resistance between soil surface and canopy air space [TL^{-1}], and σ_c is vegetation cover in fractional units.

The sensible heat fluxes from canopy, H_f, and soil surface H_g are parametrized as

$$H_f = \frac{2(T_f - T_a)}{\bar{r}_b}\rho_p c_p \tag{11.15}$$

$$H_g = \frac{(T_g - T_a)}{r_d}\rho_p c_p \tag{11.16}$$

where T_a is canopy air space temperature [θ].

Air within the canopy has negligible heat capacity, so the sensible heat flux from the canopy, H_f, and from the soil surface, H_g, must be balanced by the sensible heat flux to the atmosphere, H_t

$$H_t = H_g + H_f = \frac{(T_a - T_r)}{r_a}\rho_p c_p \tag{11.17}$$

where r_a is aerodynamic resistance [TL^{-1}], and T_r is air temperature at the reference height $z_r[\theta]$. Similarly the canopy air is assumed to have zero capacity for water storage so that the latent heat flux from canopy air space to reference height in the atmospheric boundary layer, λE_t, balances the latent heat flux from canopy vegetation to canopy air space, λE_f, and the latent heat flux from soil surface to the canopy air space, λE_g

$$\lambda E_t = \lambda E_g + \lambda E_f = \frac{\rho_p c_p}{\gamma} \frac{(e_a - e_r)}{r_a} \tag{11.18}$$

where e_r is vapour pressure of the air at reference height [ML^{-1}T^{-2}] within the atmospheric boundary layer. The canopy air space temperature, T_a, and canopy air space vapour pressure, e_a, are determined diagnostically from Eqs. (11.17) and (11.18), i.e.,

$$T_a = \frac{\dfrac{2T_f}{\bar{r}_b} + \dfrac{T_g}{r_d} + \dfrac{T_r}{r_a}}{\dfrac{2}{\bar{r}_b} + \dfrac{1}{r_d} + \dfrac{1}{r_a}} \tag{11.19}$$

and

$$e_a = \frac{\dfrac{1}{r_a} + \dfrac{\alpha_s e_*(T_g)(1 - \sigma_c)}{r_{surf} + r_d} + e_*(T_f)\left[\dfrac{w_w}{\bar{r}_b} + \dfrac{1 - w_w}{\bar{r}_b + \bar{r}_c}\right]}{\dfrac{1}{r_a} + \dfrac{1 - \sigma_c}{r_{surf} + r_d} + \left[\dfrac{w_w}{\bar{r}_b} + \dfrac{1 - w_w}{\bar{r}_b + \bar{r}_c}\right]} \tag{11.20}$$

11.4 PARAMETERIZATION OF RADIATION

The net radiation absorbed by the canopy, R_{nf}, and the soil surface, R_{ng}, $[MT^{-3}]$ is calculated as a sum of short- and long wave radiative flux,

$$R_{nf} = R_f^s + R_f^l \tag{11.21}$$

and

$$R_{ng} = R_g^s + R_g^l \tag{11.22}$$

The short-wave radiation absorbed by the canopy, R_f^s, and the soil surface, R_g^s, $[MT^{-3}]$ is

$$R_f^s = R_o^s(\sigma_f - \alpha_f)[1 + (1 - \sigma_f)\alpha_g] \tag{11.23}$$

and

$$R_g^s = R_o^s(1 - \sigma_f)(1 + \alpha_g + \alpha_f\alpha_g) \tag{11.24}$$

where R_o^s is incident downward-directed short-wave flux $[MT^{-3}]$, assumed to be known as the forcing variable, σ_f is the fractional cover of vegetation and α_g and α_f are soil-surface albedo and canopy albedo respectively. The variability of ground albedo with soil wetness is parametrized in accordance with Idso *et al.* (1975). There is no distinction between direct and diffuse radiation and it is assumed that albedo does not vary with zenith angle. Both short-wave and long-wave radiation are reflected once between the soil surface and canopy.

The long-wave radiative fluxes absorbed by the canopy, R_f^l, and the soil surface, R_g^l, $[MT^{-3}]$ are

$$R_f^l = R_o^l\sigma_f\varepsilon_f - 2\sigma_f\varepsilon_f\sigma_B + \sigma_f\varepsilon_f[R_o^l\sigma_B(1 - \varepsilon_f)T_f^4 + \varepsilon_g\sigma_B T_g^4] \tag{11.25}$$

and

$$R_g^l = \varepsilon_g[R_o^l(1 - \sigma_f) + \varepsilon_f\sigma_f\sigma_B T_f^4 + \sigma_f\varepsilon_g(1 - \varepsilon_f)\sigma_B T_g^4 - \sigma_B T_g^4] \tag{11.26}$$

where σ_B is the Stefan-Boltzman constant $[MT^{-3}\theta^{-4}]$, ε_f and ε_g are emissivities of the canopy and the soil surface respectively, and R_o^l the incident downward long-wave radiation prescribed as the forcing variable.

11.5 PARAMETERIZATION OF RESISTANCES

11.5.1 Aerodynamic resistances

The aerodynamic resistances r_a, r_b and r_d are described as

$$r_a = \int_{h_a}^{H} \frac{1}{K_s}dz + \int_{H}^{z_r} \frac{1}{K_s}dz, \tag{11.27}$$

$$r_d = \int_{z_g}^{h} \frac{1}{K_s} dz + \int_{h}^{h_a} \frac{1}{K_s} dz, \tag{11.28}$$

$$\frac{1}{\bar{r}_b} = \int_{h_a}^{H} \frac{\bar{L}_d \sqrt{u(z)}}{C_s P_s} dz, \tag{11.29}$$

where H is the canopy height [L]; K_s is turbulent transfer coefficient within and above the canopy $[L^2 T^{-1}]$ in the intervals (h_a, H) and (H, z_r) respectively; z_g is effective ground roughness length [L]; h[L] is the canopy bottom height (the height of the base of the canopy, see Fig. 11.2); $\bar{L}_d$ is the area-averaged stem and leaf area density (also called canopy density), which is related to leaf area index (LAI) as $LAI = \bar{L}_d(H - h)$; $u(h)$ is the wind speed; C_s the transfer coefficient $[L^{-1/2} T^{1/2}]$ and P_s the leaf shelter factor. According to Sellers *et al.* (1986), the position of the canopy source height, h_a, can be estimated by obtaining the centre of gravity of the $1/\bar{r}_b$ integral. Thus,

$$\int_{h}^{h_a} \frac{\bar{L}_d}{r_b} dz = \int_{h_a}^{H} \frac{\bar{L}_d}{r_b} dz = \frac{1}{2} \int_{h}^{H} \frac{\bar{L}_d}{r_b} dz = \frac{1}{2\bar{r}_b}. \tag{11.30}$$

We may obtain h_a by successive estimations until the foregoing equality is reached.

The wind speed above the canopy $u(z)$ is considered as

$$u(z) = \frac{u_*}{\kappa} \left[\ln \frac{z - d}{z_0} - \psi_m (z/L) \right], \tag{11.31}$$

where u_* is friction velocity $[LT^{-1}]$; κ is the Von Kármán constant, z_0 roughness length over the non-vegetated surface, $\psi_m(z/L)$ the stability function for momentum and L

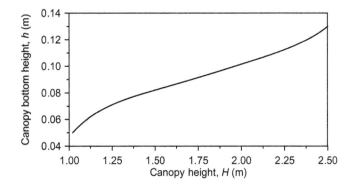

Figure 11.2. Calculated values of the canopy bottom height (h) as a function of the canopy height (H) for tall grass vegetation. The fitting curve is drawn using data from Dubov *et al.* (1978), Sellers and Dorman (1987), Mihailović and Kallos (1997), and Mihailović *et al.* (2000).

Monin-Obuhkov length. The function $\psi_m(z/L)$ is given for stable conditions $(z/L > 0)$ by $4.7z/L$ and for unstable $(z/L < 0)$ by

$$\psi_m(z/L) = -2\ln\left[\frac{(1+x)}{2}\right] - \ln\left[\frac{(1+x^2)}{2}\right] + 2\tan^{-1}(x) - \frac{\pi}{2} \tag{11.32}$$

where $x = [1 - 15z/L]^{1/4}$ (Paulson, 1970). For wind profile within short- and tall-grass canopies we used a form that approximates the wind profile within the tall-grass canopy fairly well (Brunet *et al.*, 1994; Mihailović *et al.*, 2004), i.e.,

$$u(z) = u(H)\exp\left[-\frac{1}{2}\beta\left(1 - \frac{z}{H}\right)\right], \tag{11.33}$$

where $u(H)$ is the wind speed at the canopy height $[LT^{-1}]$; and β is extinction parameter defined as

$$\beta^2 = \frac{2C_d\bar{L}_d(H-h)H}{\sigma}. \tag{11.34}$$

According to Mihailović *et al.* (2004), the value of the scaling length, σ, is defined as

$$\sigma = \frac{2C_{dg}^2 H}{C_d\bar{L}_d(H-h)}, \tag{11.35}$$

where C_{dg} is the leaf drag coefficient estimated from the size of the roughness elements of the ground (Sellers *et al.*, 1986), i.e.,

$$C_{dg} = \frac{\kappa^2}{\left[\ln\dfrac{h}{z_g}\right]^2}. \tag{11.36}$$

In Eq. (11.23) z_g is the effective roughness length. Beneath the canopy bottom height the wind speed follows a classical logarithmic profile in the form

$$u(z) = \frac{u(H)\exp\left[-\dfrac{1}{2}\beta\left(1 - \dfrac{h}{H}\right)\right]}{\ln\dfrac{h}{z_g}}\ln\frac{z}{z_g}. \tag{11.37}$$

Bearing in mind the aforementioned parameterization, the three aerodynamic resistances, r_a, r_b, and r_d, and the canopy bottom height h_a are calculated following Mihailović *et al.* (2004)

$$r_a = \frac{1}{u_*}\left\{\frac{2\kappa H}{\sigma\beta\ln\dfrac{H-d}{z_0}}\left[\exp\left[\frac{1}{2}\beta\left(1 - \frac{h_a}{H}\right)\right] - 1\right] + \frac{1}{k}\ln\frac{z_r - d}{H - d}\right\}, \tag{11.38}$$

$$r_b = \frac{1}{\sqrt{u_*}} \frac{\beta C_s P_s \sqrt{k}}{4H\bar{L}_d \sqrt{\ln \frac{H-d}{z_0}} \left[1 - \exp\left[-\frac{1}{4}\beta\left(1 - \frac{h_a}{H} \right) \right] \right]}, \tag{11.39}$$

$$r_d = \frac{1}{u_*} \left\{ \frac{2\kappa H}{\sigma\beta \ln \frac{H-d}{z_0}} \left[\exp\left[\frac{1}{2}\beta\left(1 - \frac{h}{H} \right) \right] - \exp\left[\frac{1}{2}\beta\left(1 - \frac{h_a}{H} \right) \right] - 1 \right] \right.$$

$$\left. + \frac{\exp\left[\frac{1}{2}\beta\left(1 - \frac{h}{H} \right) \right]}{\kappa \ln \frac{H-d}{z_0}} \ln^2 \frac{h}{z_g} \right\}, \tag{11.40}$$

$$h_a = H \left\{ 1 + \frac{4}{\beta} \ln \frac{1 + 2\exp\left[-\frac{1}{4}\beta\left(1 - \frac{h}{H} \right) \right]}{3} \right\}. \tag{11.41}$$

For the forest canopy the wind profile is calculated from the differential equation (Mihailović et al., 2004)

$$\frac{d}{dz}\left(K_s \frac{du}{dz} \right) = \sigma_c \frac{C_d \bar{L}_d(H - h)}{H} u^2 \tag{11.42}$$

describing the wind profile within a canopy architecture that is considered as a block of constant-density porous material placed between two heights, H and h (Sellers *et al.*, 1986; Mihailović and Kallos 1997). In this equation z is the vertical coordinate. In the case of dense vegetation ($\sigma_c = 1$), Eq. (11.42) reduces to the well-known equation for the dense vegetation. Otherwise, when $\sigma_c = 0$, Eq. (11.42) leads, by a proper choice of integration constant, to the wind profile over a bare soil. We can use Eq. (11.42) for calculating the wind speed within a vegetation canopy after we assume a functional form of K_s as it usually done. However, inadequacy of this approach lies in the fact that the behaviour of K_s must be given *a priori*, i.e. presupposed by experience (Mihailović *et al.*, 2006). After taking the derivative of Eq. (11.42) over z, we obtain a differential equation of the first order and first degree, where K_s is an unknown function, i.e.,

$$\frac{du}{dz}\frac{dK_s}{dz} + \frac{d^2u}{dz^2}K_s = \sigma_c \frac{C_d \bar{L}_d(H - h)}{H} u^2. \tag{11.43}$$

Solution to this equation can be found if the wind speed is treated as a linear combination of two terms, expressing behaviour of the wind speed over dense and sparse vegetation. Thus,

$$u(z) = \sigma_c u(H) \exp\left[-\frac{1}{2}\alpha\left(1 - \frac{z}{H} \right) \right] + (1 - \sigma_c)\frac{u_*}{\kappa}\left[\ln \frac{z}{z_b} - \psi_m(z/L) \right], \tag{11.44}$$

where α is an unknown constant to be determined, $u(H)$ the wind speed at the canopy height, u_* the friction velocity, k the Von Kármán constant, z_b the roughness length over the non-vegetated surface, $\psi_m(z/L)$ the stability function and L Monin-Obuhkov length (Paulson, 1970). The function $\psi_m(z/L)$ is given for stable conditions ($z/L > 0$) by $\psi_m(z/L) = 4.7z/L$ and for unstable ($z/L < 0$) by

$$\psi_m(z/L) = -2\ln\left[\frac{(1+x)}{2}\right] - \ln\left[\frac{(1+x^2)}{2}\right] + 2\tan^{-1}(x) - \frac{\pi}{2} \tag{11.45}$$

where $x = [1 - 15z/L]^{1/4}$. The first term in the expression (11.44) is used to approximate the wind profile within the vegetation canopy (Brunet *et al.*, 1994; Mihailović *et al.*, 2004), while the second term simulates the shape of wind profile above bare soil. After we introduce (11.44) into Eq. (11.43), and rearrange, we reach

$$\frac{dK_m}{dz} + a(z)K_m = b(z), \tag{11.46}$$

where

$$a(z) = \frac{\dfrac{1}{4H^2}\alpha^2\sigma_c u(H)e^{-\frac{1}{2}\alpha\left(1-\frac{z}{h_c}\right)} + (1-\sigma_c)\dfrac{u_*}{\kappa}\left[-\dfrac{1}{z^2} + \psi''_m(z/L)\right]}{\dfrac{1}{2H}\alpha\sigma_c u(H)e^{-\frac{1}{2}\alpha\left(1-\frac{z}{h_c}\right)} + (1-\sigma_c)\dfrac{u_*}{\kappa}\left[\dfrac{1}{z} + \psi'_m(z/L)\right]} \tag{11.47}$$

and

$$b(z) = \left[\sigma_c u(H)e^{-\frac{1}{2}\alpha\left(1-\frac{z}{h_c}\right)} + (1-\sigma_c)\dfrac{u_*}{\kappa}\left[\ln\dfrac{z}{z_b} + \psi_m(z/L)\right]\right]^2$$

$$\times \frac{\sigma_c\dfrac{C_d\bar{L}_d(H-h)}{H}}{\dfrac{1}{2H}\alpha\sigma_c u(H)e^{-\frac{1}{2}\alpha\left(1-\frac{z}{h_c}\right)} + (1-\sigma_c)\dfrac{u_*}{\kappa}\left[\dfrac{1}{z} + \psi'_m(z/L)\right]}, \tag{11.48}$$

with $\psi'_m(z/L) = d\psi_m(z/L)/dz$ and $\psi''_m(z/L) = d^2\psi_m(z/L)/dz^2$.

It is interesting to analyse the nature of the solution, K_s, of the Eq. (11.46) with the initial condition defined as $K_s(z_I) = K_s^0 > 0$, where z_I is some certain height within the canopy: (i) the solution is unique and defined over the interval $[z_I, \infty]$, that follows from the fact that the functions $a(z)$ and $b(z)$ are defined and continuous over the interval indicated; (ii) the solution is positive, that comes from the analysis of the field of directions of the given equation or more precisely due to $b(z) > 0$ and (iii) the solution is stable that can be seen from the following analysis. When $z \to \infty$ we have $a(z) \approx \alpha/(2H)$ and $b(z) \approx B\exp[\alpha z/(2H)]$. Now, Eq. (11.46) takes the form

$$\frac{dK_c}{dz} + \frac{\alpha}{2H}K_c = Be^{\alpha z/2h_c}, \tag{11.49}$$

where

$$B = \frac{2\sigma_c^2 u^2(H)C_d\bar{L}_d(H-h)}{\alpha H}. \tag{11.50}$$

The particular solution of this equation has the form $A \exp[\alpha z/(2H)]$, where A is a constant, which can be obtained after substituting the particular solution in Eq. (11.49). If we follow this procedure we get $A = BH/\alpha$. So, in this case, i.e., $z \to \infty$, the solution of Eq. (11.49) is asymptotically stable, it behaves as $A \exp[\alpha z/(2H)]$ for any given A. For the fixed α, Eq. (11.49) can be solved using the finite-difference scheme

$$K_m^{n-1} = K_m^n - \Delta z \{b^n(z) - a^n(z)K_m^n\}, \tag{11.51}$$

where n is the number of the spatial step in the numerical calculating on the interval $[H, h]$, while Δz is the grid size defined as $\Delta z = (H - h)/N$, where N is a number indicating an upper limit in number of grid size used. The calculation of the turbulent transfer coefficient for momentum starts from the canopy top with a boundary condition defined as

$$K_s^N(h_c) = \kappa^2 u(h_c) \left[\frac{\sigma_c(h_c - d)}{\ln \dfrac{h_c - d}{z_0}} + \frac{(1 - \sigma_c)h_c}{\ln \dfrac{h_c}{z_b}} \right] \tag{11.52}$$

where d is the displacement height while z_0 is the canopy roughness length calculated according to Mihailović *et al.* (1999). The procedure then goes backwards down to the canopy bottom height, h, which is defined according to Mihailović *et al.* (2004). To obtain parameter α we use an iterative procedure that does not end until the condition

$$\left| \sum_{i=1}^{N} u_i^{m+1} - \sum_{i=1}^{N} u_i^m \right| < \mu \tag{11.53}$$

is reached, where m is a number of iteration while μ is less then 0.001. Having this parameter we can calculate the wind profile on the interval $[H, h]$ according to Eq. (11.43). Beneath the canopy bottom height, the wind profile has the logarithmic shape (Sellers *et al.*, 1986; Mihailović *et al.*, 2004), i.e.,

$$u(z) = u(H) \left[\frac{\sigma_c e^{-\frac{1}{2}\alpha\left(1 - \frac{h}{H}\right)}}{\ln \dfrac{h}{z_b}} + \frac{1 - \sigma_c}{\ln \dfrac{H}{z_b}} \right] \ln \frac{z}{z_b}. \tag{11.54}$$

11.5.2 Surface, root and plant resistances

The resistances to the transport of water vapour from within the canopy and upper soil layer to the adjacent exterior air are defined as the bulk canopy stomatal resistance, $\bar{r}_c$, and soil surface resistance, r_{surf}, respectively. Combining dependence of $\bar{r}_c$ on solar radiation, air temperature, atmospheric water vapour pressure deficit and water stress (Jarvis, 1976; Dickinson *et al.*, 1986) is parametrized as

$$\bar{r}_c = \frac{r_{s\,min}}{LAI} \frac{1 + \dfrac{1.1\langle F_f \rangle}{R_0 LAI}}{\dfrac{1.1\langle F_f \rangle}{R_0 LAI} + \dfrac{r_{s\,min}}{r_{s\,max}}} [1.0 - 0.0016(298 - T_r)^2]^{-1}$$

$$\left\{1 - \eta\left[e_*\left(T_f\right) - e_r\right]\right\}^{-1}\Phi_2^{-1} \tag{11.55}$$

where $r_{s\,min}$, $r_{s\,max}$ are the minimum and maximum of stomatal resistance $[TL^{-1}]$; R_0 is limit value of 100 $[MT^{-3}]$ for canopies; and η the canopy-dependent empirical parameter that is equal to $0.025 \times 10^2 [M^{-1}LT^2]$. In this model the value of 5000 $[TL^{-1}]$ for $r_{s\,max}$ is used. The factor Φ_2 takes into account the effect of water stress on the stomatal resistance and is parametrized following Mihailović and Kallos (1997), i.e.

$$\Phi_2 = \begin{cases} 1 & \vartheta_a > \vartheta_{fc} \\ 1 - \left(\dfrac{\vartheta_{wil}}{\vartheta_a}\right)^{1.5} & \vartheta_{wil} \le \vartheta_a \le \vartheta_{fc} \\ 0 & \vartheta_a < \vartheta_{wil} \end{cases} \tag{11.56}$$

where ϑ_a is the mean volumetric soil water content in the first and second soil layers $[L^3L^{-3}]$; ϑ_{wil} is volumetric soil water content at wilting point $[L^3L^{-3}]$; and ϑ_{fc} volumetric soil water content at field capacity $[L^3L^{-3}]$.

The soil surface resistance, r_{surf}, is parametrized using the empirical expression given by Sun (1982), i.e.,

$$r_{surf} = d_1 + d_2\langle\vartheta_1\rangle^{-d_3} \tag{11.57}$$

where d_1, d_2 $[TL^{-1}]$ and d_3 are empirical constants (Mihailović, 2003), while ϑ_1 is the top layer volumetric soil water content $[L^3L^{-3}]$.

The leaf water potential ψ_l $[L]$ describing the water transfer pathway from root zone to leaf is calculated following Van der Honert (1948),

$$\psi_l = \frac{\psi_r - z_t - E_{tf}(r_{plant} + r_{soil})}{\rho} \tag{11.58}$$

where ψ_r is soil moisture potential in the root zone $[L]$, z_t is height of the transpiration source $[L]$ that is equal to canopy source height, r_{plant} is plant resistance $[T]$ imposed by the plant vascular system prescribed as a variable (Mihailović, 2003), r_{soil} is resistance of the soil and root system $[T]$, and ρ is water density $[ML^{-3}]$.

The soil water potential in the root zone, ψ_r, is parametrized as an average term obtained by summing the weighted soil water potentials of the soil layers from the surface to the rooting depth $[L]$, z_d, i.e.

$$\psi_r = \frac{\sum\limits_{0}^{z_d} \psi_i D_i}{z_d} \tag{11.59}$$

where ψ_i is soil water potential of the ith soil layer $[L]$. The soil water potential $[L]$, ψ_i, is parameterised as it is usually done, after Clapp and Hornberger (1978),

$$\psi_i = \psi_s \left(\frac{\vartheta_i}{\vartheta_s}\right)^{-B} \tag{11.60}$$

where ψ_s is soil water potential at saturation $[L]$, ϑ_i is volumetric soil moisture content of the ith soil layer $[L^3L^{-3}]$, ϑ_s is its value at saturation while B is soil type constant. The depth-averaged resistance r_{soil} to water flow from soil to roots, is parametrized according to Federer (1979)

$$r_{soil} = z_d \left(\frac{R_r}{D_d} + \frac{\alpha_j}{K_r} \right) \tag{11.61}$$

where α_j is parametrized as

$$\alpha_j = \{V_r - 3 - 2\ln[V_r/(1 - V_r)]\}/(8\pi D_d) \tag{11.62}$$

where R_r is resistance per unit root length $[\mathrm{TL}^{-1}]$, D_d is root density $[\mathrm{L}^3\mathrm{L}^{-3}]$, V_r is volume of root per unit volume of soil $[\mathrm{L}^3\mathrm{L}^{-3}]$, and K_r is mean soil hydraulic conductivity in the root zone $[\mathrm{LT}^{-1}]$ expressed as function of ψ_r

$$K_r = K_c \left(\frac{\psi_s}{\psi_r} \right)^{(2B+3)/B} \tag{11.63}$$

where K_c is saturated hydraulic conductivity $[\mathrm{LT}^{-1}]$.

11.6 PARAMETERIZATION OF HYDROLOGY

Moving from top to bottom of the soil water column, the LAPS has the three layers (Fig. 11.3). The governing equations for the three volumetric soil moisture content are given by Eqs. (11.5)–(11.7). The precipitation P_1 that infiltrates into the top soil layer is given by

$$P_1 = \begin{cases} min(P_0, K_s) & \vartheta_1 < \vartheta_s \\ 0 & \vartheta_1 < \vartheta_s \end{cases} \tag{11.64}$$

where P_0 is effective precipitation rate $[\mathrm{LT}^{-1}]$ on the soil surface given by

$$P_0 = P - (P_f - D_f), \tag{11.65}$$

P is precipitation rate above the canopy $[\mathrm{LT}^{-1}]$, P_f is rate of interception (inflow) for the canopy $[\mathrm{LT}^{-1}]$, and D_f is rate of drainage of water stored on the vegetation (outflow) for the canopy $[\mathrm{LT}^{-1}]$. P_f is given by

$$P_f = P(1 - e^{-\mu})\sigma_c \tag{11.66}$$

where μ is a constant depending on the leaf area index. It is assumed that the interception, if the rainfall can be considered via the expression describing the exponential attenuation (Sellers *et al.*, 1986), D_f is given by

$$D_f = \begin{cases} 0 & w_f < w_{max} \\ P_f & w_f < w_{max} \end{cases} \tag{11.67}$$

The transfer of water between adjacent layers $F_{i,i+1}[\mathrm{LT}^{-1}]$ is given by

$$F_{i,i+1} = K_{ef} \left[2\frac{\psi_i - \psi_{i+1}}{D_i - D_{i+1}} + 1 \right] \tag{11.68}$$

where ψ_i is soil moisture potential [L] of the ith layer, obtained by Eq. (11.60), and K_{ef} is effective hydraulic conductivity [LT^{-1}] between soil layers given by

$$K_{ef} = \frac{D_i K_i - D_{i+1} K_{i+1}}{D_i + D_{i+1}}.$$
(11.69)

In Eq. (11.69) K_i is hydraulic conductivity [LT^{-1}] of the ith soil layer determined by the empirical formula

$$K_i = K_{si} \left(\frac{\vartheta_i}{\vartheta_s} \right)^{2B+3}$$
(11.70)

where K_{si} is hydraulic conductivity at saturation [LT^{-1}] of the ith soil layer. The gravitational drainage from the bottom soil layer is defined by

$$F_3 = K_{si} \left(\frac{\vartheta_3}{\vartheta_s} \right)^{2B+3} \sin(x)$$
(11.71)

while x is mean slope angle (Sellers $et\ al.$, 1986; Abramopoulos $et\ al.$, 1988). The schematic diagram representing the drainage and run-off in the LAPS is shown in Fig. 11.3. The surface run-off R_0 [LT^{-1}] is computed as

$$R_0 = P_1 - min(P_1, K_s).$$
(11.72)

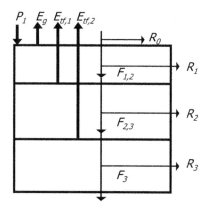

Figure 11.3. Schematic diagram of hydrology in the Land-Air Parameterization Scheme (LAPS).

The subsurface run-off R_i [LT^{-1}] is calculated for each soil layer using the expressions

$$R_1 = F_{1,2} - min(F_{1,2}, K_s)$$
(11.73)

$$R_2 = F_{2,3} - min(F_{2,3}, K_s)$$
(11.74)

$$R_3 = F_3 - min(F_3, K_s).$$
(11.75)

At the end of the time step, Δt, the value Γ_i is calculated as

$$\Gamma_i = \frac{D_i}{\Delta t}[\vartheta_i^k + A_i \Delta t - \vartheta_{fc}] \tag{11.76}$$

where ϑ_i^k is the volumetric soil moisture content at the beginning of k time step while A_i representing the terms on the right side of Eqs. (11.5)–(11.7). If the condition $\Gamma_i > 0$ is satisfied Γ_i becomes run-off, which is added to corresponding subsurface run-off R_i. Consequently, at the end of the time step, the calculated value of the volumetric soil moisture content ϑ_i^{k+1} takes the value ϑ_{fc}.

11.7 CONCLUDING REMARKS

In this chapter is given a detailed description of parameterization of the transport processes in soil-vegetation-lower atmosphere system by LAPS scheme. In designing this scheme, an effort is invested for finding a compromise between an accurate description of the main physical processes and the resolution of the number of prescribed input parameters. Land surface schemes such as LAPS aim to simulate the surface flux partitioning using an Ohm's law analogue in which surface to atmosphere fluxes are proportional to a potential difference and inversely proportional to a resistance. For sensible heat, the potential difference is the surface temperature minus the atmospheric temperature. The resistance is purely aerodynamic, and depends on the roughness of the surface, the wind speed and the atmospheric stability. For the latent heat flux the potential difference is taken as the saturated water vapour pressure at the surface temperature minus the atmospheric vapour pressure at the reference level, and the resistance depends on which moisture store is being depleted.

The hydrological state of the land surface is defined in terms of the vertical profile of soil moisture and the water lying on plant leaves or puddled on the soil surface. Evaporation from the canopy is subject to the same aerodynamic resistance as the sensible heat flux. However, evaporation from the soil and transpiration through plants is subject to an additional surface resistance. For bare soil this is related to the requirement for moisture to diffuse to the soil surface before it can evaporate. For vegetated surfaces the additional resistance represents the control that "stomata" exert over transpiration. They are open and closed in response to changes in solar radiation, temperature or soil moisture. The soil moisture that the vegetation can access for transpiration depends on the root depth and the vertical profile of soil moisture. In its configuration, LAPS updates the soil moisture in three vertical layers. The other key changes in LAPS relate directly to the surface energy balance depending on soil surface, canopy temperatures and canopy air space temperatures. The surface temperatures are calculated from the energy balance equations for bare soil and canopy surfaces, while the canopy air space temperature is calculated diagnostically from the sensible heat flux to the atmosphere balancing the sensible heat flux from the canopy and from the soil surface.

APPENDIX A – LIST OF SYMBOLS

	List of Symbols	
Symbol	Definition	Dimensions or Units
C	heat capacity	$[\text{J kg}^{-1}\,\text{K}^{-1}]$
C_s	transfer coefficient	$[\text{m}^{-1/2}\text{s}^{1/2}]$
D_d	root density	$[\text{m}^3\text{m}^{-3}]$

(Continued)

List of Symbols

Symbol	Definition	Dimensions or Units
D_f	rate of drainage of water stored on the vegetation (outflow) for the canopy	$[\text{ms}^{-1}]$
D_i	thickness of the ith soil layer	$[\text{m}]$
E	evapotranspiration rate	$[\text{kg m}^{-2}\text{s}^{-1}]$
$F_{i,i+1}$	water flux between i and $i+1$ soil layer	$[\text{ms}^{-1}]$
E_{tf}	transpiration rate from foliage	$[\text{kg m}^{-2}\text{s}^{-1}]$
$E_{tf,1}$	canopy extraction of soil moisture by transpiration from the rooted first soil layer	$[\text{kg m}^{-2}\text{s}^{-1}]$
$E_{tf,2}$	canopy extraction of soil moisture by transpiration from the rooted second soil layer	$[\text{kg m}^{-2}\text{s}^{-1}]$
E_{wf}	evaporation rate of water from the wetted fraction of canopy	$[\text{kg m}^{-2}\text{s}^{-1}]$
F_3	gravitational drainage flux from recharge soil water store	$[\text{ms}^{-1}]$
H	canopy height	$[\text{m}]$
H_f, H_g	canopy and soil sensible heat flux respectively	$[\text{W m}^{-2}]$
K_c	saturated hydraulic conductivity	$[\text{ms}^{-1}]$
K_{ef}	effective hydraulic conductivity	$[\text{ms}^{-1}]$
K_i	hydraulic conductivity	$[\text{ms}^{-1}]$
K_r	mean soil hydraulic conductivity in the root zone	$[\text{ms}^{-1}]$
K_s	turbulent transfer coefficient within and above the canopy	$[\text{m}^2\text{s}^{-1}]$
K_{si}	hydraulic conductivity at saturation	$[\text{ms}^{-1}]$
P	precipitation rate above the canopy	$[\text{ms}^{-1}]$
P_0	effective precipitation rate	$[\text{ms}^{-1}]$
P_f	water amount retained on the canopy	$[\text{ms}^{-1}]$
P_1	infiltration rate of precipitation into the upper soil moisture store	$[\text{ms}^{-1}]$
R_0	surface run-off	$[\text{ms}^{-1}]$
R_i	subsurface run-off from the ith soil layer	$[\text{ms}^{-1}]$
R_f^l	long-wave radiative fluxes absorbed by the canopy	$[\text{W m}^{-2}]$
R_g^l	long-wave radiative fluxes absorbed by the soil surface	$[\text{W m}^{-2}]$
R_n	absorbed net radiation	$[\text{W m}^{-2}]$
R_{nf}	net radiation absorbed by the canopy	$[\text{W m}^{-2}]$
R_{ng}	net radiation absorbed by the soil surface	$[\text{W m}^{-2}]$
R_f^s	short-wave radiation absorbed by the canopy	$[\text{W m}^{-2}]$
R_g^s	short-wave radiation absorbed by the soil surface	$[\text{W m}^{-2}]$
R_o^s	incident downward-directed short-wave flux	$[\text{W m}^{-2}]$
R_r	resistance per unit root length	$[\text{s m}^{-1}]$
T	surface (canopy or soil) temperature	$[\text{K}]$
T_a	canopy air space temperature	$[\text{K}]$
T_d	the deep soil temperature	$[\text{K}]$
T_f	surface canopy temperature	$[\text{K}]$
T_g	surface soil temperature	$[\text{K}]$

(Continued)

List of Symbols

Symbol	Definition	Dimensions or Units
T_r	air temperature at the reference height z_r	[K]
V_r	volume of root per unit volume of soil	[$\text{m}^3\,\text{m}^{-3}$]
$a(z), b(z)$	functions of the vertical coordinate z	
c_p	specific heat of air	[$\text{Jkg}^{-1}\text{K}^{-1}$]
d	the displacement height	[m]
d_1, d_2, d_3	empirical constants	[s m^{-1}]
$e_*(T_f)$	saturated vapour pressure at temperature T_f	[Pa]
$e_*(T_g)$	saturated vapour pressure at temperature T_g	[Pa]
e_a	canopy air space vapour pressure	[Pa]
e_r	vapour pressure of the air at reference height within the atmospheric boundary layer	[Pa]
h	the canopy bottom height (the height of the base of the canopy)	[m]
h_a	the position of the canopy source height	[m]
m	number of iteration	
n	the number of the spatial step in the numerical calculating on the interval $[H, h]$	
$\bar{r}_b$	bulk canopy boundary layer resistance	[s m^{-1}]
$\bar{r}_c$	bulk canopy stomatal resistance	[s m^{-1}]
r_a	aerodynamic resistance	[s m^{-1}]
r_d	aerodynamic resistance between soil surface and canopy air space	[s m^{-1}]
r_{plant}	plant resistance imposed by the plant vascular system	[s]
$r_{s\ min}$	minimum of stomatal resistance	[s m^{-1}]
$r_{s\ max}$	maximum of stomatal resistance	[s m^{-1}]
r_{soil}	resistance of the soil and root system	[s]
r_{surf}	soil surface resistance	[s m^{-1}]
u_*	the friction velocity	[m s^{-1}]
$u(H)$	the wind speed at the canopy height	[m s^{-1}]
$u(h)$	the wind speed at the canopy bottom height	[m s^{-1}]
$u(z)$	the wind speed above a canopy	[m s^{-1}]
w_f	the water stored on the canopy	[m]
w_w	wetted fraction of canopy	
x	mean slope angle	[$^\circ$]
z	the vertical coordinate	[m]
z_I	some certain height within the canopy	[m]
z_0, z_b	the roughness length over the non-vegetated surface	[m]
z_d	the rooting depth	[m]
z_g	effective ground roughness length	[m]
z_r	the reference height	[m]
z_t	height of the transpiration source that is equal to canopy source height	[m]
Δt	time step	[s]
Δz	the grid size	[m]
α	an unknown constant	

(Continued)

List of Symbols

Symbol	Definition	Dimensions or Units
α_f, α_g	canopy albedo and soil surface albedo, respectively	
α_s	a factor to correct for soil dryness	
β	extinction parameter	
γ	the psychrometric constant	$[\text{Pa K}^{-1}]$
$\varepsilon_f, \varepsilon_g$	emissivities of the canopy and the soil surface, respectively	
η	the canopy-dependent empirical parameter	$[\text{Pa}^{-1}]$
ϑ_1	the top layer volumetric soil water content	$[\text{m}^3\ \text{m}^{-3}]$
ϑ_a	the mean volumetric soil water content in the first and second soil layers	$[\text{m}^3\ \text{m}^{-3}]$
ϑ_{fc}	volumetric soil water content at field capacity	$[\text{m}^3\ \text{m}^{-3}]$
ϑ_i^k	the volumetric soil moisture content at the beginning of k time step	$[\text{m}^3\ \text{m}^{-3}]$
ϑ_i	volumetric soil water content in the ith layer	$[\text{m}^3\ \text{m}^{-3}]$
ϑ_s	volumetric soil moisture content at saturation	$[\text{m}^3\ \text{m}^{-3}]$
ϑ_{wil}	volumetric soil water content at wilting point	$[\text{m}^3\ \text{m}^{-3}]$
κ	Von Kármán constant	
λ	latent heat of vaporisation	$[\text{J kg}^{-1}]$
λE_f	the latent heat flux from canopy vegetation to canopy air space	$[\text{W m}^{-2}]$
λE_g	the latent heat flux from soil surface to the canopy air space	$[\text{W m}^{-2}]$
λE_t	the latent heat flux from canopy air space to reference height in the atmospheric boundary layer	$[\text{W m}^{-2}]$
μ	parameter; a constant depending on the leaf area index	
ρ	water density	$[\text{kg m}^{-3}]$
ρ_p	air density	$[\text{kg m}^{-3}]$
σ	the value of the scaling length	
σ_B	the Stefan-Boltzman constant	$[\text{W m}^{-2}\ \text{K}^{-4}]$
σ_c	vegetation cover in fractional units	
σ_f	fractional cover of vegetation	
ψ_i	soil water potential of the ith soil layer; soil moisture potential of the ith layer	$[\text{m}]$
ψ_l	the leaf water potential describing the water transfer pathway from root zone to leaf	$[\text{m}]$
$\psi_m(z/L)$	The stability function for momentum	
ψ_r	soil moisture potential in the root zone	$[\text{m}]$
ψ_s	soil water potential at saturation	$[\text{m}]$

APPENDIX B – SYNOPSIS

The interaction of the land surface and the atmosphere goes through an exchange of heat, water and momentum fluxes, such as the following: interaction of canopy vegetation with radiation, evaporation from bare soil, transpiration, evaporation of intercepted precipitation and dew, conduction of soil water through the vegetation layer, vertical movement of water in the soil, run-off, heat conduction in the soil, momentum transport and the effects of snow presence and freezing or the melting of soil moisture. Parameterization of these processes in surface schemes are divided into three groups: (i) thermal and hydraulic, (ii) bare soil transfer and (iii) canopy vegetation transport processes. The chapter briefly describes these processes through the structure of a surface scheme. The vegetation is considered to be a block of constant-density porous material that is "sandwiched" between two constant-stress layers, with an upper boundary (the height of the canopy top) and a lower boundary (the height of the canopy bottom). For a description of the transport processes in the soil, the three-soil layer approach is used. Finally, the chapter includes (i) a detailed description of governing equations and "K"-theory within and above the canopy, (ii) the resistance representation of energy fluxes and (iii) the parameterization of radiation, aerodynamic resistances and a model of hydrology.

APPENDIX C – KEYWORDS

By the end of the chapter you should have encountered the following terms. Ensure that you are familiar with them!

Surface scheme	Surface resistances	Latent heat flux
"K"-theory inside the canopy	Aerodynamic resistances	Soil heat flux
Parameterization	Canopy temperature	Intercepted water
Evaporation	Soil temperature	Surface run-off
Transpiration	Radiation in canopy	Ground run-off

APPENDIX D – QUESTIONS

What is evapotranspiration?
Which are the aerodynamic and surface resistances?
How does parameterized stomatal resistance work?
How is parameterized water intercepted by leaves?
What are surface and ground run-off?

APPENDIX E – PROBLEMS

E1. Describe the general structure of the surface scheme and its use in environmental models. more specifically, discuss the prognostic variables in the surface scheme and the corresponding equations where the variables come from.

E1. Describe the main points and discuss (i) "K"-theory within and above canopy vegetation, (ii) the resistance representation of energy fluxes and (iii) the parameterization of radiation, aerodynamic resistances and a model of hydrology.

E3. Describe how the water taken by roots is included in the parameterization of evapotranspiration and evaporation from bare soil. Consider cases when the soil moisture content of the soil is near to (i) the soil field capacity and (ii) the soil wilting point.

E4. Solve equation (42) to obtain the wind speed $u(z)$ inside the canopy, using the following assumptions: (i) turbulent transfer coefficient inside the canopy K_s is given as $K_s = \sigma u(z)$, where σ is defined by (35) and (ii) the vegetation fractional cover σ_c is given by the empirical equation for crops $\sigma_c = 1 - \exp(-kLAI)$. Discuss the solution for the values $k \in (0.2, 0.4)$.

E5. Derive an expression for the aerodynamic resistance K_s (in the case of neutral) using: (i) expression (27), (ii) turbulent transfer coefficient above the canopy given by (31) and (ii) turbulent transfer coefficient inside the canopy derived in **E4**.

REFERENCES

Abramopoulos, F., Rosenzweig, C. and Choudhury, B., 1988, Improved ground hydrology calculations for global climate models (GCMs): Soil water movement and evapotranspiration. *Journal of Climate*, 1, 921–941.

Acs, F., 1994, A coupled soil–vegetation scheme: Description, parameters, validation, and sensitivity studies. *Journal of Applied Meteorology*, **33**, 268–284.

Avissar, R., Avissar, P., Mahrer, Y. and Bravdo, B., 1985, A model to simulate response of plant stomata to environmental conditions. *Agricultural and Forest Meteorology*, **64**, 127–148.

Baron, J.S., Hartman, M.D., Kittel, T.G.F., Band, L.E., Ojima, D.S. and Lammers, R.B., 1998, Effects of land cover, water redistribution, and temperature on ecosystem processes in the South Platte Basin. *Ecological Applications*, **8**, 1037–1051.

Bosilovich, M. and Sun, W., 1995, Formulation and verification of a land surface parameterization for atmospheric models. *Boundary-Layer Meteorology*, **73**, 321–341.

Brunet, Y., Finnigan, J.J. and Raupach, M.R., 1994, A wind tunnel study of air flow in waving wheat: Single-point velocity statistics. *Boundary-Layer Meteorology*, **70**, 95–132.

Chase, T.N., Pielke, Sr.R.A., Kittel, T.G.F., Baron, J.S. and Stohlgren, T.J. 1998, Potential impacts on Colorado Rocky Mountain weather and climate due to land use changes on the adjacent Great Plains. *Journal of Geophysical Research*, **104**, 16673–16690.

Clapp, R.B. and Hornberger, G.M., 1978, Empirical equations for some soil hydraulic properties. *Water Resources Research*, **14**, 601–604.

Cox, P.M., Betts, R.A., Bunton, C.B., Essery, R.L.H., Rowntree, P.R. and Smith, J., 1999, The impact of new land surface physics on the GCM simulation of climate and climate sensitivity. *Climate Dynamics*, **15**, 183–203.

Deardorff, J.W., 1978, Efficient prediction of ground surface tem- perature and moisture, with inclusion of a layer of vegetation. *Journal of Geophysical Research*, **83**, 1889–1903.

Dickinson, R.E., 1995, Land–atmosphere interaction. *Reviews of Geophysics*, **33**, 917–922.

Dickinson, R.E., Henderson-Sellers, A., Kennedy, P.J. and Wilson, M.F., 1986, Biosphere–Atmosphere Transfer Scheme for the NCAR Community Climate Model, NCAR Tech. Rep. NCAR/TN-2751 STR, (Available from NCAR, P.O. Box 3000, Boul- der, CO 80307-3000) p. 69.

Dubov, A.S., Bikova, L.P. and Marunich, S.V., 1978, *Turbulence inside a canopy.* (Gidrometeoizdat, Leniningrad) (In Russian).

Federer, C.A., 1979, A soil-plant-atmosphere for transpiration and availability of soil water. *Water Resources Research*, **15**, 555–562.

Idso, S., Jackson, R., Kimball, B. and Nakagama, F., 1975, The dependence of bare soil albedo on soil water content. *Journal of Applied Meteorology*, **14**, 109–113.

Jarvis, P.G., 1976, The interpretation on the variations in leaf water potential and stomatal conductance found in canopies in the field. *Philosophical Transactions of the Royal Society*, **B273**, 593–610.

Lee, T.J., Pielke, R.A. and Mielke, P.W.Jr., 1995, Modeling the clear-sky surface energy budget during FIFE87. *Journal of Geophysical Research*, **100**, 25585– 25593.

Loveland, T.R., Merchant, J.W., Ohlen, D.O. and Brown, J.F., 1991, Development of a land-cover characteristics database for the conterminous U.S. *Photogrammetric Engineering and Remote Sensing*, **57**, 1453–1463.

Mihailović, D.T., 1996, Description of a land-air parameterization scheme (LAPS). *Global and Planetary Change,* **13**, 207–215.

Mihailović, D.T., 2003, Implementation of Land-Air Parameterization Scheme (LAPS) in a limited area model. *Final Report, The New York State Energy Conservation and Development Authority,* (Albany, NY).

Mihailović, D.T. and Kallos, G., 1997, A sensitivity study of a coupled soil-vegetation boundary layer scheme for use in atmospheric modeling. *Boundary-Layer Meteorology,* **82,** 283–315.

Mihailović, D.T., Rajkovic, B., Lalic, B. and Dekic, Lj., 1995, Schemes for parameterizing evaporation from a non-plant-covered surface and their impact on partitioning the sur-face energy in land-air exchange parameterization. *Journal of Applied Meteorology,* **34,** 2462–2475.

Mihailović, D.T., Pielke, R.A.Sr., Rajkovic, B., Lee, T.J. and Jeftic, M., 1993, A resistance representation of schemes for evaporation from bare and partly plant- covered surfaces for use in atmospheric models. *Journal of Applied Meteorology,* **32,** 1038–1054.

Mihailović, D.T., Kallos, G., Arsenic, I.D., Lalic, B., Rajkovic, B. and Papadopoulos, A. 1999, Sensitivity of soil surface temperature in a force-restore equation to heat fluxes and deep soil temperature. *International Journal of Climatology,* **19,** 1617– 1632.

Mihailović, D.T., Alapaty, K., Lalic, B., Arsenic, I., Rajkovic, B. and Malinovic, S., 2004, Turbulent transfer coefficients and calculation of air temperature inside the tall grass canopies in land-atmosphere schemes for environmental modeling. *Journal of Applied Meteorology,*43, 1498–1512.

Mihailović, D.T., Lalic, B., Eitzinger, J., Malinovic, S. and Arsenic I., 2006, An approach for calculation of turbulent transfer coefficient for momentum inside vegetation canopies. *Journal of Applied Meteorology and Climatology,* **45**, 348– 356.

Mihailović, D.T., Lee, T.J., Pielke, R.A., Lalic, B., Arsenic, I., Rajkovic, B. and Vidale, P.L. 2000, Comparison of different boundary layer schemes using single point micrometeorological field data. *Theorethical and Applied Climatology,* **67**, 135–151.

Noilhan, J. and Planton, S., 1989, A simple parameterization of land surface processes in meteorological models. *Monthly Weather Review,* **117**, 536–549.

Paulson, C.A., 1970, The mathematical representation of wind speed and temperature in the unstable atmospheric surface layer, *Journal of Applied Meteorology,* **9**, 857–861.

Pielke, R.A.Sr., Avissar, R., Raupach, M., Dolman, H., Zeng, X. and Denning, S., 1998, Interactions between the atmosphere and terrestrial ecosystems: Influence on weather and climate. *Global Change Biology,* **4**, 101–115.

Pielke, R.A.Sr., Walko, R.L., Steyaert, L., Vidale, P.L., Liston, G.E. and Lyons, W.A., 1999, The influence of anthropogenic landscape changes on weather in south Florida. *Monthly Weather Review,* **127,** 1663–1673.

Pleim, J. and Xiu, A., 1995, Development and testing of a surface flux and planetary bound-ary layer model for application in mesoscale models. *Journal of Applied Meteorology,* **34,** 16–32.

Sellers, P.J. and Dorman, J.L., 1987, Testing the simple biosphere model (SiB) using point micrometeorological and biophysical data. *Journal of Applied Meteorology,* **26**, 622–651.

Sellers, P.J., Mintz, Y., Sud, Y. and Dalcher, A., 1986, A simple biosphere model (SiB) for use within general circulation models. *Journal of Atmospheric Sciences,* **43**, 505–531.

Stohlgren, T.J., Chase, T.N., Pielke, R.A., Kittel, T.G.F. and Baron, J.S., 1998, Evidence that local land use practices influence regional climate, vegetation, and stream flow patterns in adjacent natural areas. *Global Change Biology,* **4**, 495–504.

Sun, S.F., 1982, *Moisture and heat transport in a soil layer forced by atmospheric conditions.* M.S. Thesis, Department of Civil Engineering, University of Connecticut.

Van der Honert, X., 1948, Water transport as a catenary process, *Discussions of the Faraday Society*, **3**, 146–153.

Viterbo, P. and Beljaars, A., 1995, An improved land-surface parameterization scheme in the ECMWF model and its validation. *Journal of Climate*, **8**, 2716– 2748.

Walko, R.L., Band, L.E., Baron, J., Kittel, T.G.F., Lammers, R., Lee, T.J., Ojima, D., Pielke, R.A.Sr., Taylor, C., Tague, C., Tremback, C.J., Vidale, P.L., 2000, Coupled atmosphere-biophysics-hydrology models for environmental modeling. *Journal of Applied Meteorology*, **39**, 931–944.

CHAPTER TWELVE

Turbulence and wind above and within the forest canopy

Branislava Lalic & Dragutin T. Mihailović
Faculty of Agriculture, University of Novi Sad, Novi Sad, Serbia

ABSTRACT

The forest has a strong influence on vertical profiles of micrometeorological variables within and above the canopy. Especially pronounced variations of all variables between ground level and crown top are primarily generated by the forest architecture. When wind encounters forest canopy, the drag of the foliage removes mean momentum of wind producing turbulent eddies. Dissipation of mean flow kinetic energy within and below the forest crown usually has been described through vertical gradient of wind speed. The accuracy of within-canopy wind profile calculation is related to assumed forest architecture and to adopted approach for parameterization of momentum turbulent fluxes. This chapter is focused on forest architecture and on turbulence produced by friction exerted when air flow encounters forest canopy. An overview of different approaches oriented towards their parameterization (forest architecture) and modelling (turbulence) is presented.

12.1 INTRODUCTION

The definition of the lower boundary condition is of great importance in dynamic environmental models (atmospheric, hydrological and ecological), especially in the presence of vegetation. Forest is a vegetation system covering more than 20% of land-based globe. Also, the atmosphere 'feels' the presence of trees up to a few hundred meters from the ground, depending on tree height. Therefore, forest as an underlying surface is often met in atmospheric and environmental models of different scales. As a dynamical source and sink of momentum, heat, water (vapour) and pollution, forest plays a crucial role in land-atmosphere-interaction modelling. To describe that role it is important to understand mechanism of forest canopy – atmosphere interaction processes. Key element of these processes is the turbulent transfer above and within the forest canopy strongly affected by forest architecture, its thermal characteristics and significant drag of foliage. Consequently, many current vegetation-atmosphere as well as the environmental models require more specific information about the forest structure describing the leaf area density variation with height in order to provide a better estimate of energy, mass and momentum exchange (Mix *et al.*, 1994; Zeng and Takahashi, 2000). In the past decades, a fair amount of literature has been accumulated that deals with the closuring problem and values of the various coefficients that must be specified in order to solve equations of motion for turbulent flow above and within the canopy. This chapter describes different approaches in designing forest canopy architecture based on leaf area index, LAI or leaf area density, LAD, as key structural characteristics. Vertical transfer of momentum, considered in this chapter, is restricted to horizontally homogeneous, extensive forest over which the mean wind is steady and unidirectional. We have selected here the parameterizations of turbulent transfer above and within the forest canopy based on first order closure model, i.e. modified K-theory.

12.2   MODELLING THE FOREST ARCHITECTURE

The forest architecture is most commonly quantified by the amount of leaves and stems, and their spatial distribution represented by leaf area index, LAI [$L^2 L^{-2}$] and leaf area density, LAD [$L^2 L^{-3}$], respectively. Following the definitions of these two characteristic quantities, the relation between them could be written in the form:

$$LAI = \int_0^{h_c} LAD(z)dz, \qquad\qquad (12.1)$$

where h_c[L] is the forest height.

However, it is extremely difficult to measure in practice these quantities inside the forest canopy. Some authors try either to provide alternative methods for measuring (Meir *et al.*, 2000), or for estimating (Law *et al.*, 2001b) the leaf area index, LAI and leaf area density, LAD, inside the different forest communities. Levi and Jarvis (1999) suggested an empirical relation for the leaf area index, LAI based on an inclusion of the forest optical characteristics,

$$LAI = -\frac{\ln \tau_H(\theta_s)}{K(\theta_s)} \qquad\qquad (12.2)$$

where τ_H is the transmittance of whole canopy ("bulk" transmittance) for radiation in photosynthetic waveband and K is an extinction coefficient which is a function of solar zenith angle, θ_s and leaf inclination angle distribution. Unfortunately, calculation of leaf area index, LAI using Eq. (12.2) is restricted to homogeneous forest and low values (less than 6 m^2 m^{-2}) of LAI. In contrast to this and other similarly established approaches, Gower (Gower *et al.*, 1999) emphasized that the direct measurement is the only reliable method for dense forest canopies having high values of LAI ($LAI > 6$ m^2 m^{-2}).

Simplest parameterization of leaf area density, LAD, related to an ideal canopy with a homogeneous crown and negligible amount of vegetation below it, could be expressed in the form

$$LAD(z) = \begin{cases} LAD_0 & h_c/2 \leq z \leq h_c \\ 0 & 0 \leq z \leq h_c/2 \end{cases} \qquad\qquad (12.3)$$

where LAD_0 [$L^2 L^{-3}$] is the leaf area density of forest crown (Watanabe and Kondo, 1990). Recently, the scientific community dealing with the environmental problems tends to derive physically more realistic empirical expressions for leaf area density, LAD, based on available observational data archives. One of the expressions among their limited collection, based on photographic method, is suggested by Meir *et al.* (2000). The photographs are being taken horizontally from the tower at different heights, using as a target a white meteorological balloon raised into the canopy at the known distance, l [L]. From these hemispherical photographs, one could determine the fraction of transmitted light through the canopy layer, τ_z as an estimate of probability, P_z, of a beam of light passing through a horizontal plane of leaves at height z within a forest canopy. Taking into account the relation between probability P_z and path length L_z through which the light comes to level z,

$$L_z = -\ln P_z, \qquad\qquad (12.4)$$

leaf area density, *LAD* for each canopy layer could be calculated as

$$LAD(z) = \frac{L_z}{l}.$$ (12.5)

Unfortunately, this method for leaf area density, *LAD* calculation can be used only in a limited number of situations when hemispherical photographs for different canopy layers are made.

A more sophisticated approach to leaf area density *LAD*, parameterization was suggested by Lalic and Mihailović (2004). On the basis of measured spatial distribution of leaves and stems, they derived the relation for *LAD(z)* taking into account tree height h_c, maximum value of leaf area density L_m and corresponding height z_m as key parameters of the forest canopy structural characteristics (Kolic, 1978; Mix *et al.*, 1994; Law *et al.*, 2001a) in the form:

$$LAD(z) = L_m \left(\frac{h_c - z_m}{h_c - z} \right)^n \exp\left[n \cdot \left(1 - \frac{h_c - z_m}{h_c - z} \right) \right],$$

$$\text{where } n = \begin{cases} 6 & 0 \leq z < z_m \\ 1/2 & z_m \leq z \leq h_c \end{cases}$$ (12.6)

Parameter *n* was found from analysis of minimum root-mean-square error (RMSE) for different measured leaf area density distribution data sets. Results of these analyses pointed out that the best choice is $n = 0.5$ for range $z \geq z_m$ and $n = 6$ for $z < z_m$. According to the classification based on z_m and h_c parameters (Kolic, 1978), all forest canopies can be divided into the three groups: 1) $z_m = 0.2h_c$ (oak and silver birch), 2) $0.2h_c < z_m < 0.4h_c$ (common maple) and 3) $z_m = 0.4h_c$ (pine), where in the bracket is a typical representative. Following this classification, empirical relation for leaf area density *LAD* described by Eq. (12.6) could be applied in the broad range of forest canopies.

12.3 TURBULENCE AND WIND ABOVE THE FOREST

In atmospheric models for different scales the underlying surface consists of patches of bare soil and plant communities with different morphological parameters. Experimental evidence indicates that there is a significant departure of the wind profile above a vegetative surface from that predicted by the logarithmic relationship, which gives the values which are greater than the observed. This situation can seriously disturb the real physical picture concerning the transfer of momentum, heat and water vapour from the surface into the atmosphere, particularly above the forest. In this section we generalise the calculation of exchange of momentum between the atmosphere and non-homogeneous vegetative surface and derive a general equation for the wind speed profile in a roughness sublayer under neutral conditions. Furthermore, these results are extended to non-neutral cases.

12.3.1 Definition of problem and motivations

Under thermally neutral conditions, steady-state flow over horizontally bare soil can be described by the well-known logarithmic law (e.g., Monin and Yaglom, 1971)

$$u(z) = \frac{u_{*g}}{\kappa} \ln \frac{z}{z_{0g}}$$ (12.7)

where $u(z)$ [L T^{-1}] is the horizontal velocity at height z[L], u_{*g}[L T^{-1}] is the friction velocity for a bare soil, which, physically, represents the shear stress $\tau = \rho_a u_{*g}$ where ρ_a [M L^{-3}] is the air density, κ is the Von Kármán's constant taken to be 0.41 (Högström, 1985) and z_{0g} [L] the roughness length of a bare soil. For vegetative surfaces, where the obstacle size has the same order of magnitude as a measuring height, Eq. (12.7) is modified as

$$u(z) = \frac{u_*}{\kappa} \ln \frac{z - d}{z_0} \qquad (12.8)$$

where u_* [L T^{-1}] is the friction velocity over the vegetation surface, d [L] the displacement height – the mean height in the vegetation on which the bulk aerodynamic drag acts (Thom, 1971) and z_0 [L] the roughness length. According to this expression, the wind speed is zero at height $d + z_0$, but the logarithmic profile cannot be extrapolated so far downwards. When the quantities d and z_0 are known the whole profile above a vegetative surface can be obtained if the wind at a single level as well as the ratio u_*/κ are known. For the non-neutral atmosphere, Eqs. (12.7) and (12.8) have to be modified due to stability effects (Businger et al., 1971).

In order to illustrate differences between these two cases, in the treatment of the lower boundary conditions, for example in surface schemes in atmospheric models, we will form a ratio u_{*g}/u_*, which is equal to $\ln[(z - d)/z_0]/\ln[z/z_{0g}]$, at the height z where the velocities given by Eqs. (12.7) and (12.8) are the same. This ratio, for several plant communities is plotted in Fig. 12.1, where displacement heights and roughness lengths used have their standard values for corresponding plant communities. Apparently, transfer of momentum between short grass and the atmosphere does not differ so much from the corresponding exchange when a bare soil is underlying surface. However, over tall grass the transfer of momentum into the atmosphere is more intensive since the u_*, which can be identified as the velocity scale of the eddies near the surface, becomes greater than u_{*g}. Difference in these velocity scales physically can be explained by the fact that the mixing length of the eddies above a vegetative surface is shorter than the mixing length above a bare soil.

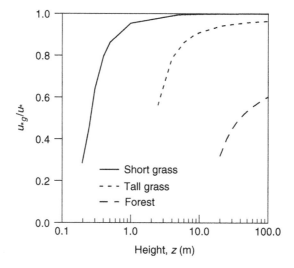

Figure 12.1. Ratio of the friction velocities over bare soil, u_{*g} and different vegetative surfaces, u_* plotted against the height z at which their wind velocities have the same value.

However, Eq. (12.8) is not valid when height z is between the vegetation (of mean height h_c) and some height z^* representing the lower limit of the roughness sublayer. Its order of magnitude can vary between $z* \cong d + 10z_0$ (De Bruin and Moore, 1985) and $z* \cong d + 20z_0$ (Tennekes, 1982). In roughness sublayer Eq. (12.8) is not valid because we are then too close to roughness elements (tall grass, trees, etc.) when the turbulence is generated by the flow around them (Garratt, 1978; De Bruin and Moore, 1985). The depth of the roughness sublayer depends on the value of displacement height d which accounts for an upward shift in the whole profile above a vegetative cover. Since z_0 is around ten percent of the canopy height then the thickness of roughness sublayer can vary between one and two canopy heights. Consequently, an improper treatment of the wind profile in roughness sublayer, systematically gives the values of shear stress and latent heat flux which can significantly deviate from observed values. In models of biosphere-atmosphere exchange when underlying vegetative surface consists of patches of bare soil and plant communities with different morphological parameters, the level of inhomogeneity in the cover has to be taken into account in addition to a spatially varying displacement height. This is of importance in the design of a new generation of land surface parameterization schemes for use in atmospheric models on scales where the patchiness of the surface is resolved (Mihailović and Kallos, 1997).

Experimental evidence indicates that in the roughness sublayer above a vegetative surface, particularly forest canopy, there is a significant departure of the wind profile from that predicted by the logarithmic relationship, giving values which are greater than observed ones (Wilson *et al.*, 1982; Shaw and Pereira, 1982; Sellers *et al.*, 1986). This problem was comprehensively considered by Garratt (1978) and Raupach and Thom (1981). They have noted that estimates of turbulent transfer coefficient K_m $[\mathrm{M L^{-1} T^{-1}}]$ above a vegetative surface h_c were 1.5–2.0 times larger than as the simple extrapolation of Eq. (12.8) would indicate. Using this estimation Eq. (12.8) can be modified, so that in roughness sublayer it takes the form

$$u(z) = \frac{u_*}{\alpha_G k}\ln\frac{z-d}{z_0} \tag{12.9}$$

where α_G is a dimensionless constant estimated to be between 1.5 and 2.0 (Raupach and Thom, 1981; Massman, 1987) resulting in 1.5–2.0 times smaller values for the wind speed than it would be expected from Eq. (12.8). Let us note that logarithmic profile given by Eq. (12.9) can only be valid for the lower part of the roughness sublayer. Some other expressions with correct matching behaviour can be found in Raupach *et al.* (1980) and Raupach (1980).

12.3.2 Exchange of momentum above a non-uniform underlying surface under neutral conditions

We will derive an expression for the turbulent transfer coefficient K_m and the wind profile, under neutral conditions, above a non-uniform underlying surface whose non-uniformity is expressed by the surface vegetation fractional cover σ_f, which takes the values from 0 (bare soil) to 1 when the ground surface is totally covered by plants. A realistic surface is rather porous, with patches of bare soil and free air spaces inside it, and vegetative portion which can produce quite different modes of turbulence in comparison with an uniform underlying surface which is either bare soil or surface covered with vegetation. Vegetative part of the underlying surface is a mosaic of patches of various size and different aerodynamic characteristics. Presumably, this mosaic will produce micro circulation with possible flow separations at leading and trailing edges setting up a highly complex dynamic flow.

In this Chapter we will not address the consequences of such non-uniformity of the vegetation part of the underlying surface. Instead, the underlying surface will be considered as a combination of the only two homogeneous portions consisting of vegetative portion, characterized with fractional cover σ_f and the bare portion, characterized with fractional cover $1 - \sigma_f$. Bearing in mind such assumption we will try to do the previously mentioned calculations.

We will start from the description of the logarithmic profile which is consistent with the following assumptions. Similarly as in the molecular gas theory, an exchange coefficient can be derived as the product of a velocity and mixing length. For molecules the mixing length can be identified with the mean free path, but for eddies above a canopy with displacement height d it is assumed that it is proportional to a corrected height $z - d$. The proportionality factor is given by Von Kármán's constant k, so the mixing height $l_m^c[L]$ is given by

$$l_m^c = \kappa(z - d) \tag{12.10}$$

which is a broadly employed expression for the mixing length in the free air above a vegetative surface in the surface layer (in further text this approach will be denoted as the "old approach"). For $d = 0$, Eq. (12.10) represents the mixing length over a bare soil, thus it becomes $l_m^b = \kappa z$ which is bigger than l_m^c. Undoubtedly, in the reality there is no situation when the underlying vegetative surface is as dense and smooth as it is assumed in deriving the mixing length given by Eq. (12.10). As we mentioned above, a natural surface is very porous and consists of vegetative surface with patches of bare soil, producing quite unpredictable mode of turbulence inside and above the vegetative surface. Experimental results by Garratt (1978) support this point. According to them, in the roughness sublayer above a vegetative surface, the mixing length l_m^α, which is bigger than l_m^c, can be written in the form $l_m^\alpha = \alpha\kappa(z - d)$ where α is a dimensionless constant representing corrected value of the mixing length in the roughness sublayer. For further consideration we will use the mixing length in the form

$$l_m^\alpha = \alpha_G\kappa(z - d) \tag{12.11}$$

where α is replaced by α_G which is defined above in Eq. (12.9). However, the eddies, with the mixing length given by Eq. (12.11), are still generated above a dense and smooth vegetative surface. In order to take into account its non-uniformity we have considered it as a block of porous material consisting of bare soil and vegetative patches which can be described by the vegetation fractional cover σ_f, with values between 0 and 1. The number of eddies generated above the underlying surface defined in such a way consists of: 1) eddies generated above the vegetative part whose number is proportional to σ_f and 2) eddies generated above the bare soil part with the factor of proportionality $(1 - \sigma_f)$. Thus, their mixing length l_m could be used as a linear combination of mixing lengths l_m^α and l_m^b, i.e., $l_m = \sigma_f l_m^\alpha + (1 - \sigma_f)l_m^b$. Let us note that a linear combination of the single lengths is not the only way of deriving a mixing length l_m accounting for the non-uniformity of the surface. Consequently, mixing lengths over a non-uniform surface would almost be different and it seems that the suggestion for a linear aggregation scheme for an effective mixing length is a simplified assumption. However, from a practical and a physical point of view this assumption might be acceptable because it is more complex than the commonly used one. After setting $l_m^\alpha = \alpha\kappa(z - d)$, the mixing lengths l_m takes the form

$$l_m = \sigma\alpha\kappa(z - d) + (1 - \sigma)\kappa z \tag{12.12}$$

where α is the dimensionless constant introduced above which depends on morphological and aerodynamic characteristics of the vegetative cover whose value varies depending on the type of vegetative cover. In this study α is considered as a function of leaf drag coefficient C_{dg}, and leaf area index LAI, i.e., $\alpha = \alpha(C_{dg} \cdot LAI)$. For $\sigma_f = 1$, Eq. (12.12) becomes Eq. (12.11) while for $\sigma_f = 0$ it reduces to the expression for the mixing length for a bare soil. The turbulent transfer coefficient K_m for the non-homogenous vegetative cover is

$$K_m = l_m U_*$$

(12.13)

where U_* [L T^{-1}] is a friction velocity above non-homogeneously covered surface. Replacing l_m, in this equation, by the expression (12) we get

$$K_m = \kappa\{[\sigma(\alpha - 1) + 1]z - \sigma\alpha d\}U_*.$$

(12.14)

The functional form of the parameter α was derived empirically by Lalic (1997). More details about this parameter can be found in Mihailović *et al.* (1999) and Chapter 3 of this book. Mihailović *et al.* (1999) found that this parameter has a typical value for forest about 1.6 while other vegetation communities have the values closer to 1.

Another characteristic of the family of lines representing the mixing length l_m, is that they cross each other at a single point, at height z_l [L], where this height does not depend on the vegetation fractional cover σ_f. The height z_l can be calculated from the condition

$$[\sigma_1(\alpha - 1) + 1]z_l - \sigma_1\alpha d = [\sigma_2(\alpha - 1) + 1]z_l - \sigma_2\alpha d$$

(12.15)

where σ_1 and σ_2 indicate different vegetation fractional covers. Solving this equation for z_l we obtain

$$z_l = \frac{\alpha}{\alpha - 1}d.$$

(12.16)

This expression explicitly shows that the point where the mixing length l_m and the turbulent transfer coefficient K_m do not depend on the vegetation fractional cover σ_f, is located at infinity where the condition that $\alpha = 1$ is satisfied. Mathematically, it means that all lines, obtained for different values of α, tend towards the line representing the "old approach". Physically, it seems that the influence of surface patchiness on the mixing length vanishes at some height z_l, however, re-emerging again above it. This situation can be explained by introducing two more degrees of freedom in the expression for the mixing length [Eq. (12.12)] in addition to the ones allowed by Eq. (12.10). Consequently, combining α and σ_f parameters we can find such a combination that makes l_m independent of surface patchiness. This dependence vanishes exactly at height z_l. Replacing this height, given by Eq. (12.16), in Eq. (12.14) we obtain $l_m = \kappa z_l$. It means that, at z_l, the mixing length is only a function of the displacement height and α. The tendency of the lines representing mixing length l_m to approach the line representing the "old approach" is more emphasized for the low height vegetation than for the taller one. Since the expression (12.14), for $\alpha = 1$, is not defined, the only physical conclusion that can be derived is that Eq. (12.10) can not be obtained as a special case of Eq. (12.12). This is not surprising because Eq. (12.12) is derived by taking into account the presence of underlying surfaces with different vegetation fractional covers, while Eq. (12.10) is not based on such an assumption. For the taller vegetation the height, z_l, where the crossover point is located, becomes lower while the lines representing the different vegetation fractional covers are more apart. The lower location of z_l, in comparison with its location for the low height vegetation, can come from the fact that this height is closer to the canopy height [Eq. (12.14)] than it is in the case of the lower vegetation. Also, these lines show a tendency of shifting towards the right side of the domain bounded by the lines obtained by the "old approach" and approach suggested by Garratt (1978).

12.3.3 Wind profile above a non-uniform underlying surface under neutral conditions

Using the foregoing assumption that the friction velocity U_* is equal to $l_m \, du/dz$ yields

$$U_* = \kappa\{[\sigma(\alpha - 1) + 1]z - \sigma\alpha d\}\frac{du}{dz} \tag{12.17}$$

This equation can be integrated to

$$u(z) = \frac{U_*}{\kappa} \frac{1}{\sigma(\alpha - 1) + 1}\ln\{[\sigma(\alpha - 1) + 1]z - \sigma\alpha d\} + C_i \tag{12.18}$$

where C_i is an integration constant. This constant can be found if we introduce the assumption that the extrapolation of wind profile (12.18) gives zero wind velocity at some height z_k defined as

$$z_k = Z_0 + D \tag{12.19}$$

where

$$Z_0 = \frac{f(\alpha, m) \, z_0}{\sigma(\alpha - 1) + 1} \tag{12.20}$$

and

$$D = \frac{\sigma\alpha d}{\sigma(\alpha - 1) + 1} \tag{12.21}$$

are generalized roughness length and displacement height, respectively and $f(\alpha, m)$ is an arbitrary function representing the dependence of Z_0 on introduced aerodynamic characteristic $\alpha = \alpha \, (LAI \cdot C_{od})$, and m denotes an arbitrary constant. Since the experimental evidence indicates that the vegetative underlying surface is rougher than it is described by the classical logarithmic profile it means that Z_0 has to be higher than z_0. Below we have assumed that the function $f(\alpha, m)$ has a power form, i.e. $f(\alpha, m) = \alpha^m$ which increases monotonically with respect to α.

Then the above condition can be written as

$$0 = \frac{U_*}{\kappa} \frac{1}{\sigma(\alpha - 1) + 1}\ln\{[\sigma(\alpha - 1) + 1]z_k - \sigma\alpha d\} + C_i. \tag{12.22}$$

After substituting the expressions (12.19), (12.20) and (12.21) in the condition (12.22), we find that the constant C_i is given by

$$C_i = -\frac{U_*}{\kappa} \frac{1}{\sigma(\alpha-1)+1} \ln \alpha^m z_0. \tag{12.23}$$

Finally, combining the expressions (12.19) and (12.23) we reach a wind profile in the roughness sublayer above the non-uniform vegetative surface under neutral conditions (hereafter referred to as the "new profile"), which can be written in the form

$$u(z) = \frac{U_*}{\kappa} \frac{1}{\sigma(\alpha-1)+1} \ln \frac{z - \frac{\sigma \alpha d}{[\sigma(\alpha-1)+1]}}{\frac{\alpha^m z_0}{[\sigma(\alpha-1)+1]}} \tag{12.24}$$

or shortly

$$u(z) = \frac{U_*}{\kappa} \frac{1}{\sigma(\alpha-1)+1} \ln \frac{z - D}{Z_0} \tag{12.25}$$

if we use the definitions (12.14) and (12.15) representing the generalized roughness length and displacement height respectively.

Comparing the expressions (12.8) and (12.25) we can see that the "new profile" explicitly includes the dependence of the wind on the non-uniformity of the underlying vegetative surface while the "old logarithmic profile" or [profile given by Eq. (12.8)] does not. Moreover, the "old logarithmic profile" interprets the underlying vegetative surface as a smooth one regardless of whether the surface is uniformly covered by the vegetation or not. The same conclusion can be emphasized for the wind profile given by Eq. (12.9) which will be referred to as the "Garratt's logarithmic profile". This profile, established on the basis of the experimental evidence, is a special case of the "new profile" for $\sigma_f = 1$, $\alpha = \alpha_G$ and $m = 1$ where α_G is taken to be 1.5. Note that the profile given by Eq. (12.8) can be formally obtained from Eq. (12.24) for $\sigma_f = 1$ and $\alpha = 1$.

In the "new logarithmic profile", given by Eq. (12.24), we still have not determined the value of the constant m. So, now we are going to focus on this constant. The "old logarithmic profile" gives systematically higher values of the wind speed in comparison with the observations. It could be an indicator that the underlying surface is much rougher than it is represented by this profile. This fact can be expressed as

$$D + Z_0 \geq d + z_0, \tag{12.26}$$

which, after substituting expressions (12.9) (with $f(\alpha, m) = \alpha^m$) and (12.20), after some rearrangement, takes the form

$$\alpha^m - \sigma\alpha - (1 - \sigma)\left(1 + \frac{d}{z_0}\right) \geq 0. \tag{12.27}$$

This inequality can be used for the estimation of the value of the parameter m. First of all we may say that m should be significantly greater than 1 since for $m = 1$ the inequality (12.27) is satisfied only for $\sigma_f = 1$ and $\alpha = 1$, i.e. when the Z_0 and D reduce to the roughness length and displacement height for the "old logarithmic profile". When $\alpha \neq 1$, the lower limit of the parameter m can be estimated from this inequality using empirical profile data. Analysing the wind profiles measured above a broad range of forest we have found that an optimum value for the parameter m is 2. With this value of m, the expression (12.24) for the wind profile which will be used in this study has the form

$$u(z) = \frac{U_*}{\kappa} \frac{1}{\sigma(\alpha - 1) + 1} \ln \frac{[\sigma(\alpha - 1) + 1]z - \sigma\alpha d}{\alpha^2 z_0}. \tag{12.28}$$

The expressions for aerodynamic parameters Z_0 and D and friction velocity U_* for forest, derived from continuity conditions, can be found in Mihailović *et al.* (1999).

12.3.4   Exchange of momentum and heat above a non-uniform vegetative surface under non-neutral conditions

As mentioned before, the exchange process can be considered as a result of movement of eddies, carrying heat and momentum. It was assumed that the velocity U_* of the eddies was of the order of $l_m \, du/dz$ where l_m is a characteristic length. Looking dimensionally we can conclude that the accelerations, caused by the friction forces, are of the order U_*^2/l_m or $l_m(du/dz)^2$. Under non-neutral conditions eddies may be also generated by buoyancy, the forces caused by density differences between the air in the eddy and the surrounding air. Buoyancy acceleration is of the order of $\Delta\Theta g/T_A$, where $\Delta\Theta[\theta]$ is the difference between the potential temperatures above and inside the canopy, g [L T^{-2}] the gravity acceleration and $T_A[\theta]$ is the mean ambient temperature. Since the difference $\Delta\Theta$ is of the order $l_m \, d\Theta/dz$ then the ratio of the buoyancy and friction acceleration is given by

$$Ri_g = \frac{g \frac{d\Theta}{dz}}{T_A \left(\frac{du}{dz}\right)^2} Pr_t \tag{12.29}$$

which is commonly used expression for the gradient Richardson's number, Ri_g and where Pr_t is the turbulent Prandtl number. This number and the Monin-Obukhov length L [L], whose precise derivation can be found in Monin and Obukhov (1954) and Priestly (1959), are the most widely used parameters characterizing the degree of non-neutrality. The Monin-Obukhov length L can be considered as the height above the displacement height, where buoyancy forces and friction forces are approximately equal. L may be given as

$$L = \frac{T_A U_*^2}{\kappa g l_m \frac{d\Theta}{dz}}. \tag{12.30}$$

Following the Monin-Obukhov theory we introduced the dimensionless height parameter denoted by ζ

$$\zeta = \frac{z - D}{L}. \tag{12.31}$$

According to Mihailović *et al.* (1999), this approach causes changes in the dimensionless height parameter ζ. Furthermore, these changes cause changes in the Φ_m and Φ_h functions in the case of non-neutrality when calculating the exchange coefficients for momentum K_m and heat transfer K_h are being calculated. These coefficients may be written as

$$K_m = \frac{\kappa U_*\{\Gamma_1 z - \sigma\alpha d\}}{\Phi_m} \tag{12.32}$$

and

$$K_h = \frac{\kappa U_*\{\Gamma_1 z - \sigma\alpha d\}}{\Phi_h} \tag{12.33}$$

where

$$\Gamma_1(\sigma, \alpha) = \sigma(\alpha - 1) + 1.$$

Functions Φ_m and Φ_h are, according to Businger *et al.* (1971),

$$\Phi_m = (1 - 15\zeta)^{-0.25} \qquad \text{unstable } \zeta < 0 \tag{12.34a}$$
$$\Phi_m = (1 + 4.7\zeta) \qquad \text{stable } \quad \zeta > 0 \tag{12.34b}$$

$$\Phi_h = 0.74(1 - 9\zeta)^{-0.5} \qquad \text{unstable } \zeta < 0 \tag{12.35a}$$
$$\Phi_h = 0.74(1 + 4.7\zeta) \qquad \text{stable } \quad \zeta > 0 \tag{12.35b}$$

Relations (12.34) and (12.35) are derived for air column over very homogeneous terrain. Certainly, it does not guarantee that their form will not be unaltered over a patchy surface. We assumed that the relations (12.34) and (12.35) can be maintained. The only differences between correction factors Φ_m and Φ_h, for homogeneous and non-homogeneous underlying surface, come from different values of the parameter ζ where its dependence on the vegetation fractional cover σ_f, is implicitly incorporated. Consequently, we have adapted correction factors Φ_m and Φ_h.

In the literature, alternative solutions can be found for the formulation of the effect of non-neutrality on the profiles of the exchange coefficients. A detailed elaboration of this subject concerning its theoretical and practical aspects is given by Goudriaan (1977). Following him we have derived the expression for the Monin-Obukhov length L in the form

$$L = \frac{T_A U_*^2 \displaystyle\int_{z_1}^{z_r} \frac{\Phi_h}{z' - D}dz'}{\Gamma_1 \kappa^2 g \Delta\Theta} \tag{12.36}$$

where the difference $\Delta\Theta$ for two heights $z_1 = D + Z_0$ and z_r an arbitrary reference level above it, is taken as $\Theta(z_r) - \Theta(z_1)$, which is negative under unstable conditions and positive under stable ones. However, the value of U_*, which is needed in Eq. (12.38) must be derived from a general profile defined by

$$\frac{du}{dz} = \frac{U_* \Phi_m}{\kappa\{\Gamma_1 z - \sigma\alpha d\}}. \tag{12.37}$$

Integration of this equation gives

$$u_r = \frac{U_*}{\kappa} \int_{D+Z_0}^{z_r} \frac{\Phi_m}{\Gamma_1 z' - \sigma\alpha d}dz' \tag{12.38}$$

where ur is the wind speed at the reference level. If we want to take into account the effect of non-neutrality then Eqs. (12.31), (12.34), (12.35), (12.36), and (12.38) must be solved simultaneously. In the stable case ($\zeta > 0$) the integration in Eqs. (12.36) and (12.38) can be done analytically. Otherwise, in the unstable conditions ($\zeta < 0$), the calculations must be done iteratively.

12.4 TURBULENCE AND WIND WITHIN THE FOREST

The main motive for studying turbulent flow within the forest is to understand processes governing momentum, mass and energy exchange between the atmosphere and forest canopy. Additionally, during the XX century the scientific community emphasised the importance of wind behaviour for the movement of spores, pollen and particles within and just above the vegetation canopy (Pingtong and Hidenori, 2000; Pinard and Wilson, 2001) as well as for the forest fires spread rate. Ecological and financial effects of forest fires have revealed a definite need for better understanding of wind profiles within and above forest (Curry and Fons, 1938).

In this section we present results of turbulent transfer parameterization within the homogeneous and non-homogeneous canopy. The first-order closure techniques based on K-theory for calculating the Reynolds' stresses within the canopy are described. Limitations of the traditionally parametrized canopy structure and the turbulent transfer coefficient for the forest canopy are considered. In addition, some approaches to turbulent transfer parameterization are presented using the forest morphological characteristics. We will focus on the momentum transfer parameterization since heat and mass transfer are treated in analogous manner.

12.4.1 Short overview of turbulent transfer parameterization within the canopy

The vertical distribution of momentum within different plant communities has usually been modelled by assuming steady and unidirectional wind and negligible pressure gradient force. Under these conditions, the time- and volume-averaged equation for the mean momentum within vegetation (Raupach *et al.*, 1986) turns into a relation describing balance between the vertical shear stress change and a drag force:

$$\frac{1}{\rho_a} \frac{\partial}{\partial z}(-\overline{uw}) = C_{dg}LAD(z)u^2 \tag{12.39}$$

where $\overline{uw}$ [M L^{-1}T^{-2}] is vertical shear (Reynolds') stress describing turbulent transfer of x-component of momentum in z-direction.

Early modelling studies (Cowan, 1968; Thom, 1971) were based on K-theory supposing that the turbulent momentum flux is equal to the product of an eddy viscosity, represented by turbulent transfer coefficient K_m [M L^{-1}T^{-1}], and the local gradient of mean wind velocity. Hence Eq. (12.39) could be written in the form:

$$\frac{1}{\rho_a} \frac{\partial}{\partial z}\left(K_m \frac{du}{dz}\right) = C_{dg}LAD(z)u^2 \tag{12.40}$$

Various assumptions have been made regarding the behaviour of K_m within the canopy. They could be classified as follows:

a) K_m is proportional to wind speed, u ($K_m \propto u$) (Cowan, 1968; Denmead, 1976);
b) K_m depends on canopy height, h_c ($K_m \propto K_m(h_c)$) (Jarvis *et al.*, 1976);
c) K_m is a product of local gradient of mean wind velocity du/dz [T^{-1}] and mixing length within the canopy l_{mc} [L] (Inoue, 1963; Raupach and Thom, 1981; Baldocchi and Meyers, 1988)

$$K_m = l_{mc}^2 \frac{\partial u}{\partial z}.$$
(12.41)

During the decade of K-theory application, it become obvious that this model can not provide accurate predictions of wind velocity in lower part of plant canopy where near-zero vertical gradient wind velocity is frequently observed (Shaw, 1977). Corsin (1974) has pointed out that the application of this, also called small-eddy closure technique, (Stull, 1988) is limited to the places where the length scales of flux-carrying motions have to be much smaller than the scales associated with average gradients (Zeng and Takahashi, 2000). Unfortunately, many measurements have shown that the air flow within and just above the canopy is dominated by turbulence with vertical length scales at least as large as the vegetation height (Kaimal and Finigan, 1994). To provide a more reliable insight into the nature of momentum transfer processes within the canopy, some authors suggested higher-order closure models (Wilson and Shaw, 1977; Meyers and Paw, 1987).

As an alternative solution to not-using these closure techniques appears a non-local first-order closure model developed by Zeng and Takahashi (2000). In this model turbulent momentum flux is divided into two parts: one, diffused by the smaller-scale eddies and parametrized according to conventional K-theory; and the other, transported by large-scale eddies as a result of non-local transport caused by shear between air flows above and within canopies. However, vertical shear stress is parametrized in the form:

$$-\overline{uw} = K_m \frac{du}{dz} + C_g u_r (u_r - u) \frac{z}{h_c}$$
(12.42)

where u_r [$L\,T^{-1}$] is a wind speed at reference height above vegetation and C_g [$M\,L^{-3}$] is an coefficient.

12.4.2   Single layer approach for parameterization of turbulent transfer within the canopy

In the case of homogeneous canopy ($LAD(z) = const.$), according to Eq. (12.1), leaf area density $LAD(z)$ can be calculated as:

$$LAD(z) = \frac{LAI}{h_c}.$$
(12.43)

Substituting $LAD(z)$ from Eq. (12.43) into Eq. (12.39), balance between the vertical shear stress change and drag force takes the form:

$$\frac{1}{\rho_a} \frac{\partial}{\partial z}(-\overline{uw}) = C_{dg} \frac{LAI}{h_c} u^2.$$
(12.44)

Assumption that K_m is proportional to wind speed u and that coefficient of proportionality σ is a constant ($K_m = \sigma u$), leads to well known Cowan's profile (Cowan, 1968):

$$u(z) = u(h_c) \left[\frac{\sinh\left(\beta_c \frac{z}{h_c}\right)}{\sinh \beta_c} \right]^{1/2}, \tag{12.45}$$

where $u(h_c)$ is the wind speed at the canopy top, h_c is the canopy height and β_c is the extinction factor defined for wind profile within the canopy:

$$\beta_c^2 = \frac{2h_c C_{dg} LAI}{\sigma}. \tag{12.46}$$

Using the third assumption for K_m, defined by Eq. (12.41), and supposing that mixing length l_{mc} is a constant within the whole canopy space, Inoue (1963) derived exponential wind profile in the form:

$$u(z) = u(h_c) \exp\left[-a_c \left(1 - \frac{z}{h_c} \right) \right], \tag{12.47}$$

where a_c is the canopy coefficient.

Using the wind profile within the canopy given by Eq. (12.48) and taking into account non-uniformity of underlying surface, Mihailović *et al.* (2006) assumed wind profile within the vegetation in the form:

$$u(z) = \sigma_f u(h_c) e^{-\frac{1}{2}\beta_1 \left(1 - \frac{z}{h_c}\right)} + (1 - \sigma_f) \frac{u_*}{\kappa} \ln \frac{z}{z_0}. \tag{12.48}$$

They supposed that the first term on right-hand side of Eq. (12.48) describes well vertical transport of momentum within homogeneously vegetated part of canopy, while the second term is responsible for the turbulence above a bare soil situated within canopy space. β_1 appearing in Eq. (12.48) is the extinction factor obtained by an iterative procedure.

12.4.3 Two-layer approach for parameterization of turbulent transfer within the canopy

The assumption that canopy is a homogeneous medium could be appropriate in the case of grass and tall grass canopy space. However, forest canopy is extremely heterogeneous due to the complexity in tree structure and presence of two specific layers, crown and stands, affecting the transport of momentum into atmosphere on the following way. The absorption of momentum between the crown top and the bottom is 70–90%, depending on the crown depth and the density. The attenuation of momentum, below the bottom of the crown, is rather small up to the roughness layer, where the rest of the air momentum is transferred to the ground due to molecular transport.

According to the observations, wind profile within the forest canopy may significantly deviate from the profiles proposed by Cowan (1968) and Inoue (1963). One should not be surprised since both relations are derived supposing that plant canopy is a homogeneous one, which is not acceptable in the case of forest canopy. In order to adequately describe within-canopy vertical momentum transfer, Lalić and Mihailović suggested (Lalić, 1997; Lalić and Mihailović, 1998; Lalić and Mihailović, 2002a; Lalić and Mihailović, 2002b;

Lalić *et al.*, 2003) an empirical expression for the wind profile within the forest based on two-layer canopy model in the form:

$$
u(z) = \begin{cases} u_h \left[\dfrac{\cosh \beta_c\left(\frac{z-z_d}{h_c}\right)}{\cosh \beta_c\left(1 - \frac{z_d}{h_c}\right)} \right]^{\frac{5}{2}} & z_d < z \leq h \\[4mm] C_h u(h_c) & z_0 < z \leq z_d \end{cases}
\tag{12.49}
$$

where: z_d is the crown bottom height and C_h is a constant. According to Massman (1987), factor β_c is equal to $4C_{dg}LAI/(\alpha^2\kappa^2)$. In the case of forest canopy, bearing in mind that smoothness or roughness of canopy from atmospheric point of view is an effect of the amount of leaves and their roughness, α can be parameterized as $\alpha^2 = 4(C_{dg}LAI)^{1/4}$ (Lalić, 1997; Lalic and Mihailović, 1998). In creating the foregoing profile, the evidence that comes from the observations of the wind profile within the forest was taken into account. After the comparisons of the wind profile observed and the wind profile defined by Eq. (12.49) it becomes obvious that two-layer approach in parameterization of forest canopy structure produces minimum deviation from the observation particularly in the layer occupied by the tree crown, where the absorption of momentum is mostly emphasized.

The wind profile defined by Eq. (12.49) requires an additional assumption in defining the momentum transfer coefficient, K_m i.e. turbulent diffusivity within the forest. Instead of commonly used assumption for K_m in the form $K_m(z) = \sigma u(z)$, describing the turbulence through the whole environment occupied by plants, we have introduced another one. For simplicity, σ is often assumed to be a constant regardless of the structure of the canopy vegetation. However, in the case of the forest this idea can be applied just in some part of its environment. Thereby, we have assumed that in the crown of the forest $(h_c > z \geq z_d)\sigma$ can be considered as a function of height z, i.e. $\sigma = \sigma(z)$, while below it $(z_d > z \geq z_0)\sigma$ remains constant. Thus, the momentum transfer coefficient can be written in the form:

$$
K_m(z) = \begin{cases} \sigma(z)u(h_c) \left[\dfrac{\cosh \beta_c\left(\frac{z-z_d}{h_c}\right)}{\cosh \beta_c\left(1 - \frac{z_d}{h_c}\right)} \right]^{\frac{5}{2}} & z_d < z \leq h_c \\[4mm] \sigma_d C_h u(h_c) & z_0 < z \leq z_d \end{cases}
\tag{12.50}
$$

where σ_d is assumed to be a constant.

The functional form of $\sigma(z)$ may be found as the solution of the differential equation describing the shear stress within the canopy according to K-theory and supposing that each of two layers is a homogeneous one

$$
\frac{d}{dz}\left(K_m\frac{du}{dz}\right) = \frac{C_{dg}LAI}{h_c}u^2.
\tag{12.51}
$$

Using expressions (12.49) and (12.50) for $u(z)$ and $K_m(z)$, the solution to Eq. (12.51) has the following form:

$$
\begin{aligned}
\sigma(z) = {} & \frac{2C_{dg}LAI\,h_c}{7\beta_c^2 ch^6\beta_c\left(\frac{z-z_d}{h_c}\right)}\left[1 + \sinh^2\beta_c\left(\frac{z-z_d}{h_c}\right)\right. \\[2mm]
& \left. + \frac{3}{5}\sinh^4\beta_c\left(\frac{z-z_d}{h_c}\right) + \frac{1}{7}\sinh^6\beta_c\left(\frac{z-z_d}{h_c}\right)\right].
\end{aligned}
\tag{12.52}
$$

The quantities z_d, C_h and σ_d included in the foregoing expressions should be derived following continuity conditions (continuity of wind speed, turbulent momentum transfer coefficient and continuity of their first derivative) at forest height and crown bottom height.

APPENDIX A – LIST OF SYMBOLS

List of Symbols

Symbol	Definition	Dimension or Units
C_h	constant	
C_i	integration constant	$[\mathrm{m\,s^{-1}}]$
C_{dg}	the leaf drag coefficient estimated from the size of the roughness elements of the ground	
D	generalized displacement height	$[\mathrm{m}]$
K	is an extinction coefficient	
K_m	turbulent transfer coefficient	$[\mathrm{m^2\,s^{-1}}]$
L	Monin-Obuhkov length	$[\mathrm{m}]$
LAI	leaf area index	$[\mathrm{m^2\,m^{-2}}]$
LAD	leaf area density	$[\mathrm{m^2\,m^{-3}}]$
LAD_0	leaf area density of forest crown	$[\mathrm{m^2\,m^{-3}}]$
$\overline{L}_d$	the area-averaged stem and leaf area density (also called canopy density)	$[\mathrm{m^2\,m^{-2}}]$
L_m	maximum value of leaf area density	$[\mathrm{m^2\,m^{-3}}]$
L_z	path length through which the light comes to level z	
P_z	probability of a beam of light passing through a horizontal plane of leaves at hight z	
P_{rt}	turbulent Prandtl number	
Ri_g	gradient Richardson's number	
T	surface (canopy or soil) temperature	$[\mathrm{K}]$
T_A	mean ambient temperature	$[\mathrm{K}]$
U_*	friction velocity above non-homogeneously covered surface	$[\mathrm{m\,s^{-1}}]$
Z_0	generalized roughness length	$[\mathrm{m}]$
a_c	canopy coefficient	
c_g	coefficient	$[\mathrm{kg\,m^{-3}}]$
d	displacement height	$[\mathrm{m}]$
du/dz	local gradient of mean wind velocity	$[\mathrm{s^{-1}}]$
g	gravitational acceleration constant	$[\mathrm{m\,s^{-2}}]$
h_c	canopy height	$[\mathrm{m}]$
l	mixing length above a non-homogeneously covered surface	$[\mathrm{m}]$
l_m^b	mixing length above the bare soil	$[\mathrm{m}]$
l_m^c	mixing length above the canopy	$[\mathrm{m}]$
l_{mc}	mixing length within the canopy	$[\mathrm{m}]$
l_m^α	the mixing length in the roughness sublayer above a vegetative surface	$[\mathrm{m}]$

(Continued)

List of Symbols

Symbol	Definition	Dimension or Units
u_{*g}	friction velocity for a bare soil	[m s^{-1}]
u_*	the friction velocity	[m s^{-1}]
$u(h_c)$	the wind speed at the canopy height	[m s^{-1}]
$u(z)$	the wind speed	[m s^{-1}]
u_r	wind speed at reference height above vegetation	[m s^{-1}]
$\overline{uw}$	vertical shear stress	[kg m^{-1}s^{-2}]
z	the vertical coordinate	[m]
z_I	some certain height within the canopy	[m]
z_0	the roughness length	[m]
z_{0g}	roughness length of a bare soil	[m]
z_d	crown bottom height	[m]
z_g	effective ground roughness length	[m]
z_k	zero wind velocity height	[m]
z_m	corresponding height	[m]
z_r	the reference height above vegetation	[m]
$\Delta\Theta$	difference between the potential temperatures above and inside the canopy	[K]
Φ_h	correction factor for heat	
Φ_m	correction factor for momentum	
α	dimensionless constant representing corrected value of the mixing length in the roughness sublayer	
α_G	dimensionless constant estimated to be between 1.5 and 2.0	
β_1	extinction factor	
β_c	extinction parameter for within canopy wind profile	
ζ	dimensionless height parameter	
θ_s	solar zenith angle	[°]
κ	Von Kármán's constant	
ρ_a	the air density	[kg m^{-3}]
σ	coefficient of proportionality	
σ_d	constant	
σ_f	vegetation cover in fractional units	
τ_H	transmittance of whole canopy	
τ_z	transmittance of canopy layer at hight z	

APPENDIX B – SYNOPSIS

Forest is a vegetation system covering more than 20% of Earth's land surface. The atmosphere 'feels' the presence of trees up to a few hundred meters from the ground, depending on tree height. As a dynamical source and sink of momentum, heat, water (vapour) and pollution, forest plays a crucial role in land-atmosphere-interaction modelling. Therefore, forest as an underlying surface is often found in atmospheric and environmental models of different scales. Experimental evidence indicates that there is a significant departure of the wind profile above a vegetative surface from that predicted by the logarithmic relationship, which yielding the values greater than those observed. The source of this deviation can be found in inadequate parameterisation of forest architecture (often considered uniform) and spatial distribution of forest canopy (often non-uniform, consisting of patches of bare soil and plant elements). A new approach to parameterisation of turbulent transfer, including a detailed simulation of the forest architecture and explicit accounting of non-uniformities in plant distribution, is an important step in producing a more realistic physical picture of turbulence within the atmospheric surface layer.

APPENDIX C – KEYWORDS

By the end of the chapter you should have encountered the following terms. Ensure that you are familiar with them!

Surface layer	Logarithmic profile	Roughness length
Leaf area density	Mixing length	Turbulent eddies
Canopy architecture	Turbulent transfer coefficient	Eddy viscosity
Monin-Obukhov theory	Friction velocity	Atmospheric stability

APPENDIX D – QUESTIONS

What is the atmospheric surface layer?

How can one model the canopy architecture?

How can one model the steady-state flow over horizontally bare soil under thermally neutral conditions?

How can one parameterise the impact of non-uniformities in plant distribution on turbulent transfer above vegetated surface?

How do different parameterisation of the canopy architecture (single layer, two layer, and multilayer approach) affect modelling of the turbulent transfer within the canopy?

APPENDIX E – PROBLEMS

E1. Describe the hierarchy in modelling the canopy architecture. Conduct a computational time vs. benefit analysis of canopy models of different complexity.

E2. Describe the importance of canopy architecture modelling from the land-atmosphere-interaction modelling point of view.

E3. Explain limitations of the standard logarithmic profile described by Eq. (12.7). Compare expected effects of Eqs. (12.7) and (12.18).

E4. Make a short overview of different approaches to turbulent transfer parameterization within the canopy.

REFERENCES

Baldocchi, D.D. and Meyers T.P., 1988, Turbulence structure in a deciduous forest. *Boundary-Layer Meteorology*, **43**, pp. 345–364.

Businger, J.A., Wyngaard, J.C., Izumi, Y.I. and Bradley E.F., 1971, Flux-Profile Relationships in the Atmospheric Surface Layer. *Journal of Atmospheric Sciences*, **28**, pp. 181–189.

Corrsin, S., 1974, Limitations of gradient transport model in random walks and in turbulence. *Advances in Geophysics*, **18A**, pp. 25–60.

Cowan, I.R., 1968, Mass, heat and momentum exchange between stands of plants and their atmospheric environment. *Quarterely Journal of Royal Meteorological Society*, **94**, pp. 523–544.

Curry, J.R. and Fons, W.L., 1938, Transfer of heat and momentum in the lowest layers of the atmosphere. *Great Britain Meteorological Office Geophysics*. Memoir No. **65**, p. 66.

De Bruin, H.A.R. and Moore, C.J., 1985, Zero-Plane Displacement and Roughness Length for Tall Vegetation, Derived From a Simple Mass Conservation Hypothesis. *Boundary-Layer Meteorology*, **42**, pp. 53–62.

Denmead, O.T., 1976, Temperate cereals. In*Vegetation and the Atmosphere*, Vol. 2, Edited by Monteith, J.L. (Academic Press, New York, U.S.A.), pp. 1–31.

Garratt, J.R., 1978, Flux Profile Relations above Tall Vegetation. *Quarterely Journal of Royal Meteorological Society*, **104**, pp. 199–211.

Goudriaan, J., 1977, *Crop Micrometeorology: A Simulation Study*. (Wageningen Center for Agricultural Publishing and Documentation).

Gower, S.T., Kucharik, C.J. and Norman J.M., 1999, Direct and indirect estimation of leaf area index, f_{APAR} and net primary production of terrestrial ecosystems. *Remote Sensing of Environment*, **70**, pp. 29–51.

Högström, U., 1985, Von Kármán's Constant in Atmospheric Boundary Layer Flow: Reevaluated. *Journal of Atmospheric Sciences*, **42**, pp. 263–270.

Inoue, E., 1963, On the turbulent structure of air flow within crop canopies. *Journal of Meteorological Society of Japan*, **41**, pp. 317–325.

Jarvis, P.G., James, B. and Landsberg J.J., 1976, Coniferous forest. In *Vegetation and the Atmosphere*, Vol. 2, Edited by Monteith, J.L. (Academic Press, New York, U.S.A.), pp. 171–240.

Kaimal, J.C. and Finnigan, J.J., 1994, *Atmospheric Boundary Layer Flows. Their Structure and Measurement*, (Oxford University Press, New York, U.S.A.).

Kolic, B. 1978, *Forest ecoclimatology*. University of Belgrade, Yugoslavia, (In Serbian).

Lalić, B., 1997, *Profile of Wind Speed in Transition Layer above the Vegetation*, Master Thesis, University of Belgrade, Serbia, (in Serbian).

Lalić, B. and Mihailović, D.T., 1998, Derivation of aerodynamic characteristics using a new wind profile in the transition layer above the vegetation. *Research Activities in Atmospheric and Oceanic Modelling*, **27**, pp. 4.17–4.19.

Lalić, B. and Mihailović, D.T., 2002a, A new approach in parameterisation of momentum transport inside and above forest canopy under neutral conditions. *Intergrated Assesment and Decision Support. Proceedings of the 1st biennial meeting of the International Environmental Modelling and Software Society*, Vol. **2**, University of Lugano, Switzerland 24/27 June, pp. 436–441.

Lalić, B. and Mihailović, D.T., 2002b, New approach to parameterisation of momentum transport within tall vegetation. *Extended Abstracts, 18th International Conference on Carpathian Meteorology*, 7/11 October 2002, Belgrade (Serbia), pp. 91–93.

Lalić, B., and Mihailović, D.T. 2004, An Empirical Relation Describing Leaf Area Density inside the Forest for Environmental Modelling. *Journal of Applied Meteorology*, **43**, pp. 641–645.

Lalić, B., Mihailović, D.T., Rajkovic, B., Arsenic, I.D. and Radlovic, D. 2003, Wind profile within the forest canopy and in the transition layer above it. *Environmental Modelling & Software*, **18**, pp. 943–950.

Law, B.E., Cescatti, A. and Baldocchi, D.D., 2001a, Leaf area distribution and radiative transfer in open-canopy forests: implications for mass and energy exchange. *Tree Physiology*, **21**, pp. 777–787.

Law, B.E., Kelliher, F.M., Baldocchi, D.D., Anthoni, P.M., Irvine, J., Moore, D. and Van Tuyl, S., 2001b, Spatial and temporal variation in respiration in a young ponderosa pine forest during a summer drought. *Agricultural and Forest Meteorology*, **110**, pp. 27–43.

Levi, P.E. and Jarvis, P.G., 1999, Direct and indirect measurements of LAI in millet and fallow vegetation in HAPEX-Sahel. *Agricultural and Forest Meteorology*, **97**, pp. 199–212.

Massman, W., 1987, A Comparative Study of some Mathematical Models of the Mean Wind Structure and Aerodynamic Drag of Plant Canopies. *Boundary-Layer Meteorology*, **40**, pp. 179–197.

Meir, P., Grace, J. and Miranda, A.C., 2000, Photographic method to measure the vertical distribution of leaf area density in forests. *Agricultural and Forest Meteorology*, **102**, pp. 105–111.

Meyers, T. and Paw, K.T., 1987, Modelling the Plant Canopy Micrometeorology with Higher-Order Closure Principles. *Agricultural and Forest Meteorology*, **41**, pp. 143–163.

Mihailović, D.T. and Kallos, G., 1997, A Sensitivity Study of a Coupled Soil-Vegetation Boundary-Layer Scheme for Use in Atmospheric Modelling. *Boundary-Layer Meteorology*. **82**, pp. 283–315.

Mihailović, D.T, Lalić, B., Rajkovic, B. and Arsenic, I., 1999, A roughness sublayer wind profile above non-uniform surface. *Boundary-Layer Meteorology*. **93**, pp. 425–451.

Mihailović, D.T., Lalić, B., Eitzinger, J., Malinovic, S. and Arsenic, I., 2006, An Approach for Calculation of Turbulent Transfer Coefficient for Momentum inside Vegetation Canopies. *Journal of Applied Meteorology and Climatology*, **45**, pp. 348–356.

Mix, W., Goldberg, V. and Bernhardt, K.H., 1994, Numerical experiments with different approaches for boundary layer modelling under large-area forest canopy conditions. *Meteorologische Zeitschrift*, **3**, pp. 187–192.

Monin, A.S. and Obukhov, A.M., 1954, Basic Regularity in Turbulent Mixing in the Surface Layer in the Atmosphere. *Trudi Geofizicheskovo Instituta Akademii Nauk USSR*, No. **24**.

Monin, A.S. and Yaglom, A.M. 1971, *Statistical Fluid Mechanics: Mechanics of turbulence*. Vol. 1, (The MIT Press, Cambridge, U.S.A.).

Pinard, J.D.J. and Wilson, J.D., 2001, First and second order closure models for wind in a plant canopy. *Journal of Applied Meteorology*, **40**, pp. 1762–1768.

Pingtong, Z. and Hidenori, T., 2000, A first order closure model for the wind flow within and above vegetation canopies. *Agricultural and Forest Meteorology*, **103**, pp. 301–313.

Priestly, C.H.B., 1959, *Turbulent transfer in the lower atmosphere*, (The University Chicago Press, Chicago, U.S.A.).

Raupach, M.R., 1980, Conditional Statistics of Reynolds Stress in Rough-Wall and Smooth-Wall Turbulent Boundary Layers. *Journal of Fluid Mechanics*, **108**, pp. 363–382.

Raupach, M.R. and Thom, A.S., 1981, Turbulence in and above Plant Canopies. *Annual Review of Fluid Mechanics*, **13**, pp. 97–129.

Raupach, M.R., Thom, A.S. and Edwards, I., 1980, A Wind Tunnel Study of Turbulent Flow Close to Regularly Arrayed Rough Surfaces. *Boundary-Layer Meteorology* **18**, 373–397.

Raupach, M.R., Coppin, P.A., Legg, B.J., 1986, Experiments on scalar dispersion within a model plant canopy. Part I: The turbulence structure. *Boundary-Layer Meteorology*, **35**, 21–52.

Sellers, P.J., Mintz, Y., Sud, Y.C. and Dachler, A. 1986, A Simple Biosphere Model (SiB) for Use within General Circulation Models. *Journal of Atmospheric Sciences*, **43**, pp. 505–531.

Shaw, R.H., 1977, Secondary wind speed maxima inside plant canopies. *Journal of Applied Meteorology*, **16**, pp. 514–521.

Shaw, R.H. and Pereira, A.R., 1982, Aerodynamic Roughness of a Plant Canopy: A Numerical Experiment. Agricultural Meteorology, **26**, pp. 51–65.

Stull, R.B., 1988, *An Introduction to Boundary Layer Meteorology*, (Kluwer Academic Publishers).

Tennekes, H., 1982, Similarity Relations, Scaling Laws and Spectral Dynamics. In *Atmospheric Turbulence and Air Pollution Modelling*, Edited by Nieuwstadt, F.T.M. and van Dop, H. (D. Reidel Publ. Co., Dordecht, Holland), pp. 37–64.

Thom, A.S., 1971, Momentum Absorption by Vegetation. *Quarterly Journal of the Royal Meteorological Society*, **97**, pp. 414–428.

Watanabe, T. and Kondo, J. 1990, The Influence of canopy structure and density upon the mixing length within and above vegetation. *Journal of the Meteorological Society of Japan*, **68**, pp. 227–235.

Wilson, N.R. and Shaw, R.H., 1977, A higher order closure model for canopy flow. *Journal of Applied Meteorology*, **16**, pp. 1197–1205.

Wilson, J.D., Ward, D.P., Thurtell, G.W. and Kidd, G.E., 1982, Statistics of Atmospheric Turbulence within and above a Corn Canopy. *Boundary-Layer Meteorology*, **24**, pp. 495–519.

Zeng, P. and Takahashi, H., 2000, A first order closure model for the wind flow within and above vegetation canopies. *Agricultural and Forest Meteorology*, **103**, pp. 301–313.

CHAPTER THIRTEEN

Flow and mass transport in vegetated surface waters

Yukie Tanino

Department of Earth Science and Engineering, Imperial College London, London, United Kingdom

ABSTRACT

This chapter describes flow and mass transport under conditions relevant to surface water systems with emergent vegetation. It is convenient to conceptualize vegetated surface waters as homogeneous arrays of discrete, rigid, two-dimensional plant elements, e.g., stems. The integral length scale of turbulence within a homogeneous emergent canopy, which has traditionally been taken to be the canopy element diameter, is constrained by the interstitial pore size in dense canopies. The canopy-averaged integral length scale is well-described by the average surface-to-surface separation between a canopy element and its nearest neighbour. One proposed consequence of this reduction in the integral length scale is that turbulent diffusion decays with increasing canopy density at solid volume fractions above $O(3)\%$. This explains the intermediate regime exhibited by net lateral dispersion, in which the dispersion coefficient, normalized by the interstitial flow velocity and canopy element diameter, decreases with increasing canopy density. Net longitudinal dispersion increases with increasing canopy density in sparse ($\phi < 6.4\%$) canopies; its behaviour in more dense canopies have not been established. In sparse canopies at transitional Reynolds numbers, the dominant contributions to longitudinal dispersion are associated with transient trapping in the unsteady wake of the elements and the velocity deficit downstream of the elements.

At an interface between an emergent canopy and open water that is parallel to the mean flow, the discontinuity in drag results in shear, which in turn gives rise to Kelvin-Helmholtz billows and shear-scale turbulence. This shear-scale turbulence dominates lateral transport across the canopy-open water interface. The penetration of the KH billows into the canopy is constrained by the canopy drag.

13.1 FOREWORD

Aquatic vegetation interacts with ambient flow in many ways. First and foremost, aquatic plants introduce additional hydraulic resistance to flow, thereby contributing to wave attenuation (e.g., Möller *et al.*, 1999; Mazda *et al.*, 2006). Similarly, some studies have attributed relatively clement local damage from tsunamis (Danielsen *et al.*, 2005) and hurricanes (Das and Vincent, 2009) to the presence of mangroves. Second, aquatic plants may improve water quality through its impact on the transport of its constituents, such as nutrients, pollutants, and sediment (e.g., Knight *et al.*, 1999). They may do this directly, through uptake and transformation of nutrients and pollutants, or indirectly, by reducing bed stress (López and García, 1998) or by providing additional surface for particle deposition (Stumpf, 1983). Taking advantage of this, wetlands have been artificially constructed to provide additional treatment of wastewater. One such site is in Augusta, GA (Fig. 13.1).

Figure 13.1. Artificial treatment wetland in Augusta, GA. Photo courtesy of Anne F. Lightbody.

Several recent works have reviewed the fluid dynamics of surface waters with submerged vegetation (Nepf, 2012; Nepf and Ghisalberti, 2008). This chapter complements these works by focusing on emergent plant canopies, in which the plant stems span the water column and penetrate the free surface. Section 1.2 describes the concept of temporal and spatial averaging, and presents the governing equations commonly used to describe flow and mass transport in homogeneous canopies. Section 1.3 focuses on the salient features of turbulence and mass transport within a homogeneous canopy. Finally, Section 1.4 discusses the dominant features of flow parallel to a water-vegetation interface. Such flows occur at, for example, river-floodplain boundaries.

13.1.1   Flow and canopy characterization

An emergent plant canopy is essentially an array of numerous plant elements, with the space between them occupied by water. For example, marshes may be regarded as an array of stems, and mangroves as an array of trunks and roots (see Fig. 13.2). Then, the density of a canopy may be characterized by the mean plant number density (elements per unit

Figure 13.2. A salt marsh in the Florida Everglades (left) and a mangrove (right). Photos courtesy of Heidi M. Nepf (left) and Brian L. White (right).

horizontal area), m. The fundamental length scale of the canopy is the characteristic width of the canopy element, d. Note that another commonly used parameter, the frontal area per unit volume, is the product of these two parameters.

Field measurements demonstrate that canopy architecture varies significantly amongst real emergent canopies. In salt marshes, for example, stem densities ranging from $m = O(20)$ to $O(4000)$ m^{-2} have been measured (Tyler and Zieman, 1999; Leonard and Luther, 1995; van Hulzen *et al.*, 2007; Valiela *et al.*, 1978; Neumeier and Amos, 2006; Huang *et al.*, 2008; Bergen *et al.*, 2000). In contrast, densities of mangrove trunks and roots are relatively smaller, with $m = 0.08$–1.0 trunks m^{-2} and $m = 30$–122 roots m^{-2} (Mazda *et al.*, 1997, 2006; Furukawa *et al.*, 1997). Similarly, the characteristic plant width may be as small as $d = 1$ to 10 mm (stem diameter) in salt marshes (Valiela *et al.*, 1978; Lightbody and Nepf, 2006; Harvey *et al.*, 2009; Larsen *et al.*, 2009; James *et al.*, 2004; Huang *et al.*, 2008), or as large as $d = 4$ cm (root diameter) to 12 cm (trunk diameter) in mangroves (Mazda *et al.*, 1997, 2006; Furukawa *et al.*, 1997). Furthermore, the appropriate canopy element need not be an individual plant. For example, Oldham and Sturman (2001) modelled a *Schoenoplectus validus* canopy as an array of plant clumps, where the characteristic clump diameter was $d = 15$ cm. Similarly, shoots in the littoral region of a lake were found to be distributed roughly uniformly randomly in regions of low mean flow, but in $O(1)$ m $\times O(1)$ m clusters in regions of high mean flow (Asaeda *et al.*, 2005). The canopy element is the shoot in the former, and the cluster of shoots in the latter.

Details of the aforementioned field measurements are summarized in Table 13.1. Not included in Table 13.1 is the extensive set of measurements collected in the Everglades, FL, USA, that are already presented in tabulated form by Carter *et al.* (1999, 2003).

The values of m and d suggest that the fraction of the total (vegetation and water) volume occupied by the vegetation, or the solid volume fraction ($1 -$ porosity), is roughly $\phi \approx 0.02$ to 3% in salt marshes. In contrast, ϕ in mangroves may be as high as $\phi = 45\%$ near the bed, due to the dense network of roots (Mazda *et al.*, 1997). Artificial wetlands may be as dense as $\phi = 65\%$ (Serra *et al.*, 2004).

An emergent canopy is often conceptualized as an array of rigid, circular cylinders of uniform diameter d distributed with a constant density m; the discussion that follows assumes that this approximation is valid. The solid volume fraction of such an array is $\phi = m\pi d^2 / 4$.

The dynamics of flow and passive transport through a canopy are largely governed by two length scales. The first is the canopy element diameter, d. The second is the characteristic spacing between canopy elements defined between their surface, s; s^2 is the characteristic area (volume per unit water depth) of the void space between the canopy elements. s, given d and m, reflects the spatial distribution of the elements. In general, the choice of s depends on the flow or transport phenomenon of interest. For example, the average surface-to-surface distance between a canopy element and its nearest neighbour, $\langle s_n \rangle_A$, controls mean drag (Tanino, 2008; cf. Spielman and Goren, 1968; Koch and Ladd, 1997) and turbulent diffusion (Tanino and Nepf, 2008b) in random arrays. In a square configuration, the nearest-neighbour separation is the same for all elements and is related to d and m as $\langle s_n \rangle_A = 1/\sqrt{m} - d$ (Fig. 13.3a). In contrast, the nearest-neighbour spacing for a particular element, s_n, in a random array differs for each element (Fig. 13.3b). Nevertheless, its average over many elements can be described analytically in terms of d and m; solutions are presented in Tanino and Nepf, 2008b.

The mean flow is characterized by the element Reynolds number,

$$\mathrm{Re}_d = \frac{\langle \bar{u} \rangle d}{\nu}, \tag{13.1}$$

where $\langle \bar{u} \rangle$ is the spatially and temporally averaged interstitial flow velocity and ν is the kinematic viscosity. Measured velocities range from 0 to 90 cm s^{-1} (Asaeda *et al.*, 2005;

Table 13.1. Selected field studies reporting properties of emergent aquatic plants. Where possible, unreported values were estimated from parameters that were reported in that study.

(Dominant) Species	Canopy element	d [mm]	m [m^{-2}]	Site	Reference
Typha augustifolia	shoot	21 to 29	48 to 286	Lake Tega, Chiba, Japan	Asaeda et al., 2005
Zizania latifolia	shoot	20 to 42	46 to 721		
Spartina alterniflora	stem	n/a	380 ± 38 to 561 ± 50	Hog Island, VA, USA	Tyler and Zieman, 1999
Juncus roemerianus	stem	n/a	100 to 625	Cedar Creek, FL, USA; Bayou Chitigue and Old Oyster Bayou, LA, USA	Leonard and Luther, 1995
Spartina alterniflora	stem	n/a	128 to 370	Westerschelde, Netherlands	van Hulzen et al., 2007
Spartina anglica	shoot	n/a	16 to 3584		
Spartina alterniflora	stem	2 to 9 (peak at 5.5)	233	Great Sippewissett Marsh, Cape Cod, MA, USA	Valiela et al., 1978
Spartina anglica	shoot	3 to 6	1450	Freiston Shore, England	Neumeier and Amos, 2006
Spartina maritima	shoot		2340	Ramalhete, Portugal	
Rhizophora stylosa	trunk	57	1.0	Nakama River, Iriomote Island, Japan	Mazda et al., 1997
	root	36	72		
Rhizophora stylosa	trunk	86	0.8	Coral Creek, Hinchinbrook Island, Australia	
	root	37	122		
Sonnerita sp.	trunk	117	0.08	Vinh Quang, Vietnam	Mazda et al., 2006
	pneumatophores	7.3	131		
Rhizophora sp., Bruguiera gymnorhiza, Ceriops tagal	trunk, roots	40	30 to 40	Middle Creek, Cairns, Australia	Furukawa et al., 1997
Spartina alterniflora	stem	1.2 ± 0.6 to 1.7 ± 0.8	n/a	Plum Island Estuary, MA, USA	Lightbody and Nepf, 2006
Cladmium jamaicense	stem, leaf	6.0 to 10.4	180 to 730	Everglades, FL, USA	Harvey et al., 2009
Eleocharis cellulosa	stem	1.9 to 2.6	n/a	Everglades, FL, USA	Larsen et al., 2009
Schoenoplectus validus	clump	150	9	Perth, Australia	Oldham and Sturman, 2001

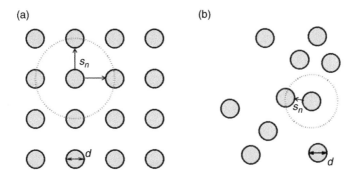

(a)   (b)

Figure 13.3. Key length scales in a canopy where the elements are (a) distributed in a periodic, square configuration or (b) distributed randomly.

Kobashi and Mazda, 2005; Mazda *et al.*, 1997; Lightbody and Nepf, 2006; Leonard and Reed, 2002; Leonard and Luther, 1995; Harvey *et al.*, 2009). These values, combined with values of d discussed above, correspond to Reynolds numbers of $Re_d = 0$ to 30000. A comparison with Re_d-dependence of flow past an isolated cylinder suggests that field conditions span the steady laminar, unsteady laminar, and turbulent flow regimes.

13.2 MATHEMATICAL FORMULATION FOR EMERGENT CANOPIES

Recall from Sec. 1.1.1 that field conditions span the steady laminar, unsteady laminar, and turbulent flow regimes. In the latter two regimes, by definition, the velocity will vary in time. For example, turbulent flow is characterized by rapid, chaotic fluctuations in velocity. In addition, canopy elements introduce *spatial* heterogeneity in the velocity field at the element-scale by, for example, introducing no-slip surfaces and spatial variations in the local interstitial cross-sectional area because of their non-uniform spacing (Fig. 13.3b). In sufficiently sparse canopies, recirculation zones may develop in the wake of the elements. Because pressure and concentration depend on the velocity, they will also exhibit temporal fluctuations and spatial heterogeneity.

In certain contexts, it may be necessary to study the instantaneous, local values of these variables. For example, many aquatic animals use chemical signals to identify and locate their prey or mate (Ferner and Weissburg, 2005; Koehl, 2006). For these animals, the details of the instantaneous concentration and its spatial gradient may be important. However, for most engineering applications, such resolution is unnecessary and inconvenient. Here, the standard approach is to consider the values of these variables averaged over element-scale fluctuations. This approach, which is an extension of the more widely used Reynolds averaging, was developed within the terrestrial canopy literature and is also referred to as *double averaging*. An excellent discussion on the physical implications of the various assumptions and mathematical operations appears in Finnigan, 1985, 2000, but the scheme was proposed even earlier (Wilson and Shaw, 1977; Raupach and Shaw, 1982). As with Reynolds averaging, a necessary condition is that there is a separation of scale between fluctuations and scales over which macroscopic properties vary (cf. Finnigan *et al.*, 2003). The key steps of the averaging scheme are outlined below.

First, the variables of interest, e.g., pressure $p(t, \mathbf{x})$, velocity $\mathbf{v}(t, \mathbf{x}) \equiv (u, v, w) \equiv (u_1, u_2, u_3)$, and solute concentration $c(t, \mathbf{x})$, where t is time, are decomposed into the local temporal average and instantaneous deviations from it. The average is taken over an interval much longer than the time scales of vortex shedding and turbulent fluctuations, but much shorter than macroscopic scales such as those of the seasonal changes in canopy characteristics (cf. Morris and Haskin, 1990), the tide, or the diurnal temperature variation. Then, the temporally-averaged quantities are decomposed into their spatial average and deviations from that average. The spatial average is taken over an infinitessimally thin volume V_f that spans many canopy elements but excludes all solid volume. The variables may now be written as

$$\mathbf{v}(t, \mathbf{x}) = \langle \overline{\mathbf{v}} \rangle (t, \mathbf{x}) + \overline{\mathbf{v}}''(t, \mathbf{x}) + \mathbf{v}'(t, \mathbf{x}), \quad \overline{\mathbf{v}'}, \langle \overline{\mathbf{v}}'' \rangle = 0 \tag{13.2}$$

$$c(t, \mathbf{x}) = \langle \overline{c} \rangle (t, \mathbf{x}) + \overline{c}''(t, \mathbf{x}) + c'(t, \mathbf{x}), \quad \overline{c'}, \langle \overline{c}'' \rangle = 0 \tag{13.3}$$

$$p(t, \mathbf{x}) = \langle \overline{p} \rangle (t, \mathbf{x}) + \overline{p}''(t, \mathbf{x}) + p'(t, \mathbf{x}), \quad \overline{p'}, \langle \overline{p}'' \rangle = 0 \tag{13.4}$$

where the overbar and $\langle \ \rangle_0$ denote the temporal and spatial averaging operations, respectively. Also, $\langle \overline{v} \rangle = \langle \overline{w} \rangle = 0$ by definition of the Cartesian coordinates, $x \equiv (x, y, z) \equiv (x_1, x_2, x_3)$, which is defined with the x-axis aligned with the mean flow $\langle \overline{u} \rangle$, the y-axis perpendicular to it in the horizontal plane, and the z-axis positive upwards. $z = 0$ at the horizontal bed.

The same temporal and spatial averaging operations are applied to all governing equations as well. Below, the averaged equations for the conservation of mass, momentum, and species are presented. Similar manipulations are applied to other governing equations, such as the turbulent kinetic energy budget.

13.2.1 Conservation of mass

Temporally averaging the continuity equation yields

$$\frac{\partial \overline{u}_j}{\partial x_j} = -\frac{\partial u'_j}{\partial x_j} = 0. \tag{13.5}$$

Spatially averaging $\partial \overline{u}_j / \partial x_j = 0$ similarly yields

$$\frac{\partial \langle \overline{u}_j \rangle}{\partial x_j} = -\frac{\partial \overline{u}''_j}{\partial x_j} = 0. \tag{13.6}$$

13.2.2 Conservation of momentum

Applying the temporal and spatial averaging operations to the Navier-Stokes equation yields

$$\frac{\partial \langle \overline{u}_i \rangle}{\partial t} + \langle \overline{u}_j \rangle \frac{\partial \langle \overline{u}_i \rangle}{\partial x_j} = g_i - \frac{1}{\rho} \frac{\partial \langle \overline{p} \rangle}{\partial x_i} - \frac{\partial \langle \overline{u'_i u'_j} \rangle}{\partial x_j} + \nu \frac{\partial}{\partial x_j} \frac{\partial}{\partial x_j} \langle \overline{u}_i \rangle - \frac{\partial \langle \overline{u}''_i \overline{u}''_j \rangle}{\partial x_j} - f_i, \tag{13.7}$$

where g_i is the ith component of the gravitational acceleration, ρ is density, and

$$f_i = -\frac{\nu}{V_f} \iint_{S_c} \frac{\partial \overline{u}_i}{\partial n} dS + \frac{1}{\rho V_f} \iint_{S_c} \overline{p} n_i \, dS \tag{13.8}$$

is the net hydrodynamic force per unit fluid mass exerted on S_c, where S_c denotes all canopy element surfaces that intersect V_f and $\mathbf{n}$ is the unit normal vector on S_c pointing out of V_f. Physically, f_i is the sum of the contribution from the viscous shear stress on the element surfaces [first term in Eq. (13.8)] and the pressure loss in the element wakes (second term). The third term on the RHS of Eq. (13.7) is the divergence of the spatial average of the familiar Reynolds stress. The fifth term, the divergence of the dispersive stress, arises from spatial correlations of the local deviations from the mean velocity.

The longitudinal ($i = 1$) component of Eqs. (13.7) and (13.8) is generally of the most interest. From f_1, the more familiar drag coefficient can be defined by

$$C_D = \frac{\langle \overline{f_D} \rangle}{\rho \langle \overline{u} \rangle^2 d / 2},$$ (13.9)

where $\langle \overline{f_D} \rangle = \rho(1 - \phi) f_1 / m$ is the mean drag per unit length of canopy element. Scaling the shear on the element surface in the first term of Eq. (13.8) as

$$-\frac{\partial \overline{u}}{\partial n} \sim \frac{\langle \overline{u} \rangle}{\langle s_n \rangle_A}$$ (13.10)

yields, for the viscous contribution to $\langle \overline{f_D} \rangle$,

$$\langle \overline{f_D} \rangle^{\text{visc}} \sim \pi \frac{d}{\langle s_n \rangle_A} \mu \langle \overline{u} \rangle.$$ (13.11)

Similarly, dimensional analysis yields, for the inertial contribution to $\langle \overline{f_D} \rangle$,

$$\langle \overline{f_D} \rangle^{\text{form}} \sim \rho d \langle \overline{u''^2} \rangle.$$ (13.12)

Clearly, $\langle s_n \rangle_A / d$ must decrease with increasing ϕ regardless of the morphology of the elements. Physical reasoning suggests that, as ϕ increases, the flow is forced to follow more tortuous paths. In turn, an increase in tortuosity will be manifested in an increase in the spatial heterogeneity of the time-averaged flow (squared), $\langle \overline{u''^2} \rangle$. Then, Eqs. (13.11) and (13.12) indicate that both $\langle \overline{f_D} \rangle^{\text{visc}}$ and $\langle \overline{f_D} \rangle^{\text{form}}$ will increase with increasing ϕ.

The canopy-averaged drag coefficient, C_D, is a function of Re_d and ϕ. In canopies where the elements may be approximated as rigid, circular cylinders, C_D decreases with decreasing Re_d and increases with increasing ϕ regardless of orientation (Koch and Ladd, 1997; Tanino and Nepf, 2008a; Ayaz and Pedley, 1999; Stone and Shen, 2002). Foliage increases ϕ, and was observed to increase the drag on an isolated reed (James et al., 2004). Real plant elements that are sufficiently flexible will reconfigure their shape to minimize drag (e.g., Shucksmith et al., 2011). Under such conditions, the element width, d, is a function of Re_d, which in turn complicates the quantification of C_D through its definition.

Also, the magnitude of drag is sensitive to the spatial distribution of the canopy elements. This is readily demonstrated in cylinder arrays, for which the normalized drag per unit length, $\langle \overline{f_D} \rangle / (\mu \langle \overline{u} \rangle)$, where μ is the dynamic viscosity, at a given Re_d and ϕ is smaller in a square configuration, irrespective of the orientation with respect to the mean flow, than in a random configuration (Koch and Ladd, 1997).

In practice, additional assumptions are made to reduce Eq. (13.7) to the more convenient depth-averaged form:

$$\langle \overline{f_D} \rangle_H m = -\rho g(1 - \phi) \frac{d \langle \overline{H} \rangle}{dx},$$ (13.13)

where $\langle\ \rangle_H$ denotes an average over the water column and $d\langle\overline{H}\rangle/dx$ is the longitudinal gradient of the water depth. Then, the mean drag of a real canopy may be estimated from the free surface gradient and the mean velocity, which are relatively easy to measure (e.g., Mazda et al. 1997; Lee et al., 2004). The conditions necessary for Eq. (13.13) to be valid are discussed in Tanino and Nepf (2008a, 2009b), and thus not included here.

13.2.3 Conservation of species

Conservation of a passive solute is described by the familiar expression

$$\frac{\partial c}{\partial t} + \mathbf{v} \cdot \nabla c = -\nabla \cdot (-D_0 \nabla c), \tag{13.14}$$

where D_0 is the molecular diffusion coefficient. Applying the same temporal and spatial averaging operations as above yields

$$\frac{\partial\langle\overline{c}\rangle}{\partial t} + \langle\overline{v_j}\rangle\frac{\partial\langle\overline{c}\rangle}{\partial x_j} = -\frac{\partial}{\partial x_j}\left\{\overline{\langle v_j'c'\rangle} + \langle\overline{v_j''}\overline{c''}\rangle - D_0\left\langle\frac{\partial}{\partial x_j}(\langle\overline{c}\rangle + \overline{c}'')\right\rangle\right\}. \tag{13.15}$$

A detailed derivation, including a physical justification for neglecting terms not retained in Eq. (13.14), is presented in Finnigan, 1985. $\overline{\langle v_j'c'\rangle}$ is the flux arising from temporal fluctuations of velocity and concentration. Under turbulent conditions, $\overline{\langle v_j'c'\rangle}$ is simply the turbulent flux, which in turn gives rise to turbulent diffusion. In unsteady laminar flow, temporal fluctuations are associated with the unsteadiness of the wake and vortex shedding. In steady laminar flow, $\overline{\langle v_j'c'\rangle} = 0$. $\langle\overline{v_j''}\overline{c''}\rangle$ is the flux arising from the spatial heterogeneity of the time-averaged velocity and concentration, and is associated with the tortuous flow path that fluid is forced to follow around the randomly-distributed plants. This mechanism gives rise to mechanical dispersion. Molecular diffusion, which is independent of the canopy architecture under typical field conditions, is negligible at the scale of the stems or of the canopy and will not be discussed further.

In general, these fluxes, like molecular diffusion, are assumed to obey the gradient-flux law, namely,

$$\overline{\langle v_j'c'\rangle} = -K_{jj}^t\frac{\partial\langle\overline{c}\rangle}{\partial x_j} \tag{13.16}$$

and

$$\langle\overline{v_j''}\overline{c''}\rangle = -K_{jj}^x\frac{\partial\langle\overline{c}\rangle}{\partial x_j} \tag{13.17}$$

where K_{jj}^t and K_{jj}^x are coefficients of diffusion and dispersion. Further, these coefficients are expected to be homogeneous within a homogeneous canopy. Then, Eq. (13.15) reduces to the more tractable form of

$$\frac{\partial\langle\overline{c}\rangle}{\partial t} + \langle\overline{v_j}\rangle\frac{\partial\langle\overline{c}\rangle}{\partial x_j} = (K_{jj}^t + K_{jj}^x + D_0)\frac{\partial^2\langle\overline{c}\rangle}{\partial x_j^2}. \tag{13.18}$$

In reference to Fick's law for the molecular diffusion of chemical species, mass transport that satisfies Eqs. (16)–(18) is often described as Fickian. A necessary criterion for Fickian transport is that the spatial scale over which the mean concentration gradient varies is much larger than the scale of the contributing processes (Corrsin, 1974). As discussed above, the characteristic length scale of turbulent eddies is d or $\langle s_n \rangle_A$ and, accordingly, $\overline{\langle v'_j c' \rangle}$ may be assumed to become Fickian within a relatively short distance $(x \gg d, \langle s_n \rangle_A)$ from the source under sufficiently turbulent conditions. Similarly, the lateral component of the time-averaged velocity varies at the element-scale of d and $\langle s_n \rangle_A$. However, under steady laminar and unsteady laminar conditions, the distance necessary for lateral dispersion, $\langle \overline{v''} \, \overline{c''} \rangle$, to become Fickian is expected to be comparable to, if not larger than, the distance over which real canopies are homogeneous (Tanino and Nepf, 2009a).

13.3 TURBULENCE AND MASS TRANSPORT IN EMERGENT CANOPIES

It was demonstrated in Sec. 1.2.2 that the temporal and spatial averaging of the species conservation equation gives rise to additional diffusion and dispersion terms. This section focuses on selected mechanisms that contribute to these terms. Specifically, we describe the two dominant mechanisms that contribute to lateral dispersion in fully turbulent flow: turbulent diffusion (Sec. 1.3.2) and dispersion associated with the tortuous flow path (Sec. 1.3.3). Then, we discuss mechanisms that contribute specifically to longitudinal dispersion: temporary trapping in the near wake and the randomly distributed velocity deficit in the far wake of the canopy elements (Sec. 1.3.5).

13.3.1 Turbulence

Obstructions in turbulent flow may alter both the structure and intensity of the turbulence. Most fundamentally, the presence of the plant elements prevents the persistence of system-scale turbulence. In sparse arrays, turbulent eddies are expected to have integral length scales of $O(d)$ everywhere (Fig. 13.4a). In dense arrays, there are regions where the separation between elements is smaller than d (Fig. 13.4b). In these regions, physical reasoning suggests that the integral length scale of turbulence is constrained by the local element spacing.

The local element spacing may be parameterized in many ways, with the distance from a cylinder to its nearest neighbour, s_n, being an obvious candidate. Clearly, s_n underestimates the maximum size of the void space around that cylinder. Nevertheless, Tanino and Nepf (2008b) have demonstrated experimentally in a random cylinder array that the mean spacing between nearest-neighbours, namely $\langle s_n \rangle_A$, accurately captures the decay of the *mean* (spatial average) integral length scale of turbulence at $d / \langle s_n \rangle_A > 1.3$ $(\phi > 9\%)$ (Fig. 13.4c). Therefore, the mean integral length scale is accurately described by

$$\langle l_t \rangle = \min \{d, \langle s_n \rangle_A\}, \tag{13.19}$$

as shown in Fig. 13.4c (red solid line).

The same physical reasoning may be used to predict the conditions under which Eq. (13.19) breaks down. If the canopy is sufficiently sparse that the distance between the canopy elements is comparable to the water depth or larger, i.e., $\langle s_n \rangle_A \geq O(\langle \overline{H} \rangle)$, the system resembles an open channel. In this limit, the integral length scale of turbulence is determined by the water depth (or channel width), and not the architecture of the canopy. Similarly, the above framework is expected to break down as the canopy becomes sufficiently dense that $\langle s_n \rangle_A$ and d approach the Kolmogorov micro-scale. However, this scenario is unlikely to occur in the field.

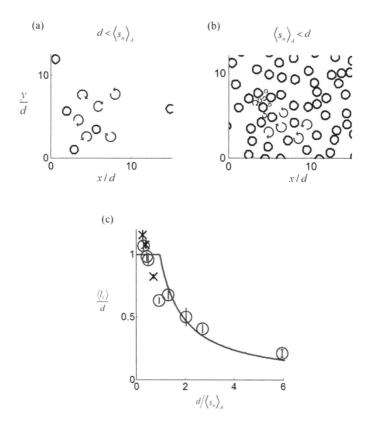

Figure 13.4. The maximum size of the turbulent eddies are O(d) in sparse canopies (a), but is constrained by the local element spacing where the surface-to-surface distance between neighboring elements is less than d (b). As a result, the mean integral length scale of turbulence is $\langle l_t \rangle \approx d$ in sparse canopies, where $d \le \langle s_n \rangle_A$, and $\langle l_t \rangle \approx \langle s_n \rangle_A$ in dense canopies, where $d > \langle s_n \rangle_A$ (c). Markers in (c) depict laboratory measurements in random cylinder arrays reported by Tanino and Nepf (2008b, circles) and White (2002, personal comm., ×); vertical bars depict standard error of the mean. The solid line is theory [Eq. (13.19)].

Note that the conceptual framework above implies that the turbulence is isotropic in the horizontal plane. Velocity measurements have confirmed this for turbulent fluctuations, i.e., $\overline{u'^2} \approx \overline{v'^2}$ (Tanino and Nepf, 2007), and for the integral length scale (Tanino and Nepf, 2007, unpublished data; White, 2002, personal comm.) in random cylinder arrays.

13.3.2 Turbulent diffusion

In general, turbulent diffusion scales with the integral length scale of turbulence and the magnitude of the turbulent velocity fluctuations:

$$K_{jj}^t \sim l_t \sqrt{k_t},$$ (13.20)

where k_t is the turbulent kinetic energy per unit mass. Canopy elements impact both parameters on the RHS of Eq. (13.20). First, they constrain the integral length scale of the turbulent eddies, as described by Eq. (13.19). Second, they enhance k_t. Like $\langle l_t \rangle$, their impact on k_t can be described in terms of the fundamental parameters of the canopy.

In two-dimensional, homogeneous canopies under macroscopically steady conditions, the viscous dissipation of turbulent kinetic energy balances the form drag (see, e.g., Tanino and Nepf, 2008b):

$$-\nu \overline{\left\langle \frac{\partial u_i'}{\partial x_j} \frac{\partial u_i'}{\partial x_j} \right\rangle} = -\langle \bar{u} \rangle \frac{\langle \overline{f_D} \rangle^{\text{form}}}{\rho(1-\phi)} \frac{\phi}{\pi d^2/4}, \tag{13.21}$$

Combining Eq. (13.21) with the classic scaling for viscous dissipation (Tennekes and Lumley, 1972),

$$\varepsilon \sim \sqrt{\bar{k_t}^3} / l_t, \tag{13.22}$$

yields the scaling for the mean turbulence intensity,

$$\left\langle \frac{\sqrt{k_t}}{\langle \bar{u} \rangle} \right\rangle \sim \left[C_D^{\text{form}} \frac{\langle l_t \rangle}{d} \frac{\phi}{(1-\phi)\pi/2} \right]^{1/3}, \tag{13.23}$$

where

$$C_D^{\text{form}} = \frac{\langle \overline{f_D} \rangle^{\text{form}}}{\rho \langle \bar{u} \rangle^2 d/2}. \tag{13.24}$$

Recall that the mean integral length scale of turbulence shifts from d to $\langle s_n \rangle_A$ as the canopy density increases [Eq. (13.19)]. In contrast, C_D^{form} increases with ϕ, at least in a cylinder array. The competition between the terms in Eq. (13.23) is such that the turbulence intensity increases with canopy density at all densities, but more gradually at higher densities ($\langle s_n \rangle_A < d$) (Fig. 13.5a). Deviations of the canopy element morphology from an idealized cylinder impact the turbulence intensity largely through C_D^{form}.

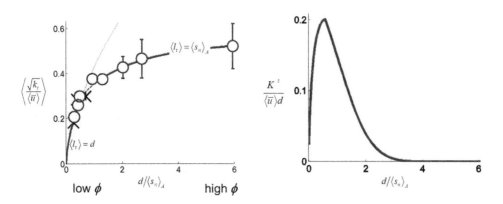

Figure 13.5. Mean turbulence intensity (a) and predicted turbulent diffusion (b) in a random cylinder array. Circles are laboratory measurements reported by Tanino and Nepf (2008b) and × are data collected by White (2002, personal comm.); vertical bars depict uncertainty. Solid lines in (a) and (b) are Eqs. (13.23) and (13.25), respectively, with scaling coefficients as reported in Tanino and Nepf, 2008b.

Recently, it has been proposed that only turbulent eddies larger than d contribute signif-icantly to net *lateral* dispersion (Tanino and Nepf, 2008b). This assumption sets the length scale of the contributing eddies to $l_t = d$, and Eq. (13.20) becomes

$$\frac{K_{yy}^{t}}{\langle \overline{u} \rangle d} \sim \left\langle \frac{\sqrt{k_t}}{\langle \overline{u} \rangle} \right\rangle n_{\text{eff}}, \tag{13.25}$$

where n_{eff} is the fractional volume of pores with dimensions larger than d. It may be inter-preted as an effective porosity. Eq. (13.25) predicts that turbulent diffusion increases rapidly with canopy density to a maximum, then decays as the canopy density increases further (Fig. 13.5b). In a random cylinder array, this maximum was observed to occur at $\phi = 3\%$ (Tanino and Nepf, 2008b).

13.3.3 Dispersion associated with the spatially heterogeneous time-averaged velocity field

As discussed in Sec. 1.3.2, the contribution of turbulent diffusion to net lateral disper-sion decays as canopy density increases above a threshold. Thus in dense canopies, net lateral dispersion is dominated by dispersion arising from the spatial heterogeneity of the time-averaged velocity field, $\langle \overline{v}'' \overline{c}'' \rangle$ (Fig. 13.6a). Recall that the corresponding Fickian coefficient is denoted by K_{yy}^{x}. Physical reasoning suggests that, in dense canopies, the time-averaged velocity field is determined primarily by the spatial distribution of the canopy elements. Therefore, its contribution to lateral dispersion is expected to depend only weakly on Re_d.

Several models for K_{yy}^{x} assume it to be Re_d-independent. The simplest describes the lateral deflection of fluid by the canopy elements as a one-dimensional random walk. In this model, a fluid particle moving through the canopy is assumed to encounter a canopy element at constant time intervals. At every encounter, it has equal probability of moving to the left or to the right of that element. Nepf (1999) proposed that the deflection induced by each element is a function of the element width, d, only, and is independent of Re_d or ϕ. With these

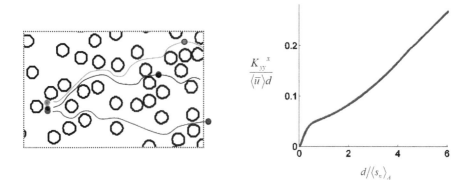

Figure 13.6. Mechanical dispersion associated with the tortuous flow path (a) and the associated lateral dispersion coefficient predicted for a random cylinder array, with the scaling coefficient as reported in Tanino and Nepf, 2008b (b).

assumptions, the asymptotic dispersion of fluid particles undergoing such encounters is described by a normalized coefficient, $K_{yy}^x/(\langle\overline{u}\rangle d)$, that is a function only of ϕ.

An alternative approach uses an analytical solution for Stokes flow through a random cylinder array derived by Koch and Brady (1986). The derivation of the solution is beyond the scope of this chapter, and will not be addressed here. In summary, the solution predicts that

$$\frac{K_{yy}^x}{\langle\overline{u}\rangle d} = \frac{1}{512\sqrt{\pi}} \frac{1}{\sqrt{\phi(1-\phi)}} \left(\frac{\langle\overline{f_D}\rangle}{\mu\langle\overline{u}\rangle}\right)^{3/2},\tag{13.26}$$

where the mean drag in Stokes flow may be approximated by (Tanino, 2008)

$$\frac{\langle\overline{f_D}\rangle}{\mu\langle\overline{u}\rangle} \approx \frac{\pi}{4}\left(\frac{1-\phi}{\phi}\right)\left(\frac{d}{\langle s_n\rangle_A}\right)^2.\tag{13.27}$$

Note that Eq. (13.26) blows up as ϕ approaches zero. In contrast, laboratory observation demonstrates that net lateral dispersion decreases as ϕ approaches zero. Combined with the anticipated decay of turbulent diffusion as ϕ approaches zero [Eq. (13.25)], this suggests that $K_{yy}^x/(\langle\overline{u}\rangle d)$ must also decay to zero. To capture this, Tanino and Nepf (2008b) adapted Koch and Brady's solution by introducing a factor to only account for elements with a neighbour close enough to permit element-element interaction; they defined the critical separation as $s_n = 4d$. Then,

$$\frac{K_{yy}^x}{\langle\overline{u}\rangle d} \sim P_{s_n<4d}\underbrace{\frac{1}{512\sqrt{\pi}} \frac{1}{\sqrt{\phi(1-\phi)}} \left(\frac{\langle\overline{f_D}\rangle}{\mu\langle\overline{u}\rangle}\right)^{3/2}}_{\substack{\text{Koch \& Brady (1986)'s}\\\text{solution for Stokes flow}}},\tag{13.28}$$

where $P_{s_n<4d}$ is the fraction of elements with a neighbour within $s_n = 4d$. If we apply values of $P_{s_n < 4d}$ and mean drag for random cylinder arrays, Eq. (13.28) predicts a monotonically increasing $K_{yy}^x/(\langle\overline{u}\rangle d)$ with canopy density (Fig. 13.6b). Such dependence is consistent with the random walk model.

13.3.4 Net lateral dispersion

Net lateral dispersion is simply the linear sum of turbulent diffusion, dispersion due to the spatial heterogeneity of the time-averaged flow, and molecular diffusion [Eq. (13.18)]. As an example, net lateral dispersion for a random cylinder array is presented in Fig. 13.7. Contrary to suggestions in earlier literature, net lateral dispersion exhibits three distinct regimes. In sparse canopies, turbulent diffusion dominates and $K_{yy}/(\langle\overline{u}\rangle d)$, where $K_{yy} = K_{yy}^x + K_{yy}^t$ represents net lateral dispersion, increases rapidly with canopy density $(d/\langle s_n\rangle_A)$. At intermediate canopy densities, turbulent diffusion decreases as $(d/\langle s_n\rangle_A)$ increases, but remains larger than dispersion associated with the spatially heterogeneous velocity field. Combined, $K_{yy}/(\langle\overline{u}\rangle d)$ also decreases. In sufficiently dense canopies, net dispersion is dominated by the contribution from the spatial heterogeneity of the velocity field, and $K_{yy}/(\langle\overline{u}\rangle d)$ once again increases with ϕ.

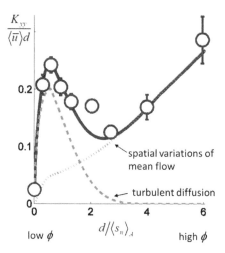

Figure 13.7. Asymptotic lateral dispersion coefficient in a random cylinder array normalized by the mean interstitial velocity, $\langle \bar{u} \rangle$, and cylinder diameter, d. The red solid line is the predicted net dispersion, which is a linear sum of the contribution from turbulent diffusion (dashed line) and dispersion arising from the spatial heterogeneity of the time-averaged velocity (dotted line). Circles depict measurements by Tanino and Nepf (2008b, 2009a, $\phi > 0$) and Nepf *et al.* (1997, $\phi = 0$).

13.3.5 Longitudinal dispersion

White and Nepf (2003) identified four mechanisms that contribute to net longitudinal dispersion in unsteady and turbulent flow. Two are molecular diffusion and element-scale turbulent diffusion, as discussed above. However, contributions of these mechanisms to longitudinal dispersion are expected to be minor because of their small length scale relative to the other two mechanisms:

(i)   dispersion associated with the spatial heterogeneity of the time-averaged velocity field, and

(ii) dispersion arising from the temporary retention of solute in the recirculation zones that constitute the near wake of each canopy element.

Note that the former may be interpreted as the longitudinal component of mechanical dispersion discussed in Sec. 1.3.3. In the context of longitudinal dispersion in sparse canopies, the spatial heterogeneity of interest is associated with the deficit in the time-averaged longitudinal velocity downstream of each canopy element, and is sometimes referred to as *wake shear dispersion* (e.g., Nepf, 2004), or *secondary wake dispersion* (White and Nepf, 2003). The latter is referred to as *near wake dispersion* or *wake trapping dispersion*.

First, we consider wake shear dispersion. The deviation of the time-averaged velocity field downstream of a cylindrical element from the mean flow was derived to be [Eq. (13.35), White and Nepf, 2003].

$$\frac{\bar{u}''_{\text{far wake}}(\tilde{x}, \tilde{y})}{\langle \bar{u} \rangle} = -\underbrace{\frac{C_D}{4\sqrt{\pi}} \sqrt{\frac{\langle \bar{u} \rangle d}{(\nu + \nu_t)}} \sqrt{\frac{d}{\tilde{x}}} \exp\left\{ -\frac{\tilde{y}^2}{4\tilde{x}} \frac{\langle \bar{u} \rangle}{(\nu + \nu_t)} \right\}}_{\text{velocity deficit in the wake of a single element}} \exp\{-C_D m\, d\tilde{x}\} \qquad (13.29)$$

where ν_t is the eddy viscosity. $(\tilde{x}, \tilde{y})$ is the Cartesian coordinate system defined relative to the particular element; the centre of the canopy element is at $(\tilde{x}, \tilde{y}) = (0, 0)$. Eq. (13.29) is the product of the velocity deficit in the wake of a single cylinder (p.189, Schlichting and Gersten, 2000) and $\exp\{-C_D m\, d\tilde{x}\}$, which describes the attenuation by the canopy drag. Note that the expression above holds for any isolated, two-dimensional body; the morphology of the canopy element is specified by C_D and d (Schlichting and Gersten, 2000).

This velocity deficit in the far wake of the elements [Eq. (13.29)] is compensated by elevated velocity elsewhere. The deviation of this elevated velocity from the mean flow may be described by [Eqs. (41, 42), White and Nepf, 2003]

$$\frac{\bar{u}''_{\text{gap}}(\tilde{x}, \tilde{y})}{\langle \bar{u} \rangle} = \begin{cases} \dfrac{C_D m\, d^2}{2(1 - md^2)} \exp\{-C_D m\, d\tilde{x}\} & |\tilde{y} - \tilde{y}_g| \le \dfrac{w}{2} \\ 0 & |\tilde{y} - \tilde{y}_g| > \dfrac{w}{2} \end{cases} , \tag{13.30}$$

where w is the mean spacing between the canopy element at $(\tilde{x}, \tilde{y}) = (0, 0)$ and its laterally adjacent neighbour, and $\tilde{y} = \tilde{y}_g$ denotes the midpoint between the two elements.

The net velocity field, $\bar{u}''(x, y)$, is described by a linear superposition of $\bar{u}''_{\text{far wake}}(\tilde{x}, \tilde{y})$ and $\bar{u}''_{\text{gap}}(\tilde{x}, \tilde{y})$ for all elements in the canopy. White and Nepf (2003) derived their contributions to the dispersion coefficient to be, to leading order,

$$\frac{K_{xx}^{\text{wake shear}}}{\langle \bar{u} \rangle d} = \frac{1}{8}\sqrt{C_D^3 \text{Re}_t} + \frac{C_D m d^2}{4(1 - md^2)}. \tag{13.31}$$

Next, we consider wake trapping dispersion, which is conceptualized as follows. A fluid particle that encounters a canopy element is detained in the recirculation zone for a period τ, then released. This particle is now displaced by a longitudinal distance of $-\langle \bar{u} \rangle \tau$ relative to a particle that was not entrained. The probability density function of τ is described by an exponential decay of the form $\exp\{-\tau/\bar{\tau}\}$, where the characteristic decay time, $\bar{\tau}$, is the mean residence time of a particle in a near wake (MacLennan and Vincent, 1982; Tanino, 2003). The probability that a particle encounters a near wake, p_w, is equal to the fractional volume of the fluid occupied by recirculation zones. Then, a scaling for longitudinal dispersion may be defined using $\langle \bar{u} \rangle \bar{\tau}$ as the length scale, $\bar{\tau}$ as the time scale, and $p_w(1 - \phi)$ as the effective porosity, i.e., the fractional volume of the system that contributes to this mechanism. Then,

$$K_{xx}^{\text{trappping}} \sim \frac{(\langle \bar{u} \rangle \bar{\tau})^2}{\bar{\tau}} p_w(1 - \phi) \tag{13.32}$$

which, in non-dimensionalized form, is

$$\frac{K_{xx}^{\text{trappping}}}{\langle \bar{u} \rangle d} \sim \frac{\langle \bar{u} \rangle \bar{\tau}}{d} p_w(1 - \phi). \tag{13.33}$$

A formal derivation is presented by White and Nepf (2003).

If the canopy is sufficiently sparse that element wakes are independent, they may be modelled as a random distribution of identical wakes, each resembling that of an isolated element. Then, p_w is a product of m (wakes per unit area) and the volume per unit depth occupied by an isolated wake. For smooth cylindrical elements, values of p_w and $\bar{\tau}$ are available in the literature. For example, in the vortex shedding regime, the length of the near wake of an isolated circular cylinder decreases as Re_d increases such that the area of

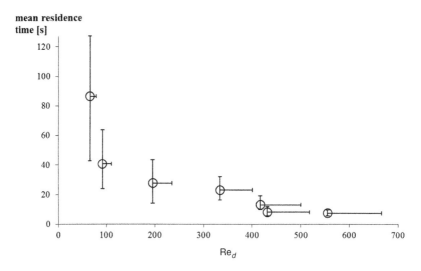

Figure 13.8. Laboratory measurements of the mean residence time in the near wake of an isolated cylinder (redrawn from Tanino, 2003).

a near wake is estimated to decrease over the range $O(2)d^2$ to $O(1)d^2$ (Paranthoen *et al.,* 1999; Tanino, 2003). Similarly, $\overline{\tau}$ in the near wake of an isolated cylinder was found to decay roughly exponentially with Re_d (Fig. 13.8; Tanino, 2003). In dense canopies, p_w and $\overline{\tau}$ will also depend on ϕ.

Net longitudinal dispersion is, then,

$$\frac{K_{xx}}{\langle \overline{u} \rangle d} = \frac{K_{xx}^{\text{wake shear}}}{\langle \overline{u} \rangle d} + \frac{K_{xx}^{\text{trappping}}}{\langle \overline{u} \rangle d}$$

$$\approx \frac{1}{8}\sqrt{C_D^3 Re_t} + \frac{C_D md^2}{4(1 - md^2)} + \frac{\langle \overline{u} \rangle \overline{\tau}}{d} p_w(1 - \phi)$$

(13.34)

Laboratory measurements of longitudinal dispersion in random cylinder arrays exhibit two distinct trends: net dispersion increases monotonically with increasing canopy density (ϕ), and decreases monotonically as Re_d increases from $Re_d = 100$ to 600 in sparse ($\phi < 6.4\%$) canopies (White and Nepf, 2003). The theory predicts correctly the qualitative dependence of net dispersion on Re_d and ϕ. Indeed, taking values from the literature, the theory predicts the ϕ-dependence to be due to the rapid increase of wake trapping dispersion. Starting from $K_{xx}^{\text{trappping}} = 0$ at $\phi = 0$, it exceeds wake shear dispersion at $\phi = 8\%$.

13.3.6   Comparison with field measurements

As expected, field measurements are limited; selected values are presented in Table 13.2. Broadly, the field measurements are larger than predictions. Such discrepancy is expected, as the theory only takes into account selected mechanisms. For example, processes other than element-scale turbulence and molecular diffusion that may enhance vertical transport include secondary flow in the near wake of the elements (Nepf *et al.*, 1997), convection, and wind-induced mixing. Contributions from these processes are negligible in the laboratory

Table 13.2. Selected field measurements of Fickian transport in emergent canopies. Equivalent solid volume fraction, ϕ, is estimated from reported values of m and d by modelling the plant elements as circular cylinders.

$\dfrac{K^t_{zz}}{\langle \overline{u} \rangle d}$	$\dfrac{K^{net}_{xx}}{\langle \overline{u} \rangle d}$	$\dfrac{K^{net}_{yy}}{\langle \overline{u} \rangle d}$	Re_d	Tracer type	Site	Reference
0.17 ± 0.08	not Fickian	n/a	15–60	Solute	Plum Island Estuary, Rowley, MA, USA	Lightbody and Nepf, 2006
0.5–0.7 0.6–0.9	1–2 14–19	n/a	20 200–300	Particulate	Water Conservation Area 3A, Everglades, FL, USA	Huang et al., 2008
9×10^{-2}	15	14	3	Particulate	Shark River Slough, Everglades, FL, USA	Saiers et al., 2003

but may be significant in the field, and may account for the observed vertical diffusion of $K^t_{zz}/(\langle \overline{u} \rangle d) = O(0.1 - 1)$, which is well above that of molecular diffusion, even where Re_d suggests laminar conditions. Similarly, contributions to longitudinal dispersion that may be significant in the field include velocity heterogeneity associated with the boundary layer at the canopy element surfaces (Koch and Brady, 1986), vertical heterogeneity in canopy properties (Lightbody and Nepf, 2006), and the shear at the bed or at the free surface. Also, the bed in a real system is permeable, and solute may diffuse into it and be retained for some time before being released into the open water. Such retention will, like wake trapping, enhance longitudinal dispersion.

In addition, the number of mechanisms that contribute to net dispersion increases as the spatial scale at which the problem is defined increases. In terms of the mathematical formulation, increasing the scale of interest corresponds to enlarging the averaging volume associated with the spatial averaging operation, $\langle \ \rangle$. This effect is best illustrated by considering the spatial heterogeneity of the time-averaged velocity field. At the canopy-scale, the spatial heterogeneity arises primarily from the presence of individual canopy elements, and thus the variations are at the scale of individual elements and element spacing. At the scale of an entire wetland, however, spatial heterogeneity in vegetation properties must be considered. For example, a natural wetland may comprise open channels with no vegetation and dense patches of vegetation. Then, large-scale shear develops between regions of different vegetation density, which contributes to longitudinal dispersion defined at the scale of the wetland (or larger). Accordingly, longitudinal dispersion in a natural system depends on the spatial scale of the experiments, with measured coefficients being larger in experiments at larger scales (Variano et al., 2009).

13.4   FLOW NEAR AN OPEN WATER-CANOPY INTERFACE

In the field, plant canopies are restricted to finite horizontal area and their boundaries with open water may play an important role in determining flow patterns and mass transport at the system-scale (Fig. 13.9). For example, field studies report that some species preferentially reside near the boundary between a salt marsh and open water (e.g., Peterson and Turner, 1994). Similarly, gradients in various chemical constituents and properties have been reported between adjacent littoral and pelagic zones in lakes (e.g., Cardinale et al., 1997; James and Barko, 1991).

Figure 13.9. Tidal marsh at Plum Island Estuary, Rowley, MA. Photo courtesy of Anne F. Lightbody.

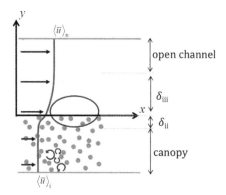

Figure 13.10. Two-layer shear flow at the interface between a homogeneous, emergent canopy and open water. White and Nepf (2008) define four distinct regions: (i) canopy outside the shear layer, (ii) canopy-side of the shear layer, (iii) open water-side of the shear layer, and (iv) open water outside the shear layer. Subscripts denote the four regions. Not to scale. Plan view.

In this chapter, we consider flow in the vicinity of the interface of a homogeneous, emergent canopy and open water, with the mean flow oriented parallel to the interface (Fig. 13.10). Such a configuration characterizes, for example, river-floodplain boundaries.

13.4.1 Mean flow

The flow configuration of interest falls into a broader set of systems that are characterized by a steady flow parallel to an infinitely long interface between a porous layer and open water, e.g., flow over and within submerged canopies (cf. Ghisalberti and Nepf, 2005; Nepf 2012). As shown in Sec. 13.2.1, the canopy elements introduce additional resistance to flow, hence a discontinuity in drag develops at an open channel-canopy interface. This discontinuity in turn generates shear at the interface, with the mean flow within the canopy, $\langle \overline{u} \rangle_1$, being smaller than that in the open water, $\langle \overline{u} \rangle_4$. This shear gives rise to Kelvin-Helmholtz (KH) billows (White and Nepf, 2007), frequently referred to as *coherent vortices* in canopy literature. Because the KH billows are larger than the element-scale turbulence within the canopy, they dominate transport across the interface. This section summarizes the four-layer model for mean flow (Fig. 13.10) proposed by White and Nepf (2008).

Sufficiently far within the emergent canopy, the mean flow is described by the balance between canopy drag and free surface gradient [Eq. (13.13)] (region i, Fig. 13.10). Sufficiently far from the interface in the open water, the flow resembles that of an open channel (region iv). Accordingly, the flow is described by the balance between the bed stress and the free surface gradient:

$$\frac{\nu}{\langle \overline{H} \rangle} \left. \frac{\partial \langle \overline{u} \rangle}{\partial z} \right|_{z=0} = -g \frac{d\langle \overline{H} \rangle}{dx}. \tag{13.35}$$

Eqs. (13.13) and (13.35) may be expressed using the more familiar drag coefficient and bed friction factor, $C_f = \mu \partial \langle \overline{u} \rangle / \partial z|_{z=0} / (\rho \langle \overline{u} \rangle^2)$:

$$\langle \overline{u} \rangle_{i,H} = \sqrt{-2g \frac{(1-\phi)}{C_D md} \frac{d\langle \overline{H} \rangle}{dx}} \tag{13.36}$$

and

$$\langle \overline{u} \rangle_{iv,H} = \sqrt{-\frac{g \langle \overline{H} \rangle}{C_f} \frac{d\langle \overline{H} \rangle}{dx}}. \tag{13.37}$$

If the coefficients C_D and C_f are known *a priori*, $\langle \overline{u} \rangle_i$ and $\langle \overline{u} \rangle_{iv}$ may be predicted from Eqs. (13.36) and (13.37) for a given free surface gradient (or bed slope).

The shear layer at the interface is also classified into two regions. On the side of the canopy (region ii), the normalized mean velocity profile is well-described by a hyperbolic tangent distribution of the form:

$$\frac{\langle \overline{u} \rangle_{ii}(y) - \langle \overline{u} \rangle_i}{\langle \overline{u} \rangle(y_0) - \langle \overline{u} \rangle_i} = 1 + \tanh \left\{ \frac{y - y_0}{\delta_{ii}} \right\}. \tag{13.38}$$

Physically, $y = y_0$ is the offset of the inflection point relative to the canopy-open water interface ($y = 0$), and δ_{ii} is the distance that the KH vortices penetrate into the canopy. The hyperbolic tangent dependence is a salient feature of mean flow at an obstructed – open flow interface, and has been reported in submerged canopies of both flexible (e.g., Ghisalberti and Nepf, 2002) and rigid elements (e.g., Huq *et al.*, 2007; Katul *et al.*, 2002). The inflection point y_0 occurs just outside the canopy in both submerged canopies and in the present configuration (Nepf, 2004; Ghisalberti and Nepf, 2002; White and Nepf, 2008). Above the canopy (region iii), the mean shear flow is well-described by a quadratic profile of the form [White and Nepf, 2008, Eq. (13.21)]

$$\frac{\langle \overline{u} \rangle_{iii}(y) - \langle \overline{u} \rangle_{iii}(y_m)}{\langle \overline{u} \rangle_{iv} - \langle \overline{u} \rangle_{iii}(y_m)} = \frac{y - y_m}{\delta_{iii}} - \frac{1}{4} \left(\frac{y - y_m}{\delta_{iii}} \right)^2. \tag{13.39}$$

δ_{iii} represents the protrusion of the KH billows into the open water beyond $y = y_m$, where $y = y_m$ is where the shear of the mean flow in the hyperbolic tangent profile region [Eq. (13.38)] matches that in the quadratic profile region [Eq. (13.39)].

The scale for δ_{ii} and δ_{iii} can be determined from the conservation of momentum. In analogy with submerged canopies (Raupach *et al.*, 1991), the momentum equation

[Eq. (13.7)] may be simplified to a balance between the canopy drag and the turbulent shear stress within the shear layer on the canopy side (region ii):

$$0 = -\frac{\partial \langle \overline{u'v'} \rangle}{\partial y} - f_1, \tag{13.40}$$

which may be rewritten in terms of the drag coefficient [Eq. (13.9)] as

$$-\frac{\partial \langle \overline{u'v'} \rangle}{\partial y} = \frac{\langle \overline{u} \rangle^2}{2} \frac{C_D m d}{(1 - \phi)}. \tag{13.41}$$

Physically, the lateral distance over which the turbulent stress decays to zero scales with δ_{ii}. Then, Eq. (13.41) yields the scaling:

$$\delta_{ii} \sim \frac{1 - \phi}{C_D m d}. \tag{13.42}$$

The same scaling, often described as the attenuation scale, applies to submerged canopies (e.g., Nepf, 2004). The scaling yields the intuitive result that the shear layer penetrates deeper into sparser canopies, which has also been verified in the laboratory (Zong and Nepf, 2011).

Within the shear layer in open water (region iii), the momentum equation reduces to a balance between turbulent shear stress and the pressure gradient:

$$0 = -\frac{\partial \langle \overline{u'v'} \rangle}{\partial y} - g \frac{d \langle \overline{H} \rangle}{dx}. \tag{13.43}$$

The free surface gradient in turn is controlled by the balance with the bed stress in the open water (region iv) [Eq. (13.35]. Rewriting the RHS of Eq. (13.43) in terms of the bed friction factor yields the scaling for the protrusion of the KH billows into the open water (White and Nepf, 2007):

$$\delta_{iii} \sim \langle \overline{H} \rangle / C_f. \tag{13.44}$$

13.4.2   Mass exchange between the canopy and the open water

The flux associated with the KH billows may be parameterized by an exchange coefficient across the canopy interface, k_e, defined by the boundary condition (Deen, 1998):

$$K_{yy} \frac{\partial \langle \overline{c} \rangle}{\partial y} \bigg|_{y=0^-} \equiv k_e (. \langle \overline{c} \rangle |_{y=0^-} - \langle \overline{c} \rangle_i), \tag{13.45}$$

where $y = 0^-$ denotes the canopy-side of the canopy interface, $\langle \overline{c} \rangle_i$ is the bulk concentration within the canopy, and K_{yy} is the net lateral dispersion coefficient of the homogeneous canopy. Selected solutions for $\langle \overline{c} \rangle$ are presented in White, 2006.

13.5   SUMMARY

The integral length scale of turbulence within a homogeneous emergent canopy, which has traditionally been taken to be the canopy element (e.g., stem) diameter, is constrained by the interstitial pore size in dense canopies. The canopy-averaged integral length scale is well-described by the average surface-to-surface separation between a canopy element and its nearest neighbour. Net lateral dispersion exhibits an intermediate regime, in which the dispersion coefficient, normalized by the interstitial flow velocity and canopy element diameter, decreases with increasing canopy density. Net longitudinal dispersion increases with increasing canopy density in sparse ($\phi < 6.4\%$) canopies; its behaviour in more dense canopies have not been established.

At an interface between an emergent canopy and open water that is parallel to the mean flow, the discontinuity in drag results in shear, which in turn gives rise to Kelvin-Helmholtz billows and shear-scale turbulence. The shear-scale turbulence dominates lateral transport across the canopy-open water interface. The penetration of the KH billows into the canopy is constrained by the canopy drag and, accordingly, its characteristic thickness is $(1 - \phi)/(C_D md)$.

APPENDIX A – LIST OF SYMBOLS

<table>
<tr><th colspan="3">List of Symbols</th></tr>
<tr><th>Symbol</th><th>Definition</th><th>Dimension or Units</th></tr>
<tr><td>ϕ</td><td>solid volume fraction</td><td>$[\cdot]$</td></tr>
<tr><td>$c(t, \mathbf{x})$</td><td>concentration</td><td>$[\text{M L}^{-3}]$</td></tr>
<tr><td>C_D</td><td>mean drag coefficient</td><td>$[\cdot]$</td></tr>
<tr><td>D_0</td><td>molecular diffusion coefficient</td><td>$[\text{L}^2\,\text{T}^{-1}]$</td></tr>
<tr><td>d</td><td>canopy element (root / stem / plant) diameter</td><td>$[\text{L}]$</td></tr>
<tr><td>f_i</td><td>net hydrodynamic drag per unit fluid mass</td><td>$[\text{L T}^{-2}]$</td></tr>
<tr><td>$\langle \overline{f_D} \rangle$</td><td>mean drag per unit length of canopy element</td><td>$[\text{M T}^{-2}]$</td></tr>
<tr><td>$\langle \overline{f_D} \rangle^{\text{form}}$</td><td>Inertial contribution to $\langle \overline{f_D} \rangle$</td><td>$[\text{M T}^{-2}]$</td></tr>
<tr><td>$\langle \overline{f_D} \rangle^{\text{visc}}$</td><td>Viscous contribution to $\langle \overline{f_D} \rangle$</td><td>$[\text{M T}^{-2}]$</td></tr>
<tr><td>g</td><td>gravitational acceleration</td><td>$[\text{L T}^{-2}]$</td></tr>
<tr><td>K_{jj}^{t}</td><td>turbulent diffusion coefficient</td><td>$[\text{L}^2\,\text{T}^{-1}]$</td></tr>
<tr><td>K_{jj}^{x}</td><td>dispersion coefficient</td><td>$[\text{L}^2\,\text{T}^{-1}]$</td></tr>
<tr><td>l_t</td><td>integral length scale of turbulence</td><td>$[\text{L}]$</td></tr>
<tr><td>$\langle l_t \rangle$</td><td>spatially-averaged integral length scale of turbulence</td><td>$[\text{L}]$</td></tr>
<tr><td>m</td><td>number of canopy elements per unit horizontal area</td><td>$[\text{L}^{-2}]$</td></tr>
<tr><td>n_{eff}</td><td>Fractional volume of pores with dimensions larger than d</td><td>$[\cdot]$</td></tr>
<tr><td>p</td><td>pressure</td><td>$[\text{M L}^{-1}\,\text{T}^{-2}]$</td></tr>
<tr><td>s</td><td>characteristic spacing between elements defined between their surfaces</td><td>$[\text{L}]$</td></tr>
<tr><td>s_n</td><td>distance between nearest neighboring elements defined between surfaces</td><td>$[\text{L}]$</td></tr>
<tr><td>$\langle s_n \rangle_A$</td><td>s_n averaged over many elements in a random array</td><td>$[\text{L}]$</td></tr>
<tr><td>t</td><td>time</td><td>$[\text{T}]$</td></tr>
<tr><td>Re_d</td><td>Reynolds number, with d as the length scale</td><td>$[\cdot]$</td></tr>
<tr><td>(u, v, w)</td><td>local, instantaneous velocity in (x, y, z) direction</td><td>$[\text{L T}^{-1}]$</td></tr>
<tr><td>(u', v', w')</td><td>deviations of (u, v, w) from their local, temporal average</td><td>$[\text{L T}^{-1}]$</td></tr>
<tr><td>$(\bar{u}, \bar{v}, \bar{w})$</td><td>local, temporal average of (u, v, w)</td><td>$[\text{L T}^{-1}]$</td></tr>
</table>

(Continued)

List of Symbols

Symbol	Definition	Dimension or Units
$(\overline{u}'', \overline{v}'', \overline{w}'')$	deviations of $(\overline{u}, \overline{v}, \overline{w})$ from their spatial average	$[\text{L T}^{-1}]$
$(\langle \overline{u} \rangle, \langle \overline{v} \rangle, \langle \overline{w} \rangle)$	temporal- and spatial-average of (u, v, w)	$[\text{L T}^{-1}]$
ε	viscous dissipation	$[\text{L}^2 \text{ T}^{-3}]$
P	density	$[\text{M L}^{-3}]$
μ	dynamic viscosity	$[\text{M L}^{-1} \text{ T}^{-1}]$
ν	kinematic viscosity	$[\text{L}^2 \text{ T}^{-1}]$
$\mathbf{v}(t, x)$	velocity vector, (u, v, w)	$[\text{L T}^{-1}]$
V_f	averaging volume	$[\text{L}^3]$
x	Cartesian coordinate aligned with the mean flow	$[\text{L}]$
y	Cartesian coordinate perpendicular to the mean flow and horizontal	$[\text{L}]$
z	vertical Cartesian coordinate	$[\text{L}]$

APPENDIX B – SYNOPSIS

This chapter describes flow and mass transport under conditions relevant to surface water systems with emergent vegetation. Vegetated surface waters are modelled as homogeneous arrays of discrete, rigid, two-dimensional plant elements. The first section summarizes typical field conditions. The second section presents the standard mathematical formulation for flow through an array of elements. The third section describes turbulence and mass transport within a homogeneous canopy. The final section describes flow at the interface between an emergent canopy and open water. Specifically, we consider characteristic behaviour of flow parallel to the water-canopy interface. Such flows occur at, for example, river-floodplain boundaries.

APPENDIX C – KEYWORDS

Canopy	Turbulent diffusion	Dispersion
Shear	Fickian	Mechanical dispersion
Integral length scale	Gradient-flux model	Drag
Double averaging	Kelvin-Helmholtz	

APPENDIX D – QUESTIONS

Define mechanical dispersion.
Why is dispersion much larger in open channels than in emergent canopies?
Name two types of emergent plant canopies in the field.
What is the dominant mechanism of lateral exchange in a partially-vegetated channel?

APPENDIX E – PROBLEMS

E1. Describe the gradient-flux model. What is assumed about the length scale of the gradient and the mechanisms contributing to the flux?

E2. How does turbulent diffusion in flow through a canopy differ from that in an unobstructed flow (open channel)?

E3. Describe the method of double averaging. Describe the implicit assumption about the characteristic scales of the canopy and the flow.

REFERENCES

Asaeda, T., Takeshi, F., and Manatunge, J., 2005, Morphological adaptations of emergent plants to water flow: a case study with *Typha angustifolia*, *Zizania latifolia* and *Phragmites australis*. *Freshwater Biology*, **50**, pp. 1991–2001.

Ayaz, F. and Pedley, T. J., 1999, Flow through and particle interception by an infinite array of closely-spaced circular cylinders. *European Journal of Mechanics – B/Fluids*, **18**(2), pp. 173–196.

Bergen, A., Alderson, C., Bergfors, R., Aquila, C., and Matsil, M. A., 2000, Restoration of a *Spartina alterniflora* salt marsh following a fuel oil spill, New York City, NY. *Wetlands Ecology and Management*, **8**(2–3), pp. 185–195.

Cardinale, B. J., Burton, T. M., and Brady, V. J., 1997, The community dynamics of epiphytic midge larvae across the pelagic – littoral interface: do animals respond to changes in the abiotic environment? *Canadian Journal of Fisheries and Aquatic Sciences*, **54**(10), pp. 2314–2322.

Carter, Virginia, Ruhl, Henry A., Rybicki, Nancy B., Reel, Justin T., and Gammon, Patricia T., 1999, Vegetative resistance to flow in South Florida: summary of vegetation sampling at sites NESRS3 and P33, Shark River Slough, April, 1996. *U. S. Geological Survey Open-File Report* 99–187.

Carter, Virginia, Ruhl, Henry A., Rybicki, Nancy B., Reel, Justin T., and Gammon, Patricia T., 2003, Vegetative resistance to flow in South Florida: summary of vegetation sampling at sites NESRS3 and P33, Shark River Slough, November, 1996. *U. S. Geological Survey Open-File Report* 99–218.

Corrsin, S., 1974, Limitations of gradient transport models in random walks and in turbulence. *Advances in Geophysics*, **18**(A), pp. 25–60.

Danielsen, F., Sørensen, M. K., Olwig, M. F., Selvam, V., Parish, F., Burgess, N. D., Hiraishi, T., Karunagaran, V. M., Rasmussen, M. S., Hansen, L. B., Quarto, A., and Suryadiputra, N., 2005, The Asian tsunami: a protective role for coastal vegetation. *Science*, **310**. p. 643.

Das, S., and Vincent, J. R., 2009, Mangroves protected villages and reduced death toll during Indian super cyclone. *Proceedings of the National Academy of Sciences of the United States of America*, **106**(18), pp. 7357–7360.

Deen, W. M., 1998, *Analysis of Transport Phenomena*. Oxford University Press.

Ferner, M. C., and Weissburg, M. J., 2005, Slow-moving predatory gastropods track prey odors in fast and turbulent flow. *Journal of Experimental Biology*, **208**, pp. 809–819.

Finnigan, J. J., 1985, Turbulent transport in flexible plant canopies. In *The Forest-Atmosphere Interaction*, editors: Hutchison, B. A. and Hicks, B. B.. D. Reidel Publishing Company.

Finnigan, J., 2000, Turbulence in plant canopies. *Annual Review of Fluid Mechanics*, **32**, pp. 519–571.

Finnigan, J. J., Clement, R., Malhi, Y., Leuning, R., and Cleugh, H.A., 2003, A re-evaluation of long-term flux measurement techniques. Part 1: averaging and coordinate rotation. *Boundary-Layer Meteorology*, **107**, pp. 1–48.

Furukawa, K., Wolanski, E., and Mueller, H., 1997, Currents and sediment transport in mangrove forests. *Estuarine, Coastal and Shelf Science*, **44**, pp. 301–310.

Ghisalberti, G. and Nepf, H. M., 2002, Mixing layers and coherent structures in vegetated aquatic flows. *Journal of Geophysical Research*, **107**(C2).

Ghisalberti, G. and Nepf, H. M., 2005, Mass transport in vegetated shear flows. *Environmental Fluid Mechanics*, **5**, pp. 527–551.

Harvey, J. W., Saiers, J. E., and Newlin J. T., 2005, Solute transport and storage mechanisms in wetlands of the Everglades, south Florida. *Water Resources Research*, **41**, W05009.

Harvey, J. W., Schaffranek, R. W., Noe, G. B., Larsen, L. G., Nowacki, D. J., and O'Connor, B. L., 2009, Hydroecological factors governing surface water flow on a low-gradient floodplain. *Water Resources Research*, **45**, W03421.

Huang, Y. H., Saiers, J. E., Harvey, J. W., Noe, G. B., and Mylon, S., 2008, Advection, dispersion, and filtration of fine particles within emergent vegetation of the Florida Everglades. *Water Resources Research*, **44**, W04408.

Huq, P., White, L. A., Carrillo, A., Redondo, J., Dharmavaram, S., Hanna, S. R., 2007, The shear layer above and in urban canopies. *Journal of Applied Meteorology and Climatology*, **46**, pp. 368–376.

James, W. F. and Barko, J. W., 1991, Estimation of phosphorous exchange between littoral and pelagic zones during nighttime convective circulation. *Limnology and Oceanography*, **36**(1), pp. 179–187.

James, C. S., Birkhead, A. L., Jordanova, A. A., and O'Sullivan J. J., 2004, Flow resistance of emergent vegetation. *Journal of Hydraulic Research*, **42**(4), pp. 390–398.

Katul, G., Wiberg, P., Albertson, J., and Hornberger, G., 2002, A mixing layer theory for flow resistance in shallow streams. *Water Resources Research*, **38**(11).

Knight, R. L., Kadlec, R. H., and Ohlendorf, H. M., 1999, The use of treatment wetlands for petroleum industry effluents. *Environmental Science & Technology*, **33**(7), pp. 973–980.

Kobashi, D. and Mazda, Y., 2005, Tidal flow in riverine-type mangroves. *Wetlands Ecology and Management*, **13**, pp. 615–619.

Koch, D. L. and Brady, D. L., 1986, The effective diffusivity of fibrous media. *AIChE Journal*, **32**(4), pp. 575–591.

Koch, D. L. and Ladd, A. J. C., 1997, Moderate Reynolds number flows through periodic and random arrays of aligned cylinders. *Journal of Fluid Mechanics*, **349**, pp. 31–66.

Koehl, M. A. R., 2006, The fluid mechanics of arthropod sniffing in turbulent odor plumes. *Chemical Senses*, **31**, pp. 93–105.

Larsen, L. G., Harvey, J. W., and Crimaldi, J. P., 2009, Predicting bed shear stress and its role in sediment dynamics and restoration potential of the Everglades and other vegetated flow systems. *Ecological Engineering*, **35**(12), pp. 1773–1785.

Lee, J. K., Roig, L. C., Jenter, H. L., and Visser, H. M., 2004, Drag coefficients for modeling flow through emergent vegetation in the Florida Everglades. *Ecological Engineering*, **22**, pp. 237–248.

Leonard, L. A., and Luther, M. E., 1995, Flow hydrodynamics in tidal marsh canopies. *Limnology and Oceanography*, **40**(8), pp. 1474–1484.

Leonard, L. A., and Reed, D. J., 2002, Hydrodynamics and sediment transport through tidal marsh canopies. *Journal of Coastal Research*, special issue **36**, pp. 459–469.

Lightbody, A. F., and Nepf, H. M., 2006, Prediction of velocity profiles and longitudinal dispersion in emergent salt marsh vegetation. *Limnology and Oceanography*, **51**(1), pp. 218–228.

López, F., and García, M., 1998, Open-channel flow through simulated vegetation: suspended sediment transport modeling. *Water Resources Research*, **34**(9), pp. 2341–2352.

MacLennan, A. S. M. and Vincent, J. H., 1982, Transport in the near aerodynamic wakes of flat plates. *Journal of Fluid Mechanics*, **120**, pp. 185–197.

Mazda, Y., Magi, M., Ikeda, Y., Kurokawa, T., and Asano, T., 2006. Wave reduction in a mangrove forest dominated by *Sonneratia* sp. *Wetlands Ecology and Management*, **14**(4), pp. 365–378.

Mazda, Y., Wolanski, E., King, B., Sase, A., Ohtsuka, D., and Magi, M., 1997, Drag force due to vegetation in mangrove swamps. *Mangroves and Salt Marshes*, **1**(3), pp. 193–199.

Möller, I., Spencer, T., French, J. R., Leggett, D. J., and Dixon, M., 1999, Wave transformation over salt marshes: a field and numerical modelling study from North Norfolk, England, *Estuarine, Coastal and Shelf Science*, **49**(3), pp. 411–426.

Morris, J. T. and Haskin, B., 1990, A 5-yr record of aerial primary production and stand characteristics of *Spartina alterniflora. Ecology*, **71**(6), pp. 2209–2217.

Nepf, H. M., 1999, Drag, turbulence, and diffusion in flow through emergent vegetation. *Water Resources Research*, **35**(2), pp. 479–489.

Nepf, H. M., 2004, Vegetated flow dynamics. In *The Ecogeomorphology of Tidal Marshes, Coastal and Estuarine Studies*, **59**, editors: Fagherazzi, S., Marani, M., and Blum, L. K. pp. 137–163.

Nepf, H. M., 2012, Flow and transport in regions with aquatic vegetation. *Annual Review of Fluid Mechanics*, **44**, pp. 123–142.

Nepf, H. M., and Ghisalberti, M., 2008, Flow and transport in channels with submerged vegetation. *Acta Geophysica*, **56**(3), pp. 753–777.

Nepf, H. M., Sullivan, J. A., and Zavistoski, R. A., 1997, A model for diffusion within emergent vegetation. *Limnology and Oceanography*, **42**(8), pp. 1735–1745.

Neumeier, U., and Amos, C. L., 2006, The influence of vegetation on turbulence and flow velocities in European salt-marshes. *Sedimentology*, **53**, pp. 259–277.

Oldham, C. E., and Sturman, J. J., 2001, The effect of emergent vegetation on convective flushing in shallow wetlands: Scaling and experiments. *Limnology and Oceanography*, **46**(6), pp. 1486–1493.

Paranthoen, P., Browne, L.W.B., Le Masson, S., Dumouchel, F., and Lecordier, J.C., 1999, Characteristics of the near wake of a cylinder at low Reynolds numbers. *European Journal of Mechanics B / Fluids* **18**, pp. 659–674.

Peterson, G. W., and Turner, R. E., 1994, The value of salt marsh edge vs interior as a habitat for fish and decapods crustaceans in a Louisiana tidal marsh. *Estuaries*, **17**(1B), pp. 235–262.

Raupach, M. R., and Shaw, R. H., 1982, Averaging procedures for flow within vegetation canopies. *Boundary-Layer Meteorology*, **22**(1), pp. 79–90.

Saiers, J. E., Harvey, J. W., and Mylon, S. E., 2003, Surface-water transport of suspended matter through wetland vegetation of the Florida everglades. *Geophysical Research Letters*, **30**(19).

Serra, T., Fernando, H. J. S., and Rodriguez, R. V., 2004, Effects of emergent vegetation on lateral diffusion in wetlands. *Water Research*, **38**, pp. 139–147.

Schlichting, H. and Gersten, K., 2000, *Boundary-Layer Theory*. 8th ed., Springer.

Shucksmith, J. D., Boxall, J. B., and Guymer, I., 2011, Bulk flow resistance in vegetated channels: analysis of momentum balance approaches based on data obtained in aging live vegetation. *Journal of Hydraulic Engineering*, **137**, pp. 1624–1635.

Spielman, L. and Goren, S. L., 1968, Model for predicting pressure drop and filtration efficiency in fibrous media. *Environmental Science and Technology*, **2**(4), pp. 279–287.

Stone, B. M. and Shen, H. T., 2002, Hydraulic resistance of flow in channels with cylindrical roughness. *Journal Hydraulic Engineering*, **128**(5), pp. 500–506.

Stumpf, R. P., 1983, The process of sedimentation on the surface of a salt marsh. *Estuarine, Coastal and Shelf Science*, **17**, pp. 495–508.

Tanino, Y., 2003, Trapping of fluid in the near wake of emergent plant stems. B.S. thesis, Massachusetts Institute of Technology, Cambridge, MA.

Tanino, Y., 2008, Flow and solute transport in random cylinder arrays: a model for emergent aquatic plant canopies. Ph.D. thesis, Massachusetts Institute of Technology, Cambridge, MA.

Tanino, Y. and Nepf, H. M., 2007, Experimental investigation of lateral dispersion in aquatic canopies. In *Proceedings of the 32nd Congress of IAHR*, editors: Di Silvio, G. and Lanzoni, S.

Tanino, Y. and Nepf, H. M., 2008a, Laboratory investigation of mean drag in a random array of rigid, emergent cylinders. *Journal of Hydraulic Engineering*, **134**, pp. 34–41.

Tanino, Y. and Nepf, H. M., 2008b, Lateral dispersion in random cylinder arrays at high Reynolds number. *Journal of Fluid Mechanics*, **600**, pp. 339–371.

Tanino, Y. and Nepf, H. M., 2009a, Laboratory investigation of lateral dispersion within dense arrays of randomly distributed cylinder at transitional Reynolds number. *Physics of Fluids*, **21**, 046603.

Tanino, Y. and Nepf, H. M., 2009b, Closure to "Laboratory Investigation of Mean Drag in a Random Array of Rigid, Emergent Cylinders" by Yukie Tanino and Heidi M. Nepf. *Journal of Hydraulic Engineering*, **135**, pp. 693–694.

Tennekes, H. and Lumley, J. L., 1972, *A First Course in Turbulence*. The MIT Press, Cambridge, MA.

Tyler, A.C., and J.C. Zieman, 1999, Patterns of development in the creekbank region of a barrier island Spartina alterniflora marsh. *Marine Ecology Progress Series*, **180**, pp. 161–177.

Valiela, I., Teal, J., and Deuser, W., 1978, The nature of growth forms in the salt marsh grass *Spartina alterniflora*. *American Naturalist*, **112**(985), pp. 461–470.

van Hulzen, J. B., van Soelen, J., and Bouma, T. J., 2007, Morphological variation and habitat modification are strongly correlated for the autogenic ecosystem engineer *Spartina anglica* (common cordgrass). *Estuaries and Coasts*, **30**(1), pp. 3–11.

Variano, E. A., Ho, D. T., Engel, V. C., Schmieder, P. J., and Reid, M. C., 2009, Flow and mixing dynamics in a patterned wetland: Kilometer-scale tracer releases in the Everglades. *Water Resources Research*, **45**, W08422.

White, B. L., 2002, Transport in random cylinder arrays: A model for aquatic canopies. Master's thesis, Massachusetts Institute of Technology, Cambridge, MA.

White, B. L., 2006, Momentum and mass transport by coherent structures in a shallow vegetated shear flow. Ph.D. thesis, Massachusetts Institute of Technology, Cambridge, MA.

White, B. L. and Nepf, H. M., 2003, Scalar transport in random cylinder arrays at moderate Reynolds number. *Journal of Fluid Mechanics*, **487**, pp. 43–79.

White, B. L. and Nepf, H. M., 2007, Shear instability and coherent structures in shallow flow adjacent to a porous layer. *Journal of Fluid Mechanics*, **593**, pp. 1–32.

White, B. L. and Nepf, H. M., 2008, A vortex-based model of velocity and shear stress in a partially vegetated shallow channel. *Water Resources Research*, **44**, W01412.

Wilson, N. R. and Shaw, R. H., 1977, A higher order closure model for canopy flow, *Journal of Applied Meteorology*, **16**, pp. 1198–1205.

Zong, L., and Nepf, H. M., 2011, Spatial distribution of deposition within a patch of vegetation, *Water Resources Research*, **47**, W03516.

CHAPTER FOURTEEN

Uniform flow and boundary layers over rigid vegetation

Paola Gualtieri & Guelfo Pulci Doria

Department of Hydraulic, Geotechnical and Environmental Engineering (DIGA), University of Naples Federico II, Naples, Italy

ABSTRACT

This chapter describes the effects of rigid, submerged and equi-spaced vegetation on uniform flow and boundary layers. With reference to uniform flow, the focus is on flow resistance evaluation and flow field modelling, and theories and methodologies to evaluate flow resistance values and describe the flow field are presented. With reference to the boundary layer, its main characteristics on a smooth plate are described, experimental measurements carried out on three different types of beds (smooth, with rigid cylinders, with rigid grass carpet) are presented, and the results are compared. Finally, a methodology to evaluate flow resistance values in a vegetated uniform flow starting from experimental measurements carried out on a turbulent boundary layer flowing on the same vegetated bed is described. A comparison between numerical values and literature experimental data shows good agreement.

14.1 INTRODUCTION

In the past, vegetation on river beds was considered an unwanted source of flow resistance, and for this reason, it was removed to improve the water conveyance. Vegetation has the following effects on water flow: it decreases the water velocity and raises the water levels, i.e., there is a reduction of flow discharge capacity; a deposition of suspended sediments; an increase or decrease in local erosion; interference with the use of the water flow for conveyance, navigation, swimming and fishing; and an influence on flooding on vegetated lands. Such effects depend mainly on the height, density, distribution, stiffness and type of vegetation. These characteristics may change with the season, e.g., the flow resistance may increase in the growing season and diminish in the dormant season.

Much of the earlier studies on the hydraulic effects of vegetation focused on determining the roughness coefficient (Yen, 2002) rather than on developing a better understanding of the physical processes. Currently, vegetation is regarded as a means for providing stabilisation for banks and channels, as habitat and food for animals, and as part of a pleasing landscape for recreational use. Therefore, preservation of vegetation is of great relevance for the ecology of water systems.

The mean flow and turbulence characteristics in vegetated flows have been studied through experimental and numerical methods, in atmospheric flows to understand the turbulence characteristics and transport processes and in open channel flows to evaluate the flow resistance.

Vegetation may be classified into two models: rigid vegetation and flexible vegetation. Rigid vegetation may be modelled using wooden or metallic cylinders or stiff natural plants; flexible vegetation may be modelled using plastic strips or flexible natural plants or grass. Moreover, models of vegetation may have different densities and they may be wholly or partially submerged, with different effects on the flow.

In this chapter, only vegetation modelled through rigid, equi-spaced and completely submerged elements and, in one case, through a rigid grass carpet, will be considered. As for the water flows in channels or rivers, the hypothesis of a wide channel with a rectangular cross section and, consequently, the hydraulic radius R equal to the flow depth h will be assumed.

14.2   UNIFORM FLOW ON RIGID SUBMERGED VEGETATION – FLOW RESISTANCE AND FLOW MODELLING

14.2.1   The problem of flow resistance

14.2.1.1   Foreword

Most research on the flow resistance of a vegetated bed is based on the two-layer approach, i.e., the idea that a vegetated flow can be divided in two different layers: the vegetation layer and the surface layer (Tsujimoto and Kitamura, 1990; Tsujimoto *et al.*, 1992). In each layer, the velocity distribution is described separately and then the two distributions are matched to one another at the separation surface. From the outcome distribution, the mean velocity is obtained and, consequently, the flow resistance is evaluated.

As for the velocity distribution in the surface layer, some authors assume the logarithmic law (characteristic of flow resistance based on the Colebrook equation), whereas others use the Manning equation. For the velocity distribution in the vegetation layer, some authors consider the velocity constant across the layer and assume that its value can be obtained starting from a single element drag, as is described in fluid mechanics. This approach provides simple but accurate enough equations of the flow resistance and it has been used recently as well. Some of these models, (e.g. Stone and Tao Shen, 2002; Van Veltzen *et al.*, 2003; Huthoff *et al.*, 2006, 2007), will be described in more detail in the Paragraph 14.2.1.2.

Some researchers approach the study of the flow resistance starting from a 1D flow model and using the Reynolds Averaged Navier-Stokes (RANS) equations with a vegetation drag term. Due to the presence of the turbulent shear stresses in the RANS equations, to solve the flow field, increasing numbers of equations are required and turbulence models are employed. For flow resistance in vegetated flows, the most-used models are the turbulent viscosity model (Klopstra *et al.*, 1997; Mejier and Van Velzen, 1999) and the k-ε model (Baptist *et al.*, 2007).

Baptist *et al.* (2007) observed that the transition from a constant velocity distribution in the vegetation layer to the logarithmic or Manning-like velocity distribution in the surface layer, can be as sharp as the two-layer model supposes. Consequently, the authors proposed a four-layer model. In particular, with respect to the elements' heights, the vegetation layer stops a little lower and the surface layer starts a little higher, generating an intermediate third layer. Finally, the fourth layer is the laminar-transition sublayer near the channel bed, which takes into account the bed roughness.

Recently, a new methodology to obtain flow resistance equations based on genetic programming has been proposed (Keijzer *et al.*, 2005; Baptist *et al.*, 2007). The results are simple, interesting equations and they will be presented in the Paragraph 14.2.1.2.

Each proposed model was experimentally validated.

14.2.1.2 Flow resistance equations

As already described, some models are particularly meaningful, and they will now be more closely examined. In particular, the models of Stone and Tao Shen (2002), Van Veltzen *et al.* (2003), Huthoff *et al.* (2006, 2007) and Baptist *et al.* (2007) will be examined, which are also synthetically described and compared by Galema (2009).

Stone and Tao Shen (2002)
The Stone and Tao Shen (2002) model can be applied to the entire bulk flow. It is based on the assumption that the driving force (water weight component along the flow) is balanced by flow resistances (bed and vegetation drag, with the second one being much greater than the first one). Therefore, if k is the vegetation height and U_V is the mean velocity in the vegetation layer, referring to the unit bed area, it holds:

$$\rho g h S \left(1 - \frac{\pi D^2 m}{4} \frac{k}{h}\right) = \frac{1}{2} C_D \rho \frac{U_V^2}{\left(1 - D\sqrt{m}\right)^2} m D k \tag{14.1}$$

where ρ is the water density, g is the gravitational acceleration, h is the water depth, S is the channel slope, D is the stem diameter of the cylindrical vegetation, m is the number of cylinders per unit area and C_D is the drag coefficient of a cylinder.

Stone and Tao Shen applied this model to dense enough vegetation that they neglected the bed resistance and considered the solidity parameter. In particular, in the entire bulk flow, the water volume was is calculated by subtracting the total stick volume and the mean velocity in the cross section occupied by the sticks was is increased by through the ratio the ratio between the whole cross section and the one obtained by subtracting the stick area. At this point, they showed that the mean velocity U, evaluated as the ratio between the total flow-rate and the full cross section of the flow, can be expressed by:

$$U = U_V \sqrt{\frac{h}{k}} \tag{14.2}$$

Inserting Eq. (14.2) in Eq. (14.1), the following equation for U was obtained:

$$U = \sqrt{\frac{2g}{C_D m D}} \sqrt{S} \left(1 - D\sqrt{m}\right) \sqrt{\left(\frac{h}{k} - \frac{1}{4}\pi m D^2\right) \frac{h}{k}} \tag{14.3}$$

Van Veltzen et al. (2003)
The Van Velzen *et al.* model, which is based on Klopstra *et al.* (1997), is a two-layer model. Van Velzen *et al.* assumed a constant flow velocity in the vegetation layer that is independent of the surface layer and used Eq. (14.1) without the solidity effects and with $h = k$:

$$U_V = \sqrt{\frac{2g}{C_D m D}} \sqrt{S} \tag{14.4}$$

The mean velocity in the surface layer U_S was described by a logarithmic term (based on the Keulegan equation), superimposed on the velocity in the vegetation layer:

$$U_S = U_V + 18\sqrt{(h-k)S} \log \frac{12(h-k)}{k_N} \tag{14.5}$$

where k_N is the Nikuradse roughness height. This parameter can be obtained from a regression analysis using the data of Meijer (1998):

$$k_N = 1.6k^{0.7} \tag{14.6}$$

Finally, the average flow velocity over the entire bulk flow was calculated as the weighted mean between the vegetation and the surface layer velocities:

$$U = \sqrt{\frac{2g}{C_D m D}}\sqrt{S} + 18\frac{(h-k)^{3/2}}{h}\sqrt{S}\log\frac{12(h-k)}{1.6k^{0.7}} \tag{14.7}$$

Huthoff et al. (2006, 2007)
The Huthoff *et al.* model is also a two-layer model. The authors introduced a characteristic drag length $b = 1/C_D m D$. Moreover, they calculated the velocity in the vegetation layer through Eq. (14.1), but without the solidity effects and they inserted a term representing the bed drag, yielding:

$$U_V = \sqrt{\frac{2bgS}{1+\frac{2b}{h}f}}\sqrt{\frac{h}{k}} \tag{14.8}$$

where f is the friction factor of the Darcy-Weisbach equation. Afterwards, Huthoff approached the study of the surface layer starting from the work of Gioia and Bombardelli (2002) and Barenblatt (2003) and some experimental calibrations, thus obtaining the following equation for the mean velocity in the surface layer:

$$U_S = U_V \left(\frac{h-k}{s}\right)^{2/3[1-(k/h)^5]} \tag{14.9}$$

where s is the constant spacing among the sticks calculated without considering their diameters. Finally, they obtained the model for the average velocity in the entire bulk flow through a weighted mean between the vegetation and surface layer velocities:

$$U = \sqrt{\frac{2bgS}{1+\frac{2b}{h}f}}\left\{\sqrt{\frac{k}{h}} + \frac{h-k}{h}\left(\frac{h-k}{s}\right)^{2/3[1-(k/h)^5]}\right\} \tag{14.10}$$

Interestingly, when h becomes large, Eq. (14.10) approaches:

$$U = \sqrt{\frac{2g}{C_D m D s^{2/3}}}h^{2/3}\sqrt{S} \tag{14.11}$$

Eq. (14.11) shows the same dependency on the flow depth and the slope channel as the Manning equation.

Baptist et al. (2007)
Baptist *et al.* (2007) proposed four procedures.
The first procedure (method of effective water depth) uses a two-layer approach, assuming a constant velocity in the vegetation layer and a logarithmic law in the surface layer. Their integration leads to an easy equation for the Chézy coefficient. The second procedure

(analytical method) assumes a logarithmic law in the surface layer and a velocity distribution is obtained from the fluid mechanics in the vegetation layer. Their integration leads to a new equation of the Chézy coefficient that is much more complex, but most likely more accurate. In the third procedure (1D turbulence model) a k–ε model is applied to numerically solve different study cases.

The validity of this 1D flow model was demonstrated by comparing its results with measurements performed by Meijer and Van Velzen (1999), Nepf and Vivoni (2000) and López and García (2001). Finally, in the fourth procedure (equation built with genetic programming), which is also summarised in Keijzer *et al.* (2005), starting from the single solutions supplied by the third procedure, 990 single solutions were considered, and through genetic programming, a new equation that best fits the previous solutions was generated. Afterwards, this equation was readjusted to fit some theoretical requirements, if C_t and C_b are the Chézy coefficient for the vegetation and the bed, respectively, they obtained:

$$C_t = \sqrt{\frac{1}{1/C_b^2 + C_D m D l / 2g}} + 18 \log \frac{12h}{(12k)} \tag{14.12}$$

Consequently, the equation of the mean velocity over the entire bulk flow was:

$$U = \left[\sqrt{\frac{1}{1/C_b^2 + C_D m D l / 2g}} + 18 \log \frac{12h}{12k} \right] \sqrt{hS} \tag{14.13}$$

14.2.1.3 Use of classical flow resistance equations

As a conclusion of the previous paragraph, it is possible to observe the following.

The classical equations of Keulegan (Colebrook-White), Manning, or Chézy-Bazin always refer to a flow without obstacles that is retarded only by the bed roughness. In contrast, the vegetation is a set of obstacles just within the flow and therefore, the classical equations cannot be usefully applied to calculate the flow resistance. This observation is expressed in Yen (2002): "*[The] presence of vegetation in the flow modifies the velocity distribution, and hence, the resistance*". In the previous four equations, this observation appears clearly. Starting from the previous equations, using the opposite procedure, the roughness coefficient values characteristic of the Keulegan (Colebrook-White), Manning, or Chézy-Bazin equations could be derived. The result would always be dependent on the flow depth and, consequently, not truly representative of the bed roughness. Hence, the flow resistance due to a vegetated bed cannot be represented through any classical equation.

Augustjin *et al.* (2008) performed some comparisons between different literature models for the flow resistances and experimental literature data. The chosen models were the three classical ones (Keulegan (Colebrook-White), Manning and Chézy-Bazin); the historical model of De Bos and Bijkerk (1963) and the more recent models of Baptist *et al.* (2007), Huthoff *et al.* (2007), and Van Velzen *et al.* (2003) that were examined previously. For the three more recent models, a first comparison between predicted water depths and literature experimental data led the authors to the conclusion that "*All methods show high correlation coefficients, small biases and root mean squares differences indicating good performance*".

For all of the models, a comparison of the normalised predicted water depths led the authors to state that "*Most models provide a reasonable fit, only constant Manning and Chézy coefficients are in general not accurate as estimators for vegetation resistance*" and "*On average the Keulegan equation shows the best agreement with experimental data, although for some data sets other models may perform better*". Furthermore they noted that

"The more recent models perform equally well as do Keulegan and De Bos and Bijkerk", the only *"disadvantage [being] that they contain an empirical parameter that needs to be calibrated"*. Finally, they observed that *"In the field, vegetation is quite heterogeneous and the theoretical geometrical parameters are not easily determined"*.

After these first comparisons, Augustjin *et al.* (2008) investigated the results that the models give if the flow depth is much higher than the vegetation height (as often occurs in floodplains). They did not find experimental data published in the literature and, consequently, they extrapolated the literature data to higher flow depths, concluding that *"Based on the different predictions of vegetation roughness descriptions, there remains a large uncertainty in flow response to the presence of vegetation at high discharges, so that more data are required"*.

A further comparison among the different models were performed, considering only the three more recent models and the Keulegan one, with different values of the h/k ratio, employing three different experimental data sets (each one with different h/k ratio values) and computing for each one the value of the equivalent Manning coefficient. They concluded that *"It can be seen that for all methods the Manning coefficient decreases with increasing submergence ratio and eventually levels off. The rate by which they level off is different for different model parameters, but in general it can be said that for $h/k > 5$ the roughness coefficient can be approached by a constant Manning coefficient. Therefore, at high discharges and associated high submergence ratios calibration on constant Manning coefficient is acceptable"*. Referring to the high flow depths and to rigid vegetation, they finally noted that *"At high submergence ratios the Manning equation provides a good approximation as long as the value is calibrated on data at submergence ratios $h/k > 5$"*.

14.2.2 The problem of vegetated flow modelling

The aim of a vegetated flow model is to evaluate the effects of vegetation on the main turbulence characteristics of the flow as the distributions of mean velocities, standard deviation, skewness, kurtosis, integral length scales and shear stresses, starting from theoretical models and experimental results. Moreover, the mean velocity distribution is strictly linked to flow resistance and, as already discussed in Paragraph 14.2.1.2, some researchers find the velocity distribution in the vegetation layer through fluid mechanics equations when starting from a two-layer model.

In particular, in most studies, RANS equations are used and due to the appearance of the turbulent shear stresses in these equations, an increasing number of equations are required to solve the flow field and turbulence models are used.

In the early studies analysing canopy flow in the field of meteorology, Wilson and Shaw (1977) addressed the problem in which the flow field was strongly 3D. They transformed it in a 1D flow field, developed in the vertical direction, by horizontally averaging the equations of momentum, turbulent kinetic energy and Reynolds stresses. Moreover, they proposed two procedures to solve the flow field. In the first procedure, the instantaneous flow field is horizontally averaged over a plane large enough to eliminate temporal and spatial variations due to the turbulence and the vegetation structure and, therefore, there was no need for temporal averaging. In the second procedure, the instantaneous flow field is temporally and then horizontally averaged over a plane with dimensions comparable to the distance between individual cylinders and, therefore, large enough only to eliminate spatial variations due to the vegetation structure. In the momentum equation, with the instantaneous variables expanded in terms of mean and fluctuating components, the above averaging procedures gives rise to terms representing drag forces due to the vegetation. The authors asserted that the procedures would have been equally effective and that their results would have been identical. The predicted results of the mean velocity and the Reynolds stresses compare well with experimental measurements, while turbulence intensities are overestimated.

In Raupach and Shaw (1982), under the assumption of stationary and horizontally homogeneous flow above vegetation, the two horizontal averaging procedures, called, respectively, Scheme I and Scheme II, were better formalised, together with the horizontal averaging operator and its properties. The result was that Scheme I and II give rise to different expressions. In fact, with reference to the momentum equation, in Scheme II, differently than in Scheme I and due to the superimposition of two averaging processes, dispersive fluxes occurred due to the obstruction of the vegetation. Again, with reference to the equation for the second moments, in Scheme II, due to the superimposition of two averaging processes, an analogous dispersive turbulent kinetic flux occurred, caused by the wakes of individual canopy elements. Moreover, the authors noted that wind tunnel experiments had failed to find dispersive momentum fluxes, even in situations in which they should have been significant. Therefore, if the dispersive fluxes are negligible, the averages of Scheme I and II coincide. The authors concluded that as the dispersive fluxes complicated the second-order equation considerably, experiments to estimate their magnitude in real canopies were required to confirm their assumption that the dispersive fluxes were negligible. This averaged approach was widely used in further studies.

Among the first studies of the turbulence characteristics in vegetated surface waters, Tsujimoto *et al.* (1992) used both a two-layer approach and a 1D flow model. The velocity distribution in the vegetation layer was not calculated through the RANS equations, but rather started from the observation that both the velocity and the Reynolds stress distributions can be approximated through exponential laws. In the surface layer, where the Reynolds stress distribution is minimally affected by vegetation, applying a mixing length model for the velocity distribution yielded a modified logarithmic law.

Shimizu and Tsujimoto (1994) numerically described the flow field in the vegetation layer through the RANS equations with a k–ε model. The governing equations were spatially averaged, neglecting the geometry of individual vegetation elements, and then modified governing equation were obtained, adding terms due to the drag effect to the momentum, k and ε equations of the standard k–ε. The calculated results showed good agreement with the measurements in the flume reported in Shimizu *et al.* (1991) and Tsujimoto *et al.* (1992).

The model proposed by Kutija and Hong (1996) is particularly interesting. Starting from a two-layer approach, RANS 1D equations were applied to both the surface and vegetation layers. For the turbulence models, in the surface layer, the mixing-length model was chosen, while in the vegetation layer, the eddy viscosity model was chosen. In fact, the comparison of the numerical results with the experimental data of Tsujimoto and Kitamura (1990) relative to rigid submerged vegetation showed that the eddy viscosity model was not suitable for the whole vegetation layer as the velocities at the top of the vegetation were not as small as they were deeper in the layer. The authors concluded that the eddy viscosity model could be applied only to a distance p from the bottom that was less than the vegetation height and that p was a basic calibration parameter. It depends on the vegetation density, diameter and stiffness, and its physical background needs to be investigated more thoroughly.

Lopez and Garcia (1997, 1998, 2001), starting from the two-layer approach, analysed the ability of turbulence models based on a two-equation closure scheme (k–ε and k–ω), to describe the flow field. The 3D problem was transformed into a 1D framework by averaging the governing equations through the procedures proposed by Raupach and Shaw (1982). The numerical results were in agreement with the field data.

Neary (2003) developed and validated a two-layer model solved with RANS 1D equations and a k–ω model because it had been demonstrated to be the model of choice for predicting bed shear stress over a wide range of roughness types (Neary, 1995; Patel and Yoon, 1995). Laboratory measurements of Shimizu and Tsujimoto (1994) were used to compare with model calculations and the results showed that both the k–ω and k–ε models gave reasonable predictions of the turbulent characteristics of the flow.

More recently, some researchers have still used the two-layer approach but have modified the physical model (Huai *et al.*, 2009a, 2009b) or the mathematical model (Gao *et al.*, 2011), focusing their attention on the local mean velocity distributions.

In particular, Huai *et al.* (2009a, 2009b) divided the vegetation layer into two sublayers, adopting a three-layer model. To predict the velocity distribution, they proposed a 1D flow model with the mixing length approximation. The fit between the analytical results and the measurement data showed good agreement.

Finally, Gao *et al.* (2011) proposed a 3D flow model using a two-layer mixing length model to describe 3D velocity profiles instead of more advanced and complex two-equation turbulence models. The model-predicted velocity distributions were compared with laboratory data measurements, with very good agreement.

Some researchers stated that solving the RANS equations by using two equations for the isotropic turbulence models provides only a limited description of the turbulent field, and they proposed to apply Large Eddy Simulation (LES) modelling to vegetated flows (Cui and Neary, 2002, 2008). The authors observed that LES results were in good agreement with measurements, but there were no discernable improvements compared with RANS equation simulations. Also Zhu *et al.* (2006) and Yue *et al.* (2008) used LES modelling and verify the validity of its results though a comparison with non-intrusive experimental data obtained from a Particle Image Velocimetry (PIV).

In the meanwhile, studies of terrestrial canopies (Raupach *et al.*, 1996) established an analogy between the atmospheric flow and a plane mixing layer more than with a perturbed boundary layer. In fact, the canonical form of atmospheric flows near the land surface in the absence of a canopy resembles a rough-wall boundary layer. However, in the presence of an extensive and dense canopy, the flow within and just above the foliage behaves as a mixing layer. A free mixing layer belongs to the family of free turbulence flows that also includes free wakes and free jets. The term "free" stresses the fact that these turbulent flows are maintained by internal velocity gradients and their evolution is independent of solid boundaries. Canonical free mixing layers evolve in co-flowing liquids of different densities or flows of different mean velocities, for instance, downstream from a splitter plate. The main characteristics of a mixing layer are:

1) The velocity distribution shows a strong inflexion corresponding to the centreline where the flows meet;
2) The distributions of the horizontal and vertical standard deviations and of Reynolds stresses show a clear peak corresponding to the centreline where the flows meet;
3) The distributions of horizontal and vertical skewness show double antisymmetric peaks.

An analogy between turbulent flow in a mixing layer and near the top of a canopy was suggested, initially, by the inflectional mean velocity distributions in both flows, which shows the generation of large-scale eddies due to the Kelvin-Helmholtz instability.

Afterwards, the mixing layer analogy provides an explanation for many of the observed distinctive features of canopy turbulence, as a peak at the top of the canopies in the standard deviations and Reynolds stresses distributions and as two antisymmetric peaks in skewness distributions.

The mixing layer analogy concept has also been used in the following studies about liquid flow-vegetation interactions, as in the case of rigid submerged vegetation in surface waters in Poggi *et al.* (2004) and Nezu and Sanjou (2008).

Poggi *et al.* (2004) observed that while in the case of a dense canopy, the flow within and just above the canopy behaved as a mixing layer, no analogous formulations existed for intermediate canopy densities. Therefore, the authors deepened the connection between the canopy density and the key turbulence statistics within and just above the canopy (referred to as the canopy sublayer i.e., CSL), to obtain a phenomenological theory that described the key turbulent statistics in terms of canopy density. During the research, experimental measurements were conducted using the LDA in an open channel over a wide range of canopy densities, showing that the relative importance of the mixing layer varies with the canopy density. It is interesting to note that given the non-homogeneity in the flow statistics within the canopy, 11 measurements locations have been chosen that are not uniformly spaced, but rather are located in regions where the flow statistics exhibit the highest spatial variability. Then, when computing statistical moments, the experimental distributions have been first time-averaged then planar-averaged to yield vertical distributions, as defined by Raupach and Shaw (1982). The proposed model divided the space within the CSL into three zones. In the deep zone the flow field was dominated by vortices connected with von Kármán vortex streets, periodically interrupted according to the canopy density. In the second zone, according to the canopy density, the inflectional mean velocity distributions near the canopy top showed the generation of large-scale eddies due to the Kelvin-Helmholtz instability. In the uppermost zone the flow obeyed to the classical surface-layer similarity theory. The authors solved the flow field with a 1D flow model, using first-order closure models (eddy diffusivity and mixing length). A comparison of numerical and experimental distributions of mean velocities and Reynolds stresses showed that the proposed model accurately reproduced them for a wide range of roughness densities.

Nezu and Sanjou (2008) noted that the mixing layer analogy concept implicitly suggests that the whole flow region can be divided into three layers, i.e., the emergent layer, the mixing layer and the log-law layer. In particular, in the emergent layer ($0 < y < h_p$), the velocity is almost constant; in the mixing layer ($h_p < y < h_{log}$), the turbulent structure is analogous to a mixing layer; and in the log-law layer ($h_{log} < y < h$), the turbulence structure is analogous to structures in boundary layers and open channel flows with rough beds. The introduced h_p is the penetration depth at which the Reynolds stress had decayed to 10% of its maximum value and as h is the flow depth. Moreover, on the basis of Laser Doppler Anemometer (LDA) and PIV measurements and LES calculations, the authors investigated turbulent structures in a vegetated canopy open channel flow.

The effect of vegetation density was also studied by Luhar *et al.* (2008). They noted that sparse and dense submerged vegetation are defined based on the relative contribution of turbulent stress, among other. In fact, in a sparse canopy, turbulent stresses remained elevated within the canopy, while in a dense canopy, they were reduced.

As previously described, for submerged canopies of sufficient density, the dominant characteristic of the flow is the generation of a shear layer at the top of the canopy that generates vortices by Kelvin-Helmholtz instability. Nepf and Ghisalberti (2008) described how these vortices control the vertical exchange of mass and momentum and influence both the mean velocity distribution and turbulent diffusivity. Already in Ghisalberti and Nepf (2006), flume experiments had been conducted with rigid vegetation to study the structure of vortices generated by Kelvin-Helmholtz instability and vertical transport in vegetated shear flows.

14.3 BOUNDARY LAYER OVER RIGID SUBMERGED VEGETATION

14.3.1 The boundary layer

The boundary layer appears in various areas of the fluid mechanics. The atmospheric boundary layer (i.e., planetary boundary layer) is the air layer most influenced by the characteristics of the earth's surface (i.e., also by the vegetation). This boundary layer grows only to a finite thickness (Ghisalberti and Nepf, 2004). In air or water flows, the boundary layer is a thin layer of reduced velocity immediately adjacent to the surface of a solid body past which the fluid is flowing. In the successive sections of the fluid, the boundary layer develops, and its thickness grows. Its characteristics are described by the boundary layer theory.

14.3.1.1 The boundary layer

The boundary layer concept was first introduced by L. Prandtl (1904). When a fluid with a low viscosity, such as air or water, flows past a streamlined solid body at high Reynolds number, the effect of viscosity should be small. Therefore, the flow may be regarded as frictionless and can be studied through the theory of irrotational flows.

However, this theory cannot be used to calculate what happens in the immediate proximity of the body because the correlated phenomena are primarily due to the viscous friction. Prandtl proposed that the effect of viscosity in the flow should be confined to a very thin layer of flow, immediately near the solid surface, where the no-slip condition results in a rather high velocity gradient that generates internal friction caused by fluid viscosity. Prandtl called this thin fluid layer the boundary layer. Figure 14.1 shows a schematic diagram of the boundary layer. In this figure, the submerged body is a long, flat plate of negligible thickness, immersed in a uniform flow of velocity u_0, with zero angle of attack. In this case, the boundary layer develops on both sides of the plate. The figure shows only the upper side.

Within the boundary layer, therefore, a velocity gradient appears in the direction normal to the plate so that velocity values increase from zero at the plate up to u_0 far from the plate. This behaviour describes the velocity distribution in the normal direction.

The boundary layer thickness is another important concept. There are many definitions of such a concept. The simplest one is that the boundary layer thickness can be considered as the distance from the plate at which the velocity attains 99% of the value of u_0 (δ_{99}). It is evident that the boundary layer thickness grows with distance from the leading edge of the plate, due to the rising influence of this obstacle. This behaviour is schematised in Figure 14.1.

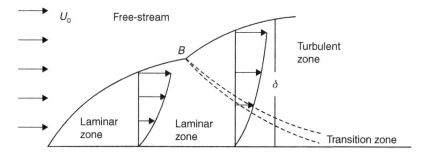

Figure 14.1. Turbulent boundary layer.

It is possible to consider the flow over the plate as being composed of the rotational boundary layer near the plate up to a distance equal to its thickness and an external irrotational flow beyond this distance. This external flow will be hereafter called the free-stream. In an ideal configuration, the free-stream has an infinite thickness, whereas in practical realisations, its thickness is a finite but very large value.

The boundary layer zone nearer the leading edge shows laminar flow, but not far from this point, the internal friction generates turbulence. Hence, there is a laminar zone and downstream a turbulent boundary layer, which is much more developed. In the turbulent zone, the boundary layer thickness grows more rapidly than in the laminar zone. In the turbulent zone, a very thin viscous sublayer is always present near the body surface.

In contrast, in the free-stream, the flow is considered to be laminar and irrotational everywhere. The described flow will hereafter be called the boundary layer.

The same characteristics are also present in the case of a lightly curved streamlined body surface, and a boundary layer arises. Consequently, the boundary layer can refer both to a flat plate and a streamlined body.

In the boundary layer over a flat plate, both u_0 and the pressure (or piezometric head, in the case of water flow) in the free-stream are constant. In contrast, in the case of a lightly curved streamlined body, u_0 and the pressure (or piezometric head) in the free-stream are often variable because of the streamlined body's surface curvature. In this second case, the u_0 and the pressure (or piezometric head) gradients in the flow direction can be either constant or variable. In particular, if these gradients are constant, the boundary layer is called an equilibrium boundary layer. The case of constant u_0 and constant pressure (or piezometric head) can be considered as the simplest case of an equilibrium boundary layer, with zero values for u_0 and pressure (or piezometric head) gradients.

A characteristic of equilibrium boundary layers is the possibility of defining a dimensionless theoretical velocity distribution that fits experimental data in all of the successive sections of the boundary layer except the viscous sublayer.

The most-used velocity distribution is the logarithmic velocity defect law, which refers to the friction velocity $u^* = (\tau/\rho)^{1/2}$ with τ shear stress at the wall and ρ fluid density. The logarithmic velocity distribution has a theoretical background, which leads to:

$$\frac{u}{u^*} = \frac{1}{0.39} \ln\left(\frac{yu^*}{\nu}\right) + 5.56 \tag{14.14}$$

with y being the distance from the plate and ν being the kinematic viscosity. The 0.39 value is an important parameter, generally defined through the symbol κ, whose name in the literature is the von Kármán Constant. This formulation was already present, for instance, in Schlichting (1955). The value of 0.4 is generally attributed to the von Kármán Constant.

Coles (1956), following some previous ideas of Clauser (1956), proposed a logarithmic corrected mean velocity law that had a new term called the Wake term. Therefore, this law is called Coles Wake law:

$$\frac{u}{u^*} = \frac{1}{0.4} \ln\left(\frac{yu^*}{\nu}\right) + 5.1 + \frac{\Pi}{0.4} W\left(\frac{y}{\delta}\right) \tag{14.15}$$

This law is considered valid for high Reynolds numbers. The W function is a universal one and is called the Wake function. Coles gave its values as a table, but these values could be calculated also as $W(y/\delta) = 1 = \sin\{\pi[2(y/\delta) - 1]/2\}$, whereas the Π parameter depends on the pressure (or piezometric head) gradient: for zero gradient, the Π parameter has a value of 0.55. Finally, it is worth noting that the δ thickness is defined as the value through which the theoretical Eq. (14.15) fits the experimental points and its value is a little more than the (δ_{99}) value. Coles Wake law is commonly presented as a velocity defect law:

$$\frac{u - u_0}{u^*} = \frac{1}{0.4} \ln\left(\frac{y}{\delta}\right) + \frac{\Pi}{0.4}\left[W\left(\frac{y}{\delta}\right) - W(1)\right] \tag{14.16}$$

The main characteristic of this equation is that it depends only on the y/δ ratio, so it is identical in all successive sections of the boundary layer: this is, consequently, the researched equilibrium distribution. The Coles Wake law fits the experimental values sufficiently well, but is affected by an important failure: its derivative where $y = \delta$ is different from zero. Dean (1976), following Granville's ideas (1976), proposed a new mean velocity distribution law, in which this Coles law failure was eliminated. His law, however, is not continuous in its second derivative, always at $y = \delta$. Gualtieri and Pulci Doria (1998) proposed a correction in the Coles law relative to the specific case of zero gradient to overcome this last failure.

14.3.1.2 The boundary layer with a turbulent free-stream

As above clarified, the hydrodynamic laws governing the boundary layer, and in particular the equilibrium boundary layer, are now reasonably well understood. In particular, for the equilibrium boundary layer, the value of the pressure (or piezometric head) gradient along the flat plate is the most important parameter; this parameter fixes the local mean velocity distribution along the direction orthogonal to the plate.

A more sophisticated type of boundary layer flow occurs when the free-stream is turbulent. This is an important and frequent type of boundary layer, and it will be hereafter defined as a boundary layer with a turbulent free-stream. If the pressure (or piezometric head) gradient of the free-stream is constant, the boundary layer is termed equilibrium boundary layer with a turbulent free-stream.

The effect of the free-stream turbulence on the turbulent boundary layer has been investigated in several experimental, analytical and computational studies, focusing on the main turbulence statistical quantities as the local mean velocity distribution (Bandypadhyay, 1992; Blair 1983a, b; Castro, 1984; Charnay et al., 1971, 1976; Evans, 1985; Evans and Horlock, 1974; Hancock and Bradshaw, 1983; Hancock and Bradshaw, 1989; Hoffmann and Mohammady, 1991; Huffman et al., 1972; Kline, 1960; McDonald, H. and Kreskowsky, 1974; Meier and Kreplin, 1980; Robertson and Holt, 1972).

Coles believed that in the case of an equilibrium boundary layer, his own Wake law could also model the effect of a turbulent free-stream, assuming that the turbulence in the free-stream would have the same effect on the velocity distribution as an adverse pressure gradient.

Pulci Doria and Taglialatela (1990) developed a velocity distribution law that can be applied to equilibrium boundary layers with turbulent free-streams and zero pressure (or piezometric head) gradients, taking into account the presence of turbulence at $y = \delta$ and also the hydrodynamic requirement of a derivative equal to zero at that point. Pulci Doria Taglialatela law (PDT) is always presented as a velocity-defect law and is characterised by the u'_0, which is the root-mean-square of the velocity fluctuations at $y = \delta$:

$$
\frac{u - u_0}{u^*} = \frac{1}{0.4}\ln\left(\frac{y}{\delta}\right) - \frac{0.0774}{0.4} - \frac{1.182}{0.4}\left(1 - \frac{u'_0}{u^+}\right) + \frac{1}{0.4}F\left(\frac{y}{\delta}\right)
$$

$$
+ \frac{1}{0.4}\left(1 - \frac{u'_0}{u^+}\right)F'\left(\frac{y}{\delta}\right) \tag{14.17}
$$

As observed, the equation holds two experimentally based functions (F and F') of the dimensionless variable y/δ. The values of these functions are shown in Table 14.1. This distribution is valid up to a value of $1.0 \div 1.2$ for the ratio u'_0/u^*. Finally, the thickness δ is defined as the value through which the theoretical Eq. (14.17) and the experimental points fit, and its value is 1.25 times as large as the δ of Coles. Experimental results of boundary layers described either through the Coles distribution or the PDT law are reported in Gualtieri et al. (2004).

Finally, it is worth noting some fundamental studies of the boundary layer: the 8th English edition of Schlichting's work (Schlichting and Gerstein, 2003), two very important reviews by Sreenivasan (1989) and Gad-el-Hak and Bandyopadhyay (1994) and the reviews by George and Castillo (1997) and Klewicki (2010). In particular, George and Castillo (1997)

Table 14.1. Functions F and F' of in PDT law (1990).

y/δ	F	F'	y/δ	F	F'	y/δ	F	F'
0.03	0.0084	0.000	0.10	0.0400	0.000	0.60	0.1272	1.008
0.05	0.0148	0.000	0.15	0.0728	0.009	0.70	1.1296	1.156
0.06	0.0180	0.000	0.20	0.0880	0.074	0.80	0.1340	1.182
0.07	0.0232	0.000	0.30	0.1080	0.277	0.90	0.1224	1.182
0.08	0.0272	0.000	0.40	0.1192	0.525	0.95	0.1056	1.182
0.09	0.220	0.000	0.50	0.1252	1.008	1.00	0.0744	1.182

proposed a theory and an experimental evaluation methodology based entirely on the averaged Navier-Stokes equations and was applied to the zero pressure gradient equilibrium boundary layer.

In terms of the literature experimental work, the greatest part of boundary layer research addresses air boundary layers developing in wind tunnels. Nonetheless, there is some research on water boundary layers generated along a flat plate inserted in a water flow (Balachandar *et al.*, 2001), or directly in an open channel (Balachandar and Ramachandran, 1999; Tachie *et al.*, 2000; Tachie *et al.*, 2001; Tachie *et al.*, 2003).

14.3.2 Effects of vegetation on the boundary layer

14.3.2.1 *Experimental research in literature*

Gualtieri and Pulci Doria (2008) and De Felice *et al.* (2009) presented experimental surveys about boundary layers developing over beds vegetated by rigid submerged elements. The density of the vegetation can be considered low with respect to the literature values for uniform flows. As in Balachandar and Ramachandran (1999), the boundary layer was obtained on the bottom of a rectangular channel, coming out of a feeding tank, through a rectangular adjustable sluice gate. The channel was 4 m long and 0.15 m wide, with plexiglass walls and bottom and a variable slope. In the first sections of the channel, beginning from the vena contracta after the sluice gate, a boundary layer was generated on the bottom. The boundary layer thickness increased in the sections along the channel, until it reached the same value as the height of the circulating flow at a distance from the inlet of the channel that depended on the dynamic characteristics of the flow itself.

Experimental measurements of the instantaneous velocity obtained via LDA had already been carried out in the same channel, but on a smooth bed (Gualtieri *et al.*, 2004). The vegetation was modelled with brass cylinders 4 mm in diameter, with three different heights (5 mm, 10 mm, 15 mm), placed according to two different geometries (rectangular and square meshes) called single-density and double-density geometries. For the rectangular meshes, the sides measurements were 5 cm long and 2.5 cm wide; for the square meshes, the sides measurements were 2.5 cm. Consequently, the number of cylinders per m^2 horizontal area was equal to 800 m^{-2} and 1600 m^{-2}; the projected area of vegetation per unit volume of water in the flow direction (Tsujimoto *et al.*, 1992) was 3.2 m^{-1} and 6.4 m^{-1}; and finally, the vegetation density, evaluated as the projected area of vegetation per m^2 horizontal area, was 0.024 and 0.048. Combinations of three different heights and two different densities produced six different vegetation models. These cylinders were glued on a plexiglass plate. The plate was then inserted onto the bottom of the experimental channel. To prevent the plate from disturbing the entry of the stream into the channel, a connecting ramp between the channel inlet and the plate itself had been installed.

To compare the experimental data with the data describing the equilibrium boundary layer on a smooth bed, some hydraulic parameters were kept constant. As a consequence, the Run 1 data of Gualtieri *et al.* (2004) was considered as a reference experiment. In particular, the height of the sluice gate was set at 7.49 cm so that the height in the vena contracta was 4.62 cm, the load on the vena contracta was at 10.34 cm, and the velocity of the free-stream was 1.424 m/s. Moreover, it was necessary to ensure the equilibrium of the boundary layer in each one of the six considered vegetation models, which means a constant piezometric head and a horizontal free surface, at least in the first 50 cm where the boundary layer develops and is measured. Therefore, it was necessary to increase the slope of the channel, to take into account the head losses due to the vegetation. The slope values are reported in Table 14.2.

Table 14.2. Slopes in different cases.

Veg. height or density	Smooth bottom	Veg. 5 mm	Veg. 10 mm	Veg. 15 mm
Single density	0.25%	0.92%	1.60%	2.27%
Double density	0.25%	1.15%	2.05%	2.95%

As in the smooth surface experiments, the test sections were set at 20, 30, 40 and 50 cm from the channel inlet. In each test section, two different measurement locations were considered. The first one was at the centre of the test section and also at the centre of a mesh. It corresponded exactly to the measurement location on a smooth bottom. The second one was laterally displaced by 1.25 cm and, consequently, was located at the centre of the cylinder row of a mesh. This last location would have been meaningless on a smooth bottom. In each test section, for each flow condition (6 cases), and for each location (2 cases), measurements of the instantaneous velocity were carried out using the LDA technique. In each location, approximately 20–30 experimental points along a vertical direction were obtained to fully describe the behaviour of the flow from the bottom to the free surface. The experimental data were processed by a Frequency Tracker. Each value of the local mean velocity was obtained using an acquisition time of 200 s, enough to reduce the turbulent velocity fluctuations. In this manner, for every location, the local mean velocity distribution was obtained.

14.3.2.2 Equilibrium boundary layer on vegetated beds

Starting from the local mean velocity distributions, the values of the δ_{99} thickness of the boundary layer were obtained and reported in Tables 14.3 and 14.4. It is useful to stress that, sometimes, the thicknesses in Tables 14.3 and 14.4 exceed the thickness of the flow, and therefore, they can be considered as virtual thickness, i.e., as the thickness the boundary layer would attain if the stream were sufficiently high. Later, a methodology to evaluate the virtual thickness will be described.

Examples of the local mean velocity distributions for the described vegetation models are represented in Figures 14.2 and 14.3 from (Gualtieri and Pulci Doria, 2008).

Tables 14.3 and 14.4 suggest some general characteristics of boundary layers on a vegetated bed:

1) Vegetation increases the boundary layer thickness.
2) The increase is greater if the vegetation is higher or denser.
3) The boundary layer thickness seems to be independent of the measurement location.

Table 14.3. Thickness of the boundary layers (single density).
Thin line: central location values
Bold line: lateral location values

B.L. thick.	Smooth bottom	Veg. 5 mm		Veg. 10 mm		Veg. 15 mm	
δ_{S1} (mm)	3.8	14.6	**16.5**	20.0	**22.8**	29.5	**27.2**
δ_{S2} (mm)	7.1	25.6	**22.7**	31.4	**30.4**	43.5	**39.5**
δ_{S3} (mm)	10.9	32.0	**33.0**	42.1	**45.0**	56.8	**51.7**
δ_{S4} (mm)	13.5	42.2	**41.3**	51.3	**53.2**	64.1	**59.2**

Table 14.4. Thickness of the boundary layers (double density).
Thin line: central location values
Bold line: lateral location values

B.L. thick.	Smooth bottom	Veg. 5 mm		Veg. 10 mm		Veg. 15 mm	
δ_{S1} (mm)	3.8	16.0	**16.3**	22.0	**25.4**	31.5	**34.2**
δ_{S2} (mm)	7.1	26.3	**28.0**	33.9	**35.4**	45.5	**46.0**
δ_{S3} (mm)	10.9	35.1	**41.0**	47.7	**49.4**	56.7	**55.9**
δ_{S4} (mm)	13.5	44.0	**52.7**	52.4	**54.7**	62.2	**57.1**

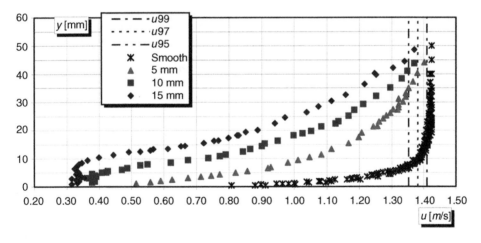

Figure 14.2. Local mean velocity distributions in test section n.1, central vertical (single density).

To stress the equilibrium characteristics of the boundary layers, the local mean velocity distributions, as functions of the distance from the bottom, were made dimensionless, respectively, through the free-stream velocity and the boundary layer thickness. A boundary layer on a flat plate is an equilibrium boundary layer when, in the free-stream, it has a streamwise zero gradient and, consequently, the dimensionless velocity distributions appear to be superimposed on one another. Examples of the dimensionless local mean velocity distributions for the described vegetation models are presented in the following figures from (Gualtieri and Pulci Doria, 2008).

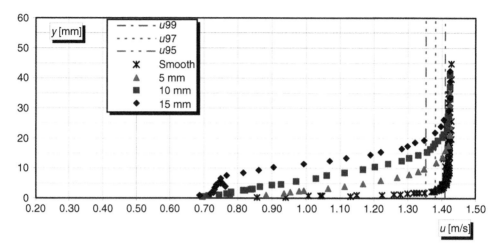

Figure 14.3. Local mean velocity distributions in test section n.4, central vertical (double density).

Some differences can be observed between Figures 14.4 and 14.5. In the first case, the dimensionless distributions are superimposed on one another, while in the second case, this superimposition is limited to the higher part of the diagrams (practically higher than the vegetation height). Therefore, when the vegetation perturbation to the stream is lower, the equilibrium characteristics completely hold, whereas, when the vegetation perturbation to the stream is higher, the equilibrium characteristics hold only in that portion of the stream that is not directly influenced by the cylinders. The authors defined these two cases, respectively, as full equilibrium and partial equilibrium. In particular, the partial equilibrium holds when the ratio between the cylinder height and the boundary layer thickness is more than 0.35 for the single density and 0.28 for the double density. These observations match the experimental data of Tsuijimoto (1990), even if relative to uniform flows, in which the ratio between the cylinder height and the uniform flow depth is 0.50, i.e., more than 0.35. However, a deeper insight in the diagrams shows that the shape of the superimposed part for the different vegetation models is not the same. To characterise this aspect, using only a single value,

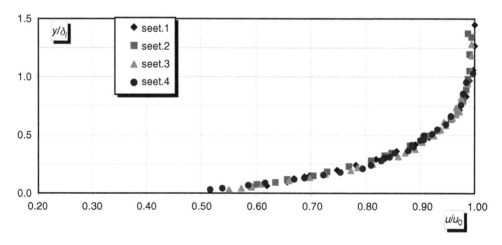

Figure 14.4. Dimensionless local mean velocity distributions, central verticals (5 mm cylinders, single density).

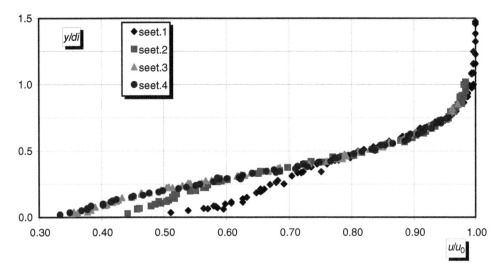

Figure 14.5. Dimensionless local mean velocity distributions, central verticals (15 mm cylinders, double density).

the authors define the ratio δ_{99}/δ_{97} as the shape factor of the distribution. The values are reported in Table 14.5. The shape factor appears as a function decreasing in a monotonic way with increasing cylinder height, cylinder density, and proximity of measurement location to a cylinder.

Table 14.5. Different Shape Factors for the different flow conditions.
Thin line: central location values
Bold line: lateral location values

Veg. height	Zero density		Single density		Double density	
0 mm	1.65	**1.65**	1.65	**1.65**	1.65	**1.65**
5 mm	1.65	**1.65**	1.50	**1.38**	1.38	**1.30**
10 mm	1.65	**1.65**	1.40	**1.27**	1.25	**1.19**
15 mm	1.65	**1.65**	1.32	**1.22**	1.20	**1.12**

14.3.2.3 Model of dense and rigid vegetation

At the beginning of Paragraph 14.3.2.1, it was stressed that the vegetation density was low with respect to literature values. Therefore, in (Gualtieri and Pulci Doria, 2008), the results of an experimental study carried out on dense, rigid, submerged vegetation made by an artificial grass carpet with a height of 18 mm are reported. To compare any different effects of low and dense rigid vegetation, some hydraulic parameters were kept constant.

In particular, the boundary layer is still a zero piezometric head gradient, obtained by assuming a channel slope of 2.25%. Moreover, the head on the vena contracta is set equal to 10.34 cm so that the velocity in the surface layer is fixed at 1.424 m/s. However, experimental measurements were carried out only in the surface layer (due to the vegetation thickness) and with two different sluice gate openings of, respectively, 8 and 10 centimetres.

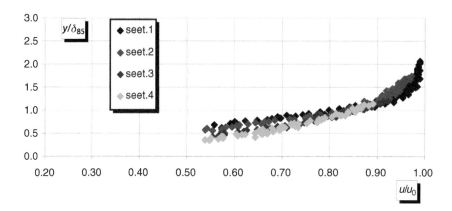

Figure 14.6. Dimensionless velocity distributions, sluice gate opening 8 cm.

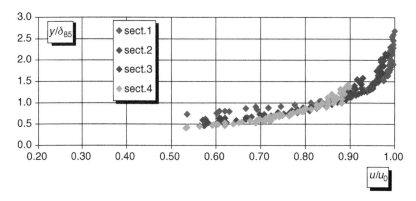

Figure 14.7. Dimensionless velocity distributions, sluice gate opening 10 cm.

The experimental results were elaborated with the methodology previously described. In particular, to stress the equilibrium characteristics of the boundary layers, the local mean velocity distributions, as functions of the distance from the bottom, were made dimensionless using the free-stream velocity and the boundary layer thickness. The results are shown in Figures 14.6 and 14.7. Based on the diagrams, the authors observed that in the case of dense vegetation, no equilibrium characteristic exists.

14.3.2.4 Conclusions about the boundary layer equilibrium problem

Based on the obtained results, the authors conclude that the effects of vegetation on the velocity distributions in boundary layers can be summarised in the following way:

First case: Vegetation very sparse and/or low – The boundary layer with a zero piezometric head gradient retains its total equilibrium characteristics, even in the lower layers of the stream, but with a velocity distribution shape that depends on the vegetation characteristics.

Second case: Vegetation denser and/or higher – The boundary layer with a zero piezometric head gradient partially holds its equilibrium characteristics, in particular, in the layers more distant from the bottom than the vegetation height, but still with a velocity distribution shape that depends on the vegetation characteristics.

Third case: Vegetation very dense and high so that the free-stream becomes very thin or no longer exists – The boundary layer with a zero piezometric head gradient does not hold its equilibrium characteristics.

14.3.2.5 Velocity distribution in the surface layer of a boundary layer

De Felice *et al.* (2009) presented a methodology to process experimental data in the surface layer of a boundary layer. First, the data analysis is restricted only to the local mean velocity distributions in the surface layer for different models of vegetation: cylindrical sticks or grass carpet. Second, velocity distribution diagrams were proposed, with starting points corresponding to the vegetation height (y_e) and to the correspondent velocity (u_e). In other words, the velocity distributions had the distance from the top of vegetation ($y - y_e$) on the ordinate and the difference of velocities ($u - u_e$) on the abscissa. These distributions were called the velocity excess in the surface layer. The obtained distributions were very different from one another, and the velocity excess attained values from 0 at the beginning up to a possible maximum equal to ($u_0 - u_e$). Once these velocity excess values in the surface layer distributions are obtained, they were transformed into dimensionless distributions using the following scaling values:

A)   For velocities, the scaling value is the difference $u_0 - u_e$;
B)   For distances from the top of the vegetation, the scaling value is a particular boundary layer thickness δ_{70}.

In fact, as the ($u_0 - u_e$) value of the velocity excess was not always attained, the authors decided to choose a scaling height δ_{70} for every distribution as the height in which the excess velocity attains 70% of ($u_0 - u_e$). A final comparison among the dimensionless distributions (blue points: central location; pink points: lateral location; yellow points: grass carpet) shows (Fig. 14.8) that they were all superimposed on one another with a good approximation.

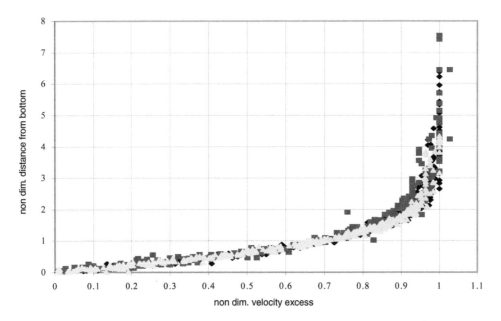

Figure 14.8. Dimensionless velocity excess in surface layer distributions.

This result showed that in a boundary layer flowing on rigid submerged vegetation, through the proposed methodology, a distribution of the local mean velocities in the surface layer, independent of the test section and of the vegetation density and height, and, moreover, of the presence or absence of equilibrium conditions, was obtained.

14.4   USE OF BOUNDARY LAYER RESULTS TO CALCULATE FLOW RESISTANCE

According to Augustijn *et al.* (2008), if the flow is 5 times higher than the vegetation height the flow resistances could be calculated using the classical equations, for instance, the Keulegan (Colebrook) or even the Manning equation.

Following the same idea, in De Felice *et al.* (2008, 2010), a methodology was proposed to evaluate the Nikuradse equivalent roughness of a vegetated bed starting from local mean velocity distributions in a boundary layer flowing on the same vegetated bottom. Therefore, using a boundary layer instead of a uniform flow, it is possible to work with shorter channels. The methodology is based on the determination of local mean velocity distributions and correspondent α Coriolis coefficients in the successive sections of a boundary layer. In particular, the α value is growing along the boundary layer until it attains its maximum value, which is higher than the correspondent one for a uniform flow with the same depth and flow-rate. Therefore, a correction coefficient for the Nikuradse equivalent roughness k_N, experimentally determined in De Felice *et al.* (2010), was applied so to better fit k_N values for the boundary layer to the correspondent ones for a uniform flow. Using this methodology, some values of Nikuradse equivalent roughness was calculated in De Felice *et al.* (2010), and to compare them with literature data (Lopez and Garcia, 1997), they were transformed into n Manning values. The result of the comparison is shown in the following Figure 14.9. In this figure, the black points represent n relative to sticks arrays of the same height but with a different density, and the line is the interpolation. Only the two red points placed along the vertical dotted lines have been calculated through the described methodology. Their correspondence to the interpolating line demonstrates the validity of the proposed methodology.

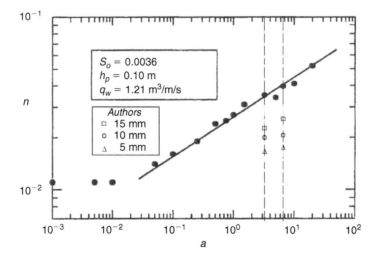

Figure 14.9. Comparison among Manning's coefficients.

14.5 CONCLUSIONS

Vegetation strongly interacts with water flows, in particular, modifying the flow resistance and turbulent characteristics. To better describe and evaluate the effects of vegetation on uniform water flows, theories and models have been developed, starting from the two-layer approach that is still used in flow resistance equations, and incorporating the mixing layer approach, derived from studies about vegetation-atmosphere interactions.

A vegetated bed influences the local mean velocity distribution and the thickness of a boundary layer, modifying its equilibrium condition. In particular, vegetation effect depends on vegetation density. For low density only the shape of the velocity distribution changes but, increasing density, the equilibrium condition of the boundary layer becomes partial and then it completely disappears.

Finally, experimental measurements of the local mean velocity distribution in a boundary layer on a vegetated bed can be used to evaluate the flow resistance in a uniform flow characterized by the same vegetated bed. This possibility is particularly interesting when the vegetation height is much lower than flow depth.

APPENDIX A – LIST OF SYMBOLS

List of Symbols		
Symbol	Definition	Dimension or Units
C_b	Chézy coefficient for the bed	$[L^{1/2}T^{-1}]$
C_D	drag coefficient of a cylinder	
C_t	Chézy coefficient for the vegetation	$[L^{1/2}T^{-1}]$
D	stem diameter of the cylindrical vegetation	$[L]$
F	first wake function in the PDT law	
F'	second wake function in the PDT law	
R	hydraulic radius	$[L]$
S	slope of the channel	
U	mean velocity in the full cross section	$[LT^{-1}]$
U_S	mean velocity in the surface layer	$[LT^{-1}]$
U_V	mean velocity in the vegetation layer	$[LT^{-1}]$
W	wake parameter in Coles law	
b	drag length	$[L]$
f	friction factor of the Darcy-Weisbach equation	
g	gravitational acceleration	$[LT^{-2}]$
h	flow depth	$[L]$
k	vegetation height	$[L]$
k_N	Nikuradse roughness	$[L]$
m	number of cylinders per m² horizontal area	$[L^{-2}]$
n	Manning resistance coefficient	$[L^{-1/6}]$
s	vegetation spacing not including sticks' diameters	$[L]$
p	maximum height validity of Kutija and Hong model	$[L]$
u	water local mean velocity	$[LT^{-1}]$
u_0	external velocity in a boundary layer flow	$[LT^{-1}]$

(Continued)

List of Symbols

Symbol	Definition	Dimension or Units
u_e	current velocity at the top of the vegetation	$[LT^{-1}]$
u_0'	root-mean-square of the velocity fluctuations at $y = \delta$	$[LT^{-1}]$
u^*	shear velocity	$[LT^{-1}]$
y	normal distance from the bed	$[L]$
y_e	normal distance from the bed to the vegetation top = vegetation height	$[L]$
Π	Coles wake law parameter	
δ	boundary layer thickness	$[L]$
ε	dissipation rate of turbulent kinetic energy per unit mass	$[L^2T^{-3}]$
κ	von Kármán Constant	
ν	kinematic viscosity	$[L^2T^{-1}]$
ρ	water density	$[ML^{-3}]$
τ	shear stress at the wall	$[ML^{-1}T^{-2}]$
ω	vorticity	$[T^{-1}]$

APPENDIX B – SYNOPSIS

This chapter describes some characteristics of the interaction between the flow and a vegetated bed. Vegetation is modelled as rigid equi-spaced elements that are completely submerged.

First, the Chapter (Section 14.3) points out two features of a uniform flow that are strongly influenced and modified by a vegetated bed, i.e. the flow resistance and the turbulence characteristics, such as the local mean velocity distribution. The evolution of theories and methodologies about flow resistance and turbulence characteristics is described. Second, Section 14.3 addresses the effects of a vegetated bed on the local mean velocity distribution and thicknesses values of a boundary layer. In particular, a description of experimental measurements on boundary layers developing on three different types of beds (smooth, with rigid cylinders, with rigid grass carpet) are described, and the results are compared. Finally, Section 14.5 presents a methodology for evaluating flow resistance values in a vegetated uniform flow starting from experimental measurements carried out on a turbulent boundary layer flowing on the same vegetated bed. A comparison between numerical values and literature experimental data showed good agreement.

APPENDIX C – KEYWORDS

By the end of the chapter, you should have encountered the following terms. Ensure that you are familiar with them!

Vegetated flow model	Flow resistance	Boundary layer
Two-layer	Velocity distribution	Boundary layer thickness
Four-layer	Average velocity in bulk flow	Dimensionless distribution
Mixing-layer	Classical flow resistance equations	Equilibrium boundary layer

APPENDIX D – QUESTIONS

Which are the main equations for the description of the flow resistance in a vegetated flow?
Which flow resistance equations are based on the two-layer approach?
Describe the evolution of the use of turbulence models in vegetated flow modelling.
What is a mixing layer, and what are its main characteristics?
What is a boundary layer?
Which characteristic has an equilibrium boundary layer?
Describe the methodology used to obtain flow resistance values in a vegetated uniform flow
  starting from turbulent boundary layer measurements.

APPENDIX E – PROBLEMS

E1. Describe a boundary layer with a turbulent free stream. How can the defect law be expressed?

E2. In a boundary layer, starting from an experimental local mean velocity distribution, describe how:
• To evaluate the boundary layer thickness
• To make the distribution dimensionless

E3. Fix some hypothetical values to the parameters in the equations for U velocity in uniform flows and compare the resulting U values.

REFERENCES

Augustijn D.C.M., Huthoff F. and van Velzen E.H., 2008, Comparison of vegetation roughness descriptions, *Proceedings of River Flow 2008*, Izmir, Turkey

Balachandar, R. and Ramachandran S., 1999, Turbulent Boundary Layers in Low Reynolds Number Shallow Open Channel Flows, *Journal of Fluids Engineering, Transactions of ASME*, **123** (2), pp. 394–400

Balachandar, R., Blakely, D., Tachie, M. and Putz, G., 2001, A Study on Turbulent Boundary Layers on a smooth Flat Plate in an Open Channel, *Journal of Fluids Engineering, Transactions of ASME*, **121**, pp. 684–689

Bandypadhyay, P.R., 1992, Reynolds Number Dependence of the Freestream Turbulence Effects on Turbulent Boundary layers, *AIAA Journal*, **30** (7), pp. 1910–1912

Baptist M.J., Babovic V., Rodriguez Uthurburu J., Keuzer M., Uittenbogaard R.E., Mynett A. and Verwey A., 2007, On inducing equations for vegetation resistance *Journal of Hydraulic Research* Vol. 45 (4), pp. 435–450

Blair, M.F., 1983a, Influence of Free-Stream Turbulence on Turbulent Boundary Layer Heat Transfer and Mean Profile Development. Part I Experimental Data, *Transactions of ASME Journal of Heat and Mass Transfer*, **105**, pp. 33–40

Blair, M.F., 1983b, Influence of Free-Stream Turbulence on Turbulent Boundary Layer Heat Transfer and Mean Profile Development. Part II Analysis of results, *Transactions of ASME Journal of Heat and Mass Transfer*, **105**, pp. 41–47

Castro, I.P., 1984, Effects of Free Stream Turbulence on Low Reynolds Number Boundary Layers, *Transactions of ASME Journal of Fluid Engineering*, **106**, pp. 298–306

Charnay, G., Mathieu, J. and Comte-Bellot, G. 1976 Response of a turbulent boundary layer to random fluctuations in an external stream, *The Physics of Fluids*, **19** (9) pp. 1261–1272

Charney, G., Comte-Bellot, G. and Mathieu J., 1971, Development of a turbulent boundary layer on a flat plate in an external turbulent flow, *AGRD CCP*, 93–71, pp. 27.1–27.10

Choi S. and Kang H., 2004, Reynolds stress modelling of vegetated open-channels flows, *Journal of Hydraulic Research*, **42** (1), pp. 3–11

Clauser, F. H., 1956, The turbulent boundary layer, *Advances in Applied Mechanics*, **4**, pp. 1–51

Coles, D., 1956, The law of the wake in the turbulent boundary layer, *Journal of Fluids Mechanics*, **1**, pp. 191–226

Cui J. and Neary V.S., 2002, Large eddy simulation (LES) of fully developed flow through vegetation, *Proceedings of the 5th Int. Conf. on Hydroinformatics*, Cardiff, UK

Cui J. and Neary V.S., 2008, LES study of turbulent flows with submerged vegetation, *Journal of Hydraulic Research*, **46** (3), pp. 307–316

De Bos W.P. and Bijkerk C., 1963, Een nieuwmonogram voor het berekenen van waterlopen *Cultuurtechnish tijdschrift*, 3, 149–155

De Felice, S., Gualtieri, P. and Pulci Doria, G., 2008, A simplified experimental method to evaluate equivalent roughness of vegetated river beds, *Proceedings of the 4th Biennial Meeting of iEMSs*, Barcelona, Spain, pp. 170–180

De Felice, S., Gualtieri, P. and Pulci Doria, G., 2009, A Universal Distribution of Local Mean Velocities in Boundary Layers Flowing Over Vegetated Bottoms, *33th IAHR Congress, Vancouver, British Columbia, Canada*, pp. 2173–2181

De Felice, S., Gualtieri, P. and Pulci Doria, G., 2010, Experimental Calibration of a Simplified Method to Evaluate Absolute Roughness of Vegetated Channels, in D.T. Mihailovic and C. Gualtieri (Eds), *Advances in Environmental Fluid Mechanics*, pp. 360, World Scientific, Singapore

Evans, R.L. and Horlock, J.K. 1974 Calculation of the Development of Turbulent Boundary Layers With a Turbulent Freestream, *Transactions of ASME Journal of Fluid Engineering*, 1974, pp. 348–352

Evans, R.L., 1985, Freestream Turbulence Effects on Turbulent Boundary Layers in an Adverse Pressure Gradient, *AIAA Journal*, **23** (11), pp. 1814–1816

Finnigan J., 2000, Turbulence in plant canopies, *Annual Review of Fluids Mechanics*, **32** (1), pp. 519–571

Gad-el-Hak, M. and Bandyopadhyay, P.R., 1994, Reynolds number effects in wall-bounded flows, *Applied Mechanics Review* **47** (8), pp. 307–365

Galema A., 2009, *Vegetation Resistance. Evaluation of vegetation resistance descriptors for flood management*, PhD Thesis, University of Twente

Gao G., Falconer R.A. and Lin B., 2011, Modelling open channel flow with vegetation using a three dimensional model, *Journal of Water Resources and Protection*, 3, 114–119

George, W.K. and Casillo, L., 1997, Zero pressure-gradient turbulent boundary layer, *Applied Mechanics Review* **50** (11), pp. 689–729

Gioia G., and F.A. Bombardelli, 2002, Scaling and similarity in rough channel flows, *Physical Review Letters*, **88** (1), pp. 14501–14504

Ghisalberti M. and Nepf H.M., 2004, The limited growth of vegetated shear layers, *Water Resources Research*, vol. 40 W07502

Ghisalberti M. and Nepf H.M., 2006, The structure of the shear layer in flows over rigid and flexible canopies, *Environmental Fluid Mechanics*, **6**, pp. 277–301

Granville, P.S., 1976, A modified law of the wake for turbulent shear layers, *ASME Journal of Fluid Engineering*, **98**, pp. 578–580

Gualtieri, P. and Pulci Doria, G. 1998, A proposal of a physically based thickness definition and of a new mean velocities distribution law in a turbulent boundary-layer on the ground of LDA measurements, *13th Australasian Fluid Mechanics Conf.*, Melbourne, Australia, pp. 845/848

Gualtieri, P., Pulci Doria, G. and Taglialatela, L., 2004, Experimental validation of turbulent boundary layers in channels *3rd HEFAT*, Cape Town, South Africa

Gualtieri, P. and Pulci Doria, G., 2008, Boundary layer development over rigid submerged vegetation, in C. Gualtieri and D.T. Mihailovic (Eds) *Fluid Mechanics of Environmental Interfaces*, Taylor & Francis, Leiden, The Netherlands, pp. 241/298

Hancock, P.E. and Bradshaw, P., 1983, Influence of Free-Stream Turbulence on Turbulent Boundary Layers, *Transactions of ASME Journal of Fluids Engineering*, **105**, pp. 284–289

Hancock, P.E. and Bradshaw, P., 1989, Turbulent structure of a boundary layer beneath a turbulent free stream, *Journal of Fluid Mechanics*, **205**, pp. 45–76

Hoffmann, J.A. and Mohammady, K., 1991, Velocity Profiles for Turbulent Boundary Layers Under Freestream Turbulence, *Transactions of ASME Journal of Fluid Engineering*, **113**, pp. 399–404

Huai W.X., Zang L.X. and Zeng Y.H., 2009a, Mathematical model for the flow with submerged and emerged vegetation, *Journal of Hydroinformatics*, **21** (5), pp. 722–729

Huai W.X., Zeng Y.H., Xu Z.g. and Yang Z.H., 2009b, Three layer model for vertical velocity distribution in open channel flow with submerged rigid vegetation, *Advances in Water Resources*, **32**, pp. 487–492

Huffman, G. D., Zimmerman, D. R. and Bennett, W. A., 1972, The effect of Free-stream turbulence leel on turbulent boundary layer behaviour, *AGARDograph 164 Paper I-5*, pp. 91–115

Huthoff F., Augustijn D.C.M. and Hulscher S.J.M.H., 2006, Depth-averaged flow in presence of submerged cylindrical elements, *Proceedings of River Flow 2006*, Lisbon, Portugal

Huthoff, F., D. C. M. Augustijn, and S. J. M. H. Hulscher, 2007, Analytical solution of the depth-averaged flow velocity in case of submerged rigid cylindrical vegetation, *Water Resources Research*, **43**, W06413, doi:10.1029/2006WR005625

Jackson P.S., 1981, On the Displacement Height in the Logarithmic Velocity Profile *Journal of Fluid Mechanics*, **111**, pp. 15–25

Keijzer M., Baptist M., Babovic V. and Uthurburu J.R., 2005, Determining equation for vegetation induced resistance using genetic programming, *Proceedings of GECCO'05*, Washington, DC, USA

Klewicki J.C., 2010, Reynolds number dependence, scaling and dynamics of turbulent boundary layers, *Journal of Fluid Engineering*, **132**, September 2010, pp. 1–47

Kline, S. J., Lisin, A.V. and Waitman, B.A., 1960 Preliminary experimental investigation of effect of free-stream turbulence on turbulent boundary-layer growth, *N.A.C.A. TN D-368*, pp. 1–60

Klopstra D., Barneveld H.J., Van Noortwijk J.M. and Van Velzen E.H., 1997, Analytical model for hydraulic roughness of submerged vegetation, *Proceedings of the 27th IAHR Congress*, San Francisco, USA, pp. 775–780

Kutija V. and Hong H.T.M., 1996, A numerical model for assessing the additional resistance to flow introduced by flexible vegetation, *Journal of Hydraulic Research*, **34** (5), pp. 99–114

Lopez F. and Garcia M., 1997, Open-Channel Flow Through Simulated Vegetation: Turbulence Modeling and Sediment Transport, *US Army of Engineers Waterway Experiment Station Wetlands Research Program Technical Report WRP-CP-10*

Lopez F. and Garcia M., 1998, Open channel flow through simulated vegetation: Suspended sediment transport modeling, *Water Resources Research*, **34** (9), pp. 2341–2352

Lopez F. and Garcia M., 2001a, Mean Flow and Turbulence Structure of Open-Channel Flow Through Non-Emergent Vegetation, *Journal of Hydraulic Engineering*, **127** (5), pp. 392–402

Luhar M., Rominger J. and Nepf H.M., 2008, Interaction between flow, transport and vegetation spatial structure, *Environmental Fluid Mechanics*, **8**, pp. 423–439

Mc Donald, H. and Kreskowsky, J.P., 1974, Effects of free stream turbulence on the turbulent boundary layer, *International Journal of Heat and Mass Transfer*, **17**, pp. 705–716

Meier H.U. and Kreplin, H.P., 1980, Influence of Freestream Turbulence on Boundary-Layer Development, *AIAA Journal*, **18** (1), pp. 11–15

Meijer, D.G., 1998, Modelproeven overstroomde vegetatie. Technical report PR121, HKV Consultants, Lelystad, The Netherlands

Mejier D.G. and Van Velzen E.H., 1999, Prototype-scale flume experiments on hydraulic roughness of submerged vegetation, *Proceedings of the 28th IAHR Congress*, Graz, Austria

Neary V.S., 1995, Numerical modeling of diversion flows, PhD Dissertation, Civil and Environmental Engineering, Univ. of Iowa, Iowa City, Iowa

Neary V.S., 2003, Numerical solution of fully developed flow with vegetative resistance, *Journal of Engineering Mechanics*, **129** (5), pp. 558–563

Nepf H.M. and Ghisalberti M., 2008, Flow and transport in channels with submerged vegetation, *Acta Geophisica*, **56** (3), pp. 753–777

Nezu I. and Nakagawa H., 1993, Turbulence in open-channel flows, Balkema, Rotterdam, The Netherlands

Nezu I., Sanijou M. and Okamoto T., 2006, Turbulent structure and dispersive properties in vegetated canopy open channel flows, *Proceedings of River Flow 2006*, Lisbon, Portugal

Nezu I. and Sanijou M., 2008, Turbulent structure and coherent motion in vegetated canopy open channel flows, *Journal of Hydro Environment Research*, **2**, pp. 62–90

Patel V.C. and Yoom J.Y., 1995, Application of turbulence models to separated flows over rough surfaces, *Journal of Fluids Engineering*, **117** (2), pp. 234–241

Poggi D., Porporato A., Ridolfi L., Albertson J.D. and Katul G.G. 2004, The effect of vegetation on canopy sub-layer turbulence, *Boundary-Layer Meteorology*, **111**, pp. 565–587

Prandtl, L., 1904, Uber Flüssigkeitsbewegungen bei sehr kleiner Reibung (On Fluid Motions with Very Small Friction), *Verhandlungen des III Internaztionalen Mathematiker Kongresses (Heidelberg 1904)*, Leipzig 1905

Pulci Doria, G. and Tagliatatela, L., 1990, Ipotesi di distribuzioni adimensionali di velocità media e di agitazione in correnti turbolente (An hypothesis about non dimensional distributions of local mean velocity and turbulent velocity fluctuations in turbulent streams) *Giornate di Studio per la celebrazione della nascita di Girolamo Ippolito*, Lacco Ameno, Italy, pp. 223–245

Raupach M.R. and Thom A.S., 1981, Turbulence in and above plant canopies, *Annual Review of Fluids Mechanics*, **13**, pp. 97–129

Raupach M.R. and Shaw R.H., 1982, Averaging procedures for flow within vegetation canopies, *Boundary Layer Meteorology*, **22**, pp. 79–90

Raupach M.R., Finnigan J.J. and Brunet Y., 1996, Coherent eddies and turbulence in vegetation: the mixing layer analogy, *Boundary Layer Metheorology*, **78**, 351–382

Robertson, J.M. and Holt, C., 1972, Stream turbulence effects on turbulent boundary layer, *Journal of Hydraulic Division Proceedings of the American Society of Civil Engineers*, **98** (HY6), pp. 1095–1099

Rodi W., 1980, Turbulence models and their application: state of art paper, Monograph, IAHR, Delft, The Netherlands

Schlichting, H. and Gersten, K., 2003, *Boundary Layer Theory (transl. by Mayes K.)*, 8th revised and Enlarged Edition 2000, Corrected Printing 2003, Springer-Verlag, Berlin-Heidelberg

Schlichting, H., 1955, *Boundary Layer Theory (transl. by Kestin J.)*, Pergamon Press LTD, London

Shimizu Y., Tsujimoto T., Nakagawa H. and Kitamura T., 1991, Experimental study of flow over rigid vegetation simulated by cylinders with equi-spacing, *Proceedings of JSCE*, **438**, pp. 31–40

Shimizu Y. and Tsujimoto T., 1994, Numerical analysis of turbulent open-channel flow over a vegetation layer using a k-ε model, *Journal of Hydroscience and Hydraulic Engineering*, **11** (2), pp. 57–67

Sreenivasan, K.R., 1989, Turbulent boundary layer, in M. Gad-el-Hak (Editor), *Frontiers in Experimental Fluid Mechanics*, pp. 159–209

Stone B.M. and Tao Shen H., 2002, Hydraulic Resistance of Flow in Channels with Cylindrical Roughness, *Journal of Hydraulic Engineering*, **128** (5), pp. 500–506

Tachie, M.F., Balachandar, R. and Bergstrom, D.J., 2003, Low Reynolds number effects in open-channel turbulent boundary layers, *Experiments in Fluids*, **35**, pp. 338–346

Tachie, M.F., Balachandar, R., Bergstrom, D.J. and Ramachandran S., 2001, Skin Friction Correlation in Open Channel Boundar Layers, *Journal of Fluids Engineering, Transactions of ASME*, **123**, pp. 953–956

Tachie, M.F., Bergstrom, D.J. and Balachandar, R., 2000, Rough Wall Turbulent Boundary Layers in Shallow Open Channel Flow, *Journal of Fluids Engineering, Transactions of ASME*, **122**, pp. 533–541

Tsujimoto T. and Kitamura T., 1990, Velocity profile of flow in vegetated-bed channels, *KHL Progressive Report, Hydraulic Laboratory, Kanazawa University*

Tsuijimoto, T., Shimizu, Y., Kitamura, T. and Okada, T., 1992, Turbulent open-channel flow over bed covered by rigid vegetation, *Journal of Hydroscience and Hydraulic Engineering*, **10** (2), pp. 13–25

Van Velzen E.H., Jesse P., Cornelissen P. and Coops H, (2003) *Stromingsweerstand vegetatie in uiterwaarden; Handbook*, Part 1 and 2 RIZA Reports 2003.028 and 2003.029, Arnhern, The Netherlands

Wilson N.R. and Shaw R.H., 1977, A higher closure model for canopy, *Journal of Applied Meteorology*, **16**, pp. 1197–1205

Yen B.C., 2002, Open channel flow resistance, *Journal of Hydraulic Engineering*, **128** (1), pp. 20–39

Yue W., Meneveau C., Parlange M.B., Zhu W., Kang H.S. and Katz J., 2008, Turbulent kinetic budget in a model of canopy: comparisons between LES and wind-tunnel experiments, *Envirommental Fluids Mechanics*, **8** (1), pp. 73–95

Zhu W., van Hout R., Luznik L., Kang H.S., Katz J. and Meneveau C., 2006, A comparison of PIV measurements of canopy turbulence performed in the field and in a wind tunnel model, *Experiments in Fluids*, **41**, pp. 308–318

CHAPTER FIFTEEN

Mass transport in aquatic environments

Gregory Nishihara

Department of Integrative Biology, University of Guelph, Guelph, Ontario, Canada

Faculty of Fisheries, Kagoshima University, Kagoshima, Japan

Josef Daniel Ackerman

Faculty of Environmental Sciences, University of Guelph, Guelph, Ontario, Canada

ABSTRACT

All forms of aquatic life rely on the surrounding fluid for the transport of resources and the products of metabolic activity. The processes that affect the transport of material to and from the surface of an organism include molecular and turbulent diffusion. However, since the viscosity of water is about 55 times that of air, the scales at which these processes occur are different and may represent considerable constraints to aquatic organisms. Transport processes in aquatic environments are considered for both pelagic (i.e., those in the water column) and benthic organisms (i.e., those at the bottom). The relevant issues related to mass transfer to and from benthic plants and animals is considered in detail.

15.1 INTRODUCTION

The concept of mass transfer is essential in aquatic environments where the fluid medium – water – serves to facilitate myriad biological processes. These include the delivery of gases used in the most basic and fundamental biochemical reactions related to the fixing of dissolved inorganic carbon (DIC; largely CO_2) via photosynthesis or chemosynthesis and the oxidation of oxygen (O_2) in respiration. They also include more complex ecological processes related to suspension feeding – the selective removal of nutritious particles from a virtual soup of material – and sexual reproduction – where sperm and eggs broadcast into a seemingly infinite spatial domain must contact one another to continue the cycle of life. All of these examples involve the physical transport of dissolved and/or particulate matter to and from aquatic organisms. The transport of these scalar quantities, whether it is generated by the organism or through environmental flows, is dictated by the principles of fluid dynamics. There is nothing magical involved, but scientific insights continue to provide intriguing examples of how aquatic organisms have evolved to use fluids.

In the parlance of fluid dynamics, it is the flux of scalar quantities (i.e., $J = UC$, where U is the velocity, and C is the concentration of the scalar) that links physics and biology in aquatic environments. It is relevant to note that it is the product of the vector and the scalar that is important rather than either the velocity or concentration alone. Too often the emphasis in many biological studies, including those of the authors, has focused on comparisons across a range of velocities; the classical experimental approach of holding one variable constant. However, a simple thought experiment will reveal that it is possible to generate the same value for J using different combinations of values for U and C. This observation should not lead the reader to think that it is physics alone that is important, rather it is evident that biology is complex and cannot be understood solely through the examination of physical principles (c.f., Pennycuick, 1992).

The focus of this chapter is mass transport in aquatic environments. From the onset this may appear to be a relatively simple task until the spatial and temporal scale and diversity of processes involved are considered. For example, aquatic organisms span eight orders of magnitude in terms of length ($10^{-7}-10^1$ m) and 21 orders of magnitude ($10^{-16}-10^5$ kg) in terms of mass (McMahon and Bonner, 1983; Niklas, 1994). It would be inconceivable to think, for example, that the same physical processes apply to a bacterium moving at μ m s^{-1} and a whale swimming at many m s^{-1}, however the truth or falsehood of this statement is scale dependent. Moreover, there are at least two scales that need to be considered, namely spatial and temporal (via velocity) scales. The familiar non-dimensional Reynolds number ($Re = 1\ U/v$; where l is the length, and v the molecular diffusivity of momentum), which relates inertial to viscous forces, provides the context by which to make this comparison (e.g., Vogel, 1994; White, 1999). A comparison of processes that occur at Re based on the scale of the whale ($Re \gg 10^3$) are turbulent in nature whereas those that occur at Re based on the scale of the bacterium ($Re \ll 1$) are creeping. Similarly, transport in the former is advective, whereas it is diffusive for the latter. There are exchange processes at the whale's surface however, that occur at the "bacterium scale", but the bacterium will never experience processes at the "whale scale" given that its environment is circumscribed within the smallest oceanic eddies (i.e., the Kolmogorov microscale; $\eta \sim$ several mm in the upper mixed layer; see Mann and Lazier, 2006). In other words, at large spatial and velocity scales conditions are turbulent and transport is advective, and simultaneously they are laminar and diffusive when examined at very small spatial and velocity scales within the external flow environment. Dealing with this continuum of scales will appear throughout this chapter, but the majority of the examples chosen are biased toward the 10 s of cm and cm s^{-1} scales and smaller. In addition, this chapter is not meant to be an exhaustive review, rather it was written to provide some historical context along with the state of the art developments, which are leading to some new understanding of aquatic environments (see for example, Niklas, 1992; Okubo and Levin, 2002; Mann and Lazier, 2006).

15.2 AQUATIC ENVIRONMENTS

When dealing with any problem or idea, it is first necessary to define the appropriate boundary conditions. This chapter will be restricted to mass transport processes in organisms inhabiting surface waters (e.g., Fischer *et al.*, 1979), and will not deal with groundwater and other interstitial environments (e.g., transport in porous material; de Beer *et al.*, 1994; de Beer and Kühl, 2001). There are a large number of ways in which to classify the structure of surface waters due largely to the availability of light and tidal forcing, which lead to different categorizations between marine and freshwater ecosystems (Fig. 15.1). For example the regions closest to shore are referred to as intertidal zones in marine ecosystems and littoral zones in both marine and freshwater systems. This is quite reasonable in the former given that the tidal influence can be so pronounced in terms of the periodic variation of conditions imparted on the resident organisms (Ingmanson and Wallace, 1995). An analogous approach is used to discriminate among freshwater ecosystems as to whether they are lotic (flowing waters, such as streams and rivers) or lentic (standing waters, such as ponds and lakes) (Kalff, 2001; Wetzel, 2001). From the perspective of this chapter, the real issue is whether the organism is pelagic (living freely within the water column) or benthic (attached to the bottom or a surface). Several important distinctions emerge from this perspective. Pelagic organisms

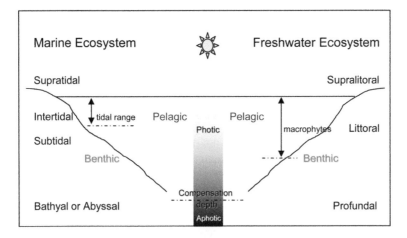

Figure 15.1. The principle classification of aquatic environments and some of the classification used in marine and freshwater ecosystems. Both ecosystems include an open water or pelagic zone (limnetic zone in freshwater) and a bottom oriented or benthic zone. The penetration of light is relevant in both cases, where the compensation depth marks the limit of the photic zone where phytoplankton can survive. In freshwater, the depth limit of rooted macrophytes delimits the bottom of the littoral zone, whereas tidal exchanges provide the various demarcations in marine systems. Note that the depth scales is much compressed in marine ecosystem.

exist in a Lagrangian reference frame as they are transported along with the water and where they may experience relative motions. This concept, which was introduced above, holds that organisms smaller than η (i.e., several mm in the upper mixed layer; e.g., many species of phyto- and zooplankton) live within the shelter of eddies. Conversely, benthic organisms, which are fixed to the bottom, inhabit an Eulerian reference frame and may experience both relative and absolute motions depending on their length. For most benthic organisms this has led to diminutive size in highly energetic environments where the whiplash-like forces of breaking waves can cause tissue damage or detachment (Denny, 1988). There are exceptions however, where long lengths can provide an escape from the breaking waves through the response known as "going with the flow" (see review in Okubo *et al.*, 2002). Recognizing these patterns and distinctions, it is possible to apply the principles of mass transport equally well to organisms living in the benthos of an estuary, a river, a coast, or a lake.

15.2.1 Aquatic Ecosystems

It would be relevant to develop further the ecosystem concept, which can be defined as the sum total of the biotic (living) and abiotic (non-living) elements and processes that occur within a particular designation (e.g., Ricklefs and Miller, 2000). In this way we can define a marine ecosystem, an estuarine ecosystem, an eelgrass ecosystem, an epiphytic ecosystem, and so forth to as many scales as one could envision (note that this example was chosen to demonstrate the hierarchy of scales; e.g., Nybakken and Bertness, 2005). The term has merit as a concept in that it is all encompassing within a system-based perspective, which provides for an understanding of the mechanistic basis of the system and allows for the comparison among ecosystems. It is also limiting because of its lack of precision of how to designate the unit, and thus avoid confusion. This is somewhat analogous to the other hierarchical scaling phenomena discussed above.

The systems analogy allows for the definition of the constituent biological components as: (1) producers (autotrophs), which are the organisms that fix chemical energy from sunlight or other sources of electron transfer (e.g., chemosynthesis), such as planktonic algae [photosynthetic protists], macroalgae, seagrasses, aquatic, marsh plants, etc.; (2) consumers (heterotrophs), which are the organisms that eat the producers such as herbivorous zooplankton, bivalves, snails, fish, turtles, sea urchins, etc.; (3) predators, which are the organisms that eat the consumers, such as large zooplankton, fish, snails, birds, etc.; and (4) detritivores, which are the organisms that utilize waste products such as fungi, bacteria, protists, and annelids. There are of course many exceptions whereby organisms may be omnivorous and feed on more than one trophic level (Ricklefs and Miller, 2000). It is also possible to define the nature of the aquatic ecosystem through its trophic status, which provides a short hand indication of the nutrient status and productivity – either as gross primary productivity (GPP) or net primary productivity (NPP) after the subtraction of the production used for respiration (Ricklefs and Miller, 2000). Productivity in this case refers to the fixation of carbon (usually $gC\ m^{-2}yr^{-1}$, i.e., flux) via the photosynthesis of cyanobacteria (blue-green algae), photosynthetic protists (algae) both planktonic (phytoplankton) and macrophytic (macroalgae or seaweeds), and nonvascular (i.e., mosses) and vascular plants (ferns, seagrasses, aquatic angiosperms or aquatic weeds).

Given that productivity is driven by nutrient levels it is not surprising that nutrient status is used to describe the trophic state of the aquatic environment (Kalff, 2001; Wetzel, 2001; Nybakken and Bertness, 2005). These states can represent the natural progression of temporal changes that occur as a newly formed water basin ages through time (note that this can also apply to a coastal embayment). In this scenario, the water body begins with relatively clear water with few nutrients, low productivity and biodiversity (i.e., oligotrophic) and through time as nutrients and sediments accumulate in the basin it progresses through mesotrophic, eutrophic and finally dystrophic where the accumulated organic matter can render the water acidic in freshwater systems (Table 15.1). Not surprisingly, excess nutrient input (principally phosphorus in freshwater and nitrogen in marine) through human activity has resulted in cultural eutrophication. Much of the past 40 years of research and engineering have been devoted to the elimination of these nutrient inputs and many researchers are describing the oligotrophication of previously culturally eutrophic environments.

Table 15.1. Trophic status of aquatic environments, which follow a continuum of sorts from low nutrient levels, productivity and biodiversity to higher levels, which can be disrupted in the extreme case of dystrophic conditions.

Condition	Trophic status			
	Oligotrophic	Mesotrophic	Eutrophic	Dystrophic
Nutrients	low	moderate	high	excess
Productivity	low	moderate	high	low
Biodiversity	low	moderate	high	low

15.3 THEORY DEVELOPMENT

As indicated above, water flowing around organisms (e.g., macrophytes, sediments, corals, and mussel beds) provides a mechanism that supplies and removes scalar quantities (e.g., dissolved gases, nutrients, seston, and gametes). Therefore, the mass transport of these scalars is an essential process for aquatic organisms (Jørgensen and Des Marais, 1990; Falter *et al.*, 2004; Larned *et al.*, 2004; Nishihara and Ackerman, 2006, 2007a). Mass transport is also a complex process involving diffusion (i.e., molecular and turbulent diffusion), advection, and boundary layer reactions, which are influenced by the properties and flow characteristics of the water, the biological and physical characteristics of the organism, the concentration and diffusional characteristics of the scalar quantity, and the kinetics and mechanism of the boundary layer reactions (Chambré and Acrivos, 1956; Acrivos and Chambré, 1957; Chambré and Young, 1958; Libby and Liu, 1966; Dang, 1983; Nishihara and Ackerman, 2006, 2007a).

15.3.1 Momentum and concentration boundary layers

Water flowing over a surface forms a momentum boundary layer (MBL) that can have laminar, transitional, or turbulent characteristics, depending on spatial and velocity scales that can be determined through the local Reynolds number ($Re = x\,U/v$, where x is the downstream distance). The MBL forms as a result of the tendency of water to adhere to surfaces (i.e., the no-slip condition; $U = 0$), which produce tangential forces (i.e., shear stresses; τ) that are greatest at the water-surface interface (i.e., wall shear stress; Ackerman and Hoover, 2001). The laminar MBL in two-dimensions can be described by solving the continuity equation and the equation of motion and their exact and approximate solutions are well known (Schlichting and Gersten, 2000). More importantly, the solutions provide a measure of the MBL thickness (δ_{MBL}), and in the case where the organisms can be approximated as a flat plate (Fig. 15.2), the δ_{MBL} at a given distance downstream (x) from the leading edge is

$$\delta_{MBL} \approx \frac{5x}{Re_x^{1/2}} \tag{15.1}$$

The turbulent MBL has vertical structure with three regions extending from the surface: (i) the viscous sublayer (VSL) where forces are largely viscous in nature; (ii) the logarithmic layer where inertial forces begin to dominate; and (iii) the outer layer where conditions approach those of the free stream (Nishihara and Ackerman, 2006, 2007b; see Table 15.2). Within the VSL, there is a very thin diffusional sublayer (DSL) where processes are largely diffusive in nature (see below).

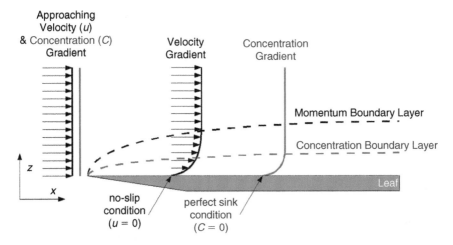

Figure 15.2. The momentum boundary layer and concentration boundary layer over a model leaf. The velocity gradient is a result of the no-slip condition at the water-surface interface and the concentration gradient occurs, given that the leaf surface acts as a sink and consumes all of the scalar arriving to the surface ($C = 0$).

By analogy, when scalars are consumed or produced at the surface of organisms, concentration boundary layers (CBL) will form. The CBL can be described in two-dimensions as

$$u\frac{\partial C}{\partial x} + w\frac{\partial C}{\partial z} = \frac{\partial}{\partial z}\left[(D + K_D)\frac{\partial C}{\partial z}\right] + R \tag{15.2}$$

where u and w are the velocities in the x and z (vertical) directions, D is the molecular diffusivity of the scalar, K_D is the turbulent analogue to D, and R is a homogeneous reaction that occurs within the CBL (Fig. 15.3; Nishihara and Ackerman, 2006, 2007a). Consider the simplest cases where the boundary conditions at the water-surface interface is constant (i.e., where the surface concentration or flux is invariant), there are no homogeneous reactions, and the turbulent diffusivity of the scalar (K_D) is much smaller than D and can be neglected (Hanratty, 1956; Shaw and Hanratty, 1977; Na and Hanratty, 2000). In this case, Eqn. (15.2) can be solved using the similarity principle (Schlichting and Gersten, 2000) and the thickness of the CBL (δ_{CBL}) is given by

$$\delta_{CBL} \approx \frac{5x}{Re_x^{1/2} Sc^{1/3}} \tag{15.3}$$

where Sc is the Schmidt number defined as the ratio of the v to D. Note that δ_{CBL} is thinner than the δ_{MBL} by a factor of $Sc^{1/3}$ (i.e., compare Eqn. (15.1) and (15.3); see Table 15.2).

Table 15.2. A comparison of momentum and concentration boundary layer definitions and theoretical values of their thickness (δ) over flat plates.

Distribution of momentum	Distribution of scalar
Momentum Boundary Layer (MBL) region from the surface to $0.99U$ Laminar[1] $\delta_{MBL} = \dfrac{5x}{Re_x^{1/2}}$ Turbulent $\delta_{MBL} = \dfrac{0.16x}{Re_x^{1/7}}$	Concentration Boundary Layer (CBL) region from the surface to $0.99C_{bulk}$ Laminar $\delta_{CBL} = \dfrac{5x}{Re_x^{1/2} Sc^{1/3}}$ Turbulent $\delta_{CBL} = \dfrac{0.16x}{Re_x^{1/7} Sc^{1/3}}$
Outer Region (Eckman Layer[2]) region where $\partial u/\partial z \to 0$ –	Outer Region region where $\partial C/\partial z \to 0$ $(K_D \approx K_v) > D$ –
Inertial Sublayer region where $\partial u/\partial z$ is exponential; v negligible $\delta_{ISL} \sim 0.15\delta_{MBL}$	Exponential Region region where $\partial C/\partial z$ is exponential; D negligible –
Viscous Sublayer (VSL) region where $K_v = v$ $\delta_{VSL} = 10\frac{v}{u_*}$	– –
Diffusional Sublayer (DSL) region where D dominates $\delta_{DSL} = \delta_{VSL} Sc^{-1/3}$	Diffusional Boundary Layer (DBL) region where $(K_D \approx K_v) < D$ depends on the nature of scalar sink/source; $\delta_{DBL} \leq \delta_{DSL}$

[1]$Re_x = 3 - 5 \times 10^5$ marks the transition to turbulence; [2]where Coriolis effects are relevant.

The CBLs in aquatic systems may have a structure similar to that of a MBL (Fig. 15.3; Levich, 1962; Nishihara and Ackerman, 2006; see Table 15.2). Adjacent to the surface there is a thin region of fluid, the diffusive boundary layer (DBL), where the $K_D < D$ and molecular diffusion is the dominant form of mass transport. The thickness of this region, which is incorrectly equated to the DSL of the MBL, extends to a height where $K_D = D$. It is relevant to note that advective transport parallel to the surface is also present in this region; therefore, diffusion is the primary mode of mass transport only in the fluid nearest to the surface where the no-slip condition is valid. Above the DSL, K_D begins to dominate D (Levich, 1962; Shaw and Hanratty, 1977; Bird *et al.*, 2002) and further from the surface, in the outer region, $K_D \gg D$, and the concentration gradient is small (Bird *et al.*, 2002; Nishihara and Ackerman, 2006).

15.3.2 Surface and CBL reactions

Unfortunately, the boundary conditions at the water-surface interface of biological systems can be complex (Nishihara and Ackerman, 2006, 2007a, b), which invalidates the assumptions of constant concentration or flux used to derive Eqn. (15.3). This is due to the variety of boundary layer reactions (i.e., heterogeneous and homogeneous boundary layer reactions) that can occur as a result of physiological activity, such as photosynthesis, nutrient uptake, and bicarbonate-carbonate chemistry (Table 15.3; Tortell *et al.*, 1997; Wolf-Gladrow and Riebesell, 1997; Martin and Tortell, 2006; Nishihara and Ackerman, 2007a). Heterogeneous reactions, occurring at the water-surface interface, serve to control the flux of material in and out of the CBL (Fig. 15.3). The kinetics can be linear or nonlinear and can vary with the concentration of the reactant and the location of the reaction (Jørgensen and Revsbech, 1985; Ploug *et al.*, 1999; Nielsen *et al.*, 2006; Nishihara and Ackerman, 2007a). Homogeneous reactions, which occur in the CBL, will also influence the concentration gradient in the CBL and violate the assumptions made to derive Eqn. (15.2) (Fig. 15.3). When reactions occur in the CBL, they serve to decrease and increase δ_{CBL} as the reaction consumes and produces the scalar, respectively (Bird *et al.*, 2002). Therefore, Eqn. (15.2) developed through the assumptions of constant concentration or flux does not apply when the heterogeneous reactions vary or when there are homogeneous reactions occurring in the CBL.

The deviations from Eqn. (15.3) have been examined in detail from a chemical engineering perspective (Chambré and Acrivos, 1956; Acrivos and Chambré, 1957; Chambré and Young, 1958; Freeman and Simpkins, 1965; Chung, 1969). Along a flat-plate undergoing a heterogeneous linear reaction where a scalar quantity is consumed, the δ_{CBL} is influenced by

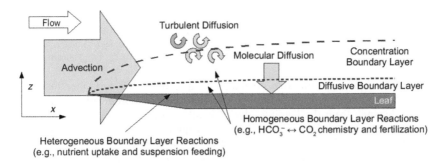

Figure 15.3. A schematic of the process occurring in the concentration boundary layer (CBL) important to mass transport in aquatic systems. Mass transfer is a function of advection, turbulent and molecular diffusion, as well as heterogeneous and homogeneous boundary layer reactions that occur at the surface and in the CBL.

the magnitude of the reaction and its proximity to the leading edge. Near the leading edge, if mass transfer rates are greater than the reaction rate, mass transport is kinetically limited (Chambré and Acrivos, 1956; Acrivos and Chambré, 1957). Therefore, the concentration at the surface ($C_{surface}$) is approximately that of the bulk concentration (C_{bulk}). Much further downstream from the leading edge where the CBL is thick, mass transfer can be slower than the reaction rate (i.e., mass transfer limitation), and the concentration at the surface is significantly lower than the bulk concentration (Chambré and Acrivos, 1956; Acrivos and Chambré, 1957). Therefore, the ratio of the CBL to the MBL may vary (Eqn. (15.3)) with regards to space, the reaction rate, and the reaction mechanism. For example, consider the case where mass transfer does not limit nutrient uptake (J) and J is not a function of the surface concentration. In this case, the surface concentration will decrease on the order of $x^{0.5}$ within a laminar CBL (Fig. 15.4). However, it has been suggested (Chambré and Acrivos, 1956; Acrivos and Chambré, 1957) that if a heterogeneous boundary layer reaction is a first-order (i.e., linear) process, a similar monotonic decrease in surface concentration will be observed, although the initial decrease of the scalar quantity will be of a smaller magnitude than for a constant reaction rate (Fig. 15.4). Moreover, it appears that for more complex reaction mechanisms (i.e., Michaelis-Menten-like reactions; Nishihara and Ackerman in review), there will be little change in surface concentrations near the leading edge, however a large decrease will be observed further downstream (Fig. 15.4) after which, the surface concentrations are predicted to asymptote to some finite value.

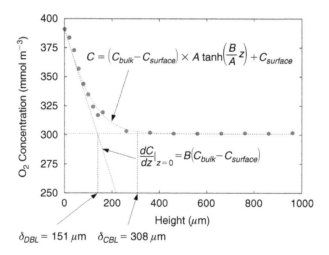

Figure 15.4. The measured and modeled oxygen gradient over a photosynthesizing leaf of the freshwater macrophyte, *Vallisneria americana* (modified after Nishihara and Ackerman, 2007b). The concentration boundary layer thickness (δ_{CBL}) and the diffusive boundary layer thickness (δ_{DBL}) are also given.

For homogeneous reactions that consume scalars, the CBL is thinner than predicted by Eqn. (15.3) (Bird *et al.*, 2002), and it will decreases monotonically in thickness with increasing distance from the leading edge (Chambré and Young, 1958). However, if the homogeneous reaction serves to produce a scalar quantity, the shape of the CBL and the concentration gradient is not as simple. Specifically, the shapes of the gradient and CBL will be similar to that of a non-homogeneous reactions near the leading edge, however further downstream, the scalar begins to accumulate in the CBL drastically altering the shape of the gradient and the characteristics of the CBL (Chambré and Young, 1958).

The combined effects of heterogeneous and homogeneous boundary layer reactions on mass transfer are not well known, however along the axial length of a pipe homogeneous reactions were suggested to have a greater influence on the consumption of scalar quantities (Dang, 1983). Biological systems involve reactions as complex as their chemical engineering analogues. Given that heterogeneous boundary layer reactions common to biology cannot always be described by simple linear or power-law functions (e.g., Michaelis-Menten kinetics), analytical solutions to determine characteristics of the CBL (e.g., Dang, 1983) are difficult to derive. Consequently, the δ_{CBL} and concentration gradients must be determined experimentally, to ensure that the fluxes and mass transfer rates derived from the concentration gradient are not based on violations of the assumptions made in Eqn. (15.3).

15.3.3 Concentration boundary layer measurements

It is relatively simple to determine the concentration gradient over the surface provided the scalar quantity can be measured easily. Unfortunately there are few scalars for which this can be achieved. The widespread use of oxygen microsensors (Glud *et al.*, 1999; Ploug *et al.*, 1999; Hondzo *et al.*, 2005; Nishihara and Ackerman, 2006, 2007a, b) has facilitated the measurement of O_2 CBLs over respiring and photosynthesizing surfaces. By appropriately positioning the microsensor, the concentration gradient over the surface can be recorded and analyzed to provide the δ_{CBL}, the thickness of the diffusive boundary layer (δ_{DBL}, where the DBL is a layer of fluid adjacent to the surface, where diffusion is the primary form of mass transfer), and the oxygen flux at the water-surface interface. It is important to make the distinction between the concept of the DSL and the DBL (Table 15.2). Whereas the DSL is a sublayer of the VSL and the ratio of the DSL to VSL is $\sim Sc^{-1/3}$ (Levich, 1962; Bird *et al.*, 2002), the DBL is a region within the CBL and will not necessarily scale with the VSL thickness uniformly over the surface of a flat plate (i.e., an organism) (Nishihara and Ackerman, 2007b, in review).

Typically, only the flux and the δ_{DBL} could be determined from the oxygen gradient due to the limited spatial resolution. Flux was evaluated by determining the slope of a line fit to the data points closest to the surface, in the region before the points became nonlinearly distributed, and multiplying the slope by the molecular diffusivity (D) for oxygen (Jørgensen and Des Marais, 1990; Ploug *et al.*, 1999; Køhler-Rink and Kühl, 2000). The δ_{DBL} was then estimated by extrapolating the line out so that it would intercept a line drawn through the data point in the bulk water (Fig. 15.4). The location of the intercept of these two lines was used to provide an estimate of the δ_{DBL}. This approach is limited because it ignores the nonlinearity observed in many of the datasets involving oxygen concentration gradients, and it is subject to errors in estimate due to the small number of data points used to determine the slope (Nishihara and Ackerman, 2007b). The solution to this problem is to select an appropriate function that can be used to model the concentration gradient determined with a microsensor.

Whereas there are a number of exponential and transcendental functions available, the hyperbolic tangent function provides an excellent solution to the problem of determining CBL properties from scalar gradients (Nishihara and Ackerman, 2006, 2007b). This function has the property that (i) far from the surface, the curve is asymptotic (i.e., models the bulk concentration) and (ii) adjacent to the surface, the first derivative of the curve is nonzero (i.e., can approximate the slope at the water-surface interface). Moreover, by making the concentration gradient dimensionless (θ) the hyperbolic tangent can be easily fit to the data regardless of whether the surface is a sink or source, by normalizing the concentration gradient [$C(z)$] to values from 0 to 1 (Fig. 15.4)

$$\theta(z) = \frac{C - C_{surface}}{C_{bulk} - C_{surface}} \tag{15.4}$$

A two-parameter hyperbolic tangent function can also be fit using nonlinear regression to the dimensionless concentration gradient (θ)

$$\theta(z) = \mathrm{Atanh}\left(\frac{B}{A}z\right) \tag{15.5}$$

where parameter A and B are constants. At the water-surface interface, the no-slip condition is valid and there is no advective component to mass transfer. Therefore, Fick's law

$$J_{(z=0)} = -D\frac{\partial C}{\partial z} \tag{15.6}$$

can be used to determine the flux through the interface by combining Eqn. (15.4) and (15.5)

$$\frac{C - C_{surface}}{C_{bulk} - C_{surface}} = \mathrm{Atanh}\left(\frac{B}{A}z\right) \tag{15.7}$$

solving for C

$$C = \mathrm{Atanh}\left(\frac{B}{A}z\right)(C_{bulk} - C_{surface}) + C_{surface} \tag{15.8}$$

and evaluating the derivative of C with respect to z

$$\frac{\partial C}{\partial z} = \mathrm{Atanh}\left(\frac{B}{A}z\right)(C_{bulk} - C_{surface}) + C_{surface} \tag{15.9}$$

Evaluating Eqn. (15.9) at $z = 0$ leads to

$$\frac{\partial C}{\partial z} = B(C_{bulk} - C_{surface}) \tag{15.10}$$

By substituting Eqn. (15.10) into Eqn. (15.6), the flux at the water-surface interface is

$$J_{(z=0)} = -DB(C_{bulk} - C_{surface}) \tag{15.11}$$

The δ_{DBL} can also be determined, by evaluating where the line with the slope determined in Eqn. (15.10) intercepts a point with a concentration of C_{bulk}. Therefore the δ_{DBL} is

$$\delta_{DBL} = \frac{C_{bulk}}{B(C_{bulk} - C_{surface})} \tag{15.12}$$

The δ_{CBL} can also be determined, following the definition for the laminar MBL, by solving for z in Eqn. (15.7), when the concentration is 99% of the bulk (i.e., when $\theta = 0.99$). Therefore, the δ_{CBL} is

$$\delta_{CBL} = \frac{A}{B}\tanh^{-1}\left(\frac{0.99}{A}\right) \tag{15.13}$$

The mass transfer coefficient (k_c) can also be determined from Eqn. (15.11)

$$J = k_c(C_{bulk} - C_{surface})$$ (15.14)

where k_c is the product of the D and parameter B.

Through k_c, a characteristic local Sherwood number, which is the ratio of the advective to diffusive flux ($Sh_x = k_c x D^{-1}$), can then be determined. In laminar flat plate boundary layer theory, Sh_x can be described by

$$Sh_x = aRe_x{}^b Sc^{0.33}$$ (15.15)

and for ideal boundary conditions, parameter a has a value of 0.339 and 0.464 for constant surface concentration and constant flux, respectively and parameter b is equal to 0.50 (Schlichting and Gersten, 2000; Bird *et al.*, 2002). For turbulent boundary layers, a has a value of 0.030 and 0.028 for constant concentration and flux, respectively and parameter b is equal to 0.80 (Schlichting and Gersten, 2000; Bird *et al.*, 2002).

Note that for laminar CBLs with the aforementioned values of a and b, Eqn. (15.15) is the dimensionless solution to the concentration boundary layer equation (Eqn. (15.2)), when $R = 0$. The parameters a and b will deviate according to the boundary conditions involved and the hydrodynamics of the system.

15.4   SETTING THE BOUNDARY CONDITIONS

As indicated above, the boundary conditions for aquatic organisms are complicated by the diverse variety of scalar quantities of interest and, most importantly, the physiological and chemical processes that create the source or sink necessary for CBL formation (Table 15.3). This is in contrast to the boundary conditions used to model purely abiotic phenomena such as water velocity or energy through the dissolution of gypsum etc. (e.g., Porter *et al.*, 2000).

Regardless, whereas it is possible to classify boundary layer reactions into two groups (heterogeneous and homogeneous boundary layer reactions) it is also possible to classify the transported material as particulate (or suspended; e.g., gametes, plankton, bacteria) and as dissolved (or in solution; e.g., gases such as oxygen and carbon dioxide and nutrients such as phosphate, nitrate and ammonium).

Heterogeneous reactions are those processes such as nutrient uptake, photosynthesis, and respiration that directly influence the flux of material through the water-surface interface (Nishihara and Ackerman, 2007a). The flux at the water-surface interface is defined by the physiological processes, which creates the sink or source. In the simplest case, the heterogeneous reaction proceeds so that the surface concentration or flux is constant and does not vary with increases in the supply or removal of the scalar (Levich, 1962). For example, marine algae in nutrient-poor water (i.e., oligotrophic conditions) suffering from nutrient limitation, will consume all the nutrients that arrive at its surface. Consequently, the concentration of nutrients at the water-surface interface would be zero (i.e., the perfect-sink condition; Vogel, 1994) and the uptake rate would be directly proportional to the flux of nutrients towards the surface. In contrast, under nutrient-rich conditions (i.e., eutrophic conditions) and where the nutrients are in excess of the alga's requirements, the uptake rate would saturate at some maximum and is invariant. The flux of nutrients at the surface would not depend on the external mass transport processes and the surface concentrations may increase if the flux through the water-surface interface is lower than the flux due to mass transport, in the case of where the surface acts as a source.

Table 15.3. Examples of heterogeneous and homogeneous boundary layer reactions that may influence the boundary conditions of the mass transport equations.

Biological processes	Type of reaction	Reference
Nutrient uptake in oligotrophic water	Constant concentration, heterogeneous	Thomas *et al.*, 1985, 1987, 2000; Sanford and Crawford, 2000; Phillips and Hurd, 2003; Larned *et al.*, 2004
Nutrient uptake in eutrophic water	Constant flux, heterogeneous	Sanford and Crawford, 2000
Ammonium uptake in *Laurencia brongniartii*	Linear flux, heterogeneous	Nishihara *et al.*, 2005
Photosynthesis in *Vallisneria americana*	Nonlinear flux, heterogeneous	Nishihara and Ackerman, 2006, 2007b
Suspension feeding activity of mussels	Nonlinear, heterogeneous	Ackerman, 1999; Ackerman *et al.*, 2001; Ackerman and Nishizaki, 2004; Tweddle *et al.*, 2005
Pollen and spore dispersal	Nonlinear, heterogeneous	Ackerman, 2000, 2006; Okubo and Levin, 2002
Fertilization by broadcast spawning	Linear/Nonlinear, homogeneous	Okubo *et al.*, 2002; Okubo and Levin, 2002
Bicarbonate – carbon dioxide chemistry above photosynthesizing organisms	Linear, homogeneous	Wolf-Gladrow and Riebesell, 1997; Tortell *et al.*, 1997; Nishihara and Ackerman, 2006, 2007a, b

As a result of heterogeneous reactions, the flux at the water-surface interface will deviate from the constant concentration and flux boundary conditions and variations are likely to occur due to the physiological characteristics of the organism. For example, the kinetics of nutrient uptake by aquatic macrophytes can vary with the nutrient (Thomas *et al.*, 1985; Thomas *et al.*, 1987; Nishihara *et al.*, 2005) as well as the spatial location on the leaf (Nishihara and Ackerman, 2007a). Moreover, uptake kinetics and the flux can be linear or nonlinear. If the nutrient can diffuse freely into the organism, the flux at the water-surface interface will be directly proportional to that of the surface (i.e., mass transfer limited). Typically however, the concentrations of nutritionally important ions (e.g., DIC, DIN) are higher in the organism than in the water (Lobban and Harrison, 1996), and thus an active uptake mechanism is required. Such active mechanisms can saturate under high nutrient concentrations (e.g., Michaelis-Menten kinetics) and hence the flux is a nonlinear function of the concentration.

Homogeneous reactions occurring in the concentration boundary layer may also alter the local concentration of material within the CBL. In the ideal case, $R = 0$ and there are no reactions present. In the natural environment however, homogeneous boundary layer reactions are likely to be relatively more common than the ideal condition. For example, the conversion of bicarbonate to CO_2 may increase the availability of CO_2 to photosynthetic organisms such as coral symbionts, algae and plants (Tortell *et al.*, 1997; Wolf-Gladrow and Riebesell, 1997; Riebesell *et al.*, 2000; Nishihara and Ackerman, 2007a). Similarly, spawning over the bed of mussels will consume eggs and sperm and produce fertilized eggs in the CBL. Ultimately these homogeneous processes affect the concentration of the material in question and clearly deviate from the ideal processes modeled by Eqn. (15.3) and Eqn. (15.14).

15.4.1 Boundary conditions for nutrient uptake

The diversity of boundary conditions discussed above is present in benthic systems, whether they are sediments, biofilms, mussel beds, or the surfaces of photosynthetic organisms. Research suggests that there is significant spatial heterogeneity in the O_2 concentrations at the water-surface interface in sediment (Jørgensen and Revsbech, 1985; Glud *et al.*, 1994; Lorke *et al.*, 2003) and biofilm systems (Nielsen *et al.*, 1990; Kuehl *et al.*, 1996). For example, the O_2 consumption in sediments were found to saturate with increasing water velocities (Jørgensen and Revsbech, 1985), indicating that the uptake mechanism is a nonlinear process. Although, there is little evidence on whether O_2 flux and nutrient flux is a linear or nonlinear process, it is more than likely that the flux is a nonlinear function. Given that the biomass in any given organism is limited, there will be some finite capacity to consume oxygen or nutrients. Therefore, for large mass transfer rates, the kinetics of the system will saturate and supply will outweigh demand. The situation is also similar in autotrophic systems, where phytoplankton, algal mats and macrophytes (e.g., macroalgae and aquatic angiosperm) consume nutrients and produce oxygen. As with sediments and biofilms, there is marked heterogeneity of flux with respect to spatial location (Nielsen *et al.*, 2006; Nishihara and Ackerman, 2007a). For example in colonies of *Phaeocystis*, oxygen flux varied along its axis (Ploug *et al.*, 1999) as was the case for algal mats, where oxygen flux was spatially heterogeneous (Glud *et al.*, 1999). Variation in oxygen flux was also evident in the leaves of the aquatic angiosperms, *Vallisneria americana*, where the flux was higher near the leading edge of the leaf than the trailing edge (Nishihara and Ackerman, 2007a). It is relevant to note that the uptake kinetics of macrophyte systems is better understood than those in multispecies arrangements (i.e., ecosystems) in sediments, biofilms, and marine aggregates (marine snow).

It is evident that uptake kinetics range from linear to nonlinear and the nonlinear behavior can be modeled as a rectangular hyperbola (i.e., Michaelis-Menten kinetics) or a more complex function as in the case of the biphasic uptake of nitrate in the diatom *Skeletonema costatum* (Serra *et al.*, 1978). The nonlinearity of the uptake kinetics can lead to spatial heterogeneity in the flux and influence the observed kinetics of the physiological process. This was the case in *Vallisneria spiralis* where the flux of oxygen saturated at both leading and trailing edges when mass transfer rates were high (i.e., high nutrient concentrations), but oxygen flux did not saturate at the trailing edge when mass transfer rates were low (i.e., low nutrient concentrations) (Nishihara and Ackerman, 2007a). Spatial heterogeneity in nutrient uptake was also observed in *Elodea canadensis*, where nutrients accumulation was highest at the edges of the leaves where mass transfer rates would be greatest (Nielsen *et al.*, 2006). It is likely that the flux through the water-surface interface is inherently nonlinear in nature with respect to concentration and space.

15.4.2 Boundary conditions for external fertilization

Reproduction involves complex boundary layer conditions with respect to the transport of gametes in and out of the momentum boundary layer, given the diversity of broadcast spawning observed in animal and macrophyte systems (see review in Okubo *et al.*, 2002). For example, the wide variety of reproductive processes seen in macroalgae limits the possibility of developing a more general theory on how mass transport processes influence reproduction. In particular, sexual reproduction in most brown algae (e.g., kelps) involves the release of male and female gametes into the water column where fertilization occurs (Lobban and Harrison, 1996). This is case of a homogeneous reaction where gametes are consumed and fertilized zygotes are produced, which would also apply to broadcast spawning invertebrates. Moreover, attractants (e.g., pheromones) are released to encourage fertilization, which would in practice enhance the production of zygotes (Lobban and Harrison, 1996). In contrast, sexual reproduction involves a heterogeneous boundary condition in the brown algal order Fucales and in red algae where the male gamete (i.e., spermatia in reds) must be transported to the female gametes that remain on the surface of the macrophyte. This is also the case in the submarine pollination of aquatic angiosperms including seagrasses (Ackerman, 2000, 2006). Similar comparisons and contrasts are also possible for the large diversity of marine and freshwater benthic animals.

15.5　SEDIMENT SYSTEMS AND BIOFILMS

Sediments and biofilms can play an important role in the exchange of dissolved organic and inorganic compounds and gases (Jørgensen and Revsbech, 1985; Glud *et al.*, 1994; Lorke *et al.*, 2003). Fortunately, microsensors have been used for some time to measure the concentration gradient and hence determine the CBL and fluxes in these systems (Jørgensen and Revsbech, 1985; Gundersen and Jørgensen, 1990; Glud *et al.*, 1994; Lorke *et al.*, 2003). There appears to be considerable the spatial heterogeneity in O_2 (Jørgensen and Revsbech, 1985; Røy *et al.*, 2002), and a recent analysis of these data revealed the nonlinear nature of the CBL of O_2 (Nishihara and Ackerman, 2007b). These studies have relied on a linear estimate of the diffusive boundary thickness (δ_{DBL}) and the assumption that advection does not occur within the DBL. In addition, the CBL thickness (δ_{CBL}) was generally not determined given that there were no objective methods to do so prior to Hondzo *et al.* (2005). As indicated above, the DBL was typically determined graphically by assuming that the oxygen gradient adjacent to the surface was linear (Jørgensen and Revsbech, 1985) and the flux through the water-surface interface was determined from the slope of the oxygen gradient. This method make two assumptions: (i) that mass transfer occurs only through diffusion in the diffusional boundary layer (DBL); and (ii) that the flux can be modeled as a one-dimensional problem (Jørgensen and Revsbech, 1985; see review in Nishihara and Ackerman, 2007b). However, it is well known that horizontal advection (e.g., mass transfer parallel to the surface and turbulent diffusion) also influences the concentration gradient near the surface (Shaw and Hanratty, 1977; Dade, 1993; Hondzo *et al.*, 2005). A power-law scaling of the concentration gradient and information on the momentum boundary layer (MBL) over the sediment revealed estimates of the δ_{DBL} that were 30% thinner than that of the linear model (Hondzo *et al.*, 2005). Moreover, a model of the CBL could be expressed in terms of the Sc, the turbulent Sc, and the MBL through the use of similarity arguments for the concentration gradient (Hondzo *et al.*, 2005). This model incorporates the fact that advective mass transport processes can be important in the CBL and discounts the notion that the DBL is a stagnant layer of water.

15.6　AUTOTROPHIC SYSTEMS

15.6.1　Pelagic producers

There has been considerable interest in the large-scale mass transport of nutrients to phytoplankton as these organisms drive pelagic ecosystems, especially in the seasonal blooms when mixing of the water column bring nutrient-rich waters to the upper water column, and in upwelling events/regions when other physical processes do the same (Ingmanson and Wallace, 1995; Kalff, 2001; Mann and Lazier, 2006). Similar interest has existed on the scale of individual cells and aggregates in an attempt to understand the mechanistic basis of blooms and hence determine parameters that can be used for modeling (e.g., Kiørboe, 2001; Kiørboe et al., 2001).

Presently, the influence of mass transport on the photosynthetic rates of phytoplankton are not well understood. However, from studies of the fluid dynamics of mass transport of sinking marine snow (Kiørboe et al., 2001; Ploug et al., 2002; see below), it is possible to infer that the local photosynthesis rate will vary spatially over the surface of the phytoplankton and produce oxygen-rich microenvironments. Regions of low CO_2 concentrations can also develop, which would influence mass transfer rates and hence, photosynthesis. For example, diatoms, which experience a Langrangian reference frame, have $Re < 10$ and are likely to produce relative thick and heterogeneous CBLs (Ploug et al., 2002). Moreover, the depletion of CO_2 within the CBL will decrease the availability of substrate for photosynthesis (Tortell et al., 1997; Wolf-Gladrow and Riebesell, 1997; Nishihara and Ackerman, 2006, 2007a). The mass transport processes dominating these microscopic organisms are believed to be primarily from diffusion (Wolf-Gladrow and Riebesell, 1997; Ploug et al., 2002). However, diatoms are able to enhance CO_2 supply by changing the CO_2 concentration through the acidification of their surrounding water thereby altering the balance of bicarbonate and CO_2 (Tortell et al., 1997; Wolf-Gladrow and Riebesell, 1997). Moreover, the biosilica (Milligan and Morel, 2002) in the cell wall of diatoms also have the ability to buffer seawater, allowing them to convert bicarbonate enzymatically to CO_2 enhancing the availability of CO_2. The magnitude of the enhancement in the supply of CO_2 through these boundary layer reactions relative to advection and diffusion are not clear. However, a numerical model of the diffusion-reaction equation (i.e., neglecting advection) suggests that 5% of the CO_2 supply is from reactions occurring in the CBL (Wolf-Gladrow and Riebesell, 1997). Given the morphological and physiological diversity of diatoms, further studies are needed to explore the relationships between their biology and physical environment, through investigations of their mass transport and fluid dynamic characteristics.

15.6.2 Benthic macrophytes

The effect of mass transport processes on aquatic macrophytes has long been recognized (e.g., Conover, 1968) and remains a topic of increased activity (see reviews in Hurd, 2000; Okubo et al., 2002). For example, mass transfer has been shown to affect the rates of photosynthesis (e.g., Sand-Jensen et al., 1985; Nishihara and Ackerman, 2006, 2007a, b), nutrient uptake (e.g., Borchardt et al., 1994; Cornelisen and Thomas, 2004), and the timing of spore release (Serrão et al., 1996; see review in Gaylord et al., 2004). Macrophytes have three-dimensional structure at a larger spatial scale than sediments and biofilms (c.f., Larned et al., 2004), and the momentum boundary layers that form around these organisms can be complex (Hurd et al., 1997; Hurd and Stevens, 1997; Stevens and Hurd, 1997). It is not surprising that most studies have simplified this complexity by configuring the macrophytes as flat plates (Wheeler, 1980; Koch, 1993; Hurd et al., 1996; Nishihara and Ackerman, 2006, 2007a, b), although there has been some efforts devoted to parameterizations and the Stanton number (*St*) analogy, where the *St* is the ratio of the flux to a surface divided by the advection past the surface (e.g., Thomas et al., 2000). Moreover, a conceptual model has been advanced to explain the relative importance of mass transport under the influence of different scales of the DBL (i.e., individual DBLs and substratum DBLs) (Larned et al., 2004). Regardless, significant details of the CBL and the mass transport properties of these organisms have been elucidated.

Most studies related to mass transport in macrophyte systems have focused their discussion using the Schmidt number scaling of the viscous sublayer thickness (δ_{VSL}) to determine the thickness of the diffusive sublayer (δ_{DSL})

$$\delta_{DSL} \approx \delta_{VSL} Sc^{-1/3} \tag{15.16}$$

which is referred to, incorrectly, as the thickness of the diffusive boundary layer (δ_{DBL}) (Larned *et al.*, 2004). Recall that the DBL is a component of the CBL and consequently

$$\delta_{DBL} \neq \delta_{DSL} \tag{15.17}$$

Moreover, when the DBL and mass flux at the water-surface interface was determined, it was by assuming a linear one-dimensional model of the concentration gradient near the surface (see above). Based on the one-dimensional model and a further simplifying assumption that the surface concentration of the nutrient was zero at the water-surface interface (i.e., a perfect-sink condition; Vogel, 1994), the δ_{DBL} of nutrients such as dissolved inorganic nitrogen (DIN) (Hurd *et al.*, 1996) and dissolved inorganic carbon (DIC) have been estimated (Wheeler, 1980). However, the concept that higher water velocities, and thus thinner δ_{DSL} and by analogy thinner δ_{DBL}, are alone responsible for increased rates of uptake or photosynthesis is false and has wasted much effort in the field. As mentioned above, it is the flux (i.e., product of the velocity and concentration) of nutrients that affects the rates of physiological processes. This was demonstrated in the case of oxygen flux in *V. americana*, which has flat ribbon-like leaves, where the effect of higher velocities (and thinner δ_{DBL}) on photosynthetic rates was observed at low nutrient concentrations and declined linearly with nutrient concentration (Nishihara and Ackerman, 2006; see below). This realization is likely one of the reasons that mass-transfer limitation has yet to be demonstrated under field conditions.

As indicated above, there has been limited success in matching predictions from linear models of the concentration gradient made using flat-plate analogies with simple boundary conditions (i.e., constant concentration and flux) and experimental results (see discussions in Hondzo *et al.*, 2005; Nishihara and Ackerman, 2006, 2007b). Several non nonlinear approximations have been used to better describe the concentration gradient measured by O_2 microsensors, and of these, the hyperbolic tangent function provides the ability to estimate both the δ_{CBL} and the δ_{DBL} of the scalar. In addition, the first derivative of the model provided estimates of the O_2 flux, which were more accurate than the typically used linear model (Nishihara and Ackerman, 2007b). The development of these techniques should provided objective methods that can be used for macrophytes as well as other organisms.

Both hydrodynamics and the concentration of DIC influenced the rates of photosynthesis in *Vallisneria americana* (Nishihara and Ackerman, 2006, 2007a, b). Increasing the DIC concentration effectively increased the mass transport of DIC to the leaf surface and decreased the saturation velocity (i.e., the velocity required to saturate photosynthesis rates). Moreover photosynthetic rates were observed to be in a state of mass transfer limitation at very low velocities even though the mass transfer of DIC through the DBL (assuming that the surface was a perfect sink) was much greater than the observed carbon uptake rates in the DBL (Nishihara and Ackerman in review). This indicates that other processes (e.g., homogeneous reactions) are likely limiting the supply of carbon.

There were also physiological differences observed between closely related species in terms of the effects of mass transfer on photosynthesis. For example, photosynthesis rates in *V. spiralis* and *V. americana* saturated at leading and trailing regions of the leaf at high DIC concentrations, however the kinetics of photosynthesis was significantly different at lower DIC concentrations (Nishihara and Ackerman, 2007a) – O_2 fluxes were much lower at the trailing edge and did not saturate with increased water velocities. There are a number of possible explanations for the differences observed at different leaf locations: (i) the physiology may differ along the leaf surface; (ii) nutrient concentrations may decline along the leaf surface; or (iii) there may be differences in the homogeneous reactions (e.g., the bicarbonate-carbon dioxide chemistry) over the leaf surface. Both of these species are known to acidify water adjacent to the surfaces, which encourages the production of CO_2 from the bicarbonate in the water (Prins *et al.*, 1980). Physiological differences may influence the rates of these homogeneous reactions and therefore affect the concentration of CO_2 that is available. It is also likely that upstream processes remove the CO_2 from the CBL and thus reduce the CO_2 availability downstream (Nishihara and Ackerman, 2007a, in review). In addition, both the δ_{CBL} and δ_{DBL} are much thinner than predictions based on Eqn. (15.3) (Nishihara and Ackerman, 2007a) and given that homogeneous reactions have non-linear responses that tend to change the thicknesses of the boundary layer, the situation is even more complicated. Evidently, more effort will be required to identify the mechanism(s) responsible for the decrease in photosynthetic rates observed downstream on the leaf surface.

The morphology of macrophytes can be quite complicated involving much branching, highly dissected leaf and frond morphologies, and surface roughness and rugosity (e.g., Sculthorpe, 1967; Lobban and Harrison, 1996). In other words, macrophytes are not simple two-dimensional organisms that can modeled as flat plates, with some obvious exception (e.g., *V. americana*). A functional explanation for this diversity is lacking, however the potential effects of some of these morphologies on the local hydrodynamic environment has long been the subject of inquiry, especially in species that have low and high energy phenotypes (i.e., smooth versus rugose and corrugated surfaces) (Wheeler, 1980; see reviews in Hurd, 2000 and Okubo *et al.*, 2002). For example, it has been long suggested that the features such as spines and corrugations along *Macrocystis* sp. (giant kelp) fronds serve to trip the boundary layer and thus periodically infuse the CBL with fresh nutrient rich water from the overlying bulk water (Hurd *et al.*, 1996). Currently, there is little evidence to support this hypothesis, and experimental results have been equivocal. For example, the twist in *V. spiralis* leaves did not appear to enhance photosynthesis rates compared to the flat leaves of *V. americana* (Nishihara and Ackerman, 2007a), but the local flow environment made it difficult to resolve O_2 measurements within the thin DBL under higher velocities. Clearly, one of the failings of this type of approach is the lack of characterization of the hydrodynamics of the flow (i.e., the MBL) and the lack of measurement of the scalar (i.e., the CBL).

It has also been suggested that the complex branching and large surface area to volume ratio enhance the ability of macrophytes to uptake nutrients (Hurd, 2000). For example, the surface area to volume ratio is large in the whorled macrophyte, *Elodea canadensis*. However at low water velocities, the boundary layers around the whorls and leaves are thick and can overlap. Most of the accumulation of carbon occurs near the edges of the leaves and whorls, where the boundary layer is thinnest and the entire leaf does not perceive the hydrodynamics in the same way (Nielsen *et al.*, 2006). Moreover, flow-induced configurational changes to the shape of macrophytes may reduce their ability to undergo photosynthesis through self-shading or the reduction of area available for nutrient uptake (Stewart and Carpenter, 2003). There are a large number of unresolved questions that remain to be answered at both small and large spatial scales around aquatic macrophytes.

15.7 HETEROTROPHIC SYSTEMS

15.7.1 Pelagic Zooplankton

Mass transport issues are relevant in pelagic environments for a number of reasons including those related to trophic and reproductive relations. In the former, there are a large number of suspension feeders and ambush predators that rely on the delivery of nutrients, chemical signals, and prey (Mann and Lazier, 2006). Chemical signals are also relevant to reproductive interactions such as mate recognition and tracking (Strickler, 1998). Both of these examples can be conceived of as encounter rate problems, which have been applied to aggregates and their formation through the application of coagulation theory (Jackson and Burd, 1998).

A key issue for pelagic mass transport at the smallest scales involves the identification of resources by small heterotrophic bacteria and protists in a well-mixed environment. In other words, this is a problem of locating a resource that has a patchy distribution in space and time. Microscale patchiness has been demonstrated to exist on the mm scale and persist on the scale of 10 min in the laboratory (Blackburn *et al.*, 1998). These patches are created by lysing cells and from sinking algal cells (which can be quite leaky) and from aggregates. Aggregates (marine snow and flocs) are composed of the lysed cells, algae, and bacteria, as well as detritus and transparent exopolymer (TEP) matter (see review in Okubo *et al.*, 2002). The chemical plumes from sinking aggregates have been examined for $Re \leq 20$ by solving the Navier-Stokes and the advection-diffusion equations numerically (Kiørboe *et al.*, 2001). Results indicate that long slender plumes, which extend from reasonably small aggregates, can have significant concentration and length depending on the flow field. An analogous phenomenon has been inferred from observations of reproductive female copepods pursed by males; the males tracking a pheromone signal released by the females (Yen *et al.*, 1998).

The case of zooplankton mass transport it is not merely an issue of flux of seston because turbulence can also affect the outcomes. As indicated above, organisms smaller than the Kolmogorov microscale are predicted to experience the relative motions within small eddies. However, this does not appear to be the case as many organisms have a dome-shaped response in which moderate levels of turbulence enhance encounter and ingestion rates for predators, whereas large levels can be inhibitory to growth in other groups, due perhaps to increased energy expenditures (Peters and Marrasé, 2000). It is evident that small-scale unsteady motion and length scales other than organism size are likely to be more relevant to these ecological processes (Peters and Marrasé, 2000).

15.7.2   Benthic animals

The diversity and ecological and economic importance of benthic suspension-feeding organisms has generated considerable interest into the biology and mechanisms of particle capture (Shimeta and Jumars, 1991; Riisgård and Larsen, 2001) and more recently the effect of suspension feeders on ecosystems (Wildish and Kristmanson, 1997; Okubo *et al.*, 2002). Concentration boundary layers have been observed over bivalves in lakes and estuaries (Dame, 1996; Ackerman *et al.*, 2001; Tweddle *et al.*, 2005) and over coral reefs (Yahel *et al.*, 1998), which indicates the important impact that suspension feeding can have on aquatic environments. Mass transport is of particular importance to suspension-feeding organisms as it is a process driven by the delivery of seston (water borne material) in the water to organisms. Suspension-feeding organisms can be classified as: (i) passive suspension feeders, such as corals, gorgonians, polychaete worms, brittle stars, sand dollars, and caddisfly and black fly larvae, which extend their feeding appendages into the water column; and (ii) active suspension feeders, such as sponges, bivalves, lophophorates, crustaceans, and ascidians, which use various pumping mechanisms to move fluid. In the former, it is the delivery of seston through horizontal advection and turbulent mixing in the vertical direction that is important – analogous to what has been described above for autotrophic organisms (see Eqn. (15.2)) – although in this case additional terms describing the settling velocity (w_s) of the seston (particulate matter) must be added

$$U\frac{\partial C}{\partial x} + w_s\frac{\partial C}{\partial z} = \frac{\partial}{\partial z}\left((K_D + D)\frac{\partial C}{\partial z}\right) + R \qquad (15.18)$$

In addition, the heterogeneous reaction of suspension feeding (ϕ) must be considered. There are many factors that can affect ϕ ranging from physical factors such as seston concentration (C), ambient velocity and the height above the bottom to biological ones including the spacing and orientation of the collecting elements (f_d; tentacles, fibers, cilia, etc.), the number (n) and numerical density (A_n) of organisms, the time (t_I) for moving the material to the site of ingestion, and the efficiency of the capture process (η_ϕ)

$$\phi = f(C, U, z, f_d, n, A_n, t_I, \eta_\phi) \qquad (15.19)$$

The circumstances are similar for active suspension feeders, but the situation requires that a term pertaining to the hydrodynamics of the pumping mechanisms used to move water through the organisms (Riisgård and Larsen, 1995) be added to ϕ in Eqn (15.19). This is of course a simplification of reality as many of the factors in Eqn. (15.19) are known to covary. For example, both the quantity and quality of the seston can affect η_ϕ, as can the product f_dA_n through the potential refiltration of water (O'Riordan *et al.*, 1995). Moreover, bivalves have behavioral responses to fluid dynamics that can affect ϕ (Ackerman, 1999).

One of the most interesting responses of suspension feeders is their unimodal response to velocity (Wildish and Kristmanson, 1997; Ackerman, 1999; Ackerman and Nishizaki, 2004). In this case, increases in velocity lead to increases in capture, clearance and/or growth rates to some peak mode after which further increases in U are inhibitory to the aforementioned rates. The phenomena has been observed in a wide variety of passive and active suspension feeders including corals, gorgonians, and bivalves (for review of bivalves see Ackerman and Nishizaki, 2004), although the mechanism responsible is not well understood. It may be, however, somewhat analogous to the model for autotrophic organisms where increased flux can saturate the physiology of the organism and other processes at high velocities can interfere with their physiology (see Fig. 15.1 in Nishihara and Ackerman, 2006). In the case of bivalves it has been suggested that it is behavioral instability due to lift and drag forcing, acting at the scale of siphons and/or shells, rather than hydrodynamic instability of the pumping mechanism or some grazing optimization that is responsible for the physiological interference (Ackerman, 1999). Regardless it indicates the importance of understanding the role of fluid dynamics at the scale of the organism for mass transport.

Environmental flows are relevant to both passive and active suspension feeding in that the formation of a CBL will be a function of the relative strength of the turbulent mixing in the water column and the sink of seston at the benthos (Fig. 15.5). CBL formation is based on the principle that the rate of seston uptake (ϕ) by suspension feeders is greater than the rate of seston delivery through mass transport. Processes such as turbulent mixing (e.g., K_D) act to obliterate the CBL through the transport of the scalar to regions where it has been depleted (Hanratty, 1956; Shaw and Hanratty, 1977). In other words, if the rate of scalar mixing (i.e., K_D) is small, it is possible that a CBL will form but if K_D is large, the CBL may not form (or may be too thin to observe) as water column mixing will eliminate it and/or cause the size of the CBL to fluctuate in thickness (Hanratty, 1956; Shaw and Hanratty, 1977). It is not surprising, therefore, that it can be difficult to detect CBLs in the field except under particular circumstances where the biomass of suspension feeders is quite large (Tweddle *et al.*, 2005) and/or mixing processes are minimized (e.g., during stratification, Ackerman *et al.*, 2001). Similar arguments can be advanced for the autotrophic systems described above.

Corals reefs represent an important component of the benthos in shallow water regions (generally < 50 m) where the average annual water temperatures are $>20°C$. This is due

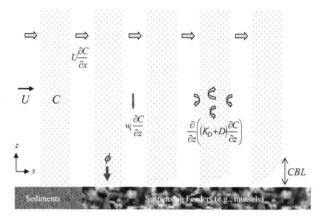

Figure 15.5. The formation of a concentration boundary layer (CBL) as seston (*C*) travels over a region of sediment to a region of suspension-feeding benthos. The stippled seston-containing region represents a slice of the water column at a particular instant in time at different downstream locations. The flux of seston to the bed is a function of advection, settling, and turbulent mixing. The CBL will be a function of the strength of the sink of seston (ϕ) and mixing of the scalar in the water column (K_D) that serves to eliminate the signal. In most cases, the time scale for turbulent mixing of the water column is faster than the time scale for benthic grazing.

in large part to their symbiotic zooxanthellae (dinoflagellate algae), which, as autotrophs, require sunlight and nutrients (e.g., DIC, DIN) for photosynthesis. The corals, being heterotrophic, also feed via the capture of particles and zooplankton on their tentacles, mucous sheets, and the extended mesenterial filaments of the gut wall. There has been considerable effort devoted to mass transport of particulate matter in terms of particle capture (Wildish and Kristmanson, 1997; Sebens *et al.*, 1998), and nutrient uptake (e.g. Atkinson and Bilger, 1992) in corals. The latter has included numerical and physical modeling as well as laboratory and field experiments (see review in Monismith, 2007). Indirect measurements of mass transfer using the dissolution of plaster has been popular recently, with experiments conducted within the skeletons of the complexly branched corals of a number of species under unidirectional and oscillatory flow in laboratory flow chambers (Reidenbach *et al.*, 2006). As might be expected, mass transfer was reduced by ~50% within the branches and mass transfer was enhanced many fold under oscillatory flow. This confirmed field results that indicated that mass transfer of gypsum blocks of various surface configurations was 30–40% higher under oscillatory flow in the field (Falter *et al.*, 2005). Importantly, the magnitude of the difference between oscillatory and linear flow declined with velocity. How these indirect measurements translate biologically for corals under natural conditions remains to be determined (c.f., Atkinson and Bilger, 1992). It has been possible, however, to ignore the intricacies of the flow-coral morphological interaction, by considering the roughness of the coral reef and parameterizing the process through the use of the Stanton number, which relates the mass transfer coefficient (k_c) to the velocity (e.g., Atkinson and Bilger, 1992). This technique has been applied recently at the scale of a reef flat community where the dissipation of waves allowed for estimates of the bottom friction (Falter *et al.*, 2004). The ability to use a measurement of the canopy friction has also made this an attractive approach for macrophyte canopies (see above).

The manner by which large-scale fluid dynamic processes in surface waters affect benthic organisms and *visa versa* is becoming better understood (Jonsson *et al.*, 2005; Loewen *et al.*, 2007). Simple measurements such as velocity are likely to provide some indication that locally depleted resources may be replenished with fresh seston. However, a description of the physical mixing processes rather than a mere reliance on simple metrics is necessary.

Unfortunately, neither the physical mixing of surface waters nor the response of suspension feeders to this mixing can be predicted or easily modeled. Additional studies at a variety of spatial scales are, therefore, warranted to better understand the role of mass transport to benthic organisms.

15.8   EMERGING PRINCIPLES

Although aquatic environments involve a great diversity of organisms, biological and ecological processes and habitats, there are some common principles that emerge when they are considered from the perspective of mass transport:

(i)   There are similarities among systems in terms of the relevance of the flux of dissolved and/or particulate scalars to the processes under consideration. Specifically flux applies equally well to the transport of nutrients to autotrophs as it is does to the flux of seston to suspension feeders. It is important to note that flux is the product of velocity and a concentration gradient, therefore, experiments should examine both the vector and the scalar.

(ii)   Concentration boundary layers (CBL) are formed when a concentration gradient forms next to a biological surface that acts as a source or sink of a scalar. CBLs are analogous to momentum boundary layers (MBL) but their structure differs. Importantly, the diffusional sublayer thickness (δ_{DSL}) defined using the MBL is not a good predictor of the much thinner diffusional boundary layer thickness (δ_{DBL}) of the CBL.

(iii)   The development and use of microsensors continues to advance our ability to examine and understand mass transport issues through the direct measurement of the concentration gradients. A hyperbolic-tangent model provides the ability to estimate the δ_{CBL}, δ_{DBL}, and the flux at the surface in a rigorous and unambiguous manner.

(iv)   CBLs can be difficult to measure under moderate and turbulent flows in the laboratory due to the small spatial scales involved and in the field due to turbulent mixing (i.e., temporal scales) that eliminate the gradients.

(v)   Both homogeneous and heterogeneous boundary layer reactions can and do occur in the CBL associated with biological and ecological processes. This realization should help to facilitate the further modeling of mass transport phenomena in aquatic systems.

(vi)   Further research into the role of physiology on the dynamics of heterogeneous reactions, especially those involving nutritiously important molecules, is required to advance our understanding of biologically relevant mass transport beyond simple physical models.

(vii)   Our current understanding of mass transport is based largely on morphological systems whereby the organism is fixed naturally to a surface or held static in experiments. Realistically, however, many biological systems are flexible and undergo complex undulations and may reconfigure morphologically under environmental flows. Unfortunately there are few techniques that can provide measurements of MBLs under these conditions, let alone characterize CBLs that vary temporally and spatially. Advances in technology and approach are needed in this area.

(viii) Although significant advances have been made with respect to mass transport there are many unresolved problems, including processes that occur under turbulent and unsteady environmental flows. This last realization provides a degree of optimism in the sense that research into mass transport will continue to be at the leading edge of aquatic research for the foreseeable future.

APPENDIX A – LIST OF SYMBOLS

<table>
<tr><td colspan="3" align="center">List of Symbols</td></tr>
<tr><td>Symbol</td><td>Definition</td><td>Units</td></tr>
<tr><td>A, B</td><td>parameters for Eqn. (15.5)</td><td>–</td></tr>
<tr><td>A_n</td><td>numerical density of organisms</td><td>indv m^{-2}</td></tr>
<tr><td>C</td><td>concentration</td><td>mol m^{-3}</td></tr>
<tr><td>C_{bulk}</td><td>bulk concentration</td><td>mol m^{-3}</td></tr>
<tr><td>$C_{surface}$</td><td>surface concentration</td><td>mol m^{-3}</td></tr>
<tr><td>CBL</td><td>concentration boundary layer</td><td>–</td></tr>
<tr><td>D</td><td>molecular diffusivity of the scalar</td><td>m^2 s^{-1}</td></tr>
<tr><td>DBL</td><td>diffusive boundary layer</td><td>m</td></tr>
<tr><td>DIC</td><td>dissolved inorganic carbon</td><td>mol m^{-3}</td></tr>
<tr><td>DIN</td><td>dissolved inorganic nitrogen</td><td>mol m^{-3}</td></tr>
<tr><td>DSL</td><td>diffusive sublayer</td><td>m</td></tr>
<tr><td>J</td><td>mass flux, scalar (e.g., nutrient) uptake rate</td><td>mol m^{-2} s^{-1}</td></tr>
<tr><td>K_D</td><td>turbulent diffusivity of the scalar</td><td>m^2 s^{-1}</td></tr>
<tr><td>K_ν</td><td>turbulent diffusivity of momentum</td><td>m^2 s^{-1}</td></tr>
<tr><td>MBL</td><td>momentum boundary layer</td><td>–</td></tr>
<tr><td>R</td><td>homogeneous boundary layer reaction</td><td>mol m^{-2} s^{-1}</td></tr>
<tr><td>Re</td><td>Reynolds number</td><td>–</td></tr>
<tr><td>Re_x</td><td>local Reynolds number</td><td>–</td></tr>
<tr><td>Sc</td><td>Schmidt number</td><td>–</td></tr>
<tr><td>Sh_x</td><td>local Sherwood number</td><td>–</td></tr>
<tr><td>St</td><td>Stanton number</td><td>–</td></tr>
<tr><td>U</td><td>freestream or bulk velocity</td><td>m s^{-1}</td></tr>
<tr><td>f_d</td><td>collecting fiber diameter</td><td>m</td></tr>
<tr><td>k_c</td><td>mass transfer coefficient</td><td>m s^{-1}</td></tr>
<tr><td>l</td><td>length</td><td>m</td></tr>
<tr><td>n</td><td>number of individuals</td><td>–</td></tr>
<tr><td>t_I</td><td>ingestion time</td><td>s</td></tr>
<tr><td>u</td><td>velocity in the x direction</td><td>m s^{-1}</td></tr>
<tr><td>w</td><td>velocity in the z direction</td><td>m s^{-1}</td></tr>
<tr><td>w_s</td><td>settling velocity</td><td>m s^{-1}</td></tr>
<tr><td>x</td><td>distance in the x (downstream) direction</td><td>m</td></tr>
<tr><td>z</td><td>distance in the z (vertical) direction</td><td>m</td></tr>
<tr><td>δ</td><td>boundary layer (BL) thickness</td><td>m</td></tr>
<tr><td>δ_{CBL}</td><td>thickness of the concentration BL</td><td>m</td></tr>
<tr><td>δ_{DBL}</td><td>thickness of the diffusive BL</td><td>m</td></tr>
<tr><td>δ_{DSL}</td><td>thickness of the diffusive sublayer</td><td>m</td></tr>
<tr><td>δ_{ISL}</td><td>thickness of the inertial sublayer</td><td>m</td></tr>
<tr><td>δ_{MBL}</td><td>thickness of the momentum BL</td><td>m</td></tr>
<tr><td>δ_{VSL}</td><td>thickness of the viscous sublayer</td><td>m</td></tr>
<tr><td>η</td><td>Kolmogorov microscale</td><td>m</td></tr>
<tr><td>η_ϕ</td><td>efficiency of suspension feeding</td><td>–</td></tr>
<tr><td>θ</td><td>dimensionless concentration gradient</td><td>–</td></tr>
<tr><td>ν</td><td>molecular diffusivity of momentum</td><td>m^2 s^{-1}</td></tr>
<tr><td>τ</td><td>shear stress</td><td>Pa</td></tr>
<tr><td>ϕ</td><td>flux due to suspension feeding</td><td>kg m^{-2} s^{-1}</td></tr>
</table>

APPENDIX B – SYNOPSIS

In aquatic environments a large number of biological processes involve the transport of scalar quantities, which is controlled by the principles of fluid dynamics. The flow of water over the surface of an organism forms a momentum boundary layer (MBL), while a concentration boundary layer (CBL) is due to a concentration gradient next to a biological surface that acts as a source or sink of a scalar. The structure of these layers is different and microsensors can be applied to a direct measurement of concentration gradients providing the basis for a rigorous estimation of the structure of the CBL and of the flux across the water-surface interface. However, these measurements could be difficult to perform under moderate and turbulent flows in the laboratory due to the small spatial scales involved and in the field due to turbulent mixing that eliminate the gradients. Both homogeneous and heterogeneous boundary layer reactions can and do occur in the CBL associated with biological and ecological processes. Further research into the role of physiology on the dynamics of heterogeneous reactions is required to advance our understanding of biologically relevant mass transport beyond simple physical models. Moreover, future analysis about mass transport should consider that many biological systems are not fixed naturally to a surface but flexible, undergoing complex undulations and may reconfigure morphologically under environmental flows.

APPENDIX C – KEYWORDS

By the end of the chapter you should have encountered the following terms. Ensure that you are familiar with them!

Autotrophic system   Ecosystem   Mass-transport
Boundary condition   Heterogeneous reaction Momentum boundary layer
Concentration boundary layer Heterotrophic system   Schmidt number
Diffusive boundary layer   Homogeneous reaction   Trophic status

APPENDIX D – QUESTIONS

Why the analysis of mass transport in the aquatic ecosystem is so important?
Which is the difference between diffusional sublayer (DSL) and diffusive boundary layer (DBL)?
How boundary layer reactions may affect boundary layer conditions for mass transport equations?
How the thickness of the concentration and the diffusive boundary layer and the flux at the surface could be estimated by using experimental data?
How suspension-feeding organisms can be classified?

APPENDIX E – PROBLEMS

E1. Introduce the concept of aquatic ecosystem and list marine and freshwater ecosystems. List and describe the constituent biological components of an ecosystem. Describe the change of the trophic status of an aquatic environment.

E2. Describe the structure of the momentum velocity layer and concentration boundary layer over a surface. Discuss the difference between the diffusional sublayer (DSL) and the diffusive boundary layer (DBL). Describe the method here presented for the measurement of the concentration and diffusive boundary layer and the flux at the water-surface interface. Point out the related problems.

E3. Compare the concepts of heterogeneous and homogeneous reaction. Provide some examples of heterogeneous and homogeneous boundary layer reactions in aquatic ecosystems.

E4. Describe main issues about mass transport in autotrophic and heterotrophic systems.

E5. Describe the main findings from the analysis of mass transport in aquatic systems presented in the chapter, point out the limits in our current understanding and list future challenges in the research on this.

REFERENCES

Ackerman J.D. 1999. Effect of velocity on the filter feeding of dreissenid mussels (*Dreissena polymorpha* and *Dreissena bugensis*): implications for trophic dynamics. *Canadian Journal of Fisheries and Aquatic Sciences* **56**:1551–61.

Ackerman, J.D. 2000. Abiotic pollen and pollination: Ecological, functional, and evolutionary perspectives. *Plant Systematics and Evolution* **222**:167–185.

Ackerman, J.D. 2006. Sexual reproduction of seagrasses: Pollination in the marine context. pp. 89–109. In: A.W.D. Larkum, J.J. Orth, and C.M. Duarte (eds.) *Seagrasses: Biology, Ecology and Their Conservation*. Springer, New York. 691 pp.

Ackerman, J.D. and Hoover, T. 2001. Measurement of local bed shear stress in streams using a Preston-static tube. *Limnology and Oceanography* **46**:2080–2087.

Ackerman, J.D., Loewen, M.R. and Hamblin, P.F. 2001. Benthic-pelagic coupling over a zebra mussel bed in the western basin of Lake Erie. *Limnology and Oceanography* **46**:892–904.

Ackerman, J.D. and Nishizaki, M.T. 2004. The effect of velocity on the suspension feeding and growth of the marine mussels *Mytilus trossulus* and *M. californianus*: Implications for competition and niche separation. *Journal of Marine Systems* **49**:195–207.

Atkinson, M. and Bilger, R.W. 1992. Effects of water velocity on phosphate uptake in coral reef-flat communities. *Limnology and Oceanography* **37**:273–279.

Acrivos, A. and Chambré, P.L. 1957. Laminar boundary layer flows with surface reactions. *Industrial and Engineering Chemistry* **59**:1025–1029.

Bird, R.B., Stewart, W. E. and Lightfoot, E.N. 2002. *Transport Phenomena*. John Wiley & Sons, New York. 895 pp.

Blackburn, N. Fenchel, T. and Mitchell, J. 1998. Microscale nutrient patches in planktonic habitats shown by chemotactic bacteria. *Science* **282**:2254–2256.

Borchardt, M.A., Hoffman, J. and Cook, P. 1994. Phosphorus uptake kinetics of *Spirogyra fluviatilis* (Charophyceae) in flowing water. *Journal of Phycology* **30**:403–417.

Chambré, P.L. and Acrivos, A. 1956. On chemical surface reactions in laminar boundary layer flows. *Journal of Applied Physics* **27**:1322–1328.

Chambré, P. L. and Young, J.D. 1958. On the diffusion of a chemically reactive species in a laminar boundary layer flow. *The Physics of Fluids* **1**: 48–54.

Chung, P. M. 1969. Periodic fluctuation of chemical species over a reactive surface. *The Physics of Fluids* **12**:53–63.

Conover JT. 1968. The importance of natural diffusion gradients and transport of substances related to benthic marine plant metabolism. *Botanica Marina* **11**:1–9

Cornelisen, C.D. and Thomas, F.I.M. 2004. Ammonium and nitrate uptake by leaves of the seagrass *Thalassia testudinum*: impact of hydrodynamic regime and epiphyte cover on uptake rates. *Journal of Marine Systems* **49**:177–194.

Dade, W. 1993. Near-bed turbulence and hydrodynamic control of diffusional mass-transfer at the sea-floor. *Limnology and Oceanography* **38**:52–69.

Dame, R. 1996. *Ecology of Marine Bivalves: An Ecosystem Approach*. CRC Marine Science Series, Boca Raton. 272 pp.

Dang, V.D. 1983. Steady-state mass transfer with homogeneous and heterogeneous reactions. *AIChE Journal* 29:19–25.

de Beer, D. and Kühl, M. 2001. Interfacial microbial mats and biofilms. 374–394. In: *The Benthic Boundary Layer*, Boudreau, B.P. and Jørgensen, B.B. (eds.). Oxford University Press, New York. 404 pp.

de Beer, D., Stoodley, P., and Lewandowski, Z. 1994. Liquid flow in heterogeneous biofilms. *Biotechnology and Bioengineering* 44:636–641.

Denny, M.W. 1988. *Biology and Mechanics of the Wave-Swept Environment*. Princeton University Press, Princeton. 344 pp.

Falter, J.L., Atkinson, M.J. and Merrifield, M.A. 2004. Mass transfer limitation of nutrient uptake by a wave-dominated reef flat community. *Limnology and Oceanography* 49:1820–1831.

Falter, J.L., Atkinson, M.J. and Coimbra, C.F.M. 2005. Effects of surface roughness and oscillatory flow on the dissolution of plaster forms: Evidence for nutrient mass transfer to coral reef communities. *Limnology and Oceanography* 50:246–254

Fischer, H., List, J., Koh, C., Imberger, J., and Brook, N. 1979. *Mixing in Inland and Coastal Waters*. Academic Press, San Diego. 302 pp.

Freeman, N.C. and Simpkins, P.G. 1965. On the diffusion of species in similar boundary layers with finite recombination rate at the wall. *Quarterly Journal of Mechanics and Applied Mathematics* 18:213–229.

Gaylord, B., Reed, D. C., Washbun, L. and Raimondi, P.T. 2004. Physical-biological coupling in spore dispersal of kelp forest macroalgae. *Journal of Marine Systems* 49:19–39.

Glud, R.N., Gundersen, J.K., Jørgensen, B.B., Revsbech, N.P. and Schulz, H.D. 1994. Diffusive and total oxygen uptake of deep-sea sediments in the eastern South Atlantic Ocean: In Situ and laboratory measurements. *Deep-Sea Research Part I Oceanographic Research Papers* 41:1767–1788.

Glud, R.N., Kühl, M., Kohls, O. and Ramsing, N.B. 1999. Heterogeneity of oxygen production and consumption in a photosynthetic microbial mat as studied by planar optodes. *Journal of Phycology* 35:270–279.

Gundersen, J.K. and Jørgensen, B.B. 1990. Microstructure of diffusive boundary-layers and the oxygen-uptake of the sea-floor. *Nature* 345:604–607.

Hanratty, T.J. 1956. Turbulent exchange of mass and momentum with a boundary. *AIChE Journal* 2:359–362.

Hondzo, M., Feyaerts, T., Donovan, R. and O'Connor, B.L. 2005. Universal scaling of dissolved oxygen distribution at the sediment-water interface: A power law. *Limnology and Oceanography* 50:1667–1676.

Hurd, C.L. and Stevens, C.L. 1997. Flow visualization around single- and multiple-bladed seaweeds with various morphologies. *Journal of Phycology* 33:360–367.

Hurd, C.L. 2000. Water motion, marine macroalgal physiology, and production. *Journal of Phycology* 36:453–472.

Hurd, C.L., Harrison, P. J. and Druehl, L. D. 1996. Effect of seawater velocity on inorganic nitrogen uptake by morphologically distinct forms of *Macrocystis integrifolia* from wave-sheltered and exposed sites. *Marine Biology* 126:205–214.

Hurd, C. L., Stevens, C.L., Laval, B., Lawrence, G. and Harrison, P.J. 1997. Visualization of seawater flow around morphologically distinct forms of the giant kelp *Macrocystis integrifolia* from wave-sheltered and exposed sites. *Limnology and Oceanography* 42:156–163.

Ingmanson, D.E., and W.J. Wallace. 1995. *Oceanography 5th Ed*. Wadsworth, Belmont, CA. 528 pp.

Jackson, G.A. and Burd, A.B. 1998. Aggregation in the marine environment. *Environmental Science and Technology* **32**:2805–2814.

Jonsson, P.R., Petersen, J.K., Karlsson, Ö., Loo, L.O. and Nilsson, S. 2005. Particle depletion above experimental bivalve beds: In situ measurements and numerical modeling of bivalve filtration in the boundary-layer. *Limnology and Oceanography* **50**:1989–1998.

Jørgensen, B.B. and Des Marais, D.J. 1990. The diffusive boundary layer of sediments: Oxygen microgradients over a microbial mat. *Limnology and Oceanography* **35**:1343–1355.

Jørgensen, B.B. and Revsbech, N.P. 1985. Diffusive boundary layers and the oxygen uptake of sediments and detritus. *Limnology and Oceanography* **30**:111–122.

Kalff, J. 2001. Limnology. Prentice Hall, Upper Saddle River, NJ. 592 pp.

Kiørboe, T. 2001. Formation and fate of marine snow: small-scale processes with large-scale implications. *Scientia Marina* **65**:57–71.

Kiørboe, T., Helle Ploug, H. and Thygesen, U.H. 2001. Fluid motion and solute distribution around sinking aggregates. I. Small-scale fluxes and heterogeneity of nutrients in the pelagic environment. *Marine Ecology Progress Series* **211**:1–13.

Koch, E.W. 1993. The effect of water flow on photosynthetic processes of the alga *Ulva lactuca* L. *Hydrobiologia* **260–261**:457–462.

Køhler-Rink, S. and Kühl, M. 2000. Microsensor studies of photosynthesis and respiration in larger symbiotic foraminifera. I The physico-chemical microenvironment of *Marginopora vertebralis, Amphistegina lobifera* and *Amphisorus hemprichii. Marine Biology* **137**: 473–486.

Kuehl, M., Glud, R.N., Ploug, H. and Ramsing, N.B. 1996. Microenvironmental control of photosynthesis and photosynthesis-coupled respiration in an epilithic cyanobacterial biofilm. *Journal of Phycology* **32**:799–812.

Larned, S.T., Nikora, V.I. and Biggs, B.J.F. 2004. Mass-transfer-limited nitrogen and phosphous uptake by stream periphyton: a conceptual model and experimental evidence. *Limnology and Oceanography* **49**:1992–2000.

Levich, V.G. 1962. *Physicochemical Hydrodynamics*. Prentice Hall. 700 pp.

Libby, P.A. and Liu, T. 1966. Laminar boundary layer with surface catalyzed reactions. *The Physics of Fluids* **9**:436–445.

Lobban, C.S. and Harrison, P.J. 1996. *Seaweed Ecology and Physiology*. Cambridge University Press, Cambridge. 376 pp.

Loewen, M.R., Ackerman, J.D. and Hamblin, P.F. 2007. Environmental implications of stratification and turbulent mixing in a shallow lake basin. *Canadian Journal of Fisheries and Aquatic Sciences* **64**:43–57.

Lorke, A., Müller, B., Maerki, M. and Wüest, A. 2003. Breathing sediments: The control of diffusive transport across the sediment-water interface by periodic boundary-layer turbulence. *Limnology and Oceanography* **48**:2077–2085.

Mann, K.H. and J.R.N. Lazier. 2006. *Dynamics of Marine Ecosystems, 3rd Ed*. Blackwell Science, Oxford. 512 pp.

Martin, C.L. and Tortell, P.D. 2006. Bicarbonate transport and extracellular carbonic anhydrase activity in Bering Sea phytoplankton assemblages: Results from isotope disequilibrium experiments. *Limnology and Oceanography* **51**:2111–2121.

McMahon, T.A. and Bonner, J.T. 1983. *On Size and Life*. W.H. Freeman, New York. 255 pp.

Milligan, A.J. and Morel, F.M.M. 2002. A proton buffering role for silica in diatoms. *Science* **297**:1848–1850.

Monismith, S.G. 2007. Hydrodynamics of coral reefs. *Annual Review of Fluid Mechanics* **39**:37–55.

Na, Y. and Hanratty, T.J. 2000. Limiting behavior of turbulent scalar transport close to a wall. *International Journal of Heat and Mass Transfer* **43**:1749–1758.

Nielsen, H.D., Nielsen, S.L. and Madsen, T.V. 2006. CO_2 uptake patterns depend on water velocity and shoot morphology in submerged stream macrophytes. *Freshwater Biology* **51**:1331–1340.

Nielsen, L.P., Christensen, P.B., Revsbech, N.P. and Sørensen, J. 1990. Denitrification and oxygen respiration in biofilms studied with a microsensor for nitrous oxide and oxygen. *Microbial Ecology* **19**:63–72.

Niklas, K.J. 1992. *Plant Biomechanics*. University of Chicago Press, Chicago. 622 pp.

Niklas, K.J. 1994. *Plant Allometry*. University of Chicago, Chicago. 395 pp.

Nishihara, G.N. and Ackerman, J.D. 2006. The effect of hydrodynamics on the mass transfer of dissolved inorganic carbon to the freshwater macrophyte *Vallisneria americana*. *Limnology and Oceanography* **51**:2734–2745.

Nishihara, G.N. and Ackerman, J.D. 2007a. The interaction of CO_2 concentration and spatial location on O2 flux and mass transport in the freshwater macrophytes *Vallisneria spiralis* and *V. americana*. *Journal of Experimental Biology* **210**:522–532.

Nishihara, G.N. and Ackerman, J.D. 2007b. On the determination of mass transfer in a concentration boundary layer. *Limnology and Oceanography Methods* **5**:88–96+ Appendix.

Nishihara, G.N., Terada, R. and Noro, T. 2005. Effect of temperature and irradiance on the uptake of ammonium and nitrate by *Laurencia brongniartii* (Rhodophyta, Ceramiales). *Journal of Applied Phycology* **17**:371–377.

Nybakken, J.W. and Bertness, M.D. 2005. *Marine Biology: An Ecological Approach (6th Edition)*. Benjamin Cummings, San Francisco. 592 pp.

Okubo, A. and Levin, S.A. 2002. *Diffusion and Ecological Problems, 2nd Ed*. Springer, New York. 488 pp.

Okubo, A., Ackerman, J.D. and Swaney, D.P. 2002. Passive Diffusion in Ecosystems. pp. 31–106 in A. Okubo and S. Levin (eds.) *Diffusion and Ecological Problems: New Perspectives, 2nd Ed*. Springer Verlag, New York. 467 pp.

O'Riordan, C.A., Monismith S.G. and Koseff , J.R. 1995. The effect of bivalve excurrent jet dynamics on mass transfer in a benthic boundary layer. *Limnology and Oceanography* **40**:330–44.

Pennycuick, C.J. 1992. *Newton Rules Biology*. Oxford University Press, Oxford. 128 pp.

Peters, F., and Marrasé, C. 2000. Effects of turbulence on plankton: an overview of experimental evidence and some theoretical considerations. *Marine Ecology Progress Series* **205**:291–306.

Phillips, J.C. and Hurd, C.L. 2003. Nitrogen ecophysiology of intertidal seaweeds from New Zealand: N uptake, storage and utilisation in relation to shore position and season. *Marine Ecology Progress Series* **264**:31–48.

Ploug, H., Stolte, W., Epping, E.H.G. and Jørgensen, B.B. 1999. Diffusive boundary layers, photosynthesis, and respiration of the colony-forming plankton algae, *Phaeocystis* sp. *Limnology and Oceanography* **44**:1949–1958.

Ploug, H. Hietanen, S., and Kuparinen, J. 2002. Diffusion and advection within and around sinking, porous diatom aggregates. *Limnology and Oceanography* **47**:1129–1136.

Porter, E.T., Sanford, L.P. and Suttles, S.E. 2000. Gypsum dissolution is not a universal integrator of 'water motion'. *Limnology and Oceanography* **45**:145–158.

Prins, H.B.A., Snel, J.F.H., Helder, R.J. and Zanstra, P.E. 1980. Photosynthetic HCO_3^- utilization and OH^- excretion in aquatic angiosperms. Plant Physiology **66**: 818–822.

Ricklefs, R.E. and Miller, G. 2000. *Ecology 4th ed*. W.H. Freeman, New York. 896 pp.

Riebesell, U., Zondervan, I., Rost, B., Tortell, P.D., Zeebe, R.E. and Morel, F.M.M. 2000. Reduced calcification of marine plankton in response to increased atmospheric CO_2. *Nature* **407**:364–367.

Reidenbach M.A., Koseff, J.R., Monismith, S.G., Steinbuck, J.V. and Genin, A. 2006. Effects of waves, unidirectional currents, and morphology on mass transfer in branched reef corals. *Limnology and Oceanography* **51**:1134–41.

Riisgård, H.U. and Larsen, P.S. 2001. Minireview: Ciliary filter feeding and bio-fluid mechanics - present understanding and unsolved problems. *Limnology and Oceanography* **46**:882–891.

Riisgård, H.U., and Larsen, P.S. 1995. Filter-feeding in marine macro-invertebrates: pump characteristics, modelling and energy cost. *Biological Reviews* **70**:67–106.

Røy, H., Hüettel, M. and Jørgensen, B.B. 2002. The role of small-scale sediment topography for oxygen flux across the diffusive boundary layer. *Limnology and Oceanography* **47**:837–847.

Sand-Jensen, K., Revsbech, N.P. and Jørgensen, B.B. 1985. Microprofiles of oxygen in epiphyte communities on submerged macrophytes. *Marine Biology* **89**:55–62.

Sanford, L.P. and Crawford, S.M. 2000. Mass transfer versus kinetic control of uptake across solid-water boundaries. *Limnology and Oceanography* **45**:1180–1186.

Schlichting, H. and Gersten, K. 2000. *Boundary-Layer Theory*. Springer, New York. 801 pp.

Sculthorpe, C.D. 1967. *The Biology of Aquatic Vascular Plants*. Edward Arnold, London. 610 pp.

Sebens, K.P., Grace, S. P., Helmuth, B., Maney, E.J. Jr. and Miles, J.S. 1998. Water flow and prey capture by three scleractinian corals, *Madracis mirabilis*, *Montastrea cavernosa* and *Porites porites*, in a field enclosure. *Marine Biology* **131**:347–360.

Serra, J.L., Llama, M.J. and Codenas, E. 1978. Nitrate utilization by the diatom *Skeletonema costatum*. 2. Regulation of nitrate uptake. *Plant Physiology* **62**:991–994.

Serrão, E.A., Pearson, G., Kautsky, L. and Brawley, S.H. 1996. Successful external fertilization in turbulent environments. *Proceedings of the National Academy of Science (USA)* **93**:5286–5290.

Shaw, D.A. and Hanratty, T.J. 1977. Turbulent mass transfer to a wall for large Schmidt numbers. *AIChE Journal* **23**: 28–37.

Shimeta, J. and Jumars, P.A. 1991. Physical mechanisms and rates of particle capture by suspension-feeders. *Oceanography and Marine Biology Annual Review* 29:191–257.

Stevens, C.L. and Hurd, C.L. 1997. Boundary-layers around bladed aquatic macrophytes. *Hydrobiologia* **346**:119–128.

Stewart, H.L. and Carpenter, R.C. 2003. The effects of morphology and water flow on photosynthesis of marine macroalgae. *Ecology* **84**:2999–3012.

Strickler, J.R. 1998. Observing free-swimming copepods mating. *Philosophical Transactions of the Royal Society (London) Series B* **353**:671–680.

Thomas F.I.M., Cornelsen, C.D. and Zande, J.M. 2000. Effects of water velocity and canopy morphology on ammonium uptake by seagrass communities. *Ecology* **81**:2704–13.

Thomas, T.E., Harrison, P.J. and Taylor, B.E. 1985. Nitrogen uptake and growth of the germlings and mature thalli of *Fucus distichus*. *Marine Biology* **84**:267–274.

Thomas, T.E., Harrison, P.J. and Turpin, D.H. 1987. Adaptations of *Gracilaria pacifica* (Rhodophyta) to nitrogen procurement at different intertidal locations. *Marine Biology* **93**:569–580.

Tortell, P.D., Reinfelder, J.R. and Morel, F.M.M. 1997. Active uptake of bicarbonate by diatoms. *Nature* **390**:243–244.

Tweddle, J.F., Simpson, J.H. and Janzen, C.D. 2005. Physical controls of food supply to benthic filter feeders in the Menai Strait, UK. *Marine Ecology Progress Series* **289**:79–88.

Vogel, S. 1994. *Life in Moving Fluids, 2nd Ed.*. Princeton University Press. pp. 476.

Wetzel, R.G. 2001 *Limnology 3rd ed*. Academic Press, San Diego 1006 pp.

Wheeler, W.N. 1980. Effect of boundary layer transport on the fixation of carbon by the giant kelp *Macrocystis pyrifera*. *Marine Biology* **56**:103–110.

White, F.M. 1999. Fluid Mechanics, 4th Ed. WCB McGraw Hill, New York.

Wildish, D. and Kristmanson, D. 1997. *Benthic Suspension Feeders and Flow.* Cambridge University Press, Cambridge. 409 pp.

Wolf-Gladrow, D. and Riebesell, U. 1997. Diffusion and reactions in the vicinity of plankton: A refined model for inorganic carbon transport. *Marine Chemistry* **59**:17–34.

Yahel, G., Post, A.F., Fabricius, K.E., Marie, D., Vaulot, D. and Genin, A.1998. Phytoplankton distribution and grazing near coral reefs. *Limnology and Oceanography* **43**:551–563.

Yen, J., Weissburg, M.J. and Doall, M.H. 1998.The fluid physics of signal perception by mate-tracking copepods. *Philosophical Transactions of the Royal Society (London) Series B* **353**:787–804.

CHAPTER SIXTEEN

Maps serving as the combined coupling between interacting environmental interfaces and their behavior in the presence of dynamical noise

Dragutin T. Mihailović
Department for Field and Vegetable Crops, Faculty of Agriculture, University of Novi Sad, Novi Sad, Serbia

Igor Balaž
Department of Physics, Faculty of Sciences, University of Novi Sad, Novi Sad, Serbia

ABSTRACT

Many physical and biological problems, in addition to environmental problems, can be described by the dynamics of driven coupled oscillators. To study their behavior as a function of coupling strength and nonlinearity, we considered the dynamics of two maps acting as the combined coupling (diffusive and linear) in the above fields. First, we considered a logistic difference equation on an extended domain that is a part of the maps, and we discuss it using its bifurcation diagram, Lyapunov exponent, sample and permutation entropy. Second, we performed the dynamical analysis of the coupled maps using Lyapunov exponents and cross-sample entropy for the dependence of two coupling parameters. Further, we investigated how dynamic noise can affect the structure of these bifurcation diagrams. This investigation was performed with noise entering in two specific ways, which disturbs either the logistic parameter on the extended domain or places an additive "shock" to the state variables. Finally, we demonstrated the effect of forcing by parametric noise on the Lyapunov exponent of coupled maps.

16.1 INTRODUCTION

16.1.1 Environmental science

To preserve this world for future generations, we must strive for sustainable development that meets the needs of the present without compromising future advances and resources. However, some scientists suspect that recent floods, heat waves and droughts in regions throughout the world are indicators of even more hazardous events to come. Increased pollution in the atmosphere is causing a gradual increase in the air temperature. The environment is becoming increasingly endangered, especially as humans encroach on and modify fragile ecosystems. Therefore, environmental issues are a primary focus of the world scientific community in the 21st century. Perhaps the most comprehensive and detailed list of these important environmental topics is outlined by the Royal Society of London for Improving Natural Knowledge (RS), the oldest national scientific society in the World and the leading national organization for the promotion of scientific research in Great Britain. The RS has emphasized the following fundamental issues for future research: (i) Are we alone in the Universe?; (ii) How can we reduce harmful greenhouse gases emissions and climate change?; (iii) What is consciousness?; (iv) How do we make decisions in an insecure world?; (v) How do we extend the average lifespan?; (vi) Is culture unique to humans?; (vii) How can we better manage the earth's resources?; (viii) What is best utilization of the internet?; (ix) How do we promote the research of stem cells?; (x) How can we best maintain biodiversity?; (xi) What is the role of geoengineering in climate change?; and (xii) How can we create new vaccines? The aforementioned topics, which are not listed in hierarchical order, highlight issues that can be linked to the environment and are primarily affected by climate change and other related issues.

A unique characteristic of these issues is that they are closely connected with the survival of of the human race on the earth. Today is the first time in the history of science that environmental problems are at the top of the scientific list of priorities, from fundamental inquiries to the development of practical applications. Environmental science can be defined from the perspective of many individual scientific disciplines. However, it is defined in the Random House Dictionary (2006) as "the branch of science concerned with the physical, chemical, and biological conditions of the environment and their effect on organisms." This definition *can be accepted as the broadest definition because it implicitly includes the main feature of environmental science, which is its multidisciplinary nature* (emphasized by D.T. Mihailović and C. Gualtieri). The field of environmental science is abundant with many interfaces and is ready for the application of new fundamental approaches that can lead to a better understanding of environmental phenomena. We have slightly evolved the previous definition by focusing on the concept of an *environmental interface,* which is defined as an interface between two abiotic or biotic environments that are in relative motion and exchange energy and substances through physical, biological and chemical processes, fluctuating temporally and spatially regardless of the space and time scale (Mihailović and Balaž, 2007; Cushman-Roisin *et al.*, 2008). We define environmental science as *"the branch of science concerned with interactions in environmental interfaces that are regarded as natural complex systems."* Environmental science encompasses issues such as climate change, conservation, biodiversity, water quality, groundwater contamination, soil contamination, use of natural resources, waste management, sustainable development, disaster reduction, air pollution, and noise pollution. The core components of environmental science are derived from the atmospheric sciences, ecology, environmental chemistry and the geosciences. *Atmospheric science* examines the earth's gaseous outer layer and emphasizes its interrelationships with other systems. It includes disciplines ranging from meteorological studies (e.g., atmospheric chemistry and atmospheric physics – Hewitt and Jackson, 2003), greenhouse gas phenomena (Bogdonoff and Rubin, 2007), atmospheric dispersion modeling of airborne contaminants, sound propagation phenomena as it relates to noise pollution (Borchgrevink, 2003), and light pollution (Welch, 1998). Studies in *Ecology* typically analyze the dynamics of biological populations and their interactions with the environment. These studies could address endangered species, predator–prey interactions, habitat integrity, the effects of environmental contaminants on populations, or the analysis of impacts of proposed land development on species viability. The interdisciplinary analysis of ecological systems that are impacted by one or more stressors related to fields in environmental science, such as water pollution, could also be evaluated. *Environmental chemistry* involves chemical alterations in the environment. Principal areas of study include soil contamination and water pollution. Topics for analysis include chemical degradation in the environment, the multi-phase transport of chemicals and chemical effects on biota. The *geosciences* include environmental geology, environmental soil science, volcanic phenomena and the evolution of the earth's crust. Hydrology and oceanography are also included as geosciences according to some classification systems (Chamley, 2003).

16.1.2 Environmental interfaces and their complexities

We previously defined the environmental interface. This definition broadly covers the unavoidable multidisciplinary approach in environmental sciences and also includes the traditional approaches in the sciences that address environmental space (Mihailović *et al.*, 2011b). For example, such interfaces can be placed between the following: human or animal bodies and surrounding air, aquatic species and their surrounding water and air systems, natural or artificially built surfaces (vegetation, ice, snow, barren soil, water, or urban communities) and the atmosphere, and cells and their surrounding environment. The environmental interface of different media has been considered in different contexts (Glazier and Graner, 1993; Nikolov *et al.*, 1995; Martins *et al.*, 2000; Niyogi and Raman, 2001; Mihailović *et al.*, 2011a, among many others). The environmental interface as a complex system is a suitable area for irregularities in temporal variations of some physical, chemical or biological quantities, describing their interactions (Rosen, 1991; Selvam and Fadnavis, 1998; Sivertsen, 2005). Complex environmental interface systems are open and hierarchically organized, and the interactions between their parts are nonlinear, while their interactions with the surrounding environment are noisy. These systems are, therefore, very sensitive to initial conditions, deterministic external perturbations and random fluctuations, which are always present in nature. Therefore, the study of noisy non-equilibrium processes is fundamental for (i) modeling the dynamics of environmental interface systems and (ii) understanding the mechanisms of spatio-temporal pattern formation in contemporary environmental sciences (Cushman-Roisin *et al.*, 2008). Recently, considerable effort has been invested in developing an understanding of how different fluctuations arise from the interplay of noise, forces, and nonlinear dynamics.

The understanding of complexity in the framework of environmental interface systems may be enhanced by starting from so-called simple systems to grasp the phenomena of interest and then add details that introduce complexity at a number of varying levels. In general, the effects of small perturbations and noise, which are ubiquitous in real systems, can be quite difficult to predict and can often yield contradictory behavior patterns. Even low-dimensional systems exhibit a enormous variety of noise-driven phenomena, ranging from less-ordered to a highly ordered system dynamics. Before proceeding further, several terms introduced in previous passages require detailed clarification. The term complex system we use in Rosen's sense (Rosen, 1991), as explicated in the following comment by Collier (2003): "*In Rosen's sense a complex system cannot be decomposed nontrivially into a set of parts for which it is the logical sum. Rosen's modelling relation requires this. Other notions of modelling would allow complete models of Rosen style complex systems, but the models would have to be what Rosen calls analytic, that is, they would have to be a logical product. Autonomous systems must be complex. Other types of systems may be complex, and some may go in and out of complex phases*". Additionally, the term complexity can entail many ambiguities because there is a large variety of uses for the concept of complexity. Sometimes (e.g., Rosen, 1991), complexity refers only to systems that cannot be modeled precisely in all respects. However, following Arshinov and Fuchs (2003), the term "complexity" has three levels of meaning: (i) there is self-organization and emergence in complex systems (Edmonds, 1999), (ii) complex systems are not organized centrally but instead are organized in a distributed manner so that there are many connections between the system's parts (Edmonds, 1999; Kauffman, 1993) and (iii) it is difficult to model complex systems and to predict their behavior, even if one knows to a large extent the parts of such systems and the connections between the parts (Edmonds, 1999; Heylighen, 1999).

In past years, the study of deterministic mathematical models of environmental systems has clearly revealed a large variety of phenomena, ranging from deterministic chaos to the presence of spatial organization. Chaos in higher dimensional systems is one of the focal subjects of physics today. Along with the approach that starts with modeling physical and biophysical systems with many degrees of freedom, there has emerged a new approach, developed by Kaneko (1983), which couples a number of one-dimensional maps to study the behavior of a system as a whole. However, this model can only be applied to study the dynamics of a single medium such as pattern formation in a fluid. What happens if two media border one another, such as in an environmental interface? One may naturally lead to the model of coupled logistic maps with different logistic parameters. Even two logistic maps coupled with each other may serve as the dynamical model of driven coupled oscillators (Midorikawa *et al.*, 1995). It has been found that two coupled identical maps possess several characteristic features that are typical for higher dimensional chaos. There exist a number of interesting physical, biological and environmental problems, such as superconducting quantum interference devices, magneto-hydrodynamics, convection in conducting fluids, chemical reactions, neurodynamics, two biological populations, coevolution of species, and substance and energy exchange between two environmental interfaces, which can be described by the dynamics of driven coupled oscillators (Hogg and Huberman, 1984; Midorikawa *et al.*, 1995; Mihailović and Balaž, 2007; Mihailović *et al.*, 2011a; Mihailović *et al.*, 2011b; Metta *et al.*, 2011, among others). In the aforementioned fields, it is of great importance to understand the global dynamics of the coupled systems as a function of both the nonlinearity and the coupling strength.

16.1.3 Maps serving as the combined coupling in environmental modeling

Consider the simplest biologically realistic model that incorporates spatial effects, which is two coupled logistic maps (Hastings, 1993; Gyllenberg *et al.*, 1993; Lloyd, 1995; Gunji and Kamiura, 2004). In terms of non-dimensional variables, these maps have the following form:

$$x_{n+1} = (1 - c_1)f(x_n) + c_1 f(y_n) \tag{16.1a}$$

$$y_{n+1} = (1 - c_1)f(y_n) + c_1 f(x_n) \tag{16.1b}$$

where $f(x) = \mu x(1 - x)$. This model supposes that the environment consists of two patches between which the entities diffuse (Fig. 16.1a). Such coupling tends to equalize the instantaneous states of the entities (*diffusive coupling*). We assume that there is a density-dependent phase followed by a dispersal phase. The density-dependent phase is modeled using the logistic map, and the dispersal phase is modeled using a simple exchange of a fixed proportion of the populations. The parameter μ is the standard logistic parameter for the logistic map, and c_1 is a measure of the diffusion of individuals between the two patches, with $0 \leq c_1 \leq 1$. We assume that the environment is homogeneous; hence, the logistic parameter μ is the same for both of the patches. This model is designed in a similar manner as that of Hassell (1991), whose host-parasitoid model consists of a pair of variables at each of at least 900 lattice sites. However, it is not necessary that μ should be taken to be constant. Following Mihailović *et al.* (2011b), the logistic parameter μ can be different in Eqs. (16.1a–16.1b), which serve as coupled maps representing energy exchange processes between two heterogeneous environmental interfaces regarded as biophysical complex systems. Notably, this coupling tends to equalize the instantaneous states of the entities. Thus, the dynamics of this simpler two-dimensional, two-parameter system will be considerably easier to understand, and the insights gained by studying it should shed light on more complex systems.

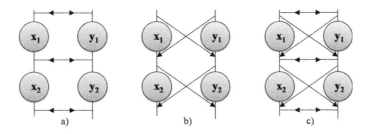

Figure 16.1. Schematic diagram of the (a) *diffusive* (b) *linear* and (c) *combined* coupling in environmental models.

There are mathematically simpler ways to couple two logistic maps: for example, we could have *linear coupling* (Fig. 16.1b)

$$x_{n+1} = f(x_n) + c_2(y_n - x_n) \tag{16.2a}$$

$$y_{n+1} = f(y_n) + c_2(x_n - y_n) \tag{16.2b}$$

Another popular form of coupling is a bilinear coupling, with the linear terms in (16.2) replaced by $\pm c_2 x_n y_n$ terms. These forms of the coupled logistic map have been studied previously, using both numerical (Kaneko, 1983; Ferretti *et al.*, 1988; Satoh and Aihara, 1990, among others) and analytic techniques (Sakaguchi and Tomita, 1990). Such forms for coupling are not biologically realistic because they involve the mixing of generations. Some of the individuals have been allowed to reproduce and die and have also been allowed to move into the other patch.

Finally, it is possible that both types of coupling are present, and this scenario will be *combined* coupling (Fig. 16.1c), which can be written in the following form:

$$x_{n+1} = (1 - c_1)\mu x_n(1 - x_n) + c_1 \mu y_n(1 - y_n) + c_2(y_n - x_n) \tag{16.3a}$$

$$y_{n+1} = (1 - c_1)\mu y_n(1 - y_n) + c_1 \mu x_n(1 - x_n) + c_2(x_n - y_n) \tag{16.3b}$$

It appears that there is no necessity to invent other types of coupling because this equation serves as a universal model of weakly coupled systems (Ivanova *et al.*, 2002), which can be broadly used in environmental modeling.

We present the main features of the function $f(x)$ with μ on an extended domain, in coupled Eqs. (16.3a)–(16.3b) in Section 16.2. Section 16.3 is devoted to the dynamical analysis of the coupled maps using the Lyapunov exponent and cross-sample entropy, while their behavior in the presence of the parametric and added noise is covered in Section 16.4. Concluding remarks are provided in Section 16.5.

16.2   LOGISTIC DIFFERENCE EQUATION ON AN EXTENDED DOMAIN

Let us consider a dynamical system

$$X_{n+1} = S(X_n) \tag{16.4}$$

and make a transformation T: $T(X) = Y$, where X and Y are vectors. If the Jacobi matrix is regular (either locally or globally), then for a transformed system,

$$Y_{n+1} = G(Y_n) \tag{16.5}$$

Information about the dynamics of this system can be obtained from the dynamics of system (16.4) and *vice versa*. In our case, we consider the following difference equation

$$x_{n+1} = \rho x_n(1 - x_n) \quad \rho < 0 \tag{16.6}$$

where the dynamics (ρ will be referred as the parameter of the difference equation) can be completely described by the dynamics of the standard logistic difference equation, as follows:

$$x_{n+1} = \mu x_n(1 - x_n) \quad 0 < \mu < 4 \tag{16.7}$$

Namely, we obtain Eq. (16.7) by making successive transformations T_1 (symmetry), T_2 (homothety) and T_3 (translation) in Eq. (16.6), where $T_1(x) = -x$, $T_2(x) = (1 - 2/\rho)x$ and $T_3(x) = x + 1 - 1/\rho$. The Jacobian for all of the transformations is globally different from zero, while r and ρ are related by the equation $\mu = 2 - \rho$. Finally, for Eq. (16.6), we have the following properties: (a) $x = 0$ is the attractive fixed point for $0 < \rho < 1$; (b) bifurcations start for $\rho < -1$ (Fig. 16.2a); (c) function $f(x) = \rho x(1 - x)$ maps interval $[1/\rho, 1 - 1/\rho]$ on itself for $-2 \le \rho < 0$; (d) the occurrence of the chaotic behavior for $-2 \le \rho < \rho_\infty$, where $\rho_\infty = 2 - \mu_\infty$ [$\mu \approx 3.56994$], while Eq. (16.6) has the same behavior for $\mu \in [r_\infty, 4]$; finally, (e) orbits tend to infinity for $\rho < -2$. Fig. 16.2a depicts the bifurcation diagram of Eq. (16.6) on the whole domain $[-2, 4]$. We now analyze the occurrence of chaos in the solution of Eq. (16.9). Because a quantitative measure for the identification of chaos is the Lyapunov exponent λ, we will calculate its spectrum for Eq. (16.9) as a function of the parameter ρ ranging from -2 to 4. Their values are seen in Fig. 16.2b. This figure depicts two features of

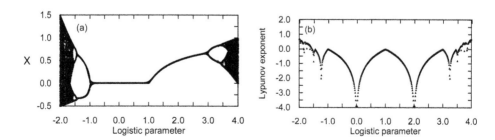

Figure 16.2. Bifurcation diagram (a) and Lyapunov exponent (b) of the difference equations (Eqs. (16.6)–(16.7)) as a function of the parameter of the difference equation ranging from -2 to 4.

the Lyapunov exponent spectrum of Eq. (16.6). They are (i) its symmetry at the point $\rho = 1$, with the exact characteristics of the logistic equation spectrum going left and right towards to values -2 and 4, respectively, and (ii) it is positive in the intervals $\rho \in [-2, 2 - \mu_\infty]$ and $\rho \in [\mu_\infty, 4]$, indicating chaotic fluctuations of x. However, inside the $\rho \in [-2, 2 - \mu_\infty]$ and $[\mu_\infty, 4]$ intervals, there are a large number of opened periodical "windows", where $\lambda < 0$. This observation implies that the dynamical system is synchronized in some regions where the chaotic regime prevails.

With increased model complexity, we are less able to manage and understand model behavior. As a result, the ability of a model to simulate complex dynamics is not an absolute value in itself, but rather a relative one: We require sufficient complexity to realistically model a process, but not so much complexity that we cannot handle it. For example, if we want to model biophysical processes over a non-uniform surface, we meet many uncertainties in the time series of the calculated temperature, energy fluxes, and other factors. Various measures of complexity were developed to compare time series and to distinguish regular (e.g., periodic), chaotic, and random behaviors. The main types of complexity parameters are entropies and Lyapunov exponents, among others. These complexity parameters are all defined for typical orbits of presumably ergodic dynamical systems, and there are profound relations between these quantities (Arshinov and Fuchs, 2003).

In this chapter, we use the sample entropy (*SampEn*) and the permutation entropy (*PermEn*) to measure the complexity and uncertainties of quantity time series described by Eq. (16.10). Sample entropy, a measure quantifying regularity and complexity, is believed to be an effective analytical method for diverse settings that include both deterministic chaotic and stochastic processes and is particularly operative in the analysis of physiological, sound, climate and environmental interface signals that involve relatively small amounts of data (Kennel *et al.*, 1992; Richman and Moorman, 2000; Lake *et al.* 2002). $SampEn(m, r, N)$ is the negative natural log of the conditional probability that two sequences similar within a tolerance r_s for m points remain similar at the next point, where N is the total number of points and self-matches are not included, i.e., $SampEn(m, r, N) = -\ln(A^m/B^m)$, where, $A^m(r) = \sum_{i=1}^{N-m} A_i^m(r)/(N-m)$ and $B^m(r) = \sum_{i=1}^{N-m} B_i^m(r)/(N-m)$. A low value of *SampEn* is interpreted as a value that results in increased regularity or order in the data series. The threshold factor, or filter, r is an important parameter. In principle, with an infinite amount of data, its value should approach zero. With finite amounts of data, or with measurement noise, the r value typically varies between 10 and 20 percent of the time series' standard deviation (Pincus, 1991).

Permutation entropy (*PermEn*) of order $n \geq 2$ is defined as $PermEn \sum p(\pi)\ln p(\pi)$, where the sum runs over all $n!$ permutations π of order n. This expression is the information that is contained in comparing n consecutive values of the time series. Consider a time series $\{x_t\}_{t=1,...,T}$. We consider all $n!$ permutations π of order n, which are considered here as possible order types of n difierent numbers. For each π, we determine the relative frequency $p(\pi) = \#\{t|0 \leq t \leq T - n, (x_{t+1}, \ldots, x_{t+n}) \text{ has type } \pi\}/(T - n + 1)$. This expression estimates the frequency of π as accurately as possible for a finite series of values. To determine $p(\pi)$ exactly, we must assume an infinite time series $\{x_1, x_2, \ldots\}$ and take the limit for $T \to \infty$ in the above formula. This limit exists with probability 1 when the underlying stochastic process fulfills a very weak stationarity condition: For $k \leq n$, the probability for $x_t < x_{t+k}$ should not depend on t. Permutation entropy as a natural complexity measure, for time series behaves similar as Lyapunov exponents and is particularly useful in the presence of dynamical or observational noise (Bandt and Pompe, 2002).

Figure 16.3 depicts *SampEn* of a single time series obtained from Eq. (16.6) as a function of the parameter ρ ranging from -2 to -1.4 (16.3a) and from 3.4 to 4 (16.3b). These two figures show the output for this equation over a range of growth values, for sample length $m = 2$. Cleary, there are a number of regions of stability around -1.83 and 3.83, respectively. We also computed the permutation entropy. Again, the test case used was

Eq. (16.6). Figs. 16.3c and 16.3d plot the computed *PermEn* versus the growth rate of parameter ρ, which is periodic for some regions and chaotic for others. We can also clearly see some regions of stability around -1.83 and 3.83, respectively. Notably, *PermEn* is very similar to the positive Lyapunov exponent (Figs. 16.3a vs. 16.3c and 16.3b vs. 16.3d).

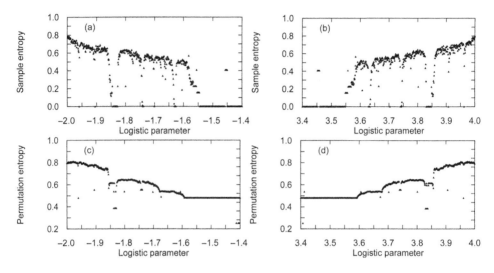

Figure 16.3. Sample (a, b) and permutation entropy (c, d) as a function of the parameter ρ, ranging from -2 to -1.4 and from 3.4 to 4.

16.3 DYNAMICAL ANALYSIS OF THE COUPLED MAPS SERVING THE COMBINED COUPLING

We will consider two parameters, i.e., the Lyapunov exponent and cross-sample entropy (Cross-SampEn). Consider the general vector mapping $\vec{x}_{n+1} = \vec{F}(\vec{x}_n), (n = 0, 1, \ldots, N)$ and its Nth iterate $\vec{F}^{(N)}(\vec{x}) \equiv \vec{F}^{(N-1)}(\vec{F}(\vec{x}))$ with $\vec{F}^{(1)}(\vec{x}) \equiv \vec{F}(\vec{x})$. The asymptotic behavior of a series of iterates of the map can be characterized by the largest Lyapunov exponent, which for an initial point $\vec{x}_0$, is an attracting region and is defined as

$$\lambda = \lim_{N \to \infty} \{\ln[\|\underline{D}^{(N)}(\vec{x}_0)\|]/N\} \qquad (16.8)$$

where $\|\,\|$ is the norm of the Jacobi matrix $\underline{D}$ and $\underline{D}^{(N)}$ for the mappings $\vec{F}(\vec{x})$ and $\vec{F}^{(N)}(\vec{x})$, respectively, and N is a large number of successive points that are used in computing the derivative matrix in Eq. (16.8). We calculated the Lyapunov exponent λ to observe the behavior of the coupled maps given by Eqs. (16.3a)–(16.3b), depending on different values of the coupling parameters c_1 and c_2. Figure 16.4 depicts the Lyapunov exponent for the coupled maps as a function of these parameters, ranging from 0 to 0.9, with an increment of 0.001, and the parameter μ ranging from -1.95 to -1.4 and from 3.4 to 3.9, with an increment of 0.01. Each point was obtained by iterating 1000 times from the initial condition to eliminate transient behavior and then averaging over another 600 iterations starting from the initial condition $x_0 = 0.20$ and $y_0 = 0.25$. This simple analysis, where we consider the Lyapunov exponent, shows very interesting features of these two coupled maps, which represent different types of interactions in environmental complex systems. From this figure, it can be seen that there exist two distinguished regions with positive as well as negative values of λ; with the property that if c_2 takes values below 0.5, then the Lyapunov exponent of the coupled maps is always negative.

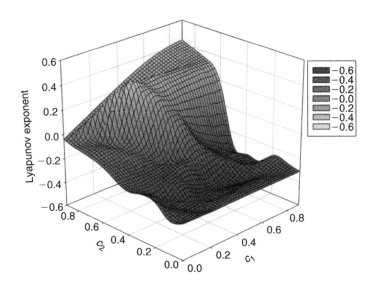

Figure 16.4. Lyapunov exponent for the combined coupling given by Eq. (16.3a)–(16.3b) for values of c_1 and c_2 that range between 0 and 0.9, with μ taken from the intervals $[-1.95, -1.4]$ and $[3.4, 3.9]$.

Cross-SampEn, a measure of asynchrony, is a recently introduced technique for comparing two different time series, to assess their degree of asynchrony or dissimilarity (Pincus and Singer, 1995; Pincus *et al.*, 1996). Let $u = [u(1), u(2), \ldots, u(N)]$ and $v = [v(1), v(2), \ldots, v(N)]$ and set the input parameters m and r_s. Given the vector sequences $x(i) = [u(i), u(i+1), \ldots, u(i+m-1)]$ and $y(j) = [v(j), v(j+1), \ldots, v(j+m-1)]$, let N be the number of data points in the time series, with $i, j = N - m + 1$. For each $i \leq N - m$, set $B_i^m(r_s)(v \parallel u) = $ (number of $j \leq N - m$ such that $d[x_m(i), y_m(j)] \leq r_s])/(N - m)$, where j ranges from 1 to $N - m$; then, $B^m(r_s)(v \parallel u) = \sum_{i=1}^{N-m} B_i^m(r_s)(v \parallel u)/(N - m)$, which is the average value of $B_i^m(v \parallel u)$. Similarly, we define A^m and A_i^m. For each $i \leq N - m$, set $A_i^m(r_s)(v \parallel u) = $ (the number of $j \leq N - m$ such that $d[x_{m+1}(i), y_{m+1}(j)] \leq r_s])/(N - m)$, where j ranges from 1 to $N - m$; then, $A^m(r_s)(v \parallel u) = \sum_{i=1}^{N-m} A_i^m(r_s)(v \parallel u)/(N - m)$ is the average value of $A_i^m(v \parallel u)$. Next, Cross-SampEn is defined as

$$Cross\text{-}SampEn = -\ln\{A^m(r)(v \parallel u)/B^m(r)(v \parallel u)\} \tag{16.9}$$

We applied *Cross-SampEn* with $m = 5$ and $r = 0.05$ for the x and y time series. Figure 5 depicts that the (c_1, c_2) phase space is covered with values of *Cross-SampEn* equal or very close to zero, corresponding to the region in Fig. 16.4 where λ is negative. It points out a high synchronization between the coupled maps in that region. In the rest of the (c_1, c_2) phase space, the entropy is greater than zero, which corresponds to positive values of λ.

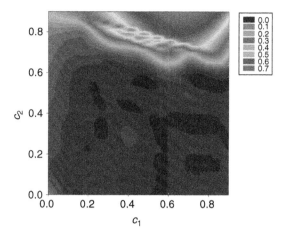

Figure 16.5. *Cross-SampEn* for the combined coupling given by Eq. (16.3a)–(16.3b) for values of parameters c_1 and c_2 ranging between 0 and 0.9 and μ taken in intervals $[-1.95, -1.4]$ and $[3.4, 3.9]$.

For the analysis of the dynamical noise in Eqs. (16.3a)–(16.3b) in the next section, we chose a system with equal values of coupling parameters, i.e., parameters $c_1 = c_2 = 0.5$. The Lyapunov exponent and *Cross-SampEn* for this choice of the parameters for the combined coupling, as a function of the logistic parameter μ, are depicted in Fig. 16.6.

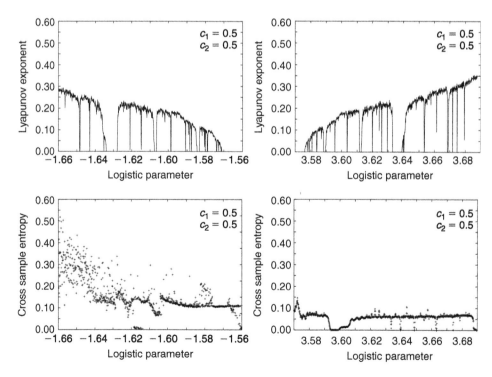

Figure 16.6. Lyapunov exponent and *Cross-SampEn* for the combined coupling given by Eq. (16.3a)–(16.3b) for $c_1 = c_2 = 0.5$ and logistic parameter μ, taken in the intervals $[-1.66, -1.56]$ and $[3.57, 3.69]$.

16.4 BEHAVIOR OF THE COUPLED MAPS SERVING AS THE COMBINED COUPLING IN THE PRESENCE OF DYNAMICAL NOISE

As Ruelle (1994, p. 27) stated, real systems can, in general, be described as deterministic systems with some added noise. This description is sufficiently vague as it appears to cover everything. In economics, for example, such a description is familiar, and the noise is called "shocks". A first remark concerning the above picture is that the separation between noise and the deterministic aspect of the evolution is ambiguous because one can always interpret "noise" as a deterministic time evolution in an infinite dimension (Serletis *et al.*, 2007b). In addition, Serletis *et al.* (2007a) argue that dynamical noise (noise that appears in systems where output becomes corrupted with noise, and noisy values are used as input during the next iteration) can dramatically change the dynamics of nonlinear dynamical systems. In fact, dynamical noise can make the detection of chaotic dynamics very difficult because it is possible to lead to the rejection of the null hypothesis of chaos. Moreover, Serletis and Shahmoradi (2006) pointed out that dynamical noise can shift bifurcation points and produce noise-induced transitions, making the determination of the bifurcation boundaries very difficult. Therefore, meaningful analyses of real systems in the environment, in terms of chaos theory, should consider the effect of dynamical noise on the system's dynamics.

In this section, we will investigate how dynamical noise can affect the structure of bifurcation diagrams of the coupled maps. Motivated by a description of spatial heterogeneity on population dynamics, many authors have considered different coupling forms for logistic maps, which is well documented by Savi (2007). This noise enters in two specific ways: It disturbs either the parameter μ (multiplicative excitation) or the deterministic law by an additive "shock" (external excitation). Specifically, we will address the coupled maps that are given by Eqs. (16.3a)–(16.3b), where $c_1 = c_2 = 0.5$.

First, the randomness influence on the following coupled maps

$$x_{n+1} = (1 - c_1)\mu x_n(1 - x_n) + c_1\mu y_n(1 - y_n) + c_2(y_n - x_n) + \Delta\xi \tag{16.10a}$$

$$y_{n+1} = (1 - c_1)\mu y_n(1 - y_n) + c_1\mu x_n(1 - x_n) + c_2(x_n - y_n) + \Delta\eta \tag{16.10b}$$

was analyzed by adding random noise. Here, $\Delta\xi_n = \mathbf{D}\delta_n^{(1)}$ and $\Delta\eta_n = \mathbf{D}\delta_n^{(2)}\Delta\eta_n$ measure the noise intensity, while $\delta_n^{(1)}$ and $\delta_n^{(2)}$ are random numbers that are uniformly distributed in the interval $[-1,1]$, and D is the amplitude of the noise. To characterize their dynamics, we plotted their bifurcation diagrams in the absence of noise (Fig. 16.7).

The influence of fluctuations in $\Delta\xi_n$ and $\Delta\eta_n$ is now the focus. By considering $\mathbf{D} = 0.1$ and $\mathbf{D} = 0.2$, the results are analyzed using the bifurcation diagrams (Fig. 16.8), which may be compared with Fig. 16.7. Notably, noise destroys some periodic windows, changing some expected behavior, already when $\mathbf{D} = 0.1$, which corresponds to a low amplitude for the noise (Serletis *et al.*, 2007b), as can be seen on the four upper panels in Fig. 16.8. Furthermore, for the doubled amplitude, i.e., $\mathbf{D} = 0.2$, the noise (the four lower panels in Fig. 16.8) significantly obscured the pictures of the bifurcation diagrams in Fig. 16.7.

As has been shown in the case of uncoupled non-linear oscillators, the addition of parametric fluctuations has a pronounced effect on the dynamics of such systems (Hogg and Huberman, 1984; Thattai and van Oudenaarden, 2001; Liu and Ma, 2005). It is therefore of interest to investigate the effect of noise on either environmental processes or events represented by the system of two maps that serve as the combined coupling. In environmental

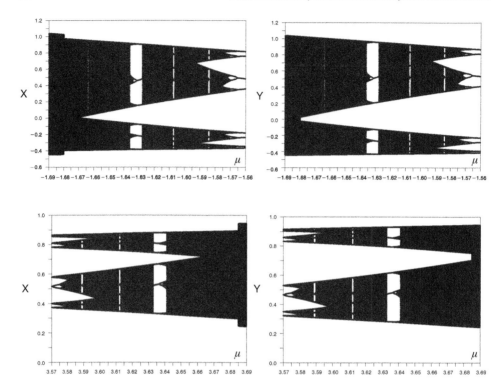

Figure 16.7. Bifurcation diagrams of the coupled maps given by Eqs. (16.3a)–(16.3b) for $c_1 = c_2 = 0.5$ in the absence of noise, $\mathbf{D} = 0$.

models, notable cases are those where the parameters of the oscillators have small, random variations that result, for example, from internal as well as external noise. These so-called parametric fluctuations can be simulated by changing the values of the nonlinearity parameters by adding uniform random numbers in a small interval.

We obtain the following map:

$$x_{n+1} = (1 - c_1)\mu_n^{(1)}x_n(1 - x_n) + c_1\mu_n^{(2)}y_n(1 - y_n) + c_2(y_n - x_n) \tag{16.11a}$$

$$y_{n+1} = (1 - c_1)\mu_n^{(2)}y_n(1 - y_n) + c_1\mu_n^{(2)}x_n(1 - x_n) + c_2(x_n - y_n) \tag{16.11b}$$

where $\mu_n^{(1)} = \mu(1 + \Delta\xi_n)$ and $\mu_n^{(1)} = \mu(1 + \Delta\eta_n)$. The bifurcation diagrams in Fig. 16.9 depict the change in their structure compared to Fig. 16.7, when the parametric noise is introduced with amplitudes $\mathbf{D} = 0.01$ and $\mathbf{D} = 0.025$, corresponding to low-intensity noise. Apparently, then, the parametric forcing produces larger changes in the bifurcation diagrams, if accomplished by adding noise. Specifically, looking at Fig. 16.8 (the four upper panels when $\mathbf{D} = 0.1$) and Fig. 16.9 (the four upper panels when $\mathbf{D} = 0.01$), we can see similar changes in the bifurcation diagrams.

To conclude considering the behavior of the maps that serve the combined coupling in the environmental modeling in the presence of dynamical noise, we will consider its behavior

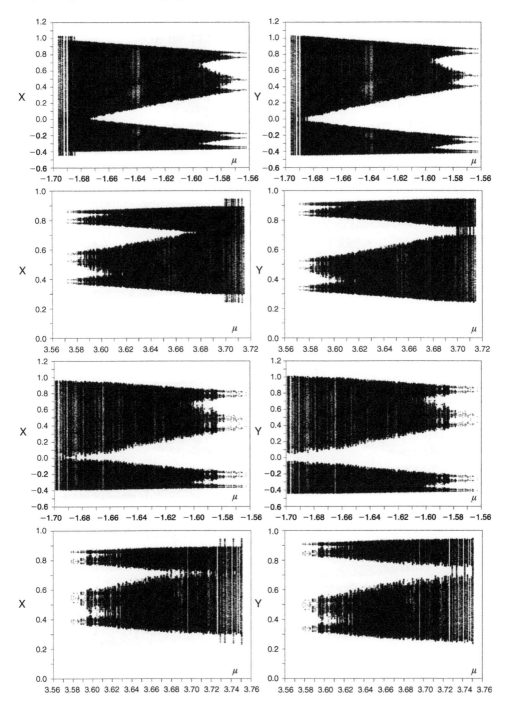

Figure 16.8. Bifurcation diagram of the coupled maps given by Eqs. (16.10a)–(16.10b) for $c_1 = c_2 = 0.5$, when forcing is performed by added noise. The first four upper panels result from $\mathbf{D} = 0.1$, while the four lower panels result from $\mathbf{D} = 0.2$.

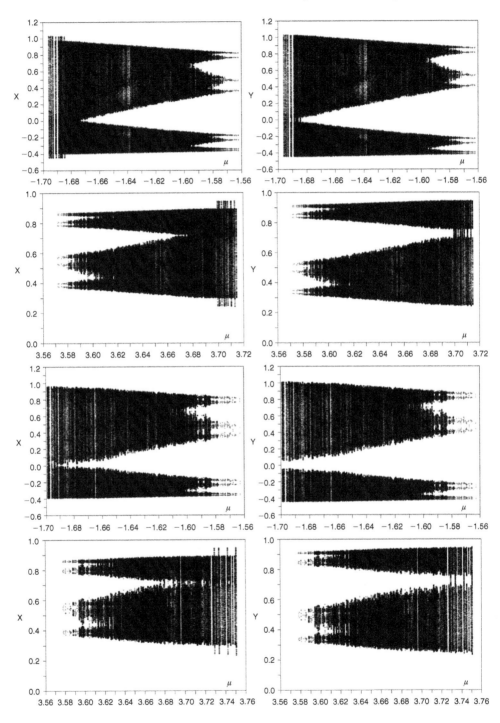

Figure 16.9. Bifurcation diagram of the coupled maps given by Eqs. (16.11a)–(16.11b) for $c_1 = c_2 = 0.5$, when forcing is performed by adding the parametric noise. The four upper panels result from $\mathbf{D} = 0.01$, while the four lower panels result from $\mathbf{D} = 0.025$.

when the parametric noise is introduced in all of the parameters in Eqs. (16.3a)–(16.3a). Thus,

$$x_{n+1} = (1 - c_{1,n})\mu_n^{(1)}x_n(1 - x_n) + c_{1,n}\mu_n^{(2)}y_n(1 - y_n) + c_{2,n}(y_n - x_n) \qquad (16.12a)$$

where $c_{1,n} = c_1(1 + \Delta\alpha_n)$, and $c_{2,n} = c_2(1 + \Delta\beta_n)$, $\Delta\alpha_n = D\delta_n^{(3)}$ and $\Delta\beta_n = D\delta_n^{(4)}$ measure the noise intensity; while $\delta_n^{(3)}$ and $\delta_n^{(4)}$ are random numbers that are uniformly distributed in the interval $[-1,1]$. Next, we focus on the changes in the Lyapunov exponent that depends on the amplitude of the noise that was introduced [29]. We calculate the

$RMSE = \left\{ \sum_{i=1}^{N} [\lambda^c(c_{1,n}, c_{2,n}, \mu_n^{(1)}, \mu_n^{(2)}) - \lambda^0(c_1, c_2, \mu)]^2 \right\}^{1/2}$ of the Lyapunov exponent, for

the coupled maps Eqs. (16.12a)–(16.12b), where λ^c and λ^0 are values calculated in the presence and absence of the noise, respectively. In calculations of the parametric noise, the amplitude D ranged from 0.0001 to 0.05, while the other parameters used were provided in Section 16.3. The results of the calculations are shown in Fig. 16.10. This figure clearly depicts that the increase of *RMSE* increases with the amplitude, in the same way as a power function. Similar results, but for the *RMSE* of the Cross-SampEn, for the maps representing biochemical substance exchange between cells were previously obtained by Mihailović and Balaz (2011).

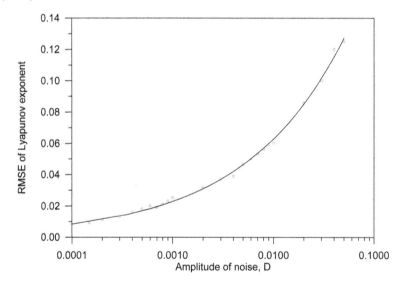

Figure 16.10. *RMSE* of the Lyapunov exponent for the coupled maps given by Eqs. (16.12a)–(16.12b) as a function of the amplitude **D** of the noise that is introduced by adding parametric forcing. The values of c_1 and c_2 and μ are the same as in Fig. 16.4.

16.5 CONCLUSIONS

We have investigated some general features of coupled maps serving as combined coupling, which can be broadly applied in environmental modeling, and their responses to additively or parametrically coupled, time-dependent fluctuating forces. Our main results are as follows.

(1) In maps 16.3(a)–16.3(b), which represent combined coupling (*diffusive* and *linear*) with two controlling parameters c_1 and c_1, we used a logistic difference equation on

an extended domain, i.e., $\mu \in [-2, 4]$, which is suitable for the modeling of environmental processes. Its features were discussed with respect to the bifurcation diagram, the Lyapunov exponent, and the sample as well as the permutation entropy. This equation shows symmetry because of the point $\mu = 1$ in the bifurcation and the Lyapunov spectrum diagrams.

(2)   In the Lyapunov coupled maps spectrum diagram, in its dependence on parameters c_1, c_1, μ (ranged in intervals $[-1.95, -1.4]$ and $[3.4$ to $3.9]$), there exist two distinguishable regions with positive as well as negative values of λ, with a feature that if c_2 takes values below 0.5, then the Lyapunov exponent is always negative. In the (c_1, c_2) phase space, the values of Cross-SampEn are mostly equal or are very close to zero, which corresponds to the region where λ is negative, indicating a high level a synchronization between the quantities x_n and y_n.

(3)   We focused on the issue of how dynamical noise can affect the bifurcation diagrams of the coupled maps that are given by Eqs. (16.3a)–(16.3b). Specifically, we considered maps for $c_1 = c_2 = 0.5$. This noise was entered in two specific ways: (1) by an additive "shock" or (2) by disturbing the parameter μ. For the forcing performed by (1), noise considerably destroys some periodic windows, changing some expected behavior when $\mathbf{D} = 0.1$, which corresponds to low-amplitude noise. Furthermore, for the doubled amplitude, i.e., $\mathbf{D} = 0.2$, the noise greatly obscured the pictures of the bifurcation diagrams with no noise. However, the forcing by (2) $\mathbf{D} = 0.01$ and $\mathbf{D} = 0.025$ produces changes that are close to the changes obtained with the forcing by (1) $\mathbf{D} = 0.1$ and $\mathbf{D} = 0.2$, respectively.

(4)   The *RMSE* of the Lyapunov exponent, when the parametric noise is introduced in all of the parameters in Eqs. (16.3a)–(16.3b), shows a growth that is similar to a power function for the amplitude of the noise $\mathbf{D}$ in the interval from 0.0001 to 0.05.

ACKNOWLEDGEMENTS

This chapter was realized as a part of the project "Studying climate change and its influence on the environment: impacts, adaptation and mitigation" (43007), which was financed by the Ministry of Education and Science of the Republic of Serbia within the framework of integrated and interdisciplinary research for the period 2011–2014.

APPENDIX A – LIST OF SYMBOLS

List of symbols		
Symbol	Definition	Dimensions or Units
c_1	Measure of diffusion	
Cross-SampEn	Cross-Sample entropy	
$\mathbf{D}$	Amplitude of the noise	
$\underline{D}$	Jacobi matrix	
PermEn	Permutation entropy	
SampEn	Sample entropy	
$\delta_n^{(1)}, \delta_n^{(2)}$	Random numbers	
λ	Lyapunov exponent	
μ	Logistic parameter	
ρ	Parameter of difference equation	

APPENDIX B – SYNOPSIS

A large number of environmental processes can be described by the dynamics of driven coupled oscillators. Most of them appear on environmental interfaces that interact with each other. That interaction is a multiple interaction because we defined *environmental interface* as an interface between two abiotic or biotic environments that are in relative motion and that exchange energy and substances through physical, biological and chemical processes, fluctuating temporally and spatially regardless of the space and time scale. In modeling these processes, the maps serving as coupled maps between the interacting interfaces must be combined when describing their complexity. In the study of map behavior as a function of coupling strength and nonlinearity, usually the dynamics of the two maps is considered, serving as the combined coupling (diffusive and linear). This type of study is performed through the analysis of (i) the logistic difference equation on extended domains that is a part of the maps, using its bifurcation diagram, Lyapunov exponent, and sample as well as the permutation entropy, and (ii) the cross-sample entropy, which depends on two coupling parameters. Dynamical noise can affect the structure of coupled map bifurcation diagrams. This analysis is provided by the noise entering in two specific ways, which disturbs either the logistic parameter on an extended domain or by an additive "shock" to the state variables.

APPENDIX C – KEYWORDS

By the end of the chapter, you should have encountered the following terms. Ensure that you are familiar with them!

Bifurcation map	Diffusive coupling	Logistic equation
Complexity	Dynamical noise	Parametric noise
Cross-sample entropy	Environmental interface	Permutation entropy
Combined coupling	Lyapunov exponent	Sample entropy
Complex systems	Linear coupling	Time series

APPENDIX D – QUESTIONS

What is the complexity?
What is the environmental interface?
Which coupling tends to equalize the instantaneous states of the interacting entities?
What is the cross-sample entropy?
What is the dynamical noise?

APPENDIX E – PROBLEMS

E1. Describe the term complex system used in Rosen's sense (Rosen, 1991), as explicated in the comment by Colier (2003), and specifically discuss the decomposition of the complex system. In addition, describe the meaning of the term complexity.

E2. Describe the environmental interface from an environmental science point of view. Following that description, count examples of this interface, taking into account different spatial and temporal scales and making its connection with complex systems.

E3. Describe how the meaning of dynamical noise can affect the dynamics of low dimensional systems and explain in what sense the term is used.

E4. Starting from maps given by Eqs. (16.3a)–(16.3b), describe the combined coupling, and calculate the Lyapunov exponent [Eq. (16.8)] for (i) diffusive coupling for $c_1 = 0.3$, $c_2 = 0$ and $\mu = 3.5$ and (ii) linear coupling for $c_1 = 0$, $c_2 = 0.4$ and $\mu = 3.5$. Plot the graph of the Lyapunov exponent for the diffusive coupling with following parameters: $\mu = 3.5$, $c_2 = 0$, while c_1 is in the interval [0.2, 0.4].

E5. Starting from maps given by Eqs. (16.3a)–(16.3b), calculate the cross-sample entropy (section 2) for diffusive coupling for $c_1 = 0.3$, $c_2 = 0$ and $\mu = 3.5$. Plot the graph of the Lyapunov exponent for the diffusive coupling for $\mu = 3.5$, $c_2 = 0$, while c_1 is in the interval [0.2, 0.4].

REFERENCES

Arshinov, V. and Fuchs, C., 2003, Preface. In: Arshinov, V. and Fuchs, C. Eds., *Causality, Emergence, Self-Organisation*, NIA-Priroda, Moscow, 1–18.

Bandt C. and Pompe B., 2002, Permutation entropy: A natural complexity measure for time series. *Physical Review Letters* **88**, 174102.

Balaz, I. and Mihailović, D.T., 2008, Evolvable biological interfaces: Outline of the new computing system. *4th Biennial Meeting: International Congress on Environmental Modelling and Software (iEMSs 2008)*, Barcelona, 7–10 July 2008, 104–113.

Bogdonoff, S. and Rubin, J., 2007, The regional greenhouse regional initiative taking action in mine. *Environment*, **49**, 9–16.

Borchgrevink, H.M., 2003, Does health promotion work in relation to noise. *Noise & Health*. **5**, 25–30.

Chamley, H., 2003, *Geosciences, Environment and Man*, Volume 1, Elsevier.

Collier, J.D., 2003, Fundamental properties of self-organization. In: Arshinov, V. and Fuchs, C. Eds., *Causality, Emergence, Self-Organisation*, NIA-Priroda, Moscow, 150–166.

Cushman-Roisin, B., Gualtieri, C. and Mihailović, D.T., 2008, Environmental Fluid Mechanics: Current issues and future outlook. In: *Fluid mechanics of environmental interfaces*, C. Gualtieri and D.T. Mihailović (Eds.), 1–16, Taylor & Francis Ltd, London, 332 p.

Edmonds, B., 1999, What is Complexity? The philosophy of complexity per se with application to some examples in evolution. In: Heylighen F., Bollen, J. and Riegler, A. Eds., *The Evolution of Complexity*, Kluwer, Dordrecht, 1–17.

Glazier, J. and Graner, F., 1993, Simulation of the differential adhesion driven rearrangement of biological cells. *Physical Review E*, **47**, 2128–2154.

Ferretti, A. and Rahman, N.K. 1998, A study of coupled logistic map and its applications in chemical physics. *Chemical Physics* **119**, 275–288.

Gunji, Y.-P. and Kamiura, M. 2004, Observational heterarchy enhancing active coupling. *Physica D* **198**, 74–105.

Gyllenberg, M., Soderbacka, G. and Ericsson, S., 1993, Does migration stabilize local population dynamics? Analysis of a discrete metapopulation model. *Mathematical Biosciences* **118**, 25–49.

Ivanova, A.S. and Kuznetsov, S.P. 2002, Scaling at the onset of chaos in a network of logistic maps with two types of global coupling. *Nonlinear Phenomena in Complex Systems* **5**, 151–154.

Hastings, A., 1993, Complex interactions between dispersal and dynamics: lessons from coupled logistic equations. *Ecology* **63**, 1362–1372.

Hassell, M.P., Comins, H.N. and May, R.M., 1991, Spatial structure and chaos in insect population dynamics. *Nature*, **353**, 255–258.

Heylighen, F., 1999, The growth of structural and functional complexity during evolution. In: Heylighen F., Bollen, J. and Riegler, A. Eds., *The Evolution of Complexity*, Kluwer, Dordrecht, 17–47.

Hewitt, C.N. and Jackson, A., 2003, *Handbook of atmospheric science principles and applications*. Blackwell, Oxford.

Hogg, T. and Huberman, B.A., 1984, Generic behavior of coupled oscillators. *Phys Rev A*, **29**, 275–281.

Kaneko, K., 1983, Transition from torus to chaos accompanied by the frequency locking with symmetry breaking – In connection with the coupled logistic map. *Progress of Theoretical Physics*, **69**, 1427–1442.

Kennel, M.B., Brown, R. and Abarbanel, H.D.I., 1992, Determining embedding dimension for phase-space reconstruction using a geometrical construction. *Phys Rev A*, **45**, 3403–3411.

Lake, D.E., Richman, J.S., Griffin, M.P. and Moorman, J.R., 2002, Sample entropy analysis of neonatal heart rate variability. *American Journal of Physiology: Regulatory Integrative and Comparative Physiology*, **283**, R789–R797.

Liu, Z. and Ma, W., 2005, Noise induced destruction of zero Lyapunov exponent in coupled chaotic systems. *Physics Letters A*, **343**, 300–305.

Lloyd, A.L. 1995, The coupled logistic map: A simple model for effects of spatial heterogeneity on population dynamics. *Journal of Theoretical Biology*, **173**, 217–230.

Martins, M.L., Ceotto, G., Alves, G., Bufon, C.C.B., Silva, J.M. and Laranjeira, F.F., 2000, Cellular automata model for citrus variegated chlorosis. *Physical Review E*, **62**, 7024–7030.

Metta, S., Provenzale, A. and Spiegel, E.A. 1995, On-off intermittency and coherent bursting in stochastically-driven coupled maps. *Chaos Soliton and Fractals*, **43**, 8–14.

Midorikawa, S., Takayuki, K. and Taksu, C., 1995, Folded bifurcation in coupled asymmetric logistic maps. *Progress of Theoretical Physics*, **94**, 571–575.

Mihailović, D.T. and Balaz, I., 2007, An essay about modeling problems of complex systems in Environmental Fluid Mechanics. *Idojaras*, **111**, 209–220.

Mihailović D.T. and Balaz, I. 2011, A model representing biochemical substances exchange between cells. Part II: effect of fluctuations of environmental parameters to behavior of the model. *Journal of Applied and Functional Analysis*, **6**, 77–84.

Mihailović, D.T. Budincevic, M., Balaž, I. and Mihailović, A. (2011a), Stability of inter-cellular exchange of biochemical substances affected by variability of environmental parameters. *Modern Physics Letters*, **25**, 1–11.

Mihailović, D.T. Budincevic, M., Kapor, D. and Balaž, I. (2011b), A numerical study of coupled maps representing energy exchange processes between two environmental interfaces regarded as biophysical complex systems. *Natural Science*, **3**, 75–84.

Kauffman, S., 1993, *The origins of order: Self-organization and selection in evolution*. Oxford University Press, Oxford.

Nikolov, N., Massman, W. and Schoettle, A., 1995, Coupling biochemical and biophysical processes at leaf level: an equilibrium photosynthesis model for leaves of C3 plants. *Ecological Modelling*, **80**, 205–235.

Niyogi, D.S. and Raman, S., 2001, Numerical modelling of gas deposition and bidirectional surface–atmosphere exchanges in mesoscale air pollution systems. In: Boybeyi, Z. Ed., *Advances in Air Pollution*, WIT Publications, Southampton, 1–51.

Pincus, S.M. 1991, Approximate entropy as a measure of system complexity. *Proceedings of the National Academy of Sciences of the United States*, **88**, 2297–2301.

Pincus, S. and Singer, B.H. 1995, Randomness and degrees of irregularity. *Proceedings of the National Academy of Sciences of the United States*, **93**, 2083–2088.

Pincus, S.M., Mulligan, T., Iranmanesh, A., Gheorghiu, S., Godschalk, M. and Veldhuis, J.D., 1996, Older males secrete luteinizing hormone and testosterone more irregularly, and jointly more asynchronously, than younger males. *Proceedings of the National Academy of Sciences of the United States,* **93**, 14100–14105.

Random House Dictionary, 2006, http://dictionary.reference.com

Richman, J.S. and Moorman, J.R., 2000, Physiological time-series analysis using approximate entropy and sample entropy. *American Journal of Physiology: Heart & Circulatory Physiology,* **278**, H2039–H2049.

Rosen, R., 1991, *Life itself: A comprehensive inquiry into the nature, origin, and fabrication of life.* Columbia University Press, New York.

Ruelle, D., 1994, Where can one hope to profitably apply the ideas of chaos? *Physics Today,* **47**, 24–30.

Sakaguchi, H. and Tomita, K., 1990, Bifurcations of the coupled logistic map. *Progress of Theoretical Physics,* **78**, 305–315.

Satoh, K. and Aihara, T. 1990, Numerical study on a coupled-logistic map as a simple model for a predator-prey system. *Journal of the Physical Society of Japan,* **59**, 1184–1189.

Savi, M.A. 2007, Effects of randomness on chaos and order of coupled maps. *Physics Letters A,* **364**, 389–395.

Selvam, A.M. and Fadnavis, S., 1998, Signatures of a universal spectrum for atmospheric interannual variability in some disparate climatic regimes. *Meteorology and Atmospheric Physics,* **66**, 87–112.

Serletis, A. and Shahmoradi, A., 2006, Comment on "Singularity Bifurcations" by Yijun He and William A. Barnett. *Journal of Macroeconomy,* **28**, 23–26.

Serletis, A., Shahmoradi, A. and Serletis, D., 2007a, Effect of noise on estimation of Lyapunov exponents from a time series. *Chaos Solitons and Fractals* **32**, 883–887.

Serletis, A., Shahmoradi, A. and Serletis, D. 2007b, Effect of noise on the bifurcation behavior of nonlinear dynamical systems. *Chaos Solitons and Fractals* **33**, 914–921.

Sivertsen, T.H. 2005, Discussing the scientific method and a documentation system of meteorological and biological parameters. *Physics and Chemistry of the Earth,* **30**, 35–43.

Thattai, M. and van Oudenaarden, A., 2001, Intrinsic noise in gene regulatory networks, *Proceedings of the National Academy of Sciences of the United States,* **98**, 8614–8619.

Welch, D., 1998. Air Issues and Ecosystem Protection, A Canadian national parks perspective. *Environmental Monitoring and Assessment,* **49**, 251–262.

Author index

Subject index

Printed and bound by CPI Group (UK) Ltd, Croydon, CR0 4YY

18/10/2024

01776254-0005